Periodic Table of the Elements with the Gmelin System Numbers

1 H 2																1 H 2	2 He 1
3 Li 20	4 Be 26											5 B 13	6 C 14	7 N 4	8 O 3	9 F 5	10 Ne 1
11 Na 21	12 Mg 27											13 Al 35	14 Si 15	15 P 16	16 S 9	17 Cl 6	18 Ar 1
19 * K 22	20 Ca 28	21 Sc 39	22 Ti 41	23 V 48	24 Cr 52	25 Mn 56	26 Fe 59	27 Co 58	28 Ni 57	29 Cu 60	30 Zn 32	31 Ga 36	32 Ge 45	33 As 17	34 Se 10	35 Br 7	36 Kr 1
37 Rb 24	38 Sr 29	39 Y 39	40 Zr 42	41 Nb 49	42 Mo 53	43 Tc 69	44 Ru 63	45 Rh 64	46 Pd 65	47 Ag 61	48 Cd 33	49 In 37	50 Sn 46	51 Sb 18	52 Te 11	53 I 8	54 Xe 1
55 Cs 25	56 Ba 30	57** La 39	72 Hf 43	73 Ta 50	74 W 54	75 Re 70	76 Os 66	77 Ir 67	78 Pt 68	79 Au 62	80 Hg 34	81 Tl 38	82 Pb 47	83 Bi 19	84 Po 12	85 At 8a	86 Rn 1
87 Fr 25a	88 Ra 31	89*** Ac 40	104 71	105 71													* NH_4 23

**Lanthanides 39	58 Ce	59 Pr	60 Nd	61 Pm	62 Sm	63 Eu	64 Gd	65 Tb	66 Dy	67 Ho	68 Er	69 Tm	70 Yb	71 Lu
***Actinides	90 Th 44	91 Pa 51	92 U 55	93 Np 71	94 Pu 71	95 Am 71	96 Cm 71	97 Bk 71	98 Cf 71	99 Es 71	100 Fm 71	101 Md 71	102 No 71	103 Lr 71

A Key to the Gmelin System is given on the Inside Back Cover

Gmelin Handbook of Inorganic Chemistry

8th Edition

Gmelin Handbook of Inorganic Chemistry

8th Edition

Gmelin Handbuch der Anorganischen Chemie

Achte, völlig neu bearbeitete Auflage

Prepared and issued by	Gmelin-Institut für Anorganische Chemie der Max-Planck-Gesellschaft zur Förderung der Wissenschaften Director: Ekkehard Fluck

Founded by	Leopold Gmelin
8th Edition	8th Edition begun under the auspices of the Deutsche Chemische Gesellschaft by R. J. Meyer
Continued by	E. H. E. Pietsch and A. Kotowski, and by Margot Becke-Goehring

Springer-Verlag Berlin Heidelberg GmbH 1990

Volumes published on "Manganese" (Syst.-No. 56)

Manganese A 1 (in German)
History – 1980

Manganese B (in German)
The Element – 1973

Manganese C 1 (in German)
Compounds (Hydrides. Oxides. Oxide Hydrates. Hydroxides) – 1973

Manganese C 2 (in German)
Compounds (Oxomanganese Ions. Permanganic Acid. Compounds and Phases with Metals of the Main and Subgroups I and II) – 1975

Manganese C 3 (in German)
Compounds of Manganese with Oxygen and Metals of the Main and Subgroups III to VI. Compounds of Manganese with Nitrogen – 1975

Manganese C 4 (in German)
Compounds of Manganese with Fluorine – 1977

Manganese C 5 (in German)
Compounds of Manganese with Chlorine, Bromine, and Iodine – 1978

Manganese C 6 (in German)
Compounds of Manganese with Sulfur, Selenium, Tellurium, Polonium – 1976

Manganese C 7
Compounds of Manganese with Boron and Carbon – 1981

Manganese C 8
Compounds of Manganese with Silicon – 1982

Manganese C 9
Compounds with Phosphorus, Arsenic, Antimony – 1983

Manganese C 10
Electronic Spectra of Manganese Halides. Cumulative Substance Index of C 1 to C 10 – 1983

Manganese D 1 (in German)
Coordination Compounds 1 – 1979

Manganese D 2 (in German)
Coordination Compounds 2 – 1980

Manganese D 3
Coordination Compounds 3 – 1982

Manganese D 4
Coordination Compounds 4 – 1985

Manganese D 5
Coordination Compounds 5 – 1987

Manganese D 6
Coordination Compounds 6 – 1988

Manganese D 7
Coordination Compounds 7 – 1990 **(present volume)**

Gmelin Handbook of Inorganic Chemistry

8th Edition

Mn
Manganese

D 7

Coordination Compounds 7

With 19 illustrations

AUTHORS
L. J. Boucher, College of Arts and Sciences, Arkansas State University, Arkansas, USA

Karl Koeber, Mirjana Kotowski, Dieter Tille, Gmelin-Institut, Frankfurt am Main

FORMULA INDEX
Ursula Hettwer, Gmelin-Institut, Frankfurt am Main

EDITORS
Helga Demmer, Mirjana Kotowski, Edith Schleitzer-Rust, Dieter Tille, Gmelin-Institut, Frankfurt am Main

CHIEF EDITOR
Edith Schleitzer-Rust, Gmelin-Institut, Frankfurt am Main

System Number 56

Springer-Verlag Berlin Heidelberg GmbH 1990

LITERATURE CLOSING DATE: 1987
IN MANY CASES MORE RECENT DATA HAVE BEEN CONSIDERED

Library of Congress Catalog Card Number: Agr 25-1383

ISBN 978-3-662-07508-1 ISBN 978-3-662-07506-7 (eBook)
DOI 10.1007/978-3-662-07506-7

Originally published by Springer-Verlag, Berlin · Heidelberg · New York · London · Paris · Tokyo in 1989
Softcover reprint of the hardcover 8th edition 1989

Preface

The present volume "Manganese" D 7 continues the description of the manganese complexes. The introduction on p. 1 shows the classes of complexes that have already been described in Chapters 1 through 32 in Volumes D 1 to D 6. Complexes with nitriles, with nitro hydrocarbons, and with ligands containing sulfur, selenium, or tellurium are now described in Chapters 33 to 36 of this volume. The characteristic features of the various complex types are summarized at the beginning of the chapters.

Complexes with sulfur-containing ligands make up the majority of the present volume. Many of them are of analytical or biological interest. An extensive literature exists on the analytical application of dithiocarbamato or dithiolato complexes; these are suitable for the spectrophotometric determination of microamounts of manganese because of their intense colors. The thiolato complexes with tetrahedral metal-sulfur coordination sites are of significant interest as synthetic analogs of the active centers in metalloproteins or nucleic acids. Among the complexes with monothiols, the adamantane-like cage of the $[Mn_4(C_6H_5S)_{10}]^{2-}$ ion, containing a tetrahedron of manganese atoms and an octahedron of bridging sulfur atoms, is noteworthy, see p. 26. Interest in the coordination chemistry of sulfoxide complexes has developed as a result of the excellent solvent action of the sulfoxides and their use in the solvent extraction of metals during refining processes.

A formula index at the end of this volume listing the empirical formulas and the linearized structural formulas of the ligands is intended to expedite locating specific compounds.

Frankfurt am Main
December 1989

Edith Schleitzer-Rust

Table of Contents

Coordination Compounds of Manganese
(Continued)

Introduction

Arrangement. In Series D, coordination compounds of manganese, with the exception of the organometallic compounds, are described. The volumes "Mangan" D1, D2 and "Manganese" D3 to D6 contain the following chapters:

"Mangan" D1, 1979
1) Review
2) Complexes with H_2O
3) Complexes with Alcohols
4) Complexes and Salts with Phenols and Other Aromatic Hydroxy Compounds
5) Complexes with Aldehydes
6) Complexes with Ketones
7) Complexes with Quinones
8) Complexes with Ethers and O-Heterocycles

"Mangan" D2, 1980
9) Complexes and Salts of Carboxylic Acids and Their Derivatives
10) Cyanomanganate Complexes
11) Cyanato, Thiocyanato, and Selenocyanato Complexes

"Manganese" D3, 1982
12) Complexes with Ammonia
13) Complexes with Amines
14) Complexes with Hydrazine and its Derivatives
15) Complexes with Hydroxylamine
16) Complexes with N-Heterocycles

"Manganese" D4, 1985
16) Complexes with N-Heterocycles (Continued)
17) Complexes with Aminoalcohols, -phenols, and -naphthols
18) Complexes with Aminoethers and Aminooxo Compounds
19) Complexes with Amino Acids
20) Complexes with Peptides
21) Complexes with Proteins

"Manganese" D5, 1987
22) Complexes with Amine-N-polycarboxylic Acids
23) Complexes with Hydrazinecaboxylic Acid and Derivatives
24) Complexes with Amides and Related Compounds
25) Complexes with Hydrazides
26) Complexes with Derivatives of Hydroxylamine
27) Complexes with Oximes and Nitroso Compounds
28) Complexes with Azo Compounds
29) Complexes with Triazenes

"Manganese" D6, 1988
30) Complexes with Schiff Bases
31) Complexes with Hydrazones or Related Compounds
32) Complexes with Carbazones Thiocarbazones, and Formazans

This volume deals with manganese complexes with nitriles or related compounds (Chapter 33), with nitro hydrocarbons (Chapter 34), with ligands containing sulfur (Chapter 35), and with ligands containing selenium or tellurium (Chapter 36).

Rules and Definitions. Generally, the names of the ligands correspond to IUPAC nomenclature; trivial names are also used.

The stepwise stability (formation) constants (K_n) for the formation of the complexes in solution from a central atom (M) and ligands (L) and the cumulative constants (β_n) are defined as follows:

$K_n = [ML_n]/[MnL_{n-1}] \cdot [L]$ in L/mol for the equilibria $ML_{n-1} + L \rightleftharpoons ML_n$ (n = 1, 2, 3, ...)
$\beta_n = [ML_n]/[M] \cdot [L]^n$ in L^n/mol^n for the equilibria $M + nL \rightleftharpoons ML_n$ (n = 1, 2, 3, ...)

The formation of complexes with protonated ligands is described by:

$K^{M}_{MH_pL} = [MH_pL]/[M] \cdot [H_pL]$ for the equilibria $M + H_pL \rightleftharpoons MH_pL$ (p = 1, 2, 3, ...)

Enthalpy (ΔH), Gibbs free energy (ΔG), or entropy changes (ΔS) are given the same subscript as the corresponding K: e.g., ΔH_1 for constant K_1. For reactions represented by cumulative constants (β_n), the notation $\Delta H_{\beta n}$ is used. Ionic strengths are given in mol/L.

With respect to magnetic properties, the conventions of Carlin, R. L., Magnetochemistry, Berlin – Heidelberg – New York, 1986 are followed. Magnetic measurements were generally performed by the Gouy method. If another technique was used, that method is reported.

Abbreviations and Dimensions. Temperatures are normally given in °C; K stands for Kelvin. Abbreviations used with temperatures are m.p. for melting point and dec. for decomposition. With thermodynamic data, (s) is used to label solids, (l) is used for liquids, and (g) is used to designate the gaseous state.

The vibrational spectra are labeled as IR (infrared) or R (Raman). The symbol ν is used for stretching vibrations and δ for deformation vibrations; wavenumbers are given in cm^{-1}. The intensities are placed in parentheses (w = weak, m = medium, s = strong, vs = very strong, etc.); sh means shoulder; br means broad. The UV-visible absorption maxima of the electronic spectra are given in nm (λ_{max}) or cm^{-1} (ν_{max}), the extinction coefficient ε is given in $L \cdot mol^{-1} \cdot cm^{-1}$.

Definitions of ESR or NMR parameters: A_0 is the isotropic hyperfine coupling constant, D_0 the isotropic zero-field splitting parameter, and g_0 the isotropic g-factors.

Abbreviations for methods used in this volume are:

DTA	differential thermal analysis	ESR	electron spin resonance
TG	thermogravimetry	NMR	nuclear magnetic resonance
DTG	differential thermogravimetry	ESCA	electron spectroscopy for chemical analysis

Abbreviations for ligands are listed on p. 251, the first page of the ligand formula index.

33 Complexes with Nitriles or Related Compounds

33.1 General

This chapter first describes complexes with nitriles, chiefly of the type, $[Mn(RCN)_6]X_2$, with $X = ClO_4$, BF_4, or halometallate anions. In these compounds the nitrile ligand is coordinated to manganese through the nitrogen atom. Included also are similar complexes with hydrocyanic acid or cyanogen halides. Complexes with polycyanoalkanes or -alkenes are described in the second part of this chapter, see p. 14. These compounds reveal a salt-like character, due to the ability of the ligands to form resonance-stabilized anions.

Carbonyl complexes of manganese, containing, additionally, nitrile ligands, will be described in the series of organometallic compounds in the same way as complexes with isonitriles where manganese is bonded to the terminal carbon atom. Complexes with cyano-substituted diketones or O-heterocycles are reported in "Mangan" D 1, 1979, pp. 112 and 145, respectively; complexes with cyano-substituted N-heterocycles in "Manganese" D 3, 1982, pp. 110, 152, 183, and "Manganese" D 4, 1985, p. 76; those with cyano carboxylic acids in "Manganese" D 3, 1982, p. 100, or "Manganese" D 5, 1987, p. 10. For complexes with 1-cyanoguanidine (= dicyandiamide) see "Manganese" D 5, 1987, p. 158; with a cyano derivative of diphenylformazane, see "Manganese" D 6, 1988, p. 368. Cyanomanganate complexes have been described in "Mangan" D 2, 1980, pp. 196/281.

The chemistry of metal nitrile complexes is reviewed in [1, 2]. Interest in this group of coordination compounds is not surprising in view of the fact that alkyl cyanides, particularly methyl cyanide (acetonitrile), are widely used in inorganic chemistry as solvents for preparation of metal compounds and measurement of their physical properties. Physical data indicate the existence of $[Mn(C_2H_3N)_6]^{2+}$ ions in acetonitrile solution. Numerous $[Mn(C_2H_3N)_6][MX_4]_n$ or $[Mn(C_2H_3N)_6][MX_6]_n$ complexes have been synthesized by halide ion transfer between stoichiometric amounts of MnX_2 salts and appropriate acceptors in anhydrous acetonitrile, see p. 7. The IR spectra of the complexes reveal positive shifts of the ν(CN) vibration mode, due to coordination of the ligand through the nitrogen atom. The tetrachlorometallates are isomorphous. Similar results were observed for the tetrabromometallates. Isomorphism was also observed with corresponding halometallates of Zn^{II}, Cd^{II}, Ni^{II}, Co^{II}, or Fe^{II}.

References:

[1] Walton, R. A. (Quart. Rev. Chem. Soc. **19** [1965] 126/43).
[2] Storhoff, B. N.; Huntley, C. L. (Coord. Chem. Rev. **23** [1977] 1/29).

33.2 With Hydrocyanic Acid HCN

$[Mn^{II}(HCN)_6][InCl_4]_2$. Manganese(II) chloride, dehydrated at 150°C by a stream of dry hydrogen chloride, was stirred with 2 equivalents of $InCl_3$ in liquid hydrocyanic acid at 15°C under anhydrous conditions. After concentrating the solution by evaporation, white crystals precipitated, and were washed with sodium-dried pentane. Characteristic IR bands of the complex (in Nujol) were assigned as follows (shifts relative to the absorption bands of gaseous free HCN are given in parentheses): ν(CH) at 3160(−151), ν(CN) at 2124(+27), δ(HCN) at 776(+63), $\nu(InCl_4)$ at 328, and ν(Mn–N) <200 cm^{-1}. From the similarity in the infrared patterns and from the ligand field spectra of the corresponding Co and Ni complexes an octahedral environment of the Mn^{II} ion by six HCN ligands, coordinated through the nitrogen atom, is assumed. The complex decomposes on heating above 120°C, and fairly rapidly in contact with water.

Reference:

Eversteijn, P. L. A.; Zuur, A. P.; Driessen, W. L. (Inorg. Nucl. Chem. Letters **12** [1976] 277/84).

33.3 With Cyanogen Halides

$\mathbf{[Mn^{II}(BrCN)_6][SbCl_6]_2}$ and $\mathbf{[Mn^{II}(ICN)_6][SbCl_6]_2}$ complexes were prepared by stirring manganese(II) chloride with $SbCl_5$ (mole ratio 1:2) in an excess of nitromethane for 24 h. The filtered solution of $[Mn(CH_3NO_2)_6][SbCl_6]_2$ (see p. 21) was then combined with 6 mol of the corresponding ligand, and the reaction mixture stirred for 1 h. The yellow complexes separated, after evaporating some of the nitromethane in vacuum. The crystals were isolated by filtration in a dry atmosphere, washed with pentane, and dried. The IR spectra show increases of ν(CN), as a result of coordination, from 2190 cm^{-1} (BrCN) to 2224 cm^{-1} or from 2162 cm^{-1} (ICN) to 2186 cm^{-1}. The complexes decompose in moist air.

Reference:

Zuur, A. P.; Driessen, W. L.; Eversteijn, P. L. A.; Aarnoudse, M. A. (Inorg. Nucl. Chem. Letters **13** [1977] 81/4).

33.4 With Acetonitrile CH_3CN (= C_2H_3N)

Remark. A great number of adducts (solvates) are reported in the literature. In several cases, well-defined crystals of coordination compounds are obtained, adequate for structure determinations. Since CH_3CN is not bonded directly to manganese, these adducts are not described in this section. They will be found in the volume and chapter for the corresponding complex type, see, e.g., acetonitrile adducts of Schiff base complexes in "Manganese" D 6, 1988, pp. 107 and 138.

33.4.1 Complexes in Solution

The existence of $\mathbf{[Mn^{II}(C_2H_3N)_6]^{2+}}$ ions in acetonitrile solutions of manganese(II) salts has been demonstrated by IR, NMR, and ESR spectral data, or by polarographic and magnetic measurements. The IR spectrum of $Mn(NO_3)_2 \cdot 6H_2O$ dissolved in acetonitrile shows two bands at 2286 and 2256 cm^{-1}, which are assigned to the ν(CN) vibration of the coordinated and the neat solvent molecules, respectively. The Mn–N bond force constant, k=1.7 mdyn/Å, was evaluated from the IR data. On adding methylamine or ethylamine to the solution the band at 2286 cm^{-1} disappears, due to the formation of amine complexes [1].

The ^{14}N nuclear relaxation times of acetonitrile solutions of manganese(II) perchlorate have been studied as a function of temperature (230 to 361 K). Saturation studies have shown that a scalar coupling mechanism controls the relaxation of the ^{14}N nucleus in the coordination sphere of the Mn^{2+} ion [2]. Proton spin-lattice and spin-spin relaxation times of acetonitrile solutions of $Mn(ClO_4)_2$ between 298 and 413 K, measured over a wide frequency range (10 kHz to 90 MHz), and ^{14}N NMR line width measurements in the same temperature range are reported in [3]. The effect of pressure and temperature on the rate of exchange of acetonitrile molecules on manganese(II) perchlorate was studied in [4]. First-order rate constants, k_1, the activation parameters, ΔH^{*} and ΔS^{*}, for the exchange of one ligand molecule from the first coordination sphere of $[Mn(C_2H_3N)_6]^{2+}$ at 298 K, and the scalar electron spin-nuclear spin coupling constant, resulting from NMR studies are tabulated below:

NMR	k_1 in s^{-1}	ΔH^{*} in kJ/mol	ΔS^{*} in $J\cdot mol^{-1}\cdot K^{-1}$	A/h in MHz	Ref.
^{14}N	1.2×10^7	30.3 ± 1.1	-7.5 ± 3.4	3.2	[2, 4]
^{1}H, ^{14}N	0.85×10^7	15.0 ± 2	-63 ± 12	3.6 ± 0.4	[3, 4]
^{14}N	1.36×10^7	29.6 ± 0.5	-8.9 ± 2.0	3.6 ± 0.07	[4]
^{1}H	3.1×10^7	35.9	19.0	-0.33*)	[5]

*) From ^{1}H NMR measurements between 233 and 253 K.

Exclusion of water as concurrent solvate was achieved by the use of carefully dried $Mn(ClO_4)_2\cdot6H_2O$ (at 380 K over P_4O_{10} under reduced pressure for 24 h) [3], dehydrating the solutions of $Mn(ClO_4)_2\cdot6H_2O$ in doubly distilled acetonitrile under reflux with molecular sieves [2, 4], or by reaction with triethyl orthoformate [5]. Problems concerning the mechanistic significance of the data, and inconsistencies of the ΔH^{*} and ΔS^{*} values are discussed. The effect of pressure on the ^{14}N line width up to 110 MPa (252 to 260 K) permits evaluation of the activation volume, $\Delta V^{*}=-7$ cm^3/mol, i.e., the volume of gain (or loss) on releasing (or coordinating) a molecule of acetonitrile in $[Mn(C_2H_3N)_6]^{2+}$ on going to the transition state for solvent-molecule exchange. The value is compared with those of $[Mn(H_2O)_6]^{2+}$ (-5.4 cm^3/mol at 298 K) and $[Mn(CH_3OH)_6]^{2+}$ (-5.0 cm^3/mol at 279 K), and with those of the corresponding Ni, Co, and Fe complexes [4]. By means of ^{1}H NMR line-broadening studies the rates of exchange of CH_3CN in solutions of Mn^{2+}, Fe^{2+}, Co^{2+}, Ni^{2+}, and Cu^{2+} are compared with the solvent exchanges of these ions in solutions of H_2O, DMSO, DMF, and CH_3OH. A decrease of k_1 for the inner-sphere solvent exchange at the hexacoordinate Mn^{2+} was observed as follows: acetonitrile > water ~ dimethyl sulfoxide > dimethylformamide > methanol. The exchange of CH_3CN at Mn^{2+} was found to be faster than at Fe^{2+} ($k_1=4.3\times10^5$ s^{-1}) or Ni^{2+} ($k_1=3.15\times10^3$ s^{-1}) [5].

The ESR spectra of manganese(II) salts in acetonitrile solution have been examined as a function of manganese concentration, counter-ion concentration (ClO_4^-), halide concentration, temperature, and Larmor frequency. The ESR spectra of $[Mn(C_2H_3N)_6]^{2+}$ at X-(9 GHz) and K-(35 GHz) band frequencies reveal an increase of the resonance line width (ΔH) of the $m_I=1/2$ hyperfine component with manganese concentration, whereas the widths are essentially independent of frequency. The addition of excess ClO_4^- had little effect on the spin-Hamiltonian parameters given below [6]:

$[Mn(C_2H_3N)_6]^{2+}$ in mol/L	$(C_4H_9)_4NClO_4$ in mol/L	ΔH in G	g	A in G
0.005	0.000	8.7 ± 0.2	2.0005 ± 0.005	-93.1 ± 0.2
0.005	0.110	9.8 ± 0.2	2.0001 ± 0.005	-93.0 ± 0.2

The g value is lower than the value, $g=2.003\pm0.001$, reported earlier in [7]; the hyperfine constant, however, is within the limits of error cited there ($A=92\pm1$ G) [6].

The molar conductivities of MnX_2 salts dissolved in acetonitrile indicate the existence of 1:2 electrolytes with the ions, $[Mn(C_2H_3N)_6]^{2+}$ and $2X^-$. For values of Λ, see the corresponding complex salts in the following chapter.

Manganese(III) complexes $[Mn^{III}L_2Cl]$ with HL = $CH_3COCH_2COCH_3$, $C_6H_5COCH_2COCH_3$, or $C_6H_5COCH_2COC_6H_5$ (see "Mangan" D 1, 1978, pp. 101, 116, and 112, respectively) form acetonitrile adducts in dichloromethane solution. Formation constants, according to $MnL_2Cl + CH_3CN = MnL_2Cl \cdot CH_3CN$, were determined spectrophotometrically at 25°C: K = 0.244(15) L/mol at 480 nm, K = 0.200(16) L/mol at 520 nm, and K = 0.234(5) L/mol at 540 nm, respectively. The molar extinction coefficients of the adducts are $\varepsilon = 570$, 220, and 320 $L \cdot mol^{-1} \cdot cm^{-1}$, respectively. When increasing amounts of acetonitrile were added to the solutions, isosbestic points were observed at 450 nm (acetylacetonato complex), 426 and 468 nm (benzoylacetonato complex), and at 483 nm (dibenzoylmethanato complex) [8].

References:

[1] Kharitonov, Yu. Ya.; Tkavadse, L. M. (Koord. Khim. **7** [1981] 1751/2; C.A. **96** [1982] No. 14592).
[2] Purcell, W. L.; Marianelli, R. S. (Inorg. Chem. **9** [1970] 1724/8).
[3] von Goldammer, E.; Bassaris, C. (J. Solution Chem. **9** [1980] 237/45).
[4] Sisley, M. J.; Yano, Y.; Swaddle, T. W. (Inorg. Chem. **21** [1982] 1141/5).
[5] Vigee, G. S.; Watkins, C. L.; Harris, M. E. (J. Inorg. Nucl. Chem. **42** [1980] 1441/5).
[6] Lynds, L.; Crawford, J. E.; Lynden-Bell, R. M.; Chan, S. I. (J. Chem. Phys. **57** [1972] 5216/30).
[7] Chan, S. I.; Fung, B. M.; Lütje, H. (J. Chem. Phys. **47** [1967] 2121/30).
[8] Ito, Y.; Kawaguchi, S. (Bull. Chem. Soc. Japan **54** [1981] 150/8).

33.4.2 Isolated Compounds

$[Mn(NO)_3(C_2H_3N)]$. The complex, characterized only by spectral data, was prepared in solution as follows: the mixture of equimolar solutions of $Mn_2(CO)_{10}$ and of $[Co(NO)_2Cl]_2$ in THF was irradiated with UV light (high-pressure Hg lamp). After evaporation of the solvent and $Co(NO)(CO)_3$, the remaining $[Mn(NO)_3THF]$ was dissolved in pentane, and tetrahydrofuran was replaced by addition of a threefold molar excess of acetonitrile. Bands observed in the IR spectrum of the stable solution at 1784 and 1688 cm^{-1} are assigned to the NO vibrations. The NMR spectrum reveals the shift parameter, $\delta(^{55}Mn) = -730$ ppm (relative to saturated aqueous $KMnO_4$ solution). The low ^{55}Mn shielding, caused by the coordinated acetonitrile, is compared with that of several ligands coordinated by O, N, S, Se, Te, C, or P atoms [1].

$[Mn^{II}(C_2H_3N)_6](ClO_4)_2$ and **$[Mn^{II}(C_2H_3N)_6](BF_4)_2$**. The perchlorate was obtained by treating a suspension of nitrosyl perchlorate in acetonitrile with metallic manganese for 4 to 6 h under reduced pressure [2]. The product was purified by repeated crystallization from acetonitrile and stored in the form of moist crystals. The strict stoichiometry of the hexasolvate is guaranteed by storing the moist crystals at a constant temperature over a large quantity of partly desolvated salt, obtained by evaporation under vacuum [3]. The colorless fluoroborate complex was prepared by metathesis from anhydrous $MnBr_2$ and $AgBF_4$ in dry acetonitrile, followed by cathodic deposition of the traces of silver not consumed in the reaction. On concentrating the solution in vacuum, the complex deposited. It was recrystallized three times from anhydrous acetonitrile [4]. The standard heat of solution for moist crystals of composition $Mn(C_2H_5N)_{6.35}(ClO_4)_2$ at 25°C in acetonitrile was found to be $\Delta H_s^\circ = 10.7 \pm 0.3$ kJ/mol and

30.7 or −78.8 kJ/mol in water or dimethyl sulfoxide solution, respectively. A correcting term of 0.99 kJ/mol represents the deviation from the hexasolvate composition. The number of solvate molecules per Mn^{II} in the saturated acetonitrile solution is 13.2 [3].

Heats of solution of $[Mn(C_2H_3N)_6](ClO_4)_2$ at 25°C in methylformamide, dimethylformamide, and dimethylacetamide are $\Delta H_s^\circ = -25.9$, −82.2, and −90.5 kJ/mol, respectively [5]. Equivalent conductivities of $Mn(ClO_4)_2$ in acetonitrile are reported in [6], see "Mangan" C 5, 1978, p. 254. From the limiting equivalent conductivity of $Mn(BF_4)_2$ in acetonitrile solution at 25°C, $\Lambda_0 =$ 207.6 $cm^2 \cdot \Omega^{-1} \cdot val^{-1}$, the equivalent conductivity of $[Mn(C_2H_3N)_6]^{2+}$, $\Lambda_0 = 99.1\ cm^2 \cdot \Omega^{-1} \cdot val^{-1}$, was calculated with $\Lambda_0 = 108.5\ cm^2 \cdot \Omega^{-1} \cdot val^{-1}$ for the BF_4^- ion. For the outer sphere association of a BF_4^- ion to $[Mn(C_2H_3N)_6]^{2+}$ an association constant log K = 2.08 mol^{-1} was evaluated from conductivity data [4].

Halometallates. Complexes of composition $\mathbf{[Mn^{II}(C_2H_3N)_6][MX_4]_n}$ with $MX_4 = AlCl_4^-$, $GaCl_4^-$, $InCl_4^{2-}$, $TlCl_4^-$, $FeCl_4^-$ [7], $InBr_4^-$, $TlBr_4^-$, $FeBr_4^-$ [8], MnI_4^{2-} [9], $GaClBr_3^-$, $InClBr_3^-$ [10], $GaClI_3^-$, $InClI_3^-$ ions [11], and $\mathbf{[Mn^{II}(C_2H_3N)_6][MX_6]_n}$ complexes with $MX_6 = SnCl_6^{2-}$ [12], $SbCl_6^-$ [13], $SbCl_5Br^-$, $NbCl_5Br^-$, $TaCl_5Br^-$ [14], $TeCl_2Br_4^{2-}$ [15], $ZrCl_2Br_4^{2-}$ [25], and $Sn(NCS)_6^{2-}$ ions [16] have been synthesized. A compound of composition $MnSn_2Cl_{10} \cdot 8C_2H_3N$ was assumed to contain the anion $[SnCl_5 \cdot C_2H_3N]^-$ and not the dimeric group $Sn_2Cl_{10}^{2-}$ [12]. The formation of a chlorostannate complex of approximate composition $MnCl_2 \cdot 2SnCl_4 \cdot 10C_2H_3N$ was reported in [17].

As shown by conductometric measurements, the halometallate complexes are formed by halide ion transfer reaction between stoichiometric amounts of MnX_2 salts and the appropriate acceptor in anhydrous acetonitrile [7 to 17, 25]. The low solubilities of the halometallate complexes and the possibility of recrystallization from pure acetonitrile is of advance. Handling under nitrogen is recommended. The starting components should be anhydrous or solid acetonitrile solvate adducts. In many cases heating the solution is necessary for rapid reaction. Evaporation and/or cooling precipitated the crystalline complexes, which were washed with acetonitrile or recrystallized from it, and dried in vacuum. On reaction of $MnCl_2$ with the tribromides [10] or the triiodides [11] of gallium and indium a small excess of the halide acceptor was used. The tetrabromoferrate was prepared from $MnBr_2$, iron powder, and bromine in acetonitrile. After reaction of the iron powder, the excess bromine and acetonitrile were evaporated in vacuum, during which the complex crystallized [8]. $[Mn(C_2H_3N)_6][MnI_4]$ was obtained by refluxing manganese (2 g) and iodine (1 to 2 g) in acetonitrile (15 mL) with protection against moisture. After filtration of the hot mixture, the complex precipitated on cooling [9]. $[Mn(C_2H_3N)_6][SnCl_6]$ was prepared by refluxing stoichiometric amounts of $Mn(C_2H_3N)_2Cl_2$ (see p. 10) and $Sn(C_2H_3N)_3Cl_4$ in acetonitrile [12]. The isothiocyanato complex $[Mn(C_2H_3N)_6][Sn(NCS)_6]$ precipitated from a mixture of $Mn(NCS)_2$ (10 mmol) and $Sn(NCS)_4 \cdot 2(C_2H_5)_2O$ (11 mmol), and was dissolved in 80 mL acetonitrile at 60 to 70°C. On evaporation and/or storage at 0 to 5°C the complex precipitated. It was washed with a mixture of acetonitrile-ether, then with pentane [16]. As shown by X-ray powder diagrams, the tetrachlorometallates are isomorphous [7]. Similar results were obtained for the tetrabromometallates [8]. Isomorphism was also observed with corresponding halometallates of Zn^{II}, Cd^{II}, Ni^{II}, Co^{II}, or Fe^{II} [7, 8].

Color, melting points, and magnetic moments (μ_{eff} in μ_B) of the $[Mn(C_2H_3N)_6]^{2+}$ halometallate complexes from susceptibility measurements (Gouy method) at room temperature are tabulated on p. 8. The magnetic moments of $[Mn(C_2H_3N)_6][TlCl_4]_2$ and $[Mn(C_2H_3N)_6][SbCl_6]_2$ were found to be independent of the temperature in the range 300 to 85 K and confirm the octahedral shape of the $[Mn(C_2H_3N)_6]^{2+}$ ion [13, 18]. The ESR X- and Q-band spectra of $[Mn(C_2H_3N)_6][GaCl_4]_2$ at room temperature (frequencies 9.364 and 33.76 GHz) reveal band maxima at 3335 and 12040 G with band widths of 670 and 700 G, respectively. The g values

2.01 and 2.00 (±0.01) are very close to the spin-only value of Mn^{2+}. No nuclear hyperfine splitting was observed in the powdered sample [18].

anion	color	m.p. in °C	μ_{eff}	Ref.
$AlCl_4^-$	white	90 (dec.)	—	[7]
$GaCl_4^-$	white	148 to 152	—	[7]
$InCl_4^-$	white	138 to 142	—	[7]
$InBr_4^-$	white	—	—	[8]
$TlCl_4^-$	white	135 to 137	5.90	[7, 18]
$TlBr_4^-$	white	—	—	[8]
$FeCl_4^-$	white	147 to 149	—	[7]
$FeBr_4^-$	brown	—	—	[8]
MnI_4^{2-}	yellow	—	5.86	[9]
$GaClBr_3^-$	gray	136 to 140	5.65	[10]
$GaClI_3^-$	white	147 to 150	5.65	[11]
$InClBr_3^-$	gray	124 to 130	5.47	[10]
$InClI_3^-$	white	141 to 146	5.50	[11]
$SnCl_5^- \cdot C_2H_3N$	white	103 to 106	—	[12]
$SbCl_6^-$	white	—	5.64*)	[13]
$SbCl_5Br^-$	yellow-orange	—	—	[14]
$NbCl_5Br^-$	orange	165 to 168	—	[14]
$TaCl_5Br^-$	yellow	170 to 174	—	[14]
$TeCl_2Br_4^{2-}$	yellow	350 (dec.)	5.81	[15]
$ZrCl_2Br_4^{2-}$	white-gray	350 (dec.)	5.80	[25]
$SnCl_6^{2-}$	white	85 (dec.)	—	[12]
$Sn(NCS)_6^{2-}$	white	—	—	[16]

*) Determined by the Faraday method.

The IR spectra of the complexes in Nujol reveal shifts of the acetonitrile vibration modes, when coordinated to the metal ion. A positive shift in the ν(CN) vibration mode of CH_3CN from 2250 to 2286 cm^{-1} was observed in the spectra of $[Mn(C_2H_3N)_6][SbCl_6]_2$ and $[Mn(C_2H_3N)_6]$-$[SnCl_6]$ [19] or to 2280 cm^{-1} in the case of $[Mn(C_2H_3N)_6][MBr_4]_2$ complexes (M = In, Tl, Fe) [8], due to coordination of the ligand through the nitrogen atom. For shifts of other acetonitrile vibration modes, see [8, 19]. A comparison with corresponding complexes of other metals resulted in a sequence of metal ions, arranged with increasing shifts in ν(CN) vibrations, corresponding to the Irving-Williams sequence [19]. Far-IR spectra have been investigated in the region 450 to 150 cm^{-1}, and band assignments made in [8, 20] with regard to the octahedral group $[Mn(C_2H_3N)_6]^{2+}$. The Mn–N stretching vibration modes occur in the 350 to 200 cm^{-1} region for most bivalent metal ions. The frequency order was also found to be in agreement with the Irving-Williams stability order. Bands assigned to Mn–N stretching (ν_{14}) and Mn–NCC wagging (ν_{15}) vibration modes of the $[Mn(C_2H_3N)_6][MX_4]_2$ and $[Mn(C_2H_3N)_6][SbCl_6]_2$ complexes (in cm^{-1}) are given below:

anion	$AlCl_4^-$	$TlCl_4^-$	$FeCl_4^-$	$FeBr_4^-$	$SbCl_6^-$
ν(Mn–N)	237	237	238	237 ± 2	237
δ(Mn–NCC)	166	160	163	163 ± 3	165
Ref.	[20]	[20]	[20]	[8]	[20]

The Raman-active fundamental frequencies of $[M(C_2H_3N)_6]^{2+}$, where M = Zn, Mn, Ni, Co, Fe, and Cu, have been studied for the polycrystalline state compounds, $[Mn(C_2H_3N)_6][SbCl_6]_2$ and $[Mn(C_2H_3N)_6][SnCl_6]$, and normal coordinate analyses have been carried out to analyze the observed Raman spectral data. The most important Raman bands, ν(Mn–N), were not identified, because these bands, expected by the normal coordinate analyses, are in the same frequency regions as the intense Raman bands of $SbCl_6^-$ and $SnCl_6^{2-}$ [21].

The Mn–N bond distance was assumed to be 2.10 Å and the force constant estimated to be k_{Mn-N} = 1.289 mdyn/Å, taking into account the values obtained for $[Al(C_2H_3N)_6]^{3+}$, reported in [22]. One of the calculated ν(Mn–N) stretching frequencies of A_{1g}, E_g, and F_{1u} symmetry species, the frequency for the F_{1u} species was found to be in excellent agreement with the observed IR vibration modes of ν(Mn–N) given in the table above [21].

The octahedral geometry of the $[Mn(C_2H_3N)_6]^{2+}$ cation in the complexes $[Mn(C_2H_3N)_6][InBr_4]_2$ and $[Mn(C_2H_3N)_6](FeBr_4)_2$ [8] or in $[Mn(C_2H_3N)_6][Sn(NCS)_6]$ [16] was also deduced from the diffuse reflectance spectra of the corresponding Ni^{II}, Co^{II}, Fe^{II} complexes. Band positions (in nm) observed in the electronic reflectance spectra of several manganese complexes are shown below with their assignments:

anion	$GaClBr_3^-$	$GaClI_3^-$	$InClBr_3^-$	$InClI_3^-$
$^6A_{1g} \rightarrow {}^4T_{1g}$ (G)	580	575	580	580
$^6A_{1g} \rightarrow {}^4T_{2g}$ (G)	470	475	470	470
$^6A_{1g} \rightarrow {}^4E_g, {}^4A_{1g}$ (G)	410	405	—	400
Ref.	[10]	[11]	[10]	[11]

anion	$SnCl_6^{2-}$	$SbCl_5Br^-$	$NbCl_5Br^-$	$TaCl_5Br^-$
$^6A_{1g} \rightarrow {}^4T_{1g}$ (G)	585	549	—	—
$^6A_{1g} \rightarrow {}^4T_{2g}$ (G)	459	444	459	488
$^6A_{1g} \rightarrow {}^4E_g, {}^4A_{1g}$ (G)	408	426	405	420
Ref.	[14]	[14]	[14]	[14]

A graph of the diffuse reflectance spectrum of $[Mn(C_2H_3N)_6][TeCl_2Br_4]$ is given in [15]. The bright yellow color and intense reflectance spectrum of $MnI_2 \cdot 3C_2H_3N$ is strongly suggestive of tetrahedral manganese(II). The complex is therefore formulated as $[Mn(C_2H_3N)_6][MnI_4]$ [9].

The molar conductivities of $[Mn(C_2H_3N)_6][MX_4]_2$ solutions (10^{-3} M) in acetonitrile at 20°C are typical for 1:2 electrolytes with values of $\Lambda = 283$, 262, 276, and 250 $cm^2 \cdot \Omega^{-1} \cdot mol^{-1}$ for $[MX_4]^- = GaClBr_3^-$, $GaClI_3^-$, $InClBr_3^-$, and $InClI_3^-$, respectively [10, 11]. The electrical conductivity of a 0.1M solution of $[Mn(C_2H_3N)_6][MnI_4]$ in acetonitrile, $\Lambda = 130$ $cm^2 \cdot \Omega^{-1} \cdot mol^{-1}$, is indicative of a 1:1 electrolyte [9].

$Mn^{II}(C_2H_3N)_4(ClO_4)_2$ and **$Mn^{II}(C_2H_3N)_4(BF_4)_2$** complexes were prepared by treating a suspension of nitrosyl perchlorate [2] or nitrosyl tetrafluoroborate [23] in acetonitrile with manganese metal for 4 to 6 h under reduced pressure. The resulting solutions were then evaporated to dryness under vacuum to yield the complexes [2, 23]. The tetrafluoroborate complex was dissolved in a minimum quantity of hot acetonitrile, and hot ethylacetate was added until the solution became cloudy. White crystals separate on cooling, and, after drying in vacuum, melt at 110 to 111°C. The X-ray powder diffraction pattern differs from those of the corresponding Zn or Cu^{II} complexes. The magnetic moment, $\mu_{eff} = 5.90$ μ_B, of $Mn(C_2H_3N)_4(BF_4)_2$ results from susceptibility measurements at room temperature. The IR spectrum in Nujol mulls (2300 to 650 cm^{-1}) indicates a square-planar arrangement of the four acetonitrile molecules, and an irregular octahedral Mn geometry with the anions coordinated above and below the plane [20]. Coordination of the acetonitrile molecules through the nitrogen atom is indicated by an increase of ν(CN) from 2248 (free ligand) to 2285 cm^{-1} (perchlorate) [2] or 2282 cm^{-1} (fluoroborate) [23]. A second band at 2300 or 2307 cm^{-1}, respectively, was assigned to a combined resonance frequency of the CH_3 deformation and the C–C stretching mode of the acetonitrile. The electrical conductivity of $Mn(C_2H_3N)_4(BF_4)_2$ in 0.1 M acetonitrile solution, $\Lambda = 143$ $cm^2 \cdot \Omega^{-1} \cdot mol^{-1}$, indicates that more than 50% of the complex is dissociated. This agrees with the molar weight, 111, determined ebullioscopically in 0.1M acetonitrile solution; the calculated value is 229. TG and DTA analyses of $Mn(C_2H_3N)_4(BF_4)_2$ show simultaneous loss of solvate and decomposition of the anion. The endothermic decomposition occurs between 170 and 240°C [23].

[$Mn^{II}(C_2H_3N)_4(SO_3Cl)_2$] was obtained, when a hot concentrated solution of 0.1 mol $Mn(SO_3Cl)_2$ in ~20 mL acetonitrile was abruptly cooled below room temperature. The crystalline solid was immediately filtered in vacuum under anhydrous conditions, washed with dry ether, and dried in vacuum. The complex melts with decomposition at 340°C. The IR spectrum reveals coordination of the chlorosulfate ions to Mn^{II} in addition to the acetonitrile molecules. Observed bands (in cm^{-1}) were assigned as follows: 2290 ν(CN); 1180, 1230 $\nu_{as}(SO_3)$; 1070 $\nu_s(SO_3)$; 580, 630 $\delta_{as}(SO_3)$; 560 $\delta_s(SO_3)$; 410 ν(S–Cl); 310 ϱ(S–Cl). The $\nu_s(SO_3)$ band does not change from that observed for the corresponding chlorosulfate, which is consistent with the existence of appreciable covalent bonding in the chlorosulfate group. The observed splitting in the doubly degenerate modes further substantiates the above argument, indicating the symmetry of the anion is lowered to a reduced C_s symmetry. The results of the magnetic susceptibility measurement at 23°C, $\mu_{eff}=5.93\ \mu_B$, and the bands observed in the diffuse reflectance spectrum, are consistent with the octahedral geometry of the complex. Band maxima at 18868 and 20472 cm^{-1} were assigned to the transitions $^6A_{1g}\rightarrow{}^4T_{1g}$ (G) and $\rightarrow{}^4T_{2g}$ (G), respectively. Charge transfer bands were observed at 28986 and 38462 cm^{-1}. The low molar conductivity indicates the existence of [$Mn(C_2H_3N)_4(SO_3Cl)_2$] complexes in solution [26].

$Mn^{II}(C_2H_3N)_2Cl_2$ and **$Mn^{II}(C_2H_3N)_2Br_2$**. The chloro complex was formed extremely slowly on passing chlorine into a suspension of manganese (2 g) in acetonitrile (50 mL), protected from moisture. It was redissolved by heating the solution. After hot filtration and cooling, the pink complex precipitated. It was washed with acetonitrile (1 to 2 mL) and dried in vacuum. The mixture of manganese (2 g) and bromine (2 mL) in 15 mL of acetonitrile was refluxed until the reaction was complete, then filtered and cooled, and the pink bromo complex was handled as above. Both the complexes were also obtained by Soxhlet extraction of anhydrous $MnCl_2$ or $MnBr_2$ with acetonitrile. The crystals were washed with acetonitrile and dried in vacuum. The magnetic moments from susceptibility measurements at room temperature are $\mu_{eff}=5.83$ and 5.79 μ_B, respectively. The IR spectra of both the complexes (in Nujol) show bands of coordinated acetonitrile, ν(CN) at 2275 and 2270 cm^{-1}, respectively, and a second band at 2295 cm^{-1}, assigned to a combination of ν(C–C) and $\delta(CH_3)$. Bands observed in the electronic reflectance spectra of the solid compounds (in $10^3\ cm^{-1}$) were assigned by comparison with the solution spectrum of $[Mn(H_2O)_6]^{2+}$:

assignment	$^4T_{1g}$(G)	$^4T_2$(G)	$^4E,^4A$(G)	$^4T_1$(G)	^4E(D)	$^4T_1$(P)	$^4A_2$(F)	$^4T_1$(F)
$Mn(C_2H_3N)_2Cl_2$	19.0	22.4	23.8	27.2	28.55	31.2	35.0	37.7
$Mn(C_2H_3N)_2Br_2$	19.2	22.2	23.8	27.0	28.2	31.2	36.7	38.0

The most likely structure consists of chains of octahedrally-coordinated Mn atoms with bridging halide ions in the xy plane and two acetonitrile ligands in the z direction [9].

$Mn^{II}(C_2H_3N)Br_2$ is assumed to have formed on electrolytic oxidation of manganese with bromine in unstirred acetonitrile. Yellow crystals were obtained, which redissolved in the reaction mixture, as the reaction proceeded; stirred solutions gave no precipitate. Complete removal of solvent and bromine from the reaction mixture yielded a pale pink solid, which was dried in vacuum at 40 to 60°C. This product had a bromine content close to that calculated for a monoacetonitrile adduct. Further heating to 240°C (10 min) yielded pink hygroscopic $MnBr_2$ [27].

$Mn^{III}(C_2H_3N)_n(NO_3)_3$. A manganese(III) complex with n between 1.5 and 2 was prepared as follows: the mixture of $KMnO_4$ (1 g) and N_2O_4 (20 mL) in 20 mL of acetonitrile was stirred for 2 h. After separation of the yellow $K_2Mn(NO_3)_4$, the brown filtrate was evaporated to yield a dark brown solid complex of unreproducible composition. The magnetic moment from susceptibility measurements at 22°C (Faraday method) is $\mu_{eff}=5.2\ \mu_B$, assuming the formula contains two molecules of acetonitrile. The IR spectrum (Nujol) shows ν(CN) bands of the coordinated acetonitrile at 2291 cm^{-1}. A strong band at 415 cm^{-1} may be assigned to the ν(Mn–N) vibration mode [24].

References:

[1] Rehder, D.; Ihmels, K.; Wenke, D.; Oltmanns, P. (Inorg. Chim. Acta **100** [1985] L11/L12).
[2] Hathaway, B. J.; Underhill, A. E. (J. Chem. Soc. **1961** 3091/6).
[3] Libuś, W.; Meçik, M.; Strzelecki, H. (J. Solution Chem. **9** [1980] 723/36).
[4] Libuś, W.; Chachulski, B.; Frączyk, L. (J. Solution Chem. **9** [1980] 355/69).
[5] Meçik, M.; Chudziak, A. (J. Solution Chem. **14** [1985] 653/64).
[6] Libuś, W.; Strzelecki, H. (Electrochim. Acta **16** [1971] 1749/55).
[7] Reedijk, J.; Groeneveld, W. L. (Recl. Trav. Chim. **87** [1968] 513/27).
[8] Reedijk, J.; Vervelde, J. B.; Groeneveld, W. L. (Recl. Trav. Chim. **88** [1969] 42/50).
[9] Hathaway, B. J.; Holah, D. G. (J. Chem. Soc. **1964** 2400/8).
[10] Masaguer, J. R.; Rodriguez, C.; Lopez, M. (Anales Quim. **74** [1978] 19/23).

[11] Masaguer, J. R.; Rodriguez, C.; Lopez, M. (Anales Quim. **74** [1978] 1041/4).
[12] Reedjik, J.; Groeneveld, W. L. (Recl. Trav. Chim. **86** [1967] 1103/26).
[13] Zuur, A. P.; Groeneveld, W. L. (Recl. Trav. Chim. **86** [1967] 1089/102).
[14] Masaguer, J.; Rodriguez, C.; Peña, M. C. (Acta Cient. Compostelana **15** [1978] 3/31).
[15] Masaguer, J. R.; Rodriguez, C.; Vazquez, M. V. (Acta Cient. Compostelana **14** [1977] 301/11).
[16] Brokaar, G.; Groeneveld, W. L.; Reedjik, J. (Recl. Trav. Chim. **89** [1970] 1117/20).
[17] Masaguer, J.; Coto, V. (Anales Real Soc. Espan. Fis. Quim. [Madrid] B **61** [1965] 905/12).
[18] Reedjik, J. (Recl. Trav. Chim. **88** [1969] 86/96).
[19] Reedjik, J.; Zuur, A. P.; Groeneveld, W. L. (Recl. Trav. Chim. **86** [1967] 1127/37).
[20] Reedjik, J.; Groeneveld, W. L. (Recl. Trav. Chem. **87** [1968] 1079/88).

[21] Hase, Y.; Alves, O. L. (An. Acad. Bras. Cienc. **52** [1980] 267/72).
[22] Hase, Y. (Chem. Scr. **13** [1979] 16/9; C.A. **91** [1979] No. 81025).
[23] Hathaway, B. J.; Holah, D. G.; Underhill, A. E. (J. Chem. Soc. **1962** 2444/8).
[24] Johnson, D. W.; Sutton, D. (Can. J. Chem. **50** [1972] 3326/31).
[25] Masaguer, J.; Rodriguez, C.; Arias, M. T. (Anales Quim. **73** [1977] 1455/9).
[26] Siddiqui, Z. A.; Shakir, M.; Aslam, M.; Khan, T. A.; Zaidi, S. A. A. (Syn. React. Inorg. Metal-Org. Chem. **13** [1983] 397/424, 405).
[27] Habeeb, J. J.; Neilson, L.; Tuck, D. G. (Inorg. Chem. **17** [1978] 306/10).

33.5 With Benzeneacetonitrile $C_6H_5CH_2CN$ (= C_8H_7N)

$[Mn^{II}(C_8H_7N)_6][MCl_4]_2$ complexes with M = In^{III} or Fe^{III} and **$[Mn^{II}(C_8H_7N)_6][SbCl_6]_2$** were prepared by stirring ~1.3 g of $MnCl_2$ (dehydrated in a vacuum at 150°C) in 25 mL of dry nitromethane for 24 h with the stoichiometric amount of the chloride acceptor component $InCl_3$, $FeCl_3$, or $SbCl_5 \cdot CH_3NO_2$. The filtered solutions were combined with 6 molar equivalents of benzeneacetonitrile. The yellow chloroantimonate and chloroindate complexes precipitated on cooling, the red-brown chloroferrate after the addition of pentane. The solid compounds were washed with pentane and dried in vacuum. $[Mn(C_8H_7N)_6][InCl_4]_2$ melts at 65°C, $[Mn(C_8H_7N)_6][FeCl_4]_2$ at 80°C, $[Mn(C_8H_7N)_6][SbCl_6]_2$ at 145°C. The IR spectra (Nujol) show ν(CN) shifts in a range from 2250 cm^{-1} (free ligand) to 2280 cm^{-1}. Anion bands were observed at 335 cm^{-1} (chloroindate and chloroantimonate) and at 375 cm^{-1} (chloroferrate). The X-ray powder patterns are similar to those of corresponding $[M(C_8H_7N)_6]X_2$ complexes (M = Ni^{II}, Co^{II}, Fe^{II}), for which an octahedral environment of the central metal atom has been deduced from the ligand field, ESR, and Mössbauer spectra. The ligand field strength of $C_6H_5CH_2CN$ is somewhat stronger than that of benzonitrile and equal to that of acetonitrile. The complexes are sensitive to hydrolysis.

Reference:

Jansen-Ligthelm, C. A.; Groeneveld, W. L.; Reedjik, J. (Inorg. Chim. Acta **7** [1973] 113/6).

33.6 With Propiononitrile CH_3CH_2CN (= Propanenitrile = C_3H_5N) or Acrylonitrile $CH_2{=}CHCN$ (= 2-Propenenitrile = C_3H_3N)

Remark. The polymerization of acrylonitrile in frozen aqueous solution is promoted by manganese(II) chloride. Coordination of monomeric acrylonitrile to manganese is assumed due to the observed positive shift of the CN stretching vibration (24 cm^{-1}). Cyclized manganese(II) complexes of silica-supported polyacrylonitrile are reported to catalyze the oxidation of cumene more actively than the manganese(II)phthalocyanine complex [2]. An increase in thermal stability and electrical conductivity was observed with polymeric acrylonitrile prepared in the presence of manganese(II) ions [3]. Individual Mn species are not characterized in [1 to 3].

$[Mn^{II}(C_3H_3N)_6][SbCl_6]_2$ was prepared by reaction of $MnCl_2$ (0.01 mol) with $SbCl_5 \cdot CH_3NO_2$ (0.2 mol) and acrylonitrile (0.06 mol) in 20 mL of nitromethane, or by treatment of 0.01 mol of $[Mn(CH_3NO_2)_6][SbCl_6]_2$ (see p. 21) with acrylonitrile (0.03 mol) in nitromethane (20 mL). The mixture was stirred for 3 h at room temperature. After filtration the solution was concentrated by evaporation in a vacuum, then allowed to crystallize. The white precipitate, washed with sodium-dried pentane in a moisture-free atmosphere, melts with decomposition at 142°C. The IR spectrum reveals a positive shift of ν(CN) to a higher wavenumber at 2252 cm^{-1} (free ligand 2228 cm^{-1}). There is no evidence for π bonding through the double bond: the C=C stretching vibration at 1003 cm^{-1} shows only a slight shift to lower wavenumber ($\Delta\nu = 6$ cm^{-1}). Vinyl out-of-plane hydrogen deformations are present at 981 (wagging) and 960 cm^{-1} (twisting), whereas the free ligand gives only a broad band centered at 970 cm^{-1}. A band at 270 cm^{-1} was assigned to ν(Mn–N). The IR spectrum is similar to those of $[M(C_3H_3N)_6][SbCl_6]_2$ complexes where M = Mg, Zn, Ni^{II}, Co^{II}, and Fe^{II} [4].

$Mn^{II}(C_3H_5N)Cl_2$ and **$Mn^{II}(C_3H_3N)Cl_2$** complexes are formed on agitation of anhydrous $MnCl_2$ with an excess of the corresponding nitrile under nitrogen. The white pulps were washed with benzene and dried in a stream of nitrogen. A magnetic moment of $\mu_{eff} = 6.02$ μ_B is reported for $Mn(C_3H_5N)Cl_2$. The IR spectrum of $Mn(C_3H_3N)Cl_2$ (mineral oil mull) shows a band for ν(CN) at 2267 cm^{-1} ($\Delta\nu = 35$ cm^{-1}). Vinyl CH and CH_2 out-of-plane deformation vibrations were observed at 975 and 962 cm^{-1}. The complexes are insoluble in propiononitrile or acrylonitrile, respectively [5].

References:

[1] Maekawa, T.; Ozaki, Y.; Yoshioka, H.; Okamura, S. (J. Macromol. Sci. Chem. A **12** [1978] 731/43).
[2] Bai, Ruke; Zong, Huijuan; He, Jigang; Jiang, Yingyan (Cuihua Xuebao **6** [1985] 191/4; C.A. **103** [1985] No. 76805).
[3] Rashidova, S. Sh.; Usmanova, M. M. (Uzb. Khim. Zh. **1976** No. 6, pp. 27/35; C.A. **86** [1977] No. 90789).
[4] Driessen, W. L.; Zuur, A. P.; Everstijn, P. L. A.; Monster, J. (Rev. Chim. Minerale **15** [1978] 93/8).
[5] Kern, R. J. (J. Inorg. Nucl. Chem. **25** [1963] 5/9).

33.7 With Benzonitrile C_6H_5CN ($=C_7H_5N$)

Complexes in Solution. Manganese(III) complexes [$Mn^{III}L_2Cl$] with HL = $CH_3COCH_2COCH_3$, $C_6H_5COCH_2COCH_3$, or $C_6H_5COCH_2COC_6H_5$ (see "Mangan" D 1, 1978, pp. 101, 116, and 112, respectively) form benzonitrile adducts $Mn^{III}L_2Cl \cdot C_6H_5CN$ in dichloromethane solution. Formation constants according to the process $Mn^{III}L_2Cl + C_6H_5CN = Mn^{III}L_2Cl \cdot C_6H_5CN$ were determined spectrophotometrically at 25°C: K = 0.258(20) L/mol at 480 nm, K = 0.276(66) L/mol at 500 nm, and K = 0.192(60) L/mol at 530 nm, respectively. The molar extinction coefficients of the adduct are $\varepsilon = 590$, 380, and 280 $L \cdot mol^{-1} \cdot cm^{-1}$, respectively. When increasing amounts of benzonitrile are added to the solution of the acetylacetonato complex an isosbestic point is observed at 447 nm [1].

Isolated Compounds. A complex of composition $Mn^{II}(C_7H_5N)_{2/3}Cl_2$ was obtained after refluxing anhydrous $MnCl_2$ with an excess of benzonitrile. The IR spectrum reveals a ν(CN) band at 2255 cm^{-1} (free ligand at 2235 cm^{-1}). The compound is assumed to be polymeric with bridging chlorine atoms. It decomposes in the range 460 to 570 K. The heat of decomposition, $\Delta H = 99 \pm 4$ kJ/mol, was measured by differential scanning calorimetry at 490 K, the peak temperature of decomposition. A weight loss of 34.7% represents the removal of the benzonitrile content (calculated 34.3%) [2].

Coordination of benzonitrile to manganese(II) ions retained in the interlayer space of montmorillonite, $Al_2(OH)_2Si_4O_{10}$, was studied by IR measurement. The formation of a complex is indicated by a shift of the ν(CN) vibration from 2235 to 2250 cm^{-1}, suggesting fixation of four molecules of benzonitrile per manganese atom [3].

References:

[1] Ito, Y.; Kawaguchi, S. (Bull. Chem. Soc. Japan **54** [1981] 150/8).
[2] Beech, G.; Marr, G.; Ashcroft, S. J. (J. Chem. Soc. A **1970** 2903/6).
[3] Tsunashima, A.; Tachiki, M.; Kodaira, K.; Matsushita, T. (Kogakubu Kenkyu Hokoku Hokkaido Daigaku No. 91 [1978] 87/94; C.A. **90** [1979] No. 196971).

33.8 With Substituted Benzonitriles

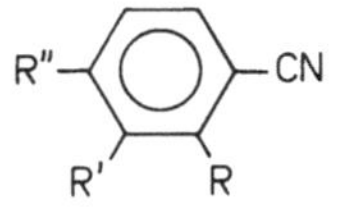

(= L)

ligand	R	R′	R″	formula
1	Cl	H	H	C_7H_4ClN
2	H	Cl	H	C_7H_4ClN
3	H	H	Cl	C_7H_4ClN
4	H	H	OCH_3	C_8H_7NO
5	H	OCH_3	OCH_3	$C_9H_9NO_2$

[$Mn^{II}L_6$][$SbCl_6$]$_2$. Complexes with ligands 1 to 5 were prepared by stirring the stoichiometric mixture of $SbCl_5 \cdot CH_3NO_2$ and $MnCl_2$ in nitromethane for 24 h. The unreacted $MnCl_2$ was then removed by centrifugation and the solution of the nitromethane complex [$Mn(CH_3NO_2)_6$][$SbCl_6$]$_2$ (see p. 21), combined with the appropriate nitrile. The complex with ligand 1 could be isolated only after the solution had been left standing for a few weeks at −5°C with a layer of dry pentane over it. The complex with ligand 2 was obtained after evaporation of part of the solvent and standing for a few hours at −5°C. The complexes with the other ligands separated as soon as a solution of the nitrile was mixed with the solution of the nitromethane complex. The crystals were washed with pentane and dried in a vacuum at room temperature. The operations were carried out in a dry atmosphere at ambient temperature or at −5°C. The X-ray powder patterns of the complexes were compared with corresponding complexes of Mg^{II}, Zn^{II}, Cd^{II}, Ni^{II}, Co^{II}, Fe^{II}, and Cu^{II}. Colors, melting points, ν(CN), and ν(Mn–N) vibration modes (in cm^{-1}) observed in the IR spectra of Nujol or KBr disks are given below:

ligand	color	m.p. in °C	ν(CN)		ν(Mn–N)
			ligand	complex	
1	white	140 to 145	2230	2256	216
2	white	153 to 155	2235	2253	205
3	white	223 to 226	2227	2248	203
4	yellow	196 to 198	2219	2240	199
5	yellow	118 to 120	2223	2240	207

Bands of the $SbCl_6^-$ ion were observed in the 344 to 339 and 149 to 175 cm^{-1} regions.

Reference:

van Driel, C. A. A.; Groeneveld, W. L. (Recl. Trav. Chim. **90** [1971] 389/404).

33.9 With Di- or Polycyano Compounds

Remark. Dicyanoamide, nitrosodicyanomethanide, and tricyanomethanide complexes of manganese(II), containing pyridine as well, have been described in "Manganese" D 3, 1982, pp. 90 and 100. For more recent data on the α- and β-form of manganese(II) dicyanoamide (preparation, IR and Raman spectra) see [1]. Complexes with 2,3-dimercapto-2-butenedinitrile are treated together with the thiolate complexes on pp. 33/9. Mixed ligand complexes containing cyano compounds and phosphines or phosphine oxides are described in "Manganese" D 8.

No.	ligand	formula
1	$CH_2(CN)_2$	$C_3H_2N_2$
2	$CH_3COCH(CN)_2$	$C_5H_4N_2O$ (= HL)
3	$NC(CH_2)_4CN$	$C_6H_8N_2$
4	$NCCH(CH_3)(CH_2)_2CN$	$C_6H_8N_2$
5	$CH(CN)_3$	C_4HN_3 (= HL)
6	$(NC)_2C{=}C(CN)_2$	C_6N_4
7	(NC)HC=(CN)CC(CN)=CH(CN)	$C_8H_2N_4$ (= HL)
8	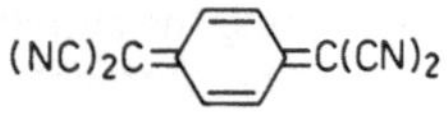	$C_{12}H_4N_4$
9	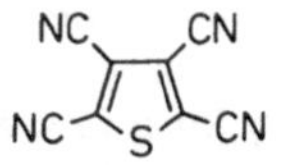	C_8N_4S
10	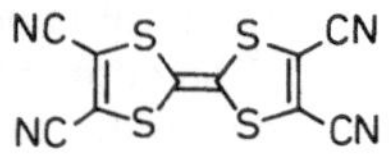	$C_{10}N_4S_4$

Complexes in Solution. Kinetic studies indicate the intermediate formation of an unspecified Mn^{III} complex with ligand 1 on polymerization of acrylonitrile and methyl methacrylate by the redox systems propanedinitrile/Mn^{III} in aqueous sulfuric acid or by propanedinitrile/ $Mn(CH_3COO)_3$ in DMF and glacial acetic acid. The irreversible decomposition yields the initiating radical [2].

$Mn^{II}(C_5H_3N_2O)_2 \cdot 4py$ and $Mn^{II}(C_5H_3N_2O)_2 \cdot 2L$ complexes with L = pyridine, 2- or 4-methylpyridine were obtained on reaction of manganese(II) nitrate (0.01 mol) dissolved in 15 to 20 mL H_2O with the appropriate amounts of the sodium salt of ligand 2 and the corresponding amine. The precipitated colorless crystals were dried in a vacuum. Decomposition points and characteristic IR bands of the complex and the sodium salt of the ligand in Nujol or KBr are shown below:

compound	t_{dec} in °C	ν(CO) in cm^{-1}	ν(CN) in cm^{-1}	
$NaC_5H_3N_2O$	—	1570	2182, 2198	2225
$Mn(C_5H_3N_2O)_2 \cdot 4py$	200	1555	2198	2220
$Mn(C_5H_3N_2O)_2 \cdot 2py$	170	1500	2190	2220
$Mn(C_5H_3N_2O)_2 \cdot 2(2\text{-}CH_3)C_5H_4N$	180	1560	2202	2220
$Mn(C_5H_3N_2O)_2 \cdot 2(4\text{-}CH_3)C_5H_4N$	200	1540 to 1550	2200	2220

The position of the ν(CO) and ν(CN) bands in comparison to the spectrum of $NaC_5H_3N_2O$ indicates coordination of the ligand anion through the enol oxygen atom. In the IR spectrum of the tetrapyridine solvate no band at 1570 cm^{-1}, indicative of noncoordinated pyridine molecules, is observed. An octahedral O_2N_4 environment at Mn is therefore assumed. The disolvate complexes are probably polymeric with bidentate bridging O,N-coordinated ligands. The complexes are slightly or not soluble in acetonitrile or dichloromethane [3].

$Mn^{II}(C_4N_3)_2$ and **$Mn^{II}(C_4N_3)_2 \cdot 0.25H_2O$.** The hydrate was obtained as a white precipitate by mixing the hot solutions of manganese(II) chloride or sulfate (25 mmol) and 50 mmol of the potassium salt of ligand 5, KC_4N_3, each dissolved in 25 mL H_2O. The complex was washed with water and dried to constant weight in a vacuum at 100°C [4]. By the same procedure (25 mmol of $MnCl_2$ in 10 mL H_2O and 50 mmol of KC_4N_3 in 25 mL H_2O), anhydrous $Mn(C_4N_3)_2$ was obtained. The elemental analysis, however, reveals only the manganese content, which is almost that of the hydrated species [5]. The anhydrous complex, $Mn(C_4N_3)_2$, was obtained from $Mn(C_4N_3)_2 \cdot 2py$ (see below) at 120°C in a vacuum [6]. The magnetic moment of $Mn(C_4N_3)_2 \cdot 0.25H_2O$ from susceptibility measurement at room temperature is $\mu_{eff} = 5.93\ \mu_B$ [4].

The infrared data indicate that the mesomeric-stabilized anion is coordinated in a tridentate fashion to manganese through the nitrile nitrogen atoms. The strong similarity between the spectrum of $C(CN)_3^-$ in its potassium salt and the spectra of the transition metal compounds provides convincing evidence that the planar structure of the ion is not significantly corrupted in the latter compounds. A polymeric structure is proposed, which consists of intersecting planes of tridentate $C(CN)_3$ moieties with Mn occupying pseudooctahedral holes. Each Mn^{2+} cation is coordinated by six nitrogen atoms. In the IR spectrum of $Mn(C_4N_3)_2 \cdot 0.25H_2O$ (Nujol and hexachlorobutadiene mulls) the ν(CN), ν(^{13}CN), and δ(H_2O) vibrations are observed at 2197, 2158, and 1630 cm^{-1}, respectively [4]. Broad bands at 275 and 230 cm^{-1} are assigned to ν(Mn–N) vibrations. The spectrum of the $C_4N_3^-$ ion shows only a weak band at 210 cm^{-1} [7]. The Raman spectrum of $Mn(C_4N_3)_2$ indicates that the D_{3h} symmetry of the free tricyanomethanide ion is essentially intact. The ν_s(CN) band of $C(CN)_3^-$ at 2225 cm^{-1} is shifted to 2256 cm^{-1} in the manganese compound. Different bands of the twice degenerate ν_e(CN) vibration mode, observed at 2209 m, 2190 vs, 2162 w, and 2155 w cm^{-1}, may be caused by the crystal structure of $Mn(C_4N_3)_2$. The spectrum of the $C(CN)_3^-$ ion shows only one band of the $\nu_e(CN)$ vibration mode at 2175 cm^{-1}. Strong bands of the $\nu(Mn–N)$ vibration mode are observed at 247, 225, and 213 cm^{-1} [8].

$Mn(C_4N_3)_2 \cdot 0.25H_2O$ is soluble only in dimethylformamide or dimethyl sulfoxide, but not in other organic solvents. The electrical conductivity of a 10^{-3} M solution of $Mn(C_4N_3)_2 \cdot 0.25H_2O$ in dimethylformamide, $\Lambda = 149\ cm^2 \cdot \Omega^{-1} \cdot mol^{-1}$, indicates the presence of a 1:2 electrolyte [4].

$Mn^{II}(C_4N_3)_2 \cdot 3py$ and **$Mn^{II}(C_4N_3)_2 \cdot 2py$**. The complex $Mn^{II}(C_4N_3)_2 \cdot 2py$ was prepared by adding 4 mL of pyridine to the mixture of 2.5 mmol of manganese(II) nitrate and 5 mmol of $K(C_4N_3)$ in 40 mL of water at 90°C. From the oil that initially separated, colorless crystals were obtained. $Mn(C_4N_3)_2 \cdot 3py$ precipitated from a solution of the disolvate (1.5 g) in 15 mL of pyridine by adding some carbon tetrachloride. A dimeric octahedral structure for the trisolvate and a polymeric structure for the disolvate, with bidentate anionic ligands, are deduced from magnetic and IR data cited for the corresponding Ni and Co species. A solution of $Mn(C_4N_3)_2 \cdot 2py$ in pyridine (8×10^{-4} M) has the electrical conductivity $\Lambda = 15.4\ cm^2 \cdot \Omega^{-1} \cdot mol^{-1}$. The complex is not soluble in chloroform, nitrobenzene, or acetonitrile. The complex decomposes on heating without melting. Maintained at 120°C under a vacuum, the complex yields $Mn(C_4N_3)_2$ [6].

$Mn^{II}(C_6N_4)$(?). The complex with ligand 6 was prepared according to the procedure given in [12] for the complex with ligand 8, $Mn(C_{12}H_4N_4)_2 \cdot 3H_2O$ (see below). An aqueous solution of the lithium salt, LiC_6N_4, was added to the aqueous solution of manganese(II) chloride (mole ratio 5:1). The precipitate was washed with water and dried in a vacuum. Elemental analysis data are not given. The IR spectrum (KBr disk) of the compound is shown in the range 3000 to 200 cm^{-1}. Characteristic bands at 1625 and 530 cm^{-1} are assigned to $\nu(C{=}C) + \nu(C{=}N)$ and $\nu(Mn–N)$ vibration modes, respectively. Additional bands are observed at 2330, 2300, 665, 650, and 275 cm^{-1}. A structure is suggested with divalent manganese bonded to the nitrogen atoms of both the 1,1-cyano groups of the dinegatively (?) charged ligand 6 [20].

A complex, **$Mn^{III}(TPP)(C_6N_4)$** containing the tetraphenylporphinato group (TPP) and the anion radical of ligand 6, is described in "Manganese" D 4, 1985, p. 117.

$Mn^{II}(C_8HN_4)_2 \cdot 3H_2O$. For preparation, 300 mg of the slightly soluble salt of ligand 7 was stirred in 100 mL of water for 30 min at 40°C together with the stoichiometric amount of hydrated manganese(II) sulfate. The mixture was filtered and the solvent evaporated below 50°C. The dry residue was extracted with acetone to yield violet crystals. A sample of $Mn(C_8HN_4)_2 \cdot 3H_2O$ pressed to 1500 kp/cm^2 shows remarkable electrical conductivity between 20 and 80°C, expressed by the volume resistivity (ϱ in $\Omega \cdot cm$): $\log \varrho = 7.7$ and is correlated with the activation energy, $E_a = 0.4$ eV [9]. The electronic spectrum of the complex dissolved in acetonitrile shows maxima at 630, 760, and 830 nm. The free ligand bands at 630 nm and at 680, 1000, and 1140 nm were assigned to the anion and the anion radical, respectively [10].

$Mn^{II}(C_{12}H_4N_4)_2$ and **$Mn^{II}(C_{12}H_4N_4)_2 \cdot nH_2O$** (n = 3, 2.5, 1). The different compounds containing the anion radical of ligand 8 were prepared as follows: A trihydrate was prepared by adding a threefold [11] or fivefold [12] excess of manganese(II) chloride to $LiC_{12}H_4N_4$, both in aqueous solution. The blue powdery solid precipitated and was washed with water and dried in a vacuum [11, 12]. $Mn(C_{12}H_4N_4)_2 \cdot 2.5H_2O$, probably identical with the trihydrate, was obtained by addition of a hot solution of 5.16 g of $MnCl_2 \cdot 4H_2O$ in 15 mL of water to a solution of 1.1 g of $LiC_{12}H_4N_4$ in 15 mL of water. The mixture was heated to boiling under nitrogen and allowed to cool slowly. The bluish solid was then washed with water and dried at room temperature. This complex was dried over P_4O_{10} in a vacuum at room temperature for about 60 h and yielded $Mn(C_{12}H_4N_4)_2 \cdot H_2O$. The monohydrate, heated at 130 to 140°C for two days under reduced pressure, yielded $Mn(C_{12}H_4N_4)_2$ [13]. $Mn(C_{12}H_4N_4)_2 \cdot 3H_2O$ decomposes at ~170°C [11, 12].

Magnetic susceptibility and electrical conductivity measurements on polycrystalline $Mn(C_{12}H_4N_4)_2$ and its hydrates are reported. Magnetic and preliminary ESR studies on the trihydrate indicate antiferromagnetic coupling of the manganese ions and an exchange interaction through delocalized anion electrons, which becomes important above 77 K. A magnetic moment of $\mu_{eff} = 4.66\ \mu_B$ results from susceptibility measurements (Faraday method) at room temperature [14]. According to [13], impurities can strongly affect the results of the susceptibility and conductivity measurements. The most likely impurity is residual $LiC_{12}H_4N_4$, which is used in the preparation of the $Mn(C_{12}H_4N_4)_2 \cdot 3H_2O$. The presence of this impurity is evident as a narrow line superimposed on the broad Mn line of the ESR spectrum. The impurity is responsible for the low value of $\mu_{eff} = 4.66\ \mu_B$ (see above). Magnetic moments and conductivities of the anhydrous compound and the hydrates $Mn(C_{12}H_4N_4)_2 \cdot 2.5H_2O$ and $Mn(C_{12}H_4N_4)_2 \cdot H_2O$ as function of temperature are shown in **Fig. 1**a and Fig. 1b [13].

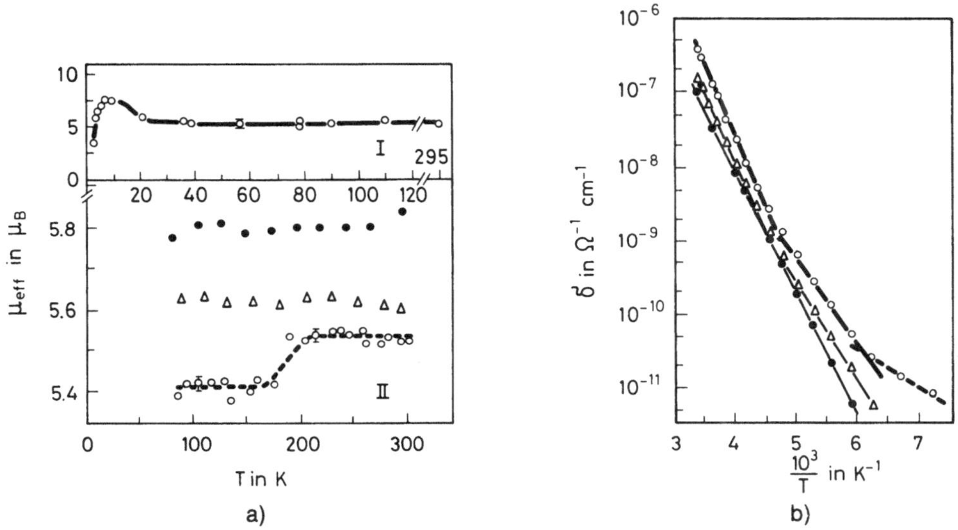

Fig. 1 Magnetic moments, μ_{eff}, (a) from ESR data (a I) and susceptibility measurements (a II) and conductivity, δ, (b) as functions of temperature; ○ $Mn(C_{12}H_4N_4)_2$, △ $Mn(C_{12}H_4N_4)_2 \cdot H_2O$, ● $Mn(C_{12}H_4N_4)_2 \cdot 2.5H_2O$ [13].

As shown in Fig. 1a the magnetic moments remain approximately constant from room temperature to about 20 K. Below this temperature the complexes show an increase in magnetic moment and in the ESR line width, due to the change from antiferromagnetic to ferromagnetic behavior [13]. At this temperature the anion electrons are localized and paired, and a strong exchange interaction (high ΔH values) occurs between Mn atoms. Above 77 K the anion electrons become delocalized, thus the Mn resonance is shifted towards the electronic g values: g = 2.025 or 2.005 at 300 K and g = 2.12 or 3.08 at 1.5 K were observed for the anhydrous compound [13] and the trihydrate [14], respectively.

The electrical conductivity of the complexes, measured between 300 and 135 K, see Fig. 1b, increases with the pressure used to make the pellets. It depends also on the amount of water present. In all cases, the conductivities follow an exp(−E/kT) law and exhibit a change in activation energies, E, to smaller values at transition temperatures below 25 K. The activation energies at higher temperatures range between 0.31 and 0.4 eV, whereas below the transition temperature the E values show only a small spread around 0.24 eV. For $Mn(C_{12}H_4N_4)_2$ a reversible change in conductivities at 210 and 165 K, corresponding to a decrease of μ_{eff} from 5.55 to 5.41 μ_B, is interpreted as a first-order structural phase transition, which results in a displacement of the radical anions parallel to the stacking axis [13]. A conductivity $\sigma=1\times10^{-5}$ $\Omega^{-1}\cdot cm^{-1}$ and the activation energy E = 0.16 eV was found for $Mn(C_{12}H_4N_4)_2\cdot3H_2O$ [15]. The complexes are insoluble in water and organic solvents [11 to 13].

$Mn^{II}(C_{12}H_4N_4)_2\cdot3CH_3CN$ was prepared by combining 0.5 mmol of $Mn(ClO_4)_3$ and 1 mmol of ligand 8 in 200 mL of acetonitrile and electrolyzing the mixture at −0.504 V (vs. SCE) for 49 h. The microcrystalline complex was removed from the cathode and compacted. The electrical conductivity is $\sigma=4.4\times10^{-3}$ $\Omega^{-1}\cdot cm^{-1}$ [16].

$[Mn^{II}phen_3](C_{12}H_4N_4)_2$ and **$[Mn^{II}phen_3](C_{12}H_4N_4)_4$**. The complex $[Mn phen_3](C_{12}H_4N_4)_2$ was prepared by mixing the boiling solution of one weight part $[Mn phen_3](ClO_4)_2$ in 11 weight parts of acetonitrile with the stoichiometric amount of the lithium salt of ligand 8 in absolute ethanol. The reaction mixture was cooled and filtered, the precipitate washed with ethanol and diethyl ether, then dried in air. The complex, $[Mn phen_3](C_{12}H_4N_4)_4$, in which two anion radicals and two neutral molecules of ligand 8 are bonded to manganese, was prepared by dissolving $[Mn phen_3]$-$(ClO_4)_2$ in acetonitrile and adding two molar equivalents of ligand 8 in boiling acetonitrile, followed by two molar equivalents $LiC_{12}H_4N_4$ in boiling absolute ethanol. Both complexes are semiconducting, their electrical resistivities are $\varrho=140$ and 1000 $\Omega\cdot cm$, respectively. The activation energies for conductivity are about 1 eV [11].

$Mn^{II}(C_{10}H_{24}N_4)_2(C_{12}H_4N_4)_6$ ($C_{10}H_{24}N_4$ = 1,4,8,11-tetraazacyclotetradecane). The complex precipitated from an aqueous solution of $Mn(C_{10}H_{24}N_4)(ClO_4)_2$ upon dropwise addition of the lithium salt of ligand 8 dissolved in H_2O. It was washed with a large amount of water and dried in a vacuum. The diffuse reflectance spectrum was investigated. The absorption band at ~7000 cm^{-1} corresponds to the charge transfer transition from the $C_{12}H_4N_4^{\overline{\cdot}}$ anion radical to incorporated (four) neutral $C_{12}H_4N_4$ molecules. An electrical resistivity, $\varrho=510$ $\Omega\cdot cm$, is observed [17].

$Mn^{II}(C_6H_8N_2)_2[NCB(C_6H_5)_3]_2$ complexes with ligand 3 or 4 (= $C_6H_8N_2$) were prepared by stirring a mixture of 51 parts of an 18% solution of $B(C_6H_5)_3$ in 3-pentenenitrile, 13 parts of 3-pentenenitrile and 2 parts of NaCN for 30 min at 50°C. The solution was then combined with 7.5 parts of $Mn(NO_3)_2\cdot6H_2O$ and the mixture held at 50°C for 3 h. After separation of solid product, 20 parts of the filtrate were combined with 20 parts of ligand 3 or 4. The solid complex with ligand 3 was washed with toluene and cyclohexane and dried. The complex with ligand 4 did not precipitate. Therefore, the mixture was extracted twice with 300 parts of cyclohexane at 20°C. The remaining solution was mixed with 300 parts of toluene and allowed to stand at room temperature for one hour. The complex separated after stirring the solution for 3 h at 50°C. It was washed and dried as above. Both white complexes are characterized by IR bands at 2275 and 2290 cm^{-1}, respectively, assigned to the ν(CN) vibration modes of nitrogen-coordinated bidentate dinitriles, which are presumed to be bridging ligands. A second band at 2195 and 2210 cm^{-1}, respectively, is assigned to the ν(CN) mode of the triphenylcyanoborate anions, also coordinated to manganese through the nitrogen atom [21].

Polymeric complexes were prepared by reaction of ligand 9 with Mn^{II} acetylacetonate [18] or ligand 10 with Mn^{II} or Mn^{III} acetylacetonate [19]. The reactants (mole ratio 2:1) were placed in a tube, the air was replaced by nitrogen, and the tube was submerged in a bath at 200 [18] or 275°C [19]. The reactions occurred spontaneously. The mixtures were then tempered in a vacuum for 4 h at ~300°C. The dark brown solids were crushed, extracted with benzene and acetone [18] or pyridine and acetone [19], then dried at 100°C. A phthalocyanine-type structure is assumed for the polymeric products. For the complex with ligand 9, the electrical conductivity, $\sigma=8.9\times10^{-9}$ $\Omega^{-1}\cdot cm^{-1}$, was measured under a pressure of 1500 kp/cm² at 300 K. The

Formulas of ligands are tabulated on p. 14

activation energy is 0.54 eV. The polymeric compound is soluble in sulfuric acid [18]. Electrical conductivities of $\sigma = 8 \times 10^{-8}$ and 3×10^{-8} $\Omega^{-1} \cdot cm^{-1}$ were measured at 298 K for the reaction products of ligand 10 with Mn^{II} or Mn^{III} acetylacetonate, respectively. The activation energy is 0.48 eV in both cases [19].

References:

[1] Civadze, A. Ju.; Köhler, H. (Z. Anorg. Allgem. Chem. **510** [1984] 25/30).
[2] Jayakrishnan, A.; Haragopal, M.; Mahadevan, V. (J. Polym. Sci. Polym. Chem. Ed. **19** [1981] 1311/8).
[3] Köhler, H.; Rassmi, K.; Skirl, R. (Z. Anorg. Allgem. Chem. **529** [1985] 173/8).
[4] Enemark, J. H.; Holm, R. H. (Inorg. Chem. **3** [1964] 1516/21).
[5] Trofimenko, S.; Little, E. L., Jr.; Mower, H. F. (J. Org. Chem. **27** [1962] 433/8).
[6] Köhler, H. (Z. Anorg. Allgem. Chem. **331** [1964] 237/52).
[7] Kolbe, A.; Köhler, H. (Z. Anorg. Allgem. Chem. **428** [1977] 113/4).
[8] Civadze, A. Ju.; Köhler, H. (Z. Anorg. Allgem. Chem. **510** [1984] 31/6).
[9] Miura, K.; Nakanome, T.; Haga, T. (Nippon Kagaku Zasshi **89** [1968] 350/6; C.A. **69** [1968] No. 26657).
[10] Miura, K.; Otaki, T.; Tsukamoto, K.; Haga, T. (Nippon Kagaku Zasshi **89** [1968] 356/62; C.A. **69** [1968] No. 23559).

[11] Acker, D. S.; Blomstrom, D. C. (U.S. 3162641 [1964]; C.A. **63** [1965] 550).
[12] Melby, L. R.; Harder, R. J.; Hertler, W. R.; Mahler, W.; Benson, R. E.; Mochel, W. E. (J. Am. Chem. Soc. **84** [1962] 3374/8).
[13] Thompson, R. C.; Gujral, V. K.; Wagner, H. J.; Schwerdtfeger, C. F. (Phys. Status Solidi A **53** [1979] 181/6).
[14] Thompson, R. C.; Hoyans, Y.; Schwerdtfeger, C. F. (Solid State Commun. **23** [1977] 633/6).
[15] Siemons, W. J.; Bierstedt, P. E.; Kepler, R. G. (J. Chem. Phys. **39** [1963] 3523/8).
[16] Kathirgamanathan, P.; Rosseinsky, D. R. (J. Chem. Soc. Chem. Commun. **1980** 839/40).
[17] Matsuoka, K.; Nogami, T.; Mikawa, H. (Mol. Cryst. Liquid Cryst. **86** [1982] 155/8).
[18] Manecke, G.; Wöhrle, D. (Makromol. Chem. **102** [1967] 1/23).
[19] Koßmehl, G.; Rohde, M. (Makromol. Chem. **178** [1977] 715/22).
[20] Shabaka, A. A.; El-Behery, K. M.; Fadly, M. (J. Mater. Sci. Letters **7** [1988] 685/7).

[21] McKay, C. M. (Ger. Offen. 3241955 [1983]; C.A. **99** [1983] No. 122640).

Formulas of ligands are tabulated on p. 14

34 Complexes with Nitro Hydrocarbons

34.1 With Nitroalkanes or Nitrocycloalkanes

ligand	formula
1	CH_3NO_2
2	$(CH_3)_2CHNO_2$ ($=C_3H_7NO_2$)
3	$C_5H_9NO_2$
4	$C_6H_{11}NO_2$
5	$C_6H_5CH_2NO_2$ ($=C_7H_7NO_2$)

Complexes in Solution. In basic solution nitrocycloalkanes (ligands 3 and 4) or phenylnitromethane (ligand 5) form the relatively stable anions $>C=\overset{+}{N}<{}^{O^-}_{O^-}$. The neutral or alkaline permanganate oxidation of these anions has been shown to give excellent yields of aldehydes and ketones. The nature of the activated manganese complexes, formed intermediately in the permanganate oxidation of the *aci*-nitrocycloalkane anions or *aci*-nitro-phenylmethane anions and substituted *aci*-nitro-phenylmethane anions is discussed in [1 to 3]. The kinetics of the oxidation reactions were investigated by means of spectrophotometric stopped-flow techniques from pH 12.5 to 13.6. The reactions are first-order in the *aci*-nitrocycloalkane or *aci*-nitro-phenylmethane anions and in permanganate and zero-order in hydroxide ion. The data are consistent with orbital rehybridization in the rate-determining step, which suggests an attack of permanganate at the carbon of the carbon-nitrogen double bond. The effect of p-substituents (Cl, Br, CH_3, OCH_3) at the phenyl ring on the rate of oxidation was studied at 1°C [2, 3].

$Mn^{II}(C_3H_6NO_2)_2$ and $Mn^{III}(C_3H_6NO_2)_2OH \cdot 0.5H_2O$. The brown manganese(II) complex and/or the deep brown manganese(III) compound were obtained by reaction of manganese atoms (formed by vaporization of the metal) with the deprotonated ligand 2 at −196°C. In the liquid state, the ligand contains the tautomeric form, $(CH_3)_2C{=}N(O)OH$, which probably participates in the initial formation reaction. During one hour at a pressure kept below 3×10^{-5} Torr, manganese (30 mmol) was condensed together with dry ligand 2 (300 mmol) onto the cold walls of the reactor. The vessel was then filled with nitrogen gas and warmed to 0°C. The solution of the reaction product in excess 2-nitropropane was filtered and the filtrate was concentrated by vacuum evaporation. The amorphous precipitates obtained on addition of hexane were dried in a vacuum. The magnetic moment, $\mu_{eff} = 5.20\ \mu_B$, results from susceptibility measurements on $Mn(C_3H_6NO_2)_2OH \cdot 0.5H_2O$ at room temperature; values of 5.91 and 5.72 μ_B were determined by ESR measurements in neat ligand solution at 29 and −80°C, respectively. The IR spectrum of the manganese(II) complex (Nujol mull) shows bands due to $\nu(OH)$ at 3300, $\nu(CN)$ at 1641, 1635, $\nu_{as}(NO_2)$ at 1159, 1130, $\nu_s(NO_2)$ at 945 cm^{-1}. Because of the split ν_{as} vibration mode, the nitro group is assumed to act as a chelating and bridging ligand. The electronic reflectance spectrum shows bands at 830, 500, 470, 360, and 300 nm. A molecular weight of ~1800 was determined by vapor pressure depression in chloroform. The different products react with acetylacetone to yield $Mn(acac)_2$ and $Mn(acac)_3$, respectively, with liberation of ligand 2 [4].

$Mn^{II}(C_7H_6NO_2)_2 \cdot 2L$ and $Mn^{II}(C_7H_6NO_2)_2 \cdot C_6H_{16}N_2$. *Aci*-nitro-phenylmethanato complexes with the additional monodentate ligands $L = H_2O$, py, or the bidentate ligand $C_6H_{16}N_2 = (CH_3)_2NCH_2CH_2N(CH_3)_2$ were prepared as follows: ligand 5 was allowed to react with an excess of aqueous 0.1M KOH solution. The pH was lowered to 7 by addition of dilute hydrochloric acid and a stoichiometric amount of an aqueous 0.1M manganese(II) nitrate solution was added. The dihydrate precipitated rapidly. The complexes with pyridine or N, N, N', N'-tetramethylethylenediamine were obtained by adding the anion solution to an aqueous solution of $Mn(NO_3)_2$

also containing a stoichiometric amount of the appropriate amine. They were also obtained from $Mn(C_7H_6NO_2)_2 \cdot 2H_2O$ by a ligand exchange reaction [5]. $Mn(C_7H_6NO_2)_2 \cdot C_6H_{16}N_2$, as shown by X-ray powder data, is isomorphous with the corresponding Ni species, the structure of which is reported. It contains chelating *aci*-nitro-phenylmethane anions. The manganese atom is in an octahedral environment of four oxygen and two nitrogen atoms. The structures of the complexes with $L = H_2O$ or py are assumed to be identical. The IR spectra of the compounds in Nujol are similar to that of $Mn(C_7H_6NO_2)_2 \cdot C_6H_{16}N_2$, which reveals bands of ν(CN) at 1578 and 1568 cm^{-1}, of $\nu_{as}(NO_2)$ at 1130, and of $\nu_s(NO_2)$ at 961 cm^{-1}. In the case of the diaqua complex the split band of the $\nu_s(NO_2)$ vibration observed at 1000 and 990 cm^{-1} may be caused by hydrogen bonding between the anions and the water molecules. The magnetic moments of the complexes, determined by susceptibility measurements at 298°C, are μ_{eff} = 5.90, 5.87, and 5.94 μ_B, respectively [5].

$[Mn^{II}(CH_3NO_2)_n][SbCl_6]_2$ (n = 6, 4). The complex with n = 6 was prepared by stirring dehydrated manganese(II) chloride with a slight excess of $SbCl_5 \cdot CH_3NO_2$ in nitromethane at room temperature. White crystals separated from the clear solution on standing at −6°C [6]. Complexes with n = 6 or 4 were prepared by adding $SbCl_5$ to the suspension of dehydrated manganese(II) chloride in carbon tetrachloride containing the corresponding amount of nitromethane. The pink microcrystalline complexes were formed in an exothermic reaction, and they were slightly contaminated with $MnCl_2$ [7]. The operations were carried out in a P_4O_{10}-dried glove box or at −6°C in a cold room [6, 7]. The IR spectrum of $[Mn(CH_3NO_2)_6][SbCl_6]_2$ in Nujol or Kel-F mulls (4000 to 200 cm^{-1}) shows no shift of the $\nu(NO_2)$ mode, but a small positive shift of the rocking mode $\varrho(NO_2)$ from 478 to 496 cm^{-1} in comparison to the free ligand [6]. Bands of the complexes with n = 6 and 4 in the range 3000 to 600 cm^{-1} are assigned in [6] as follows (free ligand bands in parentheses): $\nu_{as}(NO_2)$ at 1543, 1525 (1565); $\nu_s(NO_2)$ at 1379 (1383); $\delta(NO_2)$ at 658, 606 (659, 608). The shift and splitting of $\nu_{as}(NO_2)$ suggests an oxygen-bonded monodentate ligand [7, 8]. $[Mn(CH_3NO_2)_6][SbCl_6]_2$ decomposes at 103 to 107 [6] or 115°C [7]. The thermogravimetric analysis reveals start of decomposition at 115°C, a large loss in weight at 150 to 160°C, and the end of decomposition at 284°C [9]. The complexes decompose upon contact with water [6 to 9].

Other Nitromethane Complexes. A monosolvate, $MnCl_2 \cdot CH_3NO_2$, was obtained by refluxing anhydrous manganese(II) chloride (10 g) in nitromethane (25 mL) for 48 h. The reaction product was washed with an inert solvent. It does not melt below 250°C [10]. $MnS \cdot CH_3NO_2$ was prepared by passing dry hydrogen sulfide through the solution of manganese(II) nitrate in nitromethane. The light brown precipitate was washed with dry carbon tetrachloride and dried in a vacuum. Preparation and handling require anhydrous conditions [11].

References:

[1] Freeman, F.; Yeramyan, A.; Young, F. (J. Org. Chem. **34** [1969] 2438/40).
[2] Freeman, F.; Yeramyan, A. (Tetrahedron Leters **1968** 4783/6).
[3] Freeman, F.; Yeramyan, A. (J. Org. Chem. **35** [1970] 2061/2).
[4] Bashar, A. B. M. A.; Timms, P. L. (High Temp. Sci. **17** [1984] 417/25).
[5] Cook, J. A.; Osborne, M. J.; Rice, D. A. (J. Inorg. Nucl. Chem. **38** [1976] 711/3).
[6] Driessen, W. L.; Groeneveld, W. L. (Recl. Trav. Chim. **88** [1969] 491/8).
[7] Drăgulescu, C.; Petrovici, E.; Lupu, I. (Monatsh. Chem. **105** [1974] 1176/83).
[8] Lupu, I. (Bul. Univ. Brasov Ser. C **18** [1976] 97/102; C.A. **89** [1978] No. 119945).
[9] Lupu, I. (Bul. Univ. Brasov Ser. C **17** [1975] 77/82; C.A. **88** [1978] No. 98509).
[10] Paul, R. C.; Kaushal, R.; Pahil, S. S. (J. Indian Chem. Soc. **42** [1965] 483/5).

[11] Paul, R. C.; Kaushal, R.; Pahil, S. S. (J. Indian Chem. Soc. **44** [1967] 995/1000).

34.2 With Nitrobenzene, $C_6H_5NO_2$, or 1-Nitronaphthalene, $C_{10}H_7NO_2$

$[Mn^{II}(C_6H_5NO_2)_n][SbCl_6]_2$ (n = 6, 3). The complex with n = 6 was obtained by stirring dehydrated manganese(II) chloride with $SbCl_5 \cdot C_6H_5NO_2$ and nitrobenzene in the ratio 1:2:4 in nitromethane at room temperature. The pale yellow crystals precipitating at −6°C from the solution were washed with sodium-dried pentane [1]. The light pink complex $[Mn(C_6H_5NO_2)_3]$-$[SbCl_6]_2$ was formed in an exothermic reaction on addition of $SbCl_5$ to the mixture of dehydrated manganese(II) chloride and nitrobenzene in carbon tetrachloride. The ratio of components was 2:1:3 [2]. The complexes were handled at −6°C [1] or in a P_4O_{10}-dried glove box [2]. The magnetic moment, $\mu_{eff} = 6.35\ \mu_B$, of $[Mn(C_6H_5NO_2)_3][SbCl_6]_2$ was obtained from susceptibility measurements at 23°C [3]. The IR spectrum of $[Mn(C_6H_5NO_2)_6][SbCl_6]_2$ in Nujol is congruent with that of the ligand in the region 2000 to 700 cm^{-1}. Two bands, at ~682 and 669 cm^{-1}, appear instead of one band at 676 cm^{-1} in the free ligand spectrum, which is attributed to the symmetric NO_2 deformation mode by [4]. The band of nitrobenzene at 532 cm^{-1}, attributed to the NO_2 out-of-plane bending mode in [4], shifts to 543 cm^{-1} upon complexation. The complex bands observed at 430 and 413 cm^{-1} are probably originated from ligand vibrations at 420 and 397 cm^{-1}. A broad band at 340 ± 2 cm^{-1} is characteristic of the $\nu_3(T_{1u})$ mode of the octahedral $SbCl_6^-$ anion [1]. The spectrum of $[Mn(C_6H_5NO_2)_3][SbCl_6]_2$ shows a $\nu_{as}(NO_2)$ vibration shifted from 1527 to lower wavenumbers, but not split. This suggests that both oxygen atoms of the nitro group are bonded to Mn in an octahedral environment. A band at 339 cm^{-1} is assigned to the $SbCl_6^-$ vibration [2]. The complexes decompose upon contact with water, but they are more stable than the corresponding nitromethane species, see p. 21 [1, 2].

$Mn(C_{10}H_7NO_2)_2[SbCl_6]_2$ was prepared by adding $SbCl_5$ to the mixture of dehydrated manganese(II) chloride and 1-nitronaphthalene in carbon tetrachloride. The exothermic reaction yielded a red precipitate. In the IR spectrum of the complex, the $\nu_{as}(NO_2)$ and $\nu_s(NO_2)$ vibrations are negatively shifted, suggesting that the ligand is bidentately coordinated by the oxygen atoms [2]. The thermogravimetric analysis data reveal the start of decomposition at ~120°C. At 271°C the weight loss corresponds to the release of chlorine and 1-nitronaphthalene, according to $Mn(C_{10}H_7NO_2)_2[SbCl_6]_2 \rightarrow MnCl_2 + 2SbCl_3 + 2Cl_2 + 2C_{10}H_7NO_2$. The complex is sensitive to moisture [2, 5].

References:

[1] Driessen, W. L.; van Geldorp, L. M.; Groeneveld, W. L. (Recl. Trav. Chim. **89** [1970] 1271/5).
[2] Drăgulescu, C.; Petrovici, E.; Lupu, I. (Monatsh. Chem. **105** [1974] 1176/83).
[3] Lupu, I. (Bul. Univ. Brasov Ser. C **18** [1976] 103/8; C.A. **89** [1978] No. 52430).
[4] Green, J. H. S.; Kynaston, W.; Lindsay, A. S. (Spectrochim. Acta **17** [1961] 486/502).
[5] Lupu, I. (Bul. Univ. Brasov Ser. C **17** [1975] 77/82; C.A. **88** [1978] No. 98509).

35 Complexes with Ligands Containing Sulfur

For correlation purposes, several complexes with sulfur ligands have already been described in the "Manganese" D series, e.g., complexes with thiosemicarbazones, thiocarbazates, or thiocarbazones in "Manganese" D 6, 1988, pp. 337, 360, and 365, respectively. Remarks at the beginning of most of the sections in the present volume show where additional complexes with similar sulfur-containing ligands are to be found in volumes D 1 to D 6.

35.1 Complexes with Thiols or Thiones

General References:

Dance, I. G.; The Structural Chemistry of Metal Thiolate Complexes, Polyhedron Report Number 15, Polyhedron **5** [1986] 1037/104.

Blower, P. J.; Dilworth, J. R.; Thiolato Complexes of the Transition Metals, Coord. Chem. Rev. **76** [1987] 121/85.

Eisenberg, R.; Structural Systematics of 1,1- and 1,2-Dithiolato Chelates, Progr. Inorg. Chem. **12** [1970] 295/369.

Burns, R. P.; McCullough, F. P.; McAuliffe, C. A.; 1,1-Dithiolato Complexes of the Transition Elements, Advan. Inorg. Chem. Radiochem. **23** [1980] 211/80.

Coucouvanis, D.; The Chemistry of the Dithioacid and 1,1-Dithiolate Complexes, Progr. Inorg. Chem. **11** [1970] 233/371.

Burns, R. P.; McAuliffe, C. A.; 1,2-Dithiolene Complexes of Transition Metals, Advan. Inorg. Chem. Radiochem. **22** [1979] 303/48.

McCleverty, J. A.; Metal 1,2-Dithiolene and Related Complexes, Progr. Inorg. Chem. **10** [1968] 49/221.

Hoyer, E.; Dietzsch, W.; Dithiolenchelate, Z. Chem. [Leipzig] **11** [1971] 41/53.

Hoyer, E.; Dietzsch, W.; Heber, R.; Ungesättigte 1,2-Dithiolat-Liganden als analytische Reagenzien, Wiss. Z. Karl-Marx-Univ. Leipzig **24** Math. Naturw. Reihe [1975] 429/37.

McCleverty, J. A.; Redox Reactions and Electron Transfer Chains of Inert Transition Metal Complexes, React. Mol. Electrodes **1971** 403/60.

Cox, M.; Darken, J.; Metal Complexes of Thio-β-diketones, Coord. Chem. Rev. **7** [1971/72] 29/58.

Livingstone, S. E.; Monothio-β-diketones and Their Metal Complexes, Coord. Chem. Rev. **7** [1971/72] 59/80.

General

In the following chapter complexes with thiols and the mercapto derivatives of sulfonic or carbonic acids are described. After these, complexes with thiodiketones and mercapto ketones are treated, then complexes with heterocyclic thiols and thiones. Complexes with thiones have been included because of the thione-thiol tautomerism of these ligands.

Almost all of the thiol and thione ligands form chelate complexes, either by the use of two sulfur donor atoms, as in the case of dithiols, or by one sulfur atom and a nitrogen or an oxygen donor atom, as is the case with most of the other ligands.

Mostly, the mono- and dithiols form mononuclear tetrahedral manganese(II) complexes. Dithiols also form manganate complexes with Mn in the formal +3 and +4 oxidation states.

(A separate survey for complexes with 1,2- and 1,3-dithiols is given on p. 33.) Among the complexes with monothiols, the adamantane-like cage of the $[Mn_4(C_6H_5S)_{10}]^{2-}$ ion, containing a tetrahedron of Mn atoms and an octahedron of bridging sulfur atoms, is noteworthy. The manganate complex anions are precipitated by addition of large cation halides such as $[N(C_2H_5)_4]Br$ and $[P(C_6H_5)_4]Cl$. The complexes with thiols are extremely air-sensitive. Therefore, the manipulations of preparation and investigation were performed using standard inert-atmosphere techniques and distilled oxygen-free solvents.

Dithio-β-diketonato complexes do not form by direct reaction of the metal with the ligand. When this is attempted the ligand is first oxidized, forming a cyclic dithiolium cation with an S–S bond. On reaction with a strong reducing agent the desired Mn^{II} dithioketonate complex is formed. The heterocyclic thiols and thiones form mononuclear tetrahedral species as do the above cited mono- and dithiols. But complexes with additional coordinated halide ions or water molecules, and polymeric compounds with ligand bridges are also observed.

Because of their intense colors the thiolato and thionato complexes are suitable for the spectrophotometric determination of microamounts of manganese. Numerous studies of analytical applications have been undertaken, particularly of the complexes with 1,2-dithiols, mercapto carboxylic acids, thiodiketones, 1-hydroxy-2-pyridinethione and its derivatives, and 8-quinolinethiol and its derivatives. A separate survey is given on p. 66 for the complexes with 8-quinolinethiol and its derivatives, mainly investigated for their analytical uses.

Complexes of Mn^{II} and Mn^{III} (or other transition metals) with tetrahedral metal-sulfur coordination sites, are of significant interest as synthetic analogues for the active sites in metalloproteins or nucleic acids. Complexes of Mn^{II} with ethane- or benzenethiols were studied with a view toward their suitability as precursors for new manganese cage and cluster molecules, which itself are thought to occur as coordination units in large biomolecules.

35.1.1 **With Alkane- or Benzenethiols** RSH

ligand 1	ligand 2	ligand 3
C_2H_5SH	C_6H_5SH	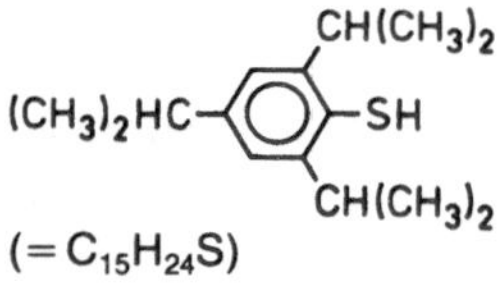
$(=C_2H_6S)$	$(=C_6H_6S)$	$(=C_{15}H_{24}S)$

$[N(CH_3)_4]_2[Mn^{II}(C_6H_5S)_4]$ was prepared by the reaction of hydrated manganese(II) chloride and the sodium salt of ligand 2 (mole ratio 1:2.5) in degassed ethanol under strictly anaerobic conditions. The reaction mixture was filtered to remove NaCl and the filtrate treated with a slight excess of the stoichiometric amount of $[N(CH_3)_4]Br$ in water. The solution was stirred, cooled to −20°C, and crystals isolated. The product was recrystallized from methanol with slow cooling to −20°C. The compound was isolated, washed with ethanol and ether, then dried in vacuum. The yield of pink crystals was 36%. They are monoclinic with a = 9.829(3), b = 14.468(5), c = 12.095(3) Å, β = 90.93(3)° [4].

$[N(C_2H_5)_4]_2[Mn^{II}(C_6H_5S)_4]$. The pale rose complex was obtained by the reaction of $[N(C_2H_5)_4]_2[MnCl_4]$ with ligand 2 and triethylamine in dimethylformamide (mole ratio 1:5:4.5) in a Schlenk tube under anaerobic conditions. The reaction mixture was stirred for 1 h, then filtered, and the deposit washed with two portions of dimethylformamide. After addition of ether, the tube was cooled to −15°C overnight. The product was isolated, washed with acetone and ether, and dried under vacuum. The complex was also prepared in 62% yield by a procedure analogous to $[N(CH_3)_4][Mn(C_6H_5S)_4]$, using $MnCl_2 \cdot 4H_2O$ and the sodium salt of ligand 2 in a mole ratio of 1:4. It was recrystallized from acetonitrile [4].

A band in the IR spectrum at 470 cm^{-1} was assigned to the ν(Mn–S) vibration. Solutions of the $[Mn^{II}(C_6H_5S)_4]^{2-}$ complex are stable in an inert atmosphere. The half-wave potential, $E_{1/2}$, for the oxidation of $[Mn^{II}(C_6H_5S)_4]^{2-}$ to $[Mn^{III}(C_6H_5S)_4]^{-}$ was measured by cyclic voltammetry at a Pt electrode vs. aqueous Ag|AgCl, 0.1 mol/L KCl. $E_{1/2} = -0.050$ V in tetrahydrofuran, using 0.1 mol/L $[(C_4H_9)_4N]BF_4$ as the supporting electrolyte, and $E_{1/2} = -0.110$ V in acetonitrile, using 0.1 mol/L $[(C_2H_5)_4N]BF_4$ as the supporting electrolyte. The electrode process is irreversible in both cases [7].

$[P(C_6H_5)_4]_2[Mn^{II}(C_6H_5S)_4]$ was prepared by the reaction of $[N(C_2H_5)_4][Mn(C_2H_5OS_2)_3]$ (see p. 131; $C_2H_5OS_2H$ = O-ethyl carbonodithioic acid), the potassium benzenethiolate, and $[P(C_6H_5)_4]Cl$ (mole ratio 1:4.7:2) in CH_3CN. The reaction was carried out in an N_2 atmosphere, using thoroughly degassed solvents. The reaction mixture was refluxed for 15 min and filtered while hot. The dark yellow filtrate was reduced in volume by evaporation under reduced pressure. Upon standing for ca. 12 h, large yellow cubic crystals formed; these were isolated, washed with ethanol, ether, and dried under vacuum at room temperature. The yield of product was 57% [1]. The complex had been previously prepared by the reaction of $[P(C_6H_5)_4]_2[Mn(C_4O_2S_2)_2]$ (see p. 40; $C_4H_2O_2S_2$ = 3,4-dimercapto-3-cyclobutene-1,2-dione) with excess C_6H_5SK in CH_3CN under rigorously oxygen-free N_2 [2]. The magnetic moment of the solid is $\mu_{eff} = 5.79\ \mu_B$ at ambient temperature. The single crystal X-ray structure determination has been carried out. The crystals are orthorhombic with a = 13.828(2), b = 17.527(4), c = 24.834(5) Å; Z = 4. The space group is $Pbc2_1$ (C^5_{2v} = No. 29). The calculated density is $D_{calc} = 1.29$ g/cm³. The structure was solved up to R = 0.055. Selected bond distances (in Å) and angles (in °) in the $[Mn(C_6H_5S)_4]^{2-}$ anion are given below (the labeling of the MnS_4 core is shown in **Fig. 2**):

Mn–S(1)	2.454(3)	S(1)–Mn–S(2)	98.4(1)	S(2)–Mn–S(4)	115.0(1)
Mn–S(2)	2.445(3)	S(3)–Mn–S(4)	101.5(1)	S(1)–Mn–S(4)	111.5(1)
Mn–S(3)	2.421(3)	S(1)–Mn–S(3)	120.0(1)	S(2)–Mn–S(3)	111.2(1)
Mn–S(4)	2.449(3)				

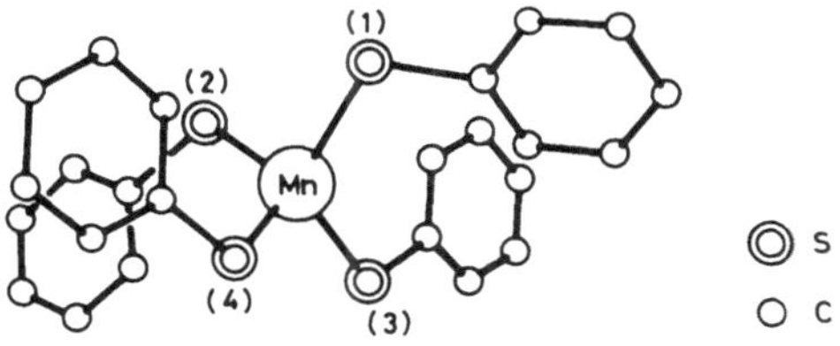

Fig. 2. Structure of the $[Mn^{II}(C_6H_5S)_4]^{2-}$ anion (H atoms omitted for clarity) [3].

The complex possesses a distorted tetrahedral geometry with a mean Mn–S distance of 2.443(12) Å [3]. It is X-ray isomorphous with the series $[P(C_6H_5)_4]_2[M(C_6H_5S)_4]$, M = Zn, Cd, Ni^{II}, Co^{II}, Fe^{II} [2, 3], for which similar distortions of the MS_4 units from T_d symmetry were observed [3]. The solid complex is stable for short periods of time, but its solutions decompose very rapidly when exposed to air [2]. $[P(C_6H_5)_4]_2[Mn(C_6H_5S)_4]$ reacts with dibenzyltrisulfane, $C_6H_5CH_2S_3CH_2C_6H_5$, in CH_3CN to give $[P(C_6H_5)_4]_2[(S_5)Mn(S_6)]$ (see p. 92) [8].

$[N(C_2H_5)_4]_2[Mn^{II}_2(C_2H_5S)_6]$ was prepared by the reaction of $MnCl_2$ in CH_3CN-DMF (3:1 v/v) with solid C_2H_5SNa (mole ratio 1:3) in an N_2 atmosphere. The reaction mixture was stirred for 4 h, treated with $[N(C_2H_5)_4]Cl$ in an amount equimolar to Mn, and stirred for an additional 2 h.

The mixture was filtered to remove NaCl and the filtrate reduced to one-fourth volume. Slow addition of ether to the filtrate was followed by slow cooling to 0°C. Maintaining the solution at 0°C for 24 h yielded extremely air-sensitive, deep red-orange crystals. The product was isolated, washed with ether, and dried in vacuum. The yield of product was 80%.

The single crystal X-ray-diffraction structure determination has been carried out. The crystal lattice is monoclinic, space group $P2_1/n$-C_{2h}^5, No. 14 (standard setting $P2_1/c$) with a = 10.993(1), b = 10.503(1), c = 18.306(2) Å, β = 93.82(1)°; Z = 2. The experimental density is D_{exp} = 1.17 g/cm³, the calculated density is D_{calc} = 1.16 g/cm³. The structure was solved up to R = 0.05. Selected bond distances (in Å) and bond angles (in °) in the $[Mn_2(C_2H_5S)_6]^{2-}$ anion (see **Fig. 3**) are:

Mn–S(1)	2.503(1)	Mn–S(2)	2.404(1)	S(1)–Mn–S(1′)	99.5(1)	S(1)–Mn–S(2)	108.6(1)
Mn–S(1′)	2.480(1)	Mn–S(3)	2.391(1)	S(2)–Mn–S(3)	112.3(1)	S(1)–Mn–S(3)	111.3(1)

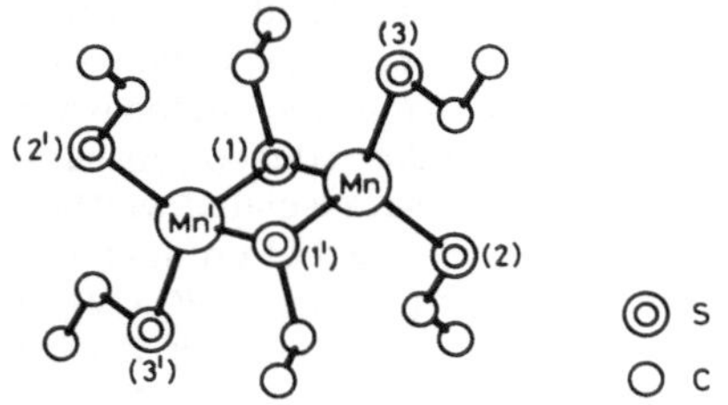

Fig. 3. Structure of the $[Mn_2^{II}(C_2H_5S)_6]^{2-}$ anion [5].

Positional parameters are presented in the paper. The anion consists of a dimeric unit in the anti configuration (C_i symmetry) with two bridging thiolate sulfur atoms and a distorted tetrahedral geometry around the Mn^{II} atoms. The mean bridge Mn–S distance is 2.492(12) Å, and the mean terminal Mn–S distance is 2.398(7) Å [5].

$[N(CH_3)_4]_2[Mn_4^{II}(C_6H_5S)_{10}]$ was prepared in an analogous manner to the tetraethylammonium salt, except the precipitant $[N(CH_3)_4]Cl$ was used. Fractional crystallization from CH_3CN-ether gave large red crystals in 10% yield. The melting point is 102°C (decomposition). The single crystal X-ray structure determination has been carried out. The complex crystallizes in the triclinic system, space group $P\bar{1}$ (C_i^1 = No. 2), with the lattice parameters a = 13.184(3), b = 23.743(4), c = 12.930(3) Å, α = 91.63(2)°, β = 113.76(1)°, γ = 79.53(2)°; Z = 2. The experimental density is D_{exp} = 1.34 g/cm³, the calculated density D_{calc} = 1.33 g/cm³. The structure was solved up to R = 0.045. Atomic positional parameters are presented in the paper. The structure of the complex anion is shown in **Fig. 4**. It consists of a tetranuclear cluster with a tetrahedral

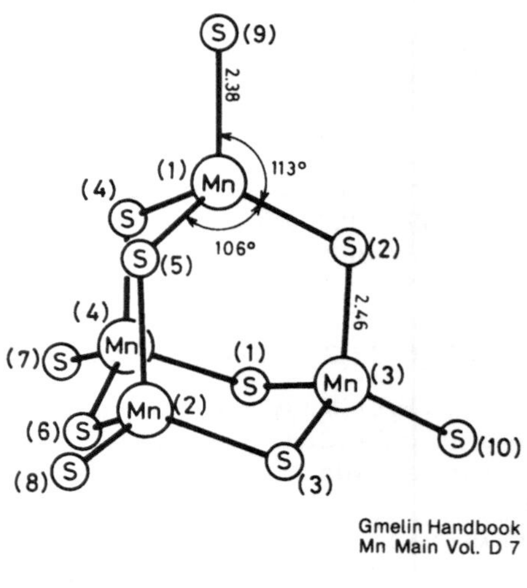

Fig. 4. Structure of the $[Mn_4^{II}(C_6H_5S)_{10}]^{2-}$ anion (phenyl groups omitted) [4].

geometry around the Mn^{II} atoms. Each metal atom is bonded to three bridging thiolate sulfur atoms, S_b, and one terminal sulfur atom, S_t. The $Mn_4(\mu\text{-}S)_6$ cage, consisting of a nearly regular Mn_4 tetrahedron and a highly distorted S_6 octahedron shows an adamantane-like stereochemistry. The Mn_3S_3 rings have chair conformation. Selected (the smallest and largest values for each type) bond distances (in Å) and bond angles (in °) are:

$Mn–S_t$		$Mn–S_b$	
Mn(4)–S(7)	2.368(7)	Mn(2)–S(6)	2.444(3)
Mn(1)–S(9)	2.382(3)	Mn(3)–S(3)	2.489(2)
$S_b–Mn–S_t$		$S_b–Mn–S_b$	
S(8)–Mn(2)–S(5)	104.2(1)	S(5)–Mn(1)–S(4)	94.3(1)
S(8)–Mn(2)–S(6)	117.1(1)	S(4)–Mn(1)–S(2)	112.9(1)

Mean values for the given ranges under idealized T_d symmetry are shown in Fig. 4. The mean values for the Mn–Mn and $S_b–S_b$ distances are 4.17(3) and 3.92(15) Å, respectively [4].

$[N(C_2H_5)_4]_2[Mn^{II}_4(C_6H_5S)_{10}]$. The complex was prepared by dropwise addition of a methanolic solution of $MnCl_2 \cdot 4H_2O$ to one of C_6H_5SNa (mole ratio 1:2.5) under strictly anaerobic conditions. The reaction mixture was stirred for 30 min, and then treated with the stoichiometric amount of $[N(C_2H_5)_4]Cl$. After stirring for 10 min, the reaction mixture was cooled to −20°C to yield an amorphous solid, which was extracted with CH_3CN. The extract was filtered, and the filtrate evaporated to a gummy residue. A methanol solution of the residue was allowed to stand, and crystals formed. The product was isolated, washed with 1:3 v/v CH_3CN-ether, ether, and then dried in vacuum. The yield of the red-orange crystals was 38%. The melting point of the solid is 75°C (decomposition). The magnetic moment of the solid is $\mu_{eff} = 4.61\ \mu_B$ per Mn atom at ambient temperature. The crystals are monoclinic with a = 15.43, b = 13.21, c = 41.52 Å, and $\beta = 96.29°$.

The solution magnetic moment of the complex in CH_3CN is temperature-dependent, increasing from $\mu_{eff} = 4.43\ \mu_B$ at 237 K to $\mu_{eff} = 4.70\ \mu_B$ at 301 K. This behavior is indicative of antiferromagnetic coupling of the Mn^{II} ions. The first reduction potential of the complex in CH_3CN (0.1 M $[N(C_4H_9)_4]ClO_4$) is −1.65 V vs. SCE (irreversible). The electronic absorption spectrum of the complex in CH_3CN shows maxima at: 490 (ε = 10), ~470 (sh), 430 (8), ~280 (sh), 258 (8460) nm [4].

$[P(C_6H_5)_4][Mn^{III}(C_{15}H_{23}S)_4]$. The purple complex was prepared by the reaction of $[N(C_2H_5)_4]_2[MnCl_5]$ and excess lithium 2,4,6-triisopropylbenzenethiolate, $Li(C_{15}H_{23}S)$, in CH_3CN-2-propanol at −20°C. A square-planar structure for the complex is likely, since its crystals are isomorphous with the analogous Co^{III} complex, which possesses a square-planar geometry. The thermal instability of the Mn compound in the solid state and in solution has prevented a complete structural characterization [6].

References:

[1] Coucouvanis, D.; Murphy, C. N.; Simhon, E.; Stremple, P.; Draganjac, M. (Inorg. Syn. **21** [1982] 23/8, 25).

[2] Holah, D. G.; Coucouvanis, D. (J. Am. Chem. Soc. **97** [1975] 6917/9).

[3] Swenson, D.; Baenziger, N. C.; Coucouvanis, D. (J. Am. Chem. Soc. **100** [1978] 1932/4).

[4] Costa, T.; Dorfman, J. R.; Hagen, K. S.; Holm, R. H. (Inorg. Chem. **22** [1983] 4091/9).

[5] Watson, A. D.; Rao, C. P.; Dorfman, J. R.; Holm, R. H. (Inorg. Chem. **24** [1985] 2820/6).

[6] Fikar, R.; Koch, S. A.; Millar, M. M. (Inorg. Chem. **24** [1985] 3311/2).

[7] Fabre, P.-L.; Poilblanc, R. (Bull. Soc. Chim. France **1986** 740/4).

[8] Coucouvanis, D.; Patil, P. R.; Kanatzidis, M. G.; Detering, B.; Baenziger, N. C. (Inorg. Chem. **24** [1985] 24/31, 25).

Formulas of ligands are tabulated on p. 24

35.1.2 With Aminoalkane- or Aminobenzenethiols NH_2RSH

No.	ligand	formula
1	$H_2N[CH_2]_2SH$	C_2H_7NS
2	$H_2N[CH_2]_2NH[CH_2]_2SH$	$C_4H_{12}N_2S$
3	$H_2N[CH_2]_3NH[CH_2]_2SH$	$C_5H_{14}N_2S$
4	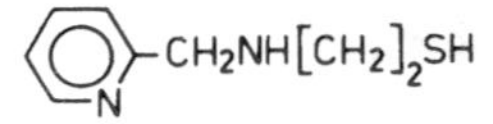	$C_8H_{12}N_2S$
5	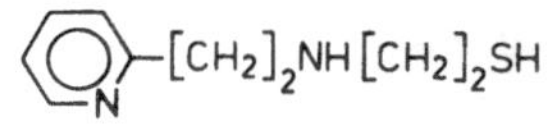	$C_9H_{14}N_2S$
6	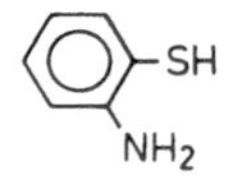	C_6H_7NS

The complex with ligand 6, $\mathbf{[Mn^{II}(C_6H_6NS)_2]}$, was prepared by the reaction of hydrated manganese(II) acetate and the ligand (mole ratio 1:2) in 75% aqueous ethanol under nitrogen at 45°C. The pale cream-colored precipitate which formed was isolated, washed with 75% aqueous ethanol, absolute ethanol, and dried under reduced pressure at about 45°C. X-ray powder photographs were recorded. (The d spacings are presented.) The magnetic moment of the solid is $\mu_{eff}=5.55\ \mu_B$ at 289 K. Variable-temperature magnetic measurements indicate a slight antiferromagnetic interaction between adjacent metal atoms, and yield the Weiss constant, $\Theta=-10$ K. The $1/\chi$ vs. T curve departs from linearity at low temperatures. The parameters $J=-1.6\ cm^{-1}$ and $g=1.92$ were obtained. The reflectance spectrum of the solid shows electronic absorption spectral maxima at 16600 and 22400 cm^{-1}, which were assigned to the high-spin octahedral Mn^{II} transitions $^6A_{1g}\rightarrow{}^4T_{1g}, {}^4T_{2g}$. The compound is insoluble in most solvents, but very slightly soluble in dimethylformamide, in which it is a nonelectrolyte. Magnetic, spectral, and solubility properties suggest a polynuclear sulfur-bridged structure [1].

Dinuclear complexes of the type $\mathbf{[Mn^{II}_2L_2X_2]}$ (see the table below) were recently prepared by the reaction of equimolar amounts of manganese(II) chloride and the respective ligand HL in methanol under nitrogen. The colorless microcrystals that separated were washed with methanol and dried in vacuum. Trinuclear complexes of the type $\mathbf{[Mn_3L_4]X_2}$ were obtained as almost colorless microcrystals in nearly the same way as described above, except for the use of an excess of the ligand (about 1.5 times). The magnetic susceptibility was measured in the 80 to 300 K range (Faraday method). The χ vs. T curves indicate intramolecular antiferromagnetic spin-exchange interactions through bridging S atoms. The magnetic moments (in μ_B) at 297.8 K and the exchange constants (in cm^{-1}, obtained by using equations based on models for dimeric or trimeric compounds) are:

HL	complex	μ_B	J
2	$[Mn_2(C_4H_{11}N_2S)_2Cl_2]$	4.88	−11.8
3	$[Mn_2(C_5H_{13}N_2S)_2Cl_2]$	5.01	−9.8
4	$[Mn_2(C_8H_{11}N_2S)_2Cl_2]$	4.94	−10.4
5	$[Mn_2(C_9H_{13}N_2S)_2Cl_2]\cdot 0.5H_2O$	5.14	−9.0
5	$[Mn_2(C_9H_{13}N_2S)_2(NO_3)_2]$	5.21	−8.5
2	$[Mn_3(C_4H_{11}N_2S)_4](ClO_4)_2$	4.71	−14.7
3	$[Mn_3(C_5H_{13}N_2S)_4](ClO_4)_2$	5.16*)	−8.9
3	$[Mn_3(C_5H_{13}N_2S)_4](NO_3)_2$	4.85	−11.1

*) At 295.4 K.

The g factors are 2.0 for the dinuclear compounds and 2.03 to 2.06 for the trinuclear compounds [2].

A light brown mixed-valence $\mathbf{Mn^{II,III}}$ **compound** was formed by the reaction of anhydrous $MnCl_2$ and ligand 1 (mole ratio 1:2) in dry ethanol in the presence of C_2H_5OK under air at 25°C for 3 h. The composition of the soluble material is $Mn_5(C_2H_6NS)_7Cl_8 \cdot 5H_2O$. Variable-temperature magnetic measurements show that the Curie-Weiss law is obeyed with the Curie constant C = 3.67 K, and the Weiss constant $\Theta = -31$ K. The ESR spectrum shows an absorption at g = 2.032. IR bands in the 3450 to 575 cm^{-1} range were assigned to OH, NH, CH, CN, or Mn–S vibrations. The experimental results suggest a polynuclear structure for the compound [3].

References:

[1] Larkworthy, L. F.; Murphy, J. M.; Phillips, D. J. (Inorg. Chem. **7** [1968] 1436/43).
[2] Handa, M.; Ōkawa, H.; Kida, S. (Bull. Chem. Soc. Japan **61** [1988] 3353/5).
[3] Kaneko, M.; Ishihara, N.; Yamada, A. (Makromol. Chem. **182** [1981] 89/99).

35.1.3 With Mercaptoalkanols, -phenols, or -benzenediols

No.	ligand	formula
1	$HSCH_2CH_2OH$	C_2H_6OS
2	$HSCH_2CH(OH)CH_2OH$	$C_3H_8O_2S$
3	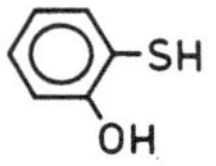	C_6H_6OS
4	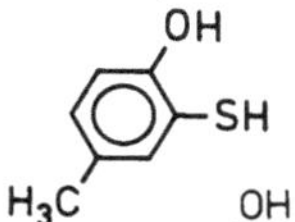	C_7H_8OS
5	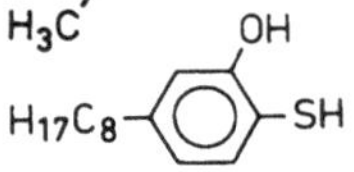	$C_{14}H_{22}OS$
6	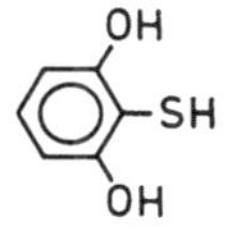	$C_6H_6O_2S$
7	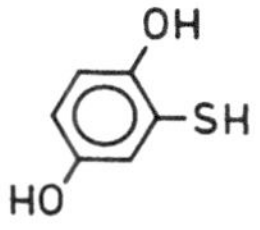	$C_6H_6O_2S$
8	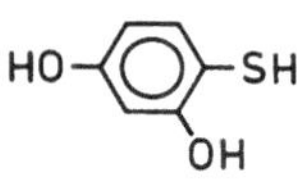	$C_6H_6O_2S$

35.1.3.1 Complexes in Solution

A soluble amber complex forms in aqueous ammonia solution between Mn^{II} and ligand 1 [1] even in the presence of sulfide ion [2]. The detection limit in aqueous ammonia was observed at a dilution of ~1:50100 [1]. An aqueous solution of ligand 2 dissolves $Mn(OH)_2$ to give a green solution, indicating complex formation between the ligand and Mn^{II} [3].

In basic, neutral, and acidic media ligand 4 and Mn^{II} give colorations or precipitates, and in dilute acid solution ternary complexes are formed with pyridine and 1,10-phenanthroline [4]. In ammoniacal or alkaline solution yellow soluble compounds are formed by ligand 6 and Mn^{II}; formation of the dianion $Mn[HOC_6H_3(O)S]_2^{2-}$ is assumed [5]. Reaction of ligand 7 [6] and ligand 8 [7] with Mn^{II} in aqueous CH_3COONa gives yellow soluble complexes which can be detected up to the dilution limit 1:5000 [6, 7].

References:

[1] Buscarons, F.; Casassas, E. (Anales Real Soc. Espan. Fis. Quim. [Madrid] B **55** [1959] 655/62).
[2] Buscarons, F.; Casassas, E. (Anales Real Soc. Espan. Fis. Quim. [Madrid] B **55** [1959] 663/8).

[3] Buscarons, F.; Casassas, E. (Anales Real Soc. Espan. Fis. Quim. [Madrid] B **51** [1955] 331/40).
[4] Dziomko, V. M.; Cherepakhin, A. I. (Vses. Zaochn. Politekhn. Inst. Sb. Statei No. 11 [1955] 37/43; C.A. **1957** 10297).
[5] Dziomko, V. M.; Cherepakhin, A. I. (Vses. Zaochn. Politekhn. Inst. Sb. Statei No. 9 [1955] 65/9; C.A. **1957** 4196).
[6] Buscarons, F.; Alsina, J. (Anales Real Soc. Espan. Fis. Quim. [Madrid] B **59** [1963] 101/8).
[7] Buscarons, F.; Alsina, J. (Inform. Quim. Anal. [Madrid] **19** [1965] 12/22).

35.1.3.2 Isolated Manganese(II) Compounds

General. A manganese(II) compound with one molecule of twice-deprotonated 2-mercaptophenol is accessible as the methanol complex, $[Mn(C_6H_4OS)(CH_3OH)_2]$, by the reaction of $MnCl_2$ with the ligand in absolute methanol in the presence of stoichiometric amounts of triethylamine under anhydrous conditions. By ligand exchange reactions with amines, the ternary amine complexes were obtained. The complex $[Mn(C_6H_4OS)]$ was obtained by heating the ternary pyridine complex. Molecular weight determinations in DMF (vapor osmometric) yielded a polymeric structure for $[Mn(C_6H_4OS)]$ and $[Mn(C_6H_4OS)(CH_3OH)_2]$ with a distorted octahedral site symmetry at the Mn atom for both compounds. The ternary amine complexes are monomeric tetrahedral. The structures of all the complexes are consistent with the magnetic moments (see below) and UV-visible spectra (not reported in the publication). All the compounds are stable to air and humidity. Thermoanalyses show the loss of solvent molecules, or amine ligands from the ternary complexes, occurring in several steps. The 2-mercaptophenol ligand is split off continuously up to 1200 K. All the complexes are insoluble in most organic solvents. Only in highly polar solvents, such as pyridine and DMF, are they more or less soluble. Some of the amine complexes, particularly the ternary complexes with pyridine and bipyridine, are also soluble in methanol and methylene chloride [1].

$[Mn(C_6H_4OS)]$. The light brown complex was prepared by suspending $[Mn(C_6H_4OS)py_2]$ in decalin and heating the suspension to boiling for 8 h. The magnetic moment is $\mu_{eff} = 5.45\ \mu_B$ at 295 K [1].

$[Mn(C_6H_4OS)(CH_3OH)_2]$ was prepared by the reaction of $MnCl_2$, 2-mercaptophenol, and triethylamine (mole ratio 1:1:2) in absolute methanol. The reaction was continued for 3 h at moderate heat. The colorless product was isolated, washed with methanol, and dried in vacuum. The magnetic moment of the solid is $\mu_{eff} = 5.84\ \mu_B$ at 295 K. Thermoanalysis of the solid shows that 0.7 mol CH_3OH per mol of complex is lost at 350 to 450 K, 0.2 mol at 600 to 650 K, and 1.1 mol at 650 to 760 K [1].

$[Mn(C_6H_4OS)\{N(C_2H_5)_3\}(CH_3OH)]$ and **$[Mn(C_6H_4OS)\{NH(C_2H_5)_2\}_2]$** were prepared by suspending $[Mn(C_6H_4OS)(CH_3OH)_2]$ in $N(C_2H_5)_3$ or $NH(C_2H_5)_2$ and heating the agitated reaction mixture to boiling for 4 h. The light yellow or light brown complexes were isolated, washed with benzene and hexane, and then dried in vacuum. The complexes can also be prepared by heating $[Mn(C_6H_4OS)(CH_3OH)_2]$ with the respective amine in benzene, as described for the following compounds. The magnetic moments of the solids are $\mu_{eff} = 5.98$ and $6.02\ \mu_B$, respectively, at 295 K. Thermoanalyses show the loss of the methanol in the interval 350 to 500 K and loss of the triethylamine in the interval 500 to 580 K. The ternary complex with diethylamine gives off one amine molecule between 400 and 580 K and the other between 680 and 790 K [1].

$[Mn(C_6H_4OS)py_2]$ and **$[Mn(C_6H_4OS)bpy]$** were prepared by suspending $[Mn(C_6H_4OS)(CH_3OH)_2]$ in benzene and heating the mixture with the stoichiometric amount of pyridine or bipyridine. The light brown or yellow solids were isolated, washed with benzene and hexane, and dried in vacuum. The magnetic moments of the solids are $\mu_{eff} = 5.17$ and $5.89\ \mu_B$, respectively, at 295 K. Thermoanalyses show that the complexes lose the amine ligands in the intervals 350 to 480 K and 480 to 690 K [1].

$Mn(C_{14}H_{21}OS)_2$ was prepared by the reaction of manganese(II) chloride and ligand 5 (mole ratio 1:2) both dissolved in alcohol or alcohol-water mixture at pH ~ 7 to 8. Coordination of the thiolate S atom and the phenolic hydroxy group to the Mn atom is assumed. The complex is effective in inhibiting the oxidation of esters of C_5 to C_9 monocarboxylic acids and pentaerythritol, which are used as lubricants [2].

References:

[1] Andrä, K.; Schmidt, K.-D. (Z. Anorg. Allgem. Chem. **498** [1983] 199/204).
[2] Kovtun, G. A.; Zhukovskaya, G. B.; Berenblyum, A. S.; Moiseev, I. I. (Neftekhimiya **22** [1982] 501/3; C.A. **97** [1982] No. 219212).

35.1.3.3 Isolated Manganese(III, IV) Compound

An insoluble solid forms in the reaction of anhydrous $MnCl_2$ and 2-mercaptoethanol (mole ratio 1:2) in dry ethanol in the presence of C_2H_5OK under air at 25°C for 3 h. The composition of the light brown material is $Mn_3(C_2H_4OS)_2O_3(OH)(H_2O)_4 \cdot 7KCl$. Variable-temperature magnetic measurements show that the Curie-Weiss law is obeyed with the Curie constant C = 2.18 K and $\Theta = -42$ K. The ESR spectrum shows an absorption at g = 2.018. IR bands in the 3400 to 550 cm^{-1} range were assigned to various groups of the coordinated ligand and to ν(Mn–S) or ν(Mn–O) vibrations. The experimental results suggest a polynuclear structure.

Reference:

Kaneko, M.; Ishihara, N.; Yamada, A. (Makromol. Chem. **182** [1981] 89/99).

35.1.4 With 1,1-Dithiols or 1,1-Dimercapto Compounds

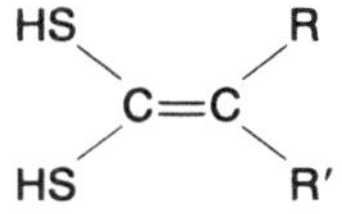

ligand 1	with R = H, R' = NO_2	(= $C_2H_3NO_2S_2$)
ligand 2	with R = R' = $C(O)CH_3$	(= $C_6H_8O_2S_2$)
ligand 3	with R = R' = $C(O)OC_2H_5$	(= $C_8H_{12}O_4S_2$)
ligand 4	with R = R' = CN	(= $C_4H_2N_2S_2$)

$Mn^{II}[Mn^{II}(C_2HNO_2S_2)]$. To an aqueous solution of the dipotassium salt of ligand 1 (0.025 mol) there was added over a few minutes an aqueous solution of 0.0125 mol Mn^{II} acetate. Thereafter a second solution of Mn^{II} acetate (0.0125 mol) was added slowly to the reaction solution. The red-brown product which precipitated during this second addition was obtained in a 50 to 60% yield [3].

$[(C_6H_5)_3P(CH_2C_6H_5)]_2[Mn^{II}(C_2HNO_2S_2)_2]$ was prepared by treating an Mn^{II} salt with the sodium salt of ligand 1 and a halide of the benzyltriphenylphosphonium cation. The X-ray powder patterns and IR spectrum (neither reported) were described as very similar to those of the corresponding Co^{II}, Ni^{II}, Cu^{II}, Pd^{II}, and Pt^{II} complexes. The IR spectrum and other physical properties are consistent with the formation of four-membered ring structures by coordination of the two sulfur atoms of the ligand. The complex undergoes reversible oxidation with iodine and reduction with borohydride in 50% (v/v) dimethylformamide-chloroform [2].

[Mn^{II}Lbpy(H_2O)$_2$] and **[MnLphen(H_2O)$_2$]** (H_2L = ligand 2 or 3). The complexes were prepared by the reaction of [Mnphen(H_2O)$_4$]$^{2+}$ or [Mnbpy(H_2O)$_4$]$^{2+}$ ions with the sodium salt of ligand 2 or 3 (mole ratio 1:1) in water at ambient temperature. After standing for 10 to 30 min, a precipitate formed. The IR spectra show bands which were assigned as follows: around 1675 cm^{-1} to ν(C=C); around 1100, 840, 610 cm^{-1} to ν(C=S), and around 390 cm^{-1} to ν(Mn–S) vibrations. The ν(Mn–N) band appears in the region of 300 cm^{-1}. Evidence for coordinated water is provided by bands around 920, 770, and 640 cm^{-1}. The IR spectra are consistent with the coordination of both sulfur atoms of the ligand to give a four-membered ring [1].

[$N(C_3H_7)_4$]$_3$[Mn^{III}($C_4N_2S_2$)$_3$]. A hot aqueous solution of the potassium salt of ligand 4 and an aqueous solution of tetrapropylammonium bromide were added to an aqueous solution of $Mn(CH_3COO)_2 \cdot 4H_2O$ with stirring. To the resulting mixture acetone was added while heating, until a homogenous green solution resulted. The solution was boiled for 1 min and filtered. On cooling for ~5 min a solid deposited and was filtered off. The remaining solution gave green plates on partial evaporation. The complex melts at 148 to 150°C (color change at 120°C). The table of the d-spacings (presented in the publication) shows that the Mn^{III} complex is not isomorphous with the corresponding Cr^{III}, Fe^{III}, and Co^{III} complexes, which are octahedral. The magnetic moment of the solid is μ_{eff} = 4.8 μ_B at 22°C. The Mn^{III} complex is a 3:1 electrolyte in nitromethane (Λ = 229 $L \cdot \Omega^{-1} \cdot mol^{-1}$ for a 10^{-3} M solution). The IR spectrum of the complex (in KBr) shows the ν(CN) band at ~2200 cm^{-1}. Other bands in the range between 3000 and 750 cm^{-1} are reported, but were not assigned [5].

[$(C_6H_5)_3P(CH_2C_6H_5)$]$_2$[Mn^{IV}($C_8H_{10}O_4S_2$)$_3$]. The purple compound was prepared by the reaction of Mn^{III} acetate monohydrate with the potassium salt of ligand 3 and [$(C_6H_5)_3P(CH_2C_6H_5)$]Cl in ethanol. The complex precipitated with the dropwise addition of water. It was collected, washed with ethanol and ether, and recrystallized from an acetonitrile-ether solution. The yield was 39%. The melting point is 147°C. The complex is X-ray powder pattern isomorphous with the corresponding Fe^{IV} complex, for which an X-ray structure determination shows S coordination and a geometry of the FeS_6 moiety intermediate between octahedral and trigonal prismatic. (The d-spacings are reported.) The magnetic moment is μ_{eff} = 3.79 μ_B at room temperature, determined by the Faraday technique. IR bands of the complex in Nujol or hexachlorobutadiene at 1708 and 1653 cm^{-1} were assigned to ν(C=O) vibrations, bands at 1455 and 1441 cm^{-1} to ν($C=CS_2$) vibrations. Other bands in the 1300 to 800 cm^{-1} range were not assigned. The IR spectrum is very similar to that of the Fe^{IV} compound. In the UV-visible spectrum of the [$Mn(C_8H_{10}O_4S_2)_3$]$^{2-}$ complex in CH_2Cl_2, three bands were observed in the 26000 to 36000 cm^{-1} range with ε between 25000 and 59500 $L \cdot mol^{-1} \cdot cm^{-1}$. Cyclic voltammetric measurements of the complex (no solvent given) with tetrabutylammonium perchlorate as the supporting electrolyte show an irreversible reduction wave at $E_{1/2}$ = 1.45 V vs. Ag|AgI [4].

References:

[1] Malik, W. U.; Bembi, R.; Bhardwaj, V. K. (J. Indian Chem. Soc. **61** [1984] 379/80).
[2] Fackler, J. P., Jr.; Coucouvanis, D. (Chem. Commun. **1965** 556/7).
[3] Lewis, S. N.; Miller, G. A. (U.S. 3449388 [1966/69]).
[4] Pedelty, R. R. (Diss. Univ. Iowa 1977; Diss. Abstr. Intern. B 38 [1977] 1708).
[5] Fackler, J. P., Jr.; Coucouvanis, D. (J. Am. Chem. Soc. **88** [1966] 3913/20).

Formulas of ligands are summarized on p. 31

35.1.5 With 1,2- or 1,3-Dithiols or Polythiols

No.	ligand		formula
1	$HSCH_2CH_2SH$		$C_2H_6S_2$
2	$HSCH_2$–CH(SH)–CH_2OH		$C_3H_8OS_2$
3	$HSCH_2$–CH(SH)–CH_2–S–S–CH_2–CH(SH)–CH_2SH		$C_6H_{14}S_6$
4	$HSCH_2$–CH_2–CH_2SH		$C_3H_8S_2$
5	$HSCH_2$–CH(SH)–CH_2SH		$C_3H_8S_3$
6	HSCH=CHSH		$C_2H_4S_2$
7	HSC(CN)=C(CN)SH		$C_4H_2N_2S_2$
8	$HSC(CF_3)$=$C(CF_3)SH$		$C_4H_2F_6S_2$
9	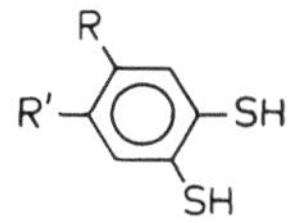	R=H, R'=CH_3	$C_7H_8S_2$
10		R=R'=CH_3	$C_8H_{10}S_2$
11	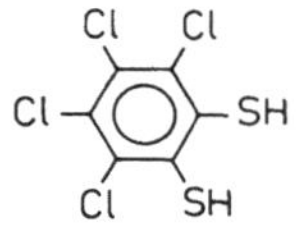		$C_6H_2Cl_4S_2$
12	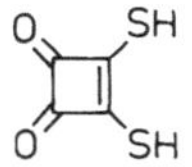		$C_4H_2O_2S_2$

Survey. Mn^{2+} ions with 1,2-dithiolates under anaerobic conditions form tetrahedral bis(dithiolato) Mn^{II} complexes in aqueous or alcoholic solution. Intensely green complexes $[MnL_2L'_n]$ with one or two additional coordinated ligands (L', such as CH_3CN, DMF, $(CH_3)_2SO$, pyridine, or imidazole), are obtained by controlled air oxidation of the Mn^{II} chelates in these solvents or in the presence of the nitrogen bases. In the case of the $[MnL_2L'_n]^-$ complexes with L = ethanedithiolate (ligand 1) partial dimerization occurs to give the dinuclear solvent-free species $[Mn_2L_4]^{2-}$ with an edge-shared tetrahedral structure. A solvent-dependent equilibrium exists between the two species. In the case of unsaturated 1,2-dithiolates (dithiolenes) with a large negative inductive (−I) effect, such as maleonitriledithiolate (ligand 7) or p-toluenedithiolate (ligand 9), controlled air oxidation of the Mn^{II} chelates yields green tris(dithiolato) complexes $[MnL_3]^{2-}$. Complex salts are precipitated by use of large cation halides. The isolated complexes with the $[MnL_2L']^-$ anions are square-pyramidal. Tris(dithiolato) complexes are octahedral or trigonal prismatic. While the $[MnL_2L'_n]^-$ complexes and the $[Mn_2L_4]^{2-}$ species are assumed, based on MO calculations, to have an Mn^{III} ion in an S = 2 state, details of the molecular and electronic structure of the tris(dithiolato) complexes are still uncertain. The complexes (except for those with ligand 12) are extremely air-sensitive, particularly the Mn^{II} chelates.

Except for the complex with ligand 12, a reversible redox equilibrium is observed between $[MnL_2]^{2-}$ and the oxidized species, $[MnL_2L'_n]^-$ or $[Mn_2L_4]^{2-}$, whereas two reversible redox equilibria exist between tris(dithiolato) complexes, $[MnL_3]^{2-}$, and the more- or less-reduced species, $[MnL_3]^{3-}$ or $[MnL_3]^-$, respectively. The intensely green colors of the oxidized species, especially $[MnL_2py]^-$ with H_2L = ligands 2 and 9 are used for the microanalysis of manganese. Mn^{II} complexes with ligands 7 and 9 show catalytic activity in the polymerization of 2,6-dimethylphenol.

Because of the apparent failure to make a traditional assignment of oxidation states to the metal atoms, particularly for the $[MnL_3]^{2-}$ complexes, the ranking in this section is by complex types, rather than (as is usual in the "Manganese" D series) by Mn oxidation states.

35.1.5.1 Complexes in Solution

Stability Data. Yellow complexes of manganese(II) and ligand 2 are formed in basic aqueous solution under a nitrogen atmosphere. The stability constant for the neutral $[Mn(C_3H_6OS_2)]$ complex is $\log K_1 = 5.23$, while for the $[Mn(C_3H_6OS_2)_2]^{2-}$ complex it is $\log \beta_2 = 10.43$ at 30°C and $I = 0.1M$ (KCl), determined by the pH method [1].

Electronic Spectra. Structure. Spectral data of dithiolato complexes in various solvents are summarized below (concentrations in mM, λ_{max} in nm, and ε in $L \cdot mol^{-1} \cdot cm^{-1}$):

ligand	complex	solvent	conc.	λ_{max} (ε)	Ref.
1	$[Mn(C_2H_4S_2)_2]^{2-}$	CH_3CN	—	~298 (sh), ~360 (sh), ~400 (sh), 460 (17)	[2]
		DMF	—	455 (14)	[2]
	$[Mn_2(C_2H_4S_2)_4]^{2-}$ a)	CH_2Cl_2	1.3	328 (sh), 356, 400 (sh)	[3]
		CH_3CN	7.02	320 (sh, 7560), 356 (10900), 388 (sh, 7850), 640 (580)	[3]
		DMF	3.6	292 (3190), 349 (9850), 392 (4480), 588 (200)	[2]
		$(CH_3)_2SO$	4.72	290 (6490), 350 (16100), 398 (10200), 600 (1250)	[3]
2	$[Mn(C_3H_6OS_2)_2(H_2O)_n]^-$ b)	C_3H_7OH–H_2O 1:1 v/v	~5	344 (7800), 390 (5300), 555 (400)	[2]
	$[Mn(C_3H_6OS_2)_2py_n]^-$ b)	pyridine-2-propanol–H_2O 4:1:1 v/v	—	353 (6300), 575 (370)	[4]
		same as above	~2	352 (7700), 392 (4200), 584 (320)	[2]
6	$[Mn(C_2H_2S_2)_2py]^-$	pyridine	—	800 (700)	[5]
7	$[Mn(C_4N_2S_2)_3]^{2-}$	CH_2Cl_2	—	313 (29800), 633 (sh, 2400), 599 (2800), 826 (2400)	[6]
		CH_2Cl_2 b)	—	308 (37000), 388 (36000), 637 (2600), 833 (2400)	[17]
		$CHCl_3$	—	581 (1800), 633 (2160), 833 (1900)	[7]
9	$[Mn(C_7H_6S_2)_2]^-$ b)	CH_3CN	—	242 (30000), 363 (11100), 510 (2300), 576 (1900), 734 (1510)	[8]
		CH_2Cl_2	—	315 (19800), 370 (40800), 410 (sh, 11800), 540 (sh, 985), 640 (sh, 385), 900 (215)	[9]
		DMF	—	327 (12600), 351 (16300), 400 (sh, 7400), 560 (2300), 570 (sh, 2200), 900 (sh, 200)	[9]

ligand	complex	solvent	conc.	λ_{max} (ε)	Ref.
9	$[Mn(C_7H_6S_2)_2py_n]^-$ c)	aqueous pyridine	—	357 (10500), 550 (550)	[4]
10	$[Mn(C_8H_8S_2)_2py]^-$	pyridine	—	365 (24700), 562 (1100)	[5]
11	$[Mn(C_6Cl_4S_2)_3]^{2-}$	CH_2Cl_2	—	272, 374, 521, 568, 714	[10]

a) Partially dissociated into solvated monomer species, $[Mn(C_2H_4S_2)_2L'_2]^-$, in coordinating solvents [2, 3], see text on structure below. – b) Containing 0.2 M $[N(C_4H_9)_4]BF_4$. – c) Probably solvated species.

The intense absorptions of the complexes in the 200 to 400 nm range are due to intra-ligand [8] or to ligand-to-Mn charge-transfer transitions [2]. The bands of the solvated species, $[MnL_2L'_n]^-$, in the range between 560 and 638 nm [2] and at 900 nm [10] with ε=190 to 330 L·mol^{-1}·cm^{-1} are probably d-d in character [2, 9].

For the $[Mn(C_2H_4S_2)_2]^{2-}$ ion in acetonitrile, the ε value of 17 L·mol^{-1}·cm^{-1} for the absorption band at 460 nm and its light yellow-green color indicate that the complex retains its tetrahedral structure in this solvent. The electronic spectrum of $[Mn_2(C_2H_4S_2)_4]^{2-}$, however, is dependent on the solvent, revealing that this complex does not retain its dimeric structure in solution [2] (see "Chemical reactions", p. 37). At the concentrations given the spectrum in dichloromethane is that of the intact dimer, whereas in dimethyl sulfoxide the maxima at 290, 398, and 600 nm are characteristic of the solvated monomer species. The spectra in dimethylformamide and acetonitrile reflect the existence of both species in these solvents [3].

Magnetic Data. Magnetic moments of dithiolato complexes in various solvents are summarized below (concentration in mmol/L, μ_{eff} = magnetic moment per Mn atom in μ_B, and γ_M = mole fraction of monomer species):

ligand	complex	solvent	concentration a)	μ_{eff} b)	γ_M	Ref.
1	$[Mn(C_2H_4S_2)_2]^{2-}$	CH_3CN	—	5.86	—	[2]
	$[Mn_2(C_2H_4S_2)_4]^{2-}$	CD_3CN	2.00	4.60	0.52	[3]
			4.01	4.31	0.28	[3]
			20.0	4.04	0.06	[3]
		$(CD_3)_2SO$	1.49	5.12	1.10	[3]
			19.5	4.88	0.77	[3]
			56.2	4.65	0.56	[3]
			127.8	4.57	0.49	[3]
	$[Mn(C_2H_4S_2)_2(DMSO)_2]^-$ c)	$(CH_3)_2SO$	2.3	5.08	—	[2]
	$[Mn(C_2H_4S_2)_2(CH_3CN)_2]^-$	CH_3CN	1.6	5.06	—	[2]
9	$[Mn(C_7H_6S_2)_2]^-$ d)	$(CH_3)_2SO$	—	5.2	—	[8]

a) Based on dimer molecular weight in the case of $[Mn_2(C_2H_4S_2)_4]^{2-}$. – b) Measurements by a standard NMR method, with $(CH_3)_4Si$ as reference [1, 9] in the 298 to 301 K temperature range [2], at 297 K [3], or by the Evans method at room temperature [8]. – c) In table in [2], $[Mn(C_2H_4S_2)_2(DMF)_2]^-$ is quoted, but $(CH_3)_2SO$ as solvent, mistake? – d) Solvated complex (see p. 37).

Formulas of ligands are tabulated on p. 33

The data for the $[Mn_2(C_2H_4S_2)_4]^{2-}$ ion in CD_3CN and $(CD_3)_2SO$ reveal the different concentration dependence of the magnetic moment per Mn atom in the two solvents, showing the partial dissociation of the dimer into the monomer species (see p. 37). The mole fractions of the monomer species were calculated from the magnetic moments [3].

IR Spectra. IR and far-IR bands were recorded for the complexes with ligand 7, $[Mn(C_4N_2S_2)_3]^{2-}$ and $[Mn(C_4N_2S_2)_3]^{3-}$, in CH_2Cl_2 (containing 0.2M $[N(C_4H_9)_4]BF_4$): ν(CN) at 2201.2 and 2211(sh) cm^{-1}, ν(Mn–S) at 348.9 cm^{-1} or ν(CN) at 2192.3 and 2205.2(sh) cm^{-1}, ν(Mn–S) at 328.1 cm^{-1}, respectively. Correlations between the ν(CN) bands and the structure, charge, and electronic configuration are discussed in [17].

Redox Reactions. The $[MnL_2]^{2-} \rightleftharpoons [MnL_2L'_n]^-$ redox reactions (or $[Mn_2L_4]^{2-}$, depending on the solvent, in the case of ligand 1) are considered by most workers to be based on reversible $Mn^{II} \rightleftharpoons Mn^{III}$ electron transfer reactions [3, 9, 11, 12]. However, from cyclovoltammetric studies of the complex with ligand 9, $[Mn(C_7H_6S_2)_2]^{2-}$, and from a comparison of its redox chemistry with that of the ligand monoanion, $C_7H_7S_2^-$, and with the catecholato, semiquinonato, and tetrachloro complexes, a ligand-centered electron transfer was inferred. For the first (reversible) oxidation step of $[Mn^{II}(C_7H_6S_2)_2]^{2-}$, formation of a radical anion of the coordinated ligand and spin-pairing with an unpaired d electron of the Mn^{II} atom is discussed. For the second (irreversible) step, formation of an oxidation product of the ligand containing S–S bonds was assumed [8]. No assignment was made for the center of electron transfer in the case of tris(dithiolato) complexes. Tris(dithiolato) complexes with ligand 11, $[Mn(C_6Cl_4S_2)_3]^z$, occur with the charges z = 0, −1, −2, −3, and −4 [10], whereas the complex with ligand 7, $[Mn(C_4N_2S_2)_3]^{2-}$, occurs only in the reduced forms with z = −2, −3, and −4 [6], the −3→−4 reactions being irreversible for both ligands. For the complex with ligand 11, the −2→−1 oxidation can also be accomplished with bromine in dichloromethane [10]. How the redox potentials reflect the greater −I effect of ligand 7, compared to ligand 11, is discussed in [10].

Redox potentials of one-electron transfer reactions of binary and ternary bis(dithiolato) and binary tris(dithiolato) manganese complexes, obtained from cyclovoltammetric or polarographic measurements, are summarized below:

H_2L No.	redox couple	solvent	conc. in mM	potential in V	ΔE in mV	Ref.
1	$[Mn(C_2H_4S_2)_2]^{2-}/[Mn(C_2H_4S_2)_2(DMSO)_n]^-$	$(CH_3)_2SO$	3.29	−0.88 b)	150	[3]
	$[Mn(C_2H_4S_2)_2]^{2-}/[Mn_2(C_2H_4S_2)_4]^{2-}$ a)	CH_3CN	2.05	−0.89 b)	120	[3]
	$[Mn_2(C_2H_4S_2)_4]^{2-}$ a)/$[Mn(C_2H_4S_2)_2]^{2-}$	$(CH_3)_2SO$	20.4	−0.86 b)	120	[3]
			7.34	−0.90 b)	190	[3]
			1.46	−0.93 b)	330	[3]
		CH_3CN	10.2	−0.94 b)	130	[3]
			4.25	−0.91 b)	140	[3]
			2.06	−0.95 b)	170	[3]
		CH_2Cl_2	~1.1	−0.92 b)	210	[3]
9	$[Mn(C_7H_6S_2)_2]^{2-}/[Mn(C_7H_6S_2)_2(DMSO)_n]^-$	$(CH_3)_2SO$	0.45	−0.76 b)	150	[3]
		CH_3CN	1	−0.67 c)	—	[8]
	$[Mn(C_7H_6S_2)_2(DMF)_n]^-/[Mn(C_7H_6S_2)_2]^{2-}$	DMF	3.33	−0.72 b)	110	[3]
			—	−1.09 d)	610	[9]
		CH_2Cl_2	—	−0.82 d)	270	[9]

H_2L No.	redox couple	solvent	conc. in mM	potential in V	ΔE in mV	Ref.
6	$[Mn(C_2H_2S_2)_2py]^{2-}/[Mn(C_2H_2S_2)_2py]^-$	py	—	−0.95[e)]	—	[11]
	$[Mn(C_2H_2S_2)_2py]^-/[Mn(C_2H_2S_2)_2py]^0$	py	—	−0.37[e)]	—	[11]
10	$[Mn(C_8H_8S_2)_2py]^{2-}/[Mn(C_8H_8S_2)_2py]^-$	py	—	−0.98[e)]	—	[11]
	$[Mn(C_8H_8S_2)_2py]^-/[Mn(C_8H_8S_2)_2py]^0$	py	—	−0.17[e)]	—	[11]
1	$[Mn(C_2H_4S_2)_2]^{2-}/[Mn(C_2H_4S_2)_2(C_3H_4N_2)]^-$ [f)]	CH_3CN	—	−0.30[g)]	1150	[12]
	$[Mn(C_2H_4S_2)_2(C_3H_4N_2)]^-/[Mn(C_2H_4S_2)_2]^{2-}$	CH_3CN	—	−1.45[h)]	1150	[12]
7	$[Mn(C_4N_2S_2)_3]^{2-}/[Mn(C_4N_2S_2)_3]^{3-}$	CH_2Cl_2	—	−0.35[i)]	—	[6]
		CH_2Cl_2	—	0.05[m)]	—	[17]
	$[Mn(C_4N_2S_2)_3]^{2-}/Mn(C_4N_2S_2)_3]^{3-}$ (irrev.)	CH_3CN	—	−0.36[k)]	—	[6]
	$[Mn(C_4N_2S_2)_3]^{3-}/[Mn(C_4N_2S_2)_3]^{4-}$ (irrev.)	CH_2Cl_2	—	−0.72[l)]	—	[10]
11	$[Mn(C_6Cl_4S_2)_3]^{2-}/[Mn(C_6Cl_4S_2)_3]^{3-}$	CH_2Cl_2	—	−0.92[l)]	—	[10]
	$[Mn(C_6Cl_4S_2)_3]^{2-}/[Mn(C_6Cl_4S_2)_3]^-$	CH_2Cl_2	—	+0.40[l)]	—	[10]
	$[Mn(C_6Cl_4S_2)_3]^{3-}/[Mn(C_6Cl_4S_2)_3]^{4-}$ (irrev.)	CH_2Cl_2	—	−1.12[l)]	—	[10]

[a)] Dimer species, $[Mn_2(C_2H_4S_2)_4]^{2-}$, in equilibrium with monomer species, $[MnL_2L'_2]^-$. – [b)] $E_{1/2}$ at ~25°C vs. SCE, glassy-carbon electrode, $[N(C_4H_9)_4]ClO_4$. – [c)] $E_{p,a}$ vs. SCE, Pt electrode, $[N(C_2H_5)_4]ClO_4$. – [d)] $E_{p,c}$ vs. SCE, Pt electrode, $[N(t\text{-}C_4H_9)_4]ClO_4$. – [e)] $E_{1/2}$ vs. SCE, dropping Hg electrode, 0.1M $NaClO_4$. – [f)] In the presence of 1H-imidazole (= $C_3H_4N_2$). – [g)] $E_{p,c}$ vs. SCE. – [h)] $E_{p,a}$ vs. SCE. – [i)] $E_{1/2}$ vs. SCE containing aqueous 1M LiCl at 25°C, Pt electrode, $[N(C_2H_5)_4]ClO_4$ or $[N(C_3H_7)_4]ClO_4$. – [k)] $E_{1/2}$ vs. normal calomel electrode at 20°C, Pt electrode, $[N(C_2H_5)_4]ClO_4$ or $[N(C_3H_7)_4]ClO_4$. – [l)] $E_{1/2}$ vs. SCE containing aqueous 1M LiCl at 25°C, Pt electrode, $[N(C_2H_5)_4]ClO_4$. – [m)] $E_{p,c}$ vs. Ag|AgCl at 20°C, Pt electrode, 0.2M$[(C_4H_9)_4N]BF_4$.

Due to adsorption phenomena the electrochemistry of acetonitrile or dimethyl sulfoxide solutions of the $[Mn_2L_4]^{2-}$ and $[MnL_2L'_n]^-$ complexes with ligand 1 shows a significant dependence on the electrode surface. The heterogeneous electron transfer rate constant depends on the electrode: $k = 4\times10^{-6}$ cm/s for the Pt electrode, $k = 3\times10^{-3}$ cm/s for the basal pyrolytic graphite electrode and $k = 2\times10^{-3}$ cm/s for the glassy-carbon electrode [3].

Chemical Reactions. The yellow $[MnL_2]^{2-}$ complexes with ligands 1, 2, and 9 in aqueous or organic aprotic solvents are extremely air-sensitive and are oxidized to the dark green $[MnL_2L'_n]^-$ species (L' = solvent, n = 1 or 2) [1, 2, 3, 9]. The complex with ligand 1, $[Mn(C_2H_4S_2)_2L'_n]^-$, partially dimerizes in organic solvents to give the dinuclear, $[Mn_2(C_2H_4S_2)_4]^{2-}$, species, in which two MnL_2^{2-} units are joined by two bridging S atoms (see **Fig. 7**, p. 43). Magnetic moments (see p. 35) and electronic spectra (see p. 34) of solutions containing the $[Mn_2(C_2H_4S_2)_4]^{2-}$ ions indicate that the equilibrium between the dimer and the monomer species is solvent-dependent. In dichloromethane and acetonitrile it is shifted to the dimer side, whereas in dimethyl sulfoxide it is shifted to the monomer side. The equilibrium constant for $[Mn_2(C_2H_4S_2)_4]^{2-} + 4\,L' \rightleftharpoons 2\,[Mn(C_2H_4S_2)_2L'_2]^-$, derived from the concentration dependence of the magnetic moment in dimethyl sulfoxide, is $K \approx 81(18)$ mmol/L. Solutions containing $[Mn_2(C_2H_4S_2)_4]^{2-}$ ions are stable in aprotic solvents (acetonitrile, dimethyl sulfoxide, or dimethylformamide) [3]. Addition of excess imidazole to the dark green solution in dimethylformamide leads to a color change to bright green and formation of the $[Mn(C_2H_4S_2)_2(C_3H_4N_2)]^-$ complex, which is converted in good donor solvents to the $[Mn(C_2H_4S_2)_2L'_n]^-$ complex by

Formulas of ligands are tabulated on p. 33

ligand exchange [12]. Axial ligation reactions were also observed with pyridine [2, 5]. In some cases (complexes with ligands 6 and 10) only the pyridine-containing ternary, but not binary, complexes were found. Solutions of the $[MnL_2py]^-$ ions in pyridine are stable for several hours. The decomposition is more rapid in dimethyl sulfoxide and dimethylformamide [5]. The tris(dithiolato) complex with ligand 7, $[Mn(C_4N_2S_2)_3]^{2-}$, forms dark green air-unstable solutions in acetone, chloroform, dichloromethane, or dimethylformamide. In aerated acetone, it slowly decomposes to give yellow solids which do not contain manganese [6].

Applications. The intensely dark green colorations produced by ligands 2 and 9 with manganese ions in aqueous ammonia [13, 14], acetate-buffered solutions [15], or aqueous pyridine [4, 16] under nitrogen [4] can be used for the detection of manganese [4, 13 to 16] down to 2 μg in aqueous pyridine [16]. The sensitivity in aqueous ammonia is 1:3000000 [14]. Color reactions were also observed under these conditions with ligands 1, 4, and 8 in contrast to monothiols which give no colorations. The reaction with Mn^{II} acetate can therefore be used as a test for vicinal dithiols [4]. Solutions containing the complexes with ligands 7 and 9, $[Mn(C_4N_2S_2)_2]^{2-}$ and $[Mn(C_8H_8S_2)_2]^{2-}$, in methanol-dichlorobenzene were found to show catalytic activity in the oxidative polymerization of 2,6-dimethylphenol [18].

References:

[1] Leussing, D. L.; Tischer, T. N. (J. Am. Chem. Soc. **83** [1961] 65/70).
[2] Costa, T.; Dorfman, J. R.; Hagen, K. S.; Holm, R. H. (Inorg. Chem. **22** [1983] 4091/9).
[3] Mukherjee, R. N.; Rao, C. P.; Holm, R. H. (Inorg. Chem. **25** [1986] 2979/89).
[4] Rosenblatt, D. H.; Jean, G. N. (Anal. Chem. **27** [1955] 951/4).
[5] Hoyer, E.; Dietzsch, W.; Müller, H. (Z. Chem. [Leipzig] **7** [1967] 354/5).
[6] McCleverty, J. A.; Locke, J.; Wharton, E. J.; Gerloch, M. (J. Chem. Soc. A **1968** 816/23).
[7] Stiefel, E. I.; Bennett, L. E.; Dori, Z.; Crawford, T. H.; Simo, C.; Gray, H. B. (Inorg. Chem. **9** [1970] 281/6).
[8] Sawyer, D. T.; Srivatsa, G. S.; Bodini, M. E.; Schaefer, W. P.; Wing, R. M. (J. Am. Chem. Soc. **108** [1986] 936/42).
[9] Henkel, G.; Greiwe, K.; Krebs, B. (Angew. Chem. **97** [1985] 113/4).

[10] Wharton, E. J.; McCleverty, J. A. (J. Chem. Soc. A **1969** 2258/66).
[11] Hoyer, E.; Dietzsch, W.; Heber, H. (Proc. 3rd Symp. Coord. Chem., Debrecen, Hung., 1970, Vol. 1, pp. 259/71).
[12] Seela, J. L.; Huffman, J. C.; Christou, G. (J. Chem. Soc. Chem. Commun. **1985** 58/60).
[13] Přibil, P.; Roubal, Z. (Collection Czech. Chem. Commun. **19** [1954] 1162/70).
[14] Buscarons, F.; Casassas, E. (Anales Real Soc. Espan. Fis. Quim. B **55** [1959] 655/62).
[15] Clark, R. E. D.; Neville, R. G. (J. Chem. Educ. **36** [1959] 390/3).
[16] Clark, R. E. D. (Analyst [London] **82** [1957] 177/82).
[17] Best, S. P.; Clark, R. J. H.; McQueen, R. C. S.; Walton, J. R. (Inorg. Chem. **17** [1988] 884/90).
[18] Kaneko, M. (Makromol. Chem. **178** [1977] 733/40).

35.1.5.2 Isolated Compounds

35.1.5.2.1 $M_2[MnL_2]$ Complexes and Their Solvates

$[N(CH_3)_4]_2[Mn^{II}(C_2H_4S_2)_2]\cdot0.5CH_3CN$ was prepared under anaerobic conditions by the dropwise addition of a methanol solution of $MnCl_2$ to sodium 1,2-ethanedithiolate in methanol (mole ratio 1:2). After stirring 2 h, a stoichiometric amount of $[N(CH_3)_4]Cl$ in methanol was added to the reaction mixture. The solution was stirred for 12 h at room temperature and the solvent removed in vacuum. Addition of CH_3CN to the residue gave a yellow suspension which was filtered. Reducing the filtrate to half its volume, then addition of ether and cooling to −20°C, gave large yellow crystals. The product was isolated, washed with ether, and dried in vacuum. The yield of product was 84%.

A single crystal X-ray structural determination of the complex was carried out. It crystallizes in the monoclinic system, space group $P2_1/n-C^5_{2h}$, No. 14 (standard setting $P2_1/c$) with the lattice parameters a = 9.833(5), b = 17.325(7), c = 13.819(6) Å, α = 93.43(2)°; Z = 4. The calculated density is D_{calc} = 1.21 g/cm³, the experimental density is D_{exp} = 1.22 g/cm³. The structure was solved up to R = 4.4%. Atomic positional parameters for the $[Mn(C_2H_4S_2)_2]^{2-}$ unit are presented in the publication. The structure of $[Mn(C_2H_4S_2)_2]^{2-}$, together with the atom-numbering scheme, is shown in **Fig. 5**. Selected bond distances (in Å) and angles are:

Mn–S(1)	2.441(2)	S(1)–Mn–S(2)	91.6(1)°	S(1)–Mn–S(4)	117.9(1)°
Mn–S(2)	2.423(2)	S(3)–Mn–S(4)	91.4(1)°	S(2)–Mn–S(3)	117.6(1)°
Mn–S(3)	2.435(2)	S(1)–Mn–S(3)	121.9(1)°	S(2)–Mn–S(4)	119.2(1)°
Mn–S(4)	2.433(2)				

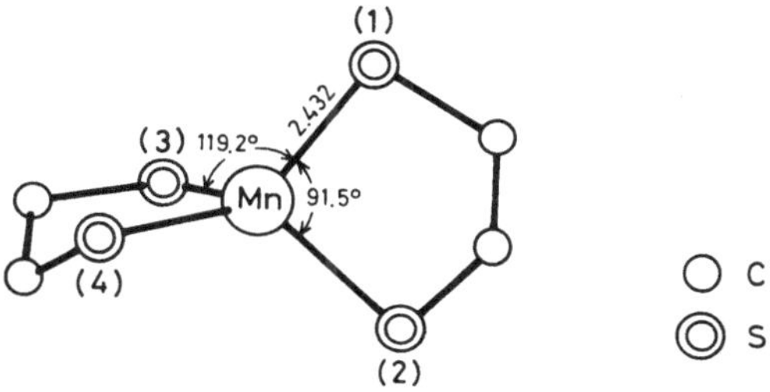

Fig. 5. Structure of the 1,2-ethanedithiolato complex, $[Mn^{II}(C_2H_4S_2)_2]^{2-}$, in $[N(CH_3)_4]_2[Mn(C_2H_4S_2)_2]\cdot 0.5\,CH_3CN$ (H atoms omitted) [1].

The Mn^{II} atom is four-coordinate with a distorted tetrahedral arrangement of the sulfur donor atoms around the metal. The mean Mn–S distance is 2.43 Å (mean bond angles are shown in the figure). The five-membered chelate rings adopt a gauche conformation [1].

$[N(C_2H_5)_4]_2[Mn^{II}(C_2H_4S_2)_2]$ and **$[P(C_6H_5)_4]_2[Mn^{II}(C_2H_4S_2)_2]$** complexes, were prepared in a manner analogous to $[N(CH_3)_4]_2[Mn(C_2H_4S_2)_2]\cdot 0.5\,CH_3CN$, but using $[N(C_2H_5)_4]Cl$ or $[P(C_6H_5)_4]Cl$ instead of $[N(CH_3)_4]Cl$. The yield of the phosphonium salt was 42%. Its magnetic moment in the solid state (orange-yellow crystals) is μ_{eff} = 6.02 μ_B at room temperature (298 to 301 K) [1].

$[N(C_2H_5)_4]_2[Mn^{II}(C_4N_2S_2)_2]$. The complex with ligand 7 was prepared by the slow addition of an N_2-degassed aqueous solution of $MnCl_2\cdot 4H_2O$ to an N_2-degassed mixture of $[N(C_2H_5)_4]Cl$ and the sodium salt of the ligand in water (mole ratio 1:2:2.5). After stirring for 6 h at ambient temperature, light brown crystals had formed. The product was isolated, washed with water, ethanol, ether, and air dried. The yield of the product was 56%. The solid complex slowly decomposes in air, less rapidly than in solution. It is sparingly soluble in degassed acetone, giving an orange-red solution. It is a 2:1 electrolyte in acetone [2].

$[P(C_6H_5)_4]_2[Mn^{II}(C_7H_6S_2)_2]\cdot 2\,CH_3OH$ was prepared by the reaction of hydrated manganese(II) chloride with sodium 1,2-p-toluenedithiolate (mole ratio ≦1:2) in methanol, in the absence of oxygen. Addition of $[P(C_6H_5)_4]Br$ to the yellowish brown reaction mixture yielded brown columnar-shaped crystals. The single crystal X-ray determination of the structure gave the crystal data: monoclinic system, space group $P2_1/c-C^5_{2h}$ (No. 14); with a = 22.378(9), b = 12.102(4), c = 21.888(8) Å, β = 110.48(3)°; Z = 4. The structure was solved up to R = 6.3%.

Formulas of ligands are tabulated on p. 33

The bidentate ligands are oriented almost perpendicularly, and a strongly distorted tetrahedral coordination of the sulfur donors is present. The Mn–S distances are: 2.395, 2.414, 2.435, and 2.434 Å ($\sigma = 0.002$ Å) with a mean value of 2.42 Å. The S–Mn–S bond angles are: 88.59°, 89.15°, 117.13°, 117.41°, 123.06°, and 124.59° ($\sigma = 0.07°$). The compound is isomorphous with $[As(C_6H_5)_4]_2[Cd(C_7H_6S_2)_2]\cdot 2C_2H_5OH$. It shows the magnetic behavior typical of a high-spin d^5 complex [3].

$H_2[Mn^{II}L_2]\cdot 4H_2O$ and $Mn^{II}[MnL_2]\cdot 4H_2O$. Compounds of composition $Mn[MnL_2]\cdot 4H_2O$ are reported to precipitate on reaction of $MnCl_2\cdot 4H_2O$ with ligand 1 or 9 in ethanol in the presence of C_2H_5OK. Precipitates obtained from the filtrate on addition of a twofold amount of water were assumed to be $H_2[MnL_2]\cdot 4H_2O$. The complex with ligand 9, $Mn[Mn(C_7H_6S_2)_2]\cdot 4H_2O$, is reported to be an active catalyst in the oxidative polymerization of 2,6-dimethylphenol, whereas the complex with ligand 1, $Mn[Mn(C_2H_4S_2)_2]\cdot 4H_2O$, was found to be almost inactive [4].

$K_2[Mn^{II}(C_4O_2S_2)_2]\cdot 4H_2O$. The air-sensitive complex was prepared by the reaction of stoichiometric quantities of manganese(II) chloride and the potassium salt of dithiosquaric acid (ligand 12) in water, under a nitrogen atmosphere. The reaction mixture was cooled to 0°C, and yellow crystals formed. The product was isolated, washed with ethanol, then ether, and dried under vacuum at room temperature. The complex is a 2:1 electrolyte in water. The solid has a magnetic moment of $\mu_{eff} = 6.13\ \mu_B$ (Faraday method) at ambient temperature. The IR spectrum of the complex in Nujol mull shows the prominent absorptions: ν(C⩦O) at 1900, 1730, 1650 cm^{-1}, ν(C⩦C⩦O) at 1420, 1360 cm^{-1}, ν(C⩦C⩦S) at 1215, 1120 cm^{-1}, and ν(C⩦S) at 950, 915 cm^{-1}. The electronic absorption spectrum of the complex in water shows maxima (ε in parentheses) at 29410 (38500), 31250 (31000), and 40820 cm^{-1}(22000). The IR spectrum and the X-ray powder patterns (d spacings are given) are different from those of either planar bis(dithio-squarato) complexes or the tetrahedral Co^{II}, Fe^{II}, or Zn^{II} complexes. From comparing the magnetic moment with that of bis(dithiophosphato) and bis(dithiophosphinato) complexes, a tetrahedral structure was suggested [5].

$[P(C_6H_5)_4]_2[Mn^{II}(C_4O_2S_2)_2]$ and $[P(C_6H_5)_4]_2[Mn^{II}(C_4O_2S_2)_2]\cdot 2L'$ ($L' = H_2O$ or C_2H_5OH). The dihydrate was prepared by the reaction of the stoichiometric ratios of hydrated manganese(II) chloride, the potassium salt of ligand 12, $K_2(C_4O_2S_2)$, and $[P(C_6H_5)_4]Cl$ in water, under a nitrogen atmosphere. The insoluble complex that formed was filtered off and dried under vacuum. All subsequent reactions were carried out in an inert atmosphere. Recrystallization of the product from wet solvents (DMF or CH_3CN) gave the pure dihydrate. Crystals of the ethanol adduct were obtained by recrystallization from ethanol. Pure crystals of $[P(C_6H_5)_4]_2[Mn(C_4O_2S_2)_2]$ were obtained a) by recrystallization of the crude dihydrate, which had been dried under vacuum for several hours, from dry CH_3CN or CH_2Cl_2, followed by addition of ether, b) by heating the ethanol solvate for 8 h at 60 to 80°C under vacuum, or c) by recrystallizing the ethanol solvate from hot CH_3CN [5].

The complex is a 2:1 electrolyte in CH_3CN. The magnetic moment of the solid anhydrous compound is $\mu_{eff} = 5.87\ \mu_B$ at ambient temperature. The electronic absorption spectrum of the anhydrous complex in CH_3CN shows maxima (ε in parentheses) at 32260 (38180), 36360 (30960), and 44440 cm^{-1} (86700). The IR spectrum of the dihydrate in Nujol mull shows the prominent bands: ν(C⩦O) at 1720, 1680, 1665, 1645 cm^{-1}, ν(C⩦C⩦O) at 1370, 1345, 1340 cm^{-1}, ν(C⩦C⩦S) at 1200, 1190, 1170 cm^{-1}, and ν(C⩦S) at 955, 905, 895 cm^{-1}. The ethanol solvate shows the bands: ν(C⩦O) at 1832, 1725, 1690, 1660 cm^{-1}, ν(C⩦C⩦O) at 1385, 1363 cm^{-1}, ν(C⩦C⩦S) at 1175, 1165 cm^{-1}, and ν(C⩦S) at 1040, 992, 915, 888, 875 cm^{-1}. A tetrahedral structure is assumed as for the potassium compound (see above). Air oxidation of $[P(C_6H_5)_4]_2[Mn(C_4O_2S_2)_2]$ in acetonitrile gave the $[C_8O_4S_3]^{2-}$ monosulfide anion [5].

$[P(C_6H_5)_4]_2[Mn^{II}(C_4O_2S_2)(C_6H_5S)_2]$ was prepared by treating a hot acetonitrile solution of $[P(C_6H_5)_4]_2[Mn(C_4O_2S_2)_2]$ with potassium benzenethiolate under rigorously oxygen-free conditions. The precipitated potassium salt, $K_2(C_4O_2S_2)$, was filtered off, and the filtrate yielded the yellow product. The magnetic moment of the solid is $\mu_{eff} = 5.84\ \mu_B$ at ambient temperature. The complex is X-ray isomorphous with the corresponding Fe^{II}, Co^{II}, and Zn compounds. The electronic spectra (not given for the Mn^{II} complex) and magnetic properties are very similar to those expected for tetrahedrally coordinated ions. The $[P(C_6H_5)_4]_2[Mn(C_4O_2S_2)(SC_6H_5)_2]$ complex reacts with further potassium benzenethiolate in CH_3CN under nitrogen to give $[P(C_6H_5)_4]_2$-$[Mn(SC_6H_5)_4]$ (see p. 25) and $K_2(C_4O_2S_2)$ [6].

References:

[1] Costa, T.; Dorfman, J. R.; Hagen, K. S.; Holm, R. H. (Inorg. Chem. **22** [1983] 4091/9).
[2] McCleverty, J. A.; Locke, J.; Wharton, E. J.; Gerloch, M. (J. Chem. Soc. A **1968** 816/23).
[3] Henkel, G.; Greiwe, K.; Krebs, B. (Angew. Chem. **97** [1985] 113/4).
[4] Kaneko, M. (Makromol. Chem. **178** [1977] 733/40).
[5] Coucouvanis, D.; Holah, D. G.; Hollander, F. J. (Inorg. Chem. **14** [1975] 2657/65).
[6] Holah, D. G.; Coucouvanis, D. (J. Am. Chem. Soc. **97** [1975] 6917/9).

35.1.5.2.2 Mixed Valence Complex $[P(C_6H_5)_4]_2[Mn_3(C_3H_6S_2)_5]$

An ethanolic solution containing $MnCl_2$, sodium 1,3-propanedithiolate, and $[P(C_6H_5)_4]Br$, in a 1:2:1 mole ratio, was cooled to ~0°C and oxidized by successive additions of small portions of air, until generation of a deep green color ceased and a microcrystalline green precipitate no longer formed. Recrystallization of the product from DMF-ether under anaerobic conditions at 25°C yielded black prisms in 40% yield.

The single crystal X-ray structure determination of the complex was carried out. The crystal data are: triclinic system, space group $P\bar{1}$-C_i^1 (No. 2) with the lattice constants a = 14.385(6), b = 23.734(11), c = 9.881(3) Å, α = 100.39(2)°, β = 93.25(2)°, γ = 107.53(2)°; Z = 2. The structure was solved up to R = 6.41% (data collected at −160°C). The structure of the anion is shown in **Fig. 6**, p. 42. It contains a nearly linear arrangement of the three Mn atoms (bonding angle Mn(1)–Mn(2)–Mn(3) = 169.7(1)°). The central Mn^{II} atom possesses a distorted octahedral geometry with each of six thiolate sulfur atoms bridging to terminal Mn^{III} atoms. The Mn^{II}–S distances range from 2.554 to 2.677 Å. The mean Mn^{II}–Mn^{III} distance is 3.11 Å. The Mn^{III} atoms are five-coordinate with a geometry intermediate between the square-pyramidal and trigonal-bipyramidal extremes. The Mn^{III}–S distances range from 2.315 to 2.327 Å. The magnetic moment of the solid is $\mu_{eff} = 6.75\ \mu_B$ at 300 K. The moment decreases to $\mu_{eff} = 3.9\ \mu_B$ at 25 K, corresponding to a spin quartet, S = 3/2, ground state. The variable temperature magnetic measurements indicate that the spin states of the three metal atoms are: Mn^{III}, S = 2, Mn^{II}, S = 5/2, Mn^{III}, S = 2. The exchange parameter between Mn^{III} and Mn^{II} is $J = -18.3\ cm^{-1}$.

Reference:

Seela, J. L.; Folting, K.; Wang, R.-J.; Huffman, J. C.; Christou, G.; Chang, H.-R.; Hendrickson, D. N. (Inorg. Chem. **24** [1985] 4454/6).

Formulas of ligands are tabulated on p. 33

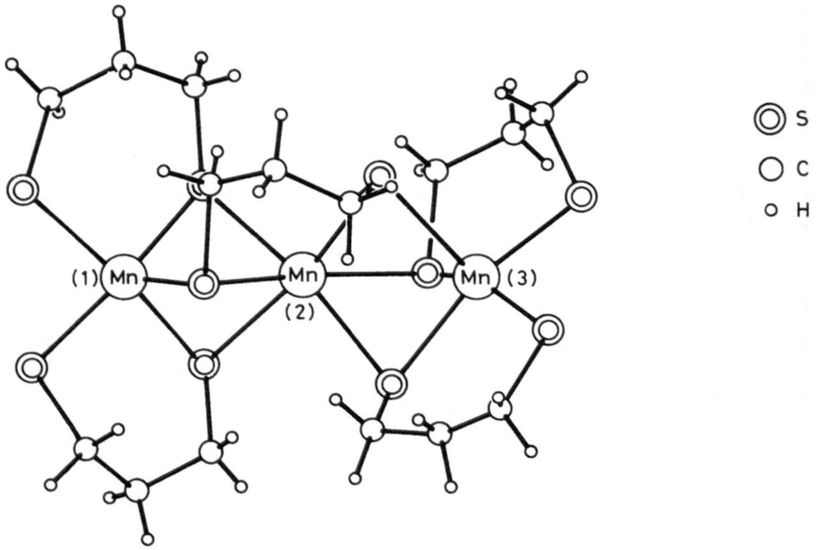

Fig. 6. Structure of the 1,3-propanedithiolato complex, $[Mn_3(C_3H_6S_2)_5]^{2-}$.

35.1.5.2.3 $M_2[Mn_2L_4]$ and $M_2[Mn_2L_2]$ Complexes

$[N(C_2H_5)_4]_2[Mn_2^{III}(C_2H_4S_2)_4]$ was prepared by the reaction of hydrated manganese(II) chloride, sodium 1,2-ethanedithiolate, and $[N(C_2H_5)_4]Br$ (mole ratio 1:2:2) in ethanol solution under nitrogen. Upon controlled exposure to air, small green-black crystals appeared. The crystals were isolated (under nitrogen) and recrystallized from CH_3CN-THF at 60°C to give black prisms in a 50% yield [1]. The complex can also be prepared by controlled air oxidation of $[N(C_2H_5)_4]_2[Mn(C_2H_4S_2)_2]$ in acetonitrile [2]. The single crystal X-ray structure of the complex was determined [1, 2], and data obtained from the two determinations are summarized below:

crystal system	monoclinic	monoclinic
space group	$P2_1/n$-C_{2h}^5, No. 14 (standard setting $P2_1/c$)	$P2_1/c$-C_{2h}^5, No. 14
lattice parameters in Å or °	a=12.039(4), b=10.699(2), c=15.012(3), β=110.89(2); Z=2	a=12.139(2), b=10.699(2), c=15.182(3), β=115.05(1); Z=2
reliability in %	R=4.4	R=4.48
density in g/cm³	d_{calc}=1.36 d_{exp}=1.37	— —
Ref.	[2]	[1]

Atomic positional parameters are presented in [2]. In the drawing of the $[Mn_2(C_2H_4S_2)_4]^{2-}$ anion structure (see **Fig. 7**) primed and unprimed atoms are related by an inversion center. The five-coordinate geometry around the metal atom can be described as distorted trigonal bipyramidal. The dimeric structure contains one terminal gauche chelate ring for each metal atom and two chelate rings with one bridging sulfur atom bonded to two metal atoms, forming an Mn_2S_2

unit [1, 2]. The Mn–S distances are shown in Fig. 7 [2]. Quite similar Mn–S distances were found by [1]. The Mn–Mn′ distance is 3.596(3) Å [2], while 3.543(2) Å was found by [1]. Bond angles around the manganese atoms (in °) are [2]:

S(1)–Mn–S(1′)	87.7(1)	S(1)–Mn–S(3)	89.7(1)	S(1′)–Mn–S(4)	95.6(1)
Mn–S(1)–Mn′	92.3(1)	S(2)–Mn–S(4)	90.6(1)	S(1)–Mn–S(4)	176.7(1)
S(1)–Mn–S(2)	88.6(1)	S(1′)–Mn–S(2)	105.7(1)	S(2)–Mn–S(3)	143.0(1)
S(3)–Mn–S(4)	89.1(1)	S(1′)–Mn–S(3)	111.2(1)		

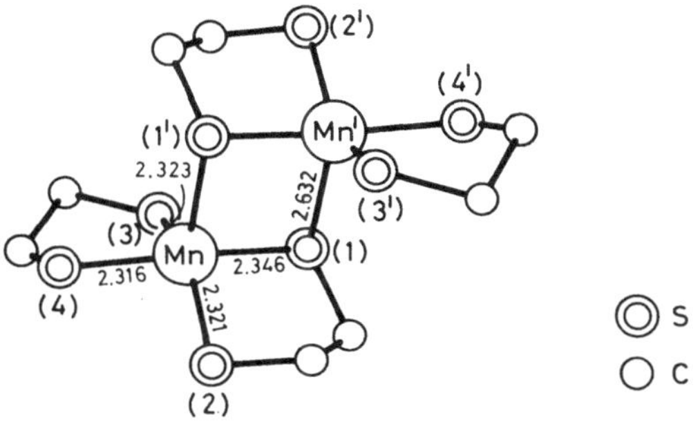

Fig. 7. Structure of the 1,2-ethanedithiolato complex, $[Mn^{III}_2(C_2H_4S_2)_4]^{2-}$, with selected bond distances and angles (H atoms omitted) [2].

The magnetic moment of the solid is $\mu_{eff}=3.96\ \mu_B$ at ambient temperature [2, 3]. It decreases to near zero with decreasing temperature. The lowering of the moment from that expected for a high-spin Mn^{III} complex is attributed to antiferromagnetic spin-spin coupling in the dimer ($J=-18.7\ cm^{-1}$) [3]. The complex is stable under anaerobic conditions [2].

$[P(C_6H_5)_4]_2[Mn^{III}_2(C_2H_4S_2)_4]$. The yellow filtrate obtained in the preparation of $[P(C_6H_5)_4]_2$-$[Mn^{II}(C_2H_4S_2)_2]$ (p. 39) was treated with ~300 mL of air injected by syringe. The reaction mixture immediately turned dark green and flaky microcrystals separated out. The product was collected, washed with diethyl ether, and dried in vacuum. Recrystallization from DMF-diethyl ether gave dark green microcrystals in 38% yield [2].

$[(C_2H_5)_3N(CH_2C_6H_5)]_2[Mn^{III}_2(C_6H_{10}S_6)_2]$. The complex with ligand 3 (formed in situ by oxidation of ligand 5) was prepared by treating an ethanolic solution containing $MnCl_2$, sodium 1,2,3-propanetrithiolate, and $[(C_2H_5)_3N(CH_2C_6H_5)]Br$ in a 1:2:4 mole ratio with air, as described for $[P(C_6H_5)_4]_2[Mn_3(C_3H_6S_2)_5]$ (p. 41). The product was isolated and recrystallized from DMF-diethyl ether to give black crystals in ~40% yield. The single crystal X-ray determination of the structure was carried out. The crystal data are: monoclinic system, space group $P2_1/c$-C^5_{2h} (No. 14) with a = 11.540(2), b = 12.115(2), c = 17.478(4) Å, β = 101.78(1)°; Z = 2. The structure was solved up to R = 3.59% (data collected at −160°C). The structure of the anion can be described as an Mn^{III} dimer containing two bridging bis-bidentate, tetrathiolate ligands, see **Fig. 8**, p. 44. As in the $[Mn_2(C_2H_4S_2)_4]^{2-}$ anion, the manganese geometries are approximately trigonal bipyramidal. Comparison of the two structures shows near congruency of the portions common to both molecules. The Mn–Mn′ distance is 3.598(2) Å. The bridging Mn–S distances are 2.3530(12) and 2.6546(12) Å. The magnetic moment of the solid is $\mu_{eff}=3.95\ \mu_B$ at 300 K. It decreases to near zero with decreasing temperature. The magnetic data indicate that the two high-spin Mn^{III} centers are antiferromagnetically coupled with $J=-18.7\ cm^{-1}$ [3].

Formulas of ligands are tabulated on p. 33

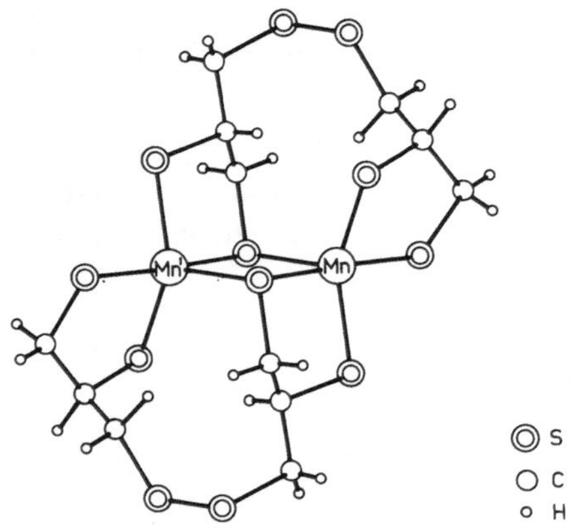

Fig. 8. Structure of the $[Mn_2^{III}(C_6H_{10}S_6)_2]^{2-}$ anion [3].

References:

[1] Christou, G.; Huffman, J. C. (J. Chem. Soc. Chem. Commun. **1983** 558/60).
[2] Costa, T.; Dorfman, J. R.; Hagen, K. S.; Holm, R. H. (Inorg. Chem. **22** [1983] 4091/9).
[3] Seela, J. L.; Folting, K.; Wang, R.-J.; Huffman, J. C.; Christou, G.; Chang, H.-R.; Hendrickson, D. N. (Inorg. Chem. **24** [1985] 4454/6).

35.1.5.2.4 $[N(C_2H_5)_4]_2[MnL_2NO]$ Complexes

Remark. Attempts to prepare the ternary complex with ligand 9, $[N(C_2H_5)_4]_2[Mn(C_7H_6S_2)_2NO]$, and the quaternary complex with ligands 7 and 11, $[N(C_2H_5)_4][Mn(C_4N_2S_2)(C_6Cl_4S_2)NO]$, by a π-cyclopentadienyl ring displacement reaction, as described below for $[N(C_2H_5)_4]$-$[Mn(C_6Cl_4S_2)_2NO]$, resulted in green unstable solids which decomposed during recrystallization [1].

$[N(C_2H_5)_4]_2[Mn(C_4N_2S_2)_2NO]$. The complex was prepared by bubbling NO gas through a degassed aqueous ethanol solution of hydrated manganese(II) chloride for 10 min, then adding a solution of the sodium salt of ligand 7 in methanol, with subsequent NO passage for an additional 10 min. The ligand-to-metal mole ratio used was 2:1. The reaction mixture was purged with N_2 gas and filtered into a methanol solution of $[N(C_2H_5)_4]Br$ (equimolar to the ligand). Partial removal of the solvent in vacuum yielded a green-black precipitate. The product was isolated, washed with 2-propanol, pentane, and dried in vacuum. The yield of the air-sensitive product was 38%. Recrystallization even under nitrogen resulted in decomposition. The melting point of the compound is 218 to 220°C. It is soluble in polar organic solvents and in dichloromethane, but the solutions decomposed in <3 min. It is a 2:1 electrolyte in acetone. The IR spectrum of the complex in KBr shows the characteristic bands: ν(CN) at 2210 and 2198 cm^{-1}, ν(NO) at 1687 cm^{-1}. Another 12 bands in the 1500 to 750 cm^{-1} range were not assigned [1].

$[N(C_2H_5)_4]_2[Mn(C_4F_6S_2)_2NO]$. The complex with ligand 8 was prepared by the reaction of excess ethanolic hydrazine with the cyclopentadienyl compound, $[(C_5H_5)Mn(C_4F_6S_2)NO]$. The reaction mixture turned green after 4 h, and was then treated with an ethanol solution of excess $[N(C_2H_5)_4]Br$. Storage of the solution for two weeks at −15°C gave green crystals. The product was isolated and dried. Recrystallization from acetone-ethanol gave the product in 23% yield. The solid melts at 165 to 171°C with decomposition. The magnetic moment of the compound is $\mu_{eff} = 1.30\ \mu_B$ at 23°C. This is intermediate between the expected diamagnetism for a structure with no unpaired electrons (singlet state) and the expected magnetic moment of 2.83 μ_B for a structure with two unpaired electrons (triplet state). The IR spectrum of the complex in a KBr disk shows the prominent bands: ν(CH) at 2950, ν(NO) at 1660, ν(C=C) at 1518, ν(CF) at 1230, 1150, and 1105 cm^{-1}. Another 12 bands in the 1500 to 680 cm^{-1} range were not assigned. The electronic absorption spectrum of the complex in 9:1 ethanol-hydrazine (pale green solution) shows a strong absorption at 725 nm ($\varepsilon = 2040\ L \cdot mol^{-1} \cdot cm^{-1}$) and an inflection point at ~303 nm ($\varepsilon = 6100\ L \cdot mol^{-1} \cdot cm^{-1}$). A square-pyramidal structure is suggested for the $[Mn(C_4F_6S_2)_2NO]^{2-}$ anion [2].

$[N(C_2H_5)_4]_2[Mn(C_6Cl_4S_2)_2NO]$. The complex was prepared by adding over 10 min potassium tetrachloro-1,2-benzenedithiolate, dissolved in ethanol, to an acetone solution of $[(C_5H_5)MnNO(CO)_2][PF_6]$ in a mole ratio of 1:2. The reaction mixture was stirred continuously for 15 min, CO evolved vigorously, and a deep green color developed gradually. The solution was filtered under N_2, and the filtrate treated with an equimolar amount of $[N(C_2H_5)_4]Br$ in water. The solution was evaporated until green crystals formed. The product was isolated, washed with 2-propanol and pentane, and dried in vacuum. The air-sensitive solid melts at 300°C. It is soluble in polar organic solvents and in dichloromethane, but the solutions decompose rapidly. The IR spectrum of the complex in a KBr disk shows the characteristic ν(NO) band at 1676 cm^{-1}. Another 14 bands in the 2950 to 680 cm^{-1} range were not assigned [1].

References:

[1] James, T. A.; McCleverty, J. A. (J. Chem. Soc. A **1970** 3318/21).
[2] King, R. B.; Bisnette, M. B. (Inorg. Chem. **6** [1967] 469/79).

35.1.5.2.5 $M[MnL_2L']$ and $M_3[Mn_2L_4L']$ Complexes

$[As(C_6H_5)_4][Mn^{III}(C_2H_2S_2)_2N_2H_4]$ was prepared by the reaction of hydrated manganese(II) chloride and the disodium salt of ligand 6 (mole ratio 1:2) in methanol with excess hydrazine at 0°C. The reaction mixture was stirred for two minutes in air. The cherry red solution was filtered under N_2, and the filtrate treated with an equimolar amount of $[As(C_6H_5)_4]Cl$ in methanol. The precipitate that formed was isolated, dissolved in hydrazine, and reprecipitated by slow addition of 1:1 isopropyl alcohol-isobutyl alcohol to give brown needles. The complex is a 1:1 electrolyte in nitromethane. The compound starts decomposing at ~123°C. The magnetic moment of the solid is $\mu_{eff} = 5.04\ \mu_B$ at 20°C [1].

$[N(C_2H_5)_4][Mn^{III}L_2py]$ for H_2L = ligand 6 and **$[As(C_6H_5)_4][Mn^{III}L_2py]$** for H_2L = ligand 6 or 10. The complexes with ligand 6 were prepared in the same way as $[As(C_6H_5)_4]$-$[Mn(C_2H_2S_2)N_2H_4]$, but using pyridine instead of hydrazine and $[N(C_2H_5)_4]Cl$ instead of $[As(C_6H_5)_4]Cl$ in the case of the tetraethylammonium salt. The complex with ligand 10 was prepared by the reaction of an ethanolic solution of hydrated manganese(II) chloride and a slight excess of the disodium salt of the ligand in pyridine-ethanol. The reaction mixture was filtered and an ethanol solution of a small excess of $[As(C_6H_5)_4]Cl$ added to the filtrate. The brown crystals that formed were isolated, dissolved in pyridine, and reprecipitated by slow addition of a 1:1 mixture of isopropyl alcohol-isobutyl alcohol. $[As(C_6H_5)_4][Mn(C_2H_2S_2)_2py]$ melts at 134 to 136°C and $[As(C_6H_5)_4][Mn(C_8H_8S_2)_2py]$ at 150°C (dec.) [1, 2]. Goniometer photographs show that $[As(C_6H_5)_4][Mn(C_2H_2S_2)_2py]$ is isomorphous with the corresponding Fe^{III} compound [1]. The effective magnetic moments of the three compounds are between 4.88 and 4.95 μ_B, i.e., near to the spin-only value (4.90 μ_B) for a high-spin d^4 complex [1, 2]. For $[As(C_6H_5)_4][Mn(C_2H_2S_2)_2py]$ the Curie-Weiss law is obeyed with the Weiss constant $\Theta = -36$ K. The IR spectrum of this compound shows the ν(C=C) band at 1515 cm^{-1} (at 1634 cm^{-1} for $Na_2(C_2H_2S_2)$) [1].

Formulas of ligands are tabulated on p. 33

The compounds are extremely sensitive to oxygen [1]. Crystalline $[N(C_2H_5)_4]$-$[Mn(C_2H_2S_2)_2py]$ was obtained as a pyrophoric material in several cases. The complexes with ligand 6 are readily soluble in pyridine. The solutions are stable for several hours. Solutions in dimethylformamide or dimethyl sulfoxide are considerably less stable. In acetone or dichloromethane the complexes dissolve with decomposition [2]. The compounds are 1:1 electrolytes in nitromethane [1].

$[N(C_2H_5)_4][Mn^{III}(C_2H_4S_2)_2(C_3H_4N_2)]$. The complex with ligand 1 and 1H-imidazole (= $C_3H_4N_2$) was prepared by the reaction of $[N(C_2H_5)_4]_2[Mn_2(C_2H_4S_2)_4]$ (see p. 42) with excess imidazole in dimethylformamide. After the addition of an equal volume of acetone to the reaction mixture, and cooling overnight, black crystals formed in 69% yield. The complex can also be conveniently obtained by controlled air oxidation of the ethanol slurry of hydrated manganese(II) chloride, $Na_2(C_2H_4S_2)$, 1H-imidazole, and $[N(C_2H_5)_4]Br$ in a 1:2:4:3 mole ratio. Upon oxidation the white solid dissolves, and a microcrystalline product precipitates. Recrystallization from acetonitrile-tetrahydrofuran gave greater than 90% overall yield.

The single crystal X-ray determination of the structure of the complex was carried out. The complex crystallizes in the monoclinic system, space group $P2_1/n-C^5_{2h}$, No. 14 (standard setting $P2_1/c$) with the lattice constants a = 13.974(5), b = 14.317(5), c = 10.564(3) Å, β = 90.13(2)°; Z = 4 at −72°C. The structure was solved up to R = 5.86%. The manganese atom is five-coordinate and is in a slightly distorted square-pyramidal environment. The apex is occupied by the nitrogen atom, and the basal positions are occupied by the sulfur donors with the metal 0.385 Å above the basal plane. The bond distances are: Mn–N = 2.224(7) Å and Mn–S from 2.319(3) to 2.334(3) Å. The N–Mn–S angles are 97.02(18)° to 102.88(18)° and the S–Mn–S angles from 88.16(12)° to 89.13(13)°. The imidazole ring is planar and makes an angle of 13.52° with the Mn–N vector [3].

The magnetic moment of the complex is $\mu_{eff} = 4.97\ \mu_B$ at 300 K. A sharp decrease in the moment below 15 K is indicative of zero field splitting with a value of 5.71 cm^{-1} [4]. While the complex can be recrystallized satisfactorily from acetonitrile or acetonitrile-acetone, attempts to recrystallize it from dimethylformamide-acetone (in the absence of excess imidazole) yielded the dinuclear complex, $[Mn_2(C_2H_2S_2)_4]^{2-}$, with the monomeric complex, $[Mn(C_2H_4S_2)_2L_n]^-$ containing solvate molecules, as an intermediate [3].

$[N(CH_3)_4]_3[Mn^{III}_2(C_2H_4S_2)_4(C_3H_3N_2)]$ and **$[N(CH_3)_4]_3[Mn^{III}_2(C_2H_4S_2)_4(C_3H_3N_2)]\cdot 2CH_3CN$.** The unsolvated complex was prepared by the reaction of $MnCl_2$, sodium 1,2-ethanedithiolate, sodium imidazolate, and $[N(CH_3)_4]Cl$ (mole ratio 1:2:4:4) in ethanol. The reaction mixture was cooled to 0°C and oxidized by successive additions of small portions of air until generation of a deep green color and precipitation of a green powder could no longer be observed. At this point anaerobic conditions were reestablished, and the product was isolated and recrystallized from warm (45°C) dimethylformamide-tetrahydrofuran to give analytically pure black prisms in 40% overall yield. Recrystallization from acetonitrile yielded the bis(acetonitrile) solvate. The single crystal X-ray structure determination of $[N(CH_3)_4]_3[Mn_2(C_2H_4S_2)_4(C_3H_3N_2)]\cdot 2CH_3CN$ was carried out. The complex crystallizes in the orthorhombic space group $Pna2_1-C^9_{2v}$ (No. 33) with the lattice constants a = 17.965(5), b = 16.094(4), and c = 14.789(3) Å; Z = 4. The structure was solved up to R = 4.58%. In the dimeric anion, the two five-coordinate Mn^{III} atoms are bridged by the imidazolate ring in a symmetric fashion (see **Fig. 9**) with the two Mn–N distances being 2.197(8) Å. The sulfur donors occupy the four basal positions of the square-pyramidal coordination sphere. The Mn–Mn distance is 6.487 Å and the Mn–S distances are in the range 2.318 to 2.357(3) Å. The N–Mn–S and S–Mn–S angles are 94.78° to 109.11(21)° and 86.60° to 88.87(11)°, respectively. The magnetic moment of the solid is $\mu_{eff} = 4.51\ \mu_B$ at 300 K and decreases to near zero as the temperature decreases. The data indicate an antiferromagnetic coupling of two high-spin Mn^{III} centers with J = −1.9 cm^{-1} [4].

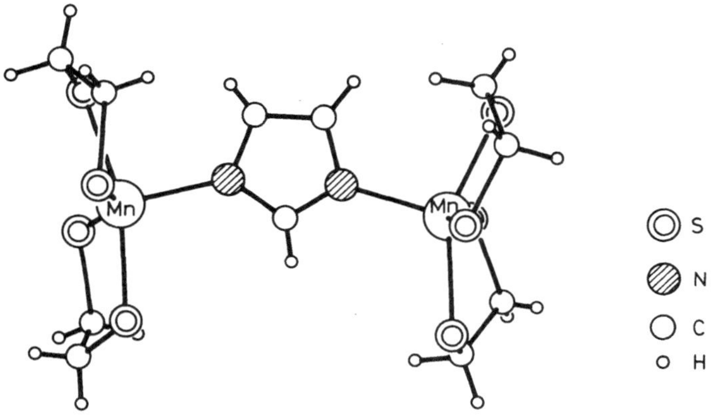

Fig. 9. Structure of the $[Mn_2^{III}(C_2H_4S_2)_4(C_3H_3N_2)]^{3-}$ anion [4] ($C_2H_6S_2$ = 1,2-ethanedithiole, $C_3H_4N_2$ = 1H-imidazole).

$[P(C_6H_5)_4]_2[Mn^{III}(C_7H_6S_2)_2]\cdot[Mn^{III}(C_7H_6S_2)_2CH_3OH]\cdot 3\,CH_3OH$. The Mn^{II} complex with ligand 9, $[Mn(C_7H_6S_2)_2]^{2-}$, was first prepared by the reaction of hydrated manganese(II) chloride and $Na_2(C_7H_6S_2)$ (mole ratio $\leqq 1:2$) in methanol in the absence of oxygen. The yellowish brown reaction mixture was brought into contact with atmospheric oxygen, and then $[P(C_6H_5)_4]Br$ was added. The color of the solution turned brown, and black crystals formed. The single crystal X-ray structure determination of the complex was carried out. It crystallizes in the triclinic system, space group $P\bar{1}-C_i^1$ (No. 2) with the lattice constants a = 10.001(4), b = 19.447(8), c = 19.814(8) Å, α = 99.64(3)°, β = 90.01(3)°, γ = 104.12(3)°; Z = 2. The structure was solved up to R = 4.1%. The complex contains the $[Mn(C_7H_6S_2)_2]^-$ anion with a four-coordinate square-planar structure and the $[Mn(C_7H_6S_2)_2CH_3OH]^-$ anion with a five-coordinate square-pyramidal structure. In $[Mn(C_7H_6S_2)_2]^-$ the Mn–S distances are 2.280 to 2.283 Å. Four S–Mn–S angles are 89.77° to 90.23°. The other two are 179.86° and 180°. In $[Mn(C_7H_6S_2)_2CH_3OH]^-$ sulfur atoms occupy the basal positions with the Mn–S distances from 2.306 to 2.314 Å. The methanol oxygen occupies the apical position with an Mn–O distance of 2.224(2) Å. Four S–Mn–S angles are from 88.16° to 89.45°, two are 161.17° and 165.78°. The S–Mn–O angles are from 95.48° to 100.65°. The magnetic moment is μ_{eff} = 5.2 μ_B per Mn atom at room temperature. The ^1H NMR spectrum of the complex in perdeuterated dimethyl sulfoxide shows aromatic proton signals shifted upfield (referred to TMS) to δ = 4.6, 9.3, and 20.8, while the methyl signal is shifted downfield to 15.7. The results indicate that only solvated $[Mn(C_7H_6S_2)_2]^-$ ions exist in solution [5].

References:

[1] Hoyer, E.; Dietzsch, W.; Heber, H. (Proc. 3rd Symp. Coord. Chem., Debrecen, Hung., 1970, Vol. 1, pp. 259/71; C.A. **78** [1973] No. 105468).

[2] Hoyer, E.; Dietzsch, W.; Müller, H. (Z. Chem. [Leipzig] **7** [1967] 354/5).

[3] Seela, J. L.; Huffman, J. C.; Christou, G. (J. Chem. Soc. Chem. Commun. **1985** 58/60).

[4] Seela, J. L.; Folting, K.; Wang, R.-J.; Huffman, J. C.; Christou, G.; Chang, H.-R.; Hendrickson, D. N. (Inorg. Chem. **24** [1985] 4454/6).

[5] Henkel, G.; Greiwe, K.; Krebs, B. (Angew. Chem. **97** [1985] 113/4).

Formulas of ligands are tabulated on p. 33

35.1.5.2.6 $M_2[MnL_3]$ Complexes

$[P(C_6H_5)_4]_2[Mn(C_4N_2S_2)_3]$ was prepared by adding $MnCl_2 \cdot 4H_2O$ in water to the sodium salt of ligand 7, $Na_2(C_4N_2S_2)$, in aqueous ethanol and shaking the mixture in air for a few minutes. The solution was filtered into ethanolic $[P(C_6H_5)_4]Br$ and the product precipitated by addition of ether. It was purified by dissolving in degassed acetone and reprecipitating with ether. The dark green crystals were washed with ethanol, diethyl ether, and air-dried. The yield of product was 47%. The melting point of the solid is 172 to 175°C [1]. Comparison of the X-ray powder patterns of the $[P(C_6H_5)_4][M'(C_4N_2S_2)_3]$ complexes with M' = V, Cr, Mn, Fe, Mo, and W (a table of d spacings is presented) indicates that the Mn compound is not isomorphous with the other compounds. The magnetic moment of the solid varies from $\mu_{eff} = 4.01\ \mu_B$ at 296.5 K to 3.94 μ_B at 86.3 K. Susceptibility data satisfy the Curie-Weiss law with the Weiss constant $\Theta = -10$ K. The magnetic results could not be related to the molecular orbital schemes suggested earlier for dithiolene complexes of the type $[M(C_2S_2R_2)_3]^{z-}$ having D_{3h} symmetry. The IR spectrum of the complex in a KBr disk shows the ν(CN) bands at 2208 and 2200 cm^{-1}. Other bands in the 3050 to 690 cm^{-1} range were not assigned. The complex is a 2:1 electrolyte in acetone [1].

$[(CH_3)P(C_6H_5)_3]_2[Mn(C_4N_2S_2)_3]$. The green complex was prepared by the air oxidation of a mixture of a manganese(II) salt and the sodium salt of ligand 7, $Na_2(C_4N_2S_2)$, in ethanol, followed by precipitation on addition of $[CH_3P(C_6H_5)_3]I$ [2]. The magnetic moment of the complex corresponds to S = 3/2 [1]. (The value of 4.97 μ_B reported for the magnetic moment in [2] was wrong [1].) The complex is a 2:1 electrolyte in nitromethane [2].

$[As(C_6H_5)_4]_2[Mn(C_4N_2S_2)_3]$ was prepared by slow addition of solid $Mn(CH_3COO)_3 \cdot 2H_2O$ to a hot ethanol solution of $Na_2(C_4N_2S_2)$ and $[As(C_6H_5)_4]Cl$ with nitrogen bubbling through the mixture, which was heated on a steam bath. The mole ratio used was 1:2:2:2. The reaction mixture was heated for an additional five minutes and then cooled. The green solid was isolated and recrystallized twice from dichloromethane-isopropyl alcohol under nitrogen. The yield was about 20%, based on the manganese used. The melting point of the solid is 176 to 181°C. Comparison of the X-ray powder patterns indicates isomorphism with $[As(C_6H_5)_4]_2$-$[Fe(C_4N_2S_2)_3]$, for which an octahedral structure can be assumed. (A table of d spacings, but not of the Fe compound, is presented in the publication.) The magnetic moment of the solid is $\mu_{eff} = 3.85\ \mu_B$ at room temperature. The complex is a 2:1 electrolyte in nitromethane [3].

$[N(C_2H_5)_4]_2[Mn(C_6Cl_4S_2)_3]$ and **$[N(C_4H_9)_4]_2[Mn(C_6Cl_4S_2)_3]$.** The complexes were prepared by reaction of an aqueous ethanol solution of hydrated manganese(II) chloride with the potassium salt of ligand 11, $K_2(C_6Cl_4S_2)$ (prepared in situ from potassium and the ligand) in the case of the tetraethylammonium compound, or with $NH(C_2H_5)_3(C_6Cl_4S_2)$ (prepared in situ from triethylamine and the ligand) in the case of the tetrabutylammonium compound in a mole ratio of 1:3. The reaction mixture was stirred for 30 min and filtered into an ethanol solution of excess $[N(C_2H_5)_4]Br$ or $[N(C_4H_9)_4]I$. Partial evaporation of the solvent under reduced pressure gave magenta precipitates. The complexes were recrystallized from acetone-ethanol to afford black crystals, which were washed with ethanol, diethyl ether, and air-dried. The yields were 84 or 44%, based on $MnCl_2 \cdot 4H_2O$. $[N(C_4H_9)][Mn(C_6Cl_4S_2)_3]$ was also prepared by the addition of $KMnO_4$, dissolved in the minimum volume of degassed water, to an ethanolic solution of $K_2(C_6Cl_4S_2)$ (mole ratio 1:3). Acetone was added and the reaction mixture refluxed for 20 min under nitrogen. Reaction of the resulting magenta solution with $[N(C_4H_9)_4]I$, and purification of the black precipitate that formed, was as described above. The yield was 61%. The tetraethylammonium salt melts at 280°C with decomposition, the tetrabutylammonium salt from 252 to 254°C. The magnetic moment of solid $[N(C_4H_9)_4][Mn(C_6Cl_4S_2)_3]$ is $\mu_{eff} = 3.97\ \mu_B$ at 20°C. The tetrabutylammonium salt is soluble in acetone and dichloromethane. It is a 2:1 electrolyte in acetone [4].

Formulas of ligands are tabulated on p. 33

References:

[1] McCleverty, J. A.; Locke, J.; Wharton, E. J. (J. Chem. Soc. A **1968** 816/23).
[2] Gerloch, M.; Kettle, S. F. A.; Locke, J.; McCleverty, J. A. (Chem. Commun. **1966** 29/31).
[3] Stiefel, E. I.; Bennett, L. E.; Dori, Z.; Crawford, T. H.; Simo, C.; Gray, H. B. (Inorg. Chem. **9** [1970] 281/6).
[4] Wharton, E. J.; McCleverty, J. A. (J. Chem. Soc. A **1969** 2258/66).

35.1.5.2.7 Polymeric Manganese Dithiolato Complexes

Polymeric manganese complexes with ligand 7 have been prepared with dyes as cations. The compositions of two black solids have been reported: $(TH)_6Mn_4(C_4N_2S_2)_5 \cdot 21\,H_2O$ [1] and $(TH)_6Mn_4(C_4N_2S_2)_5(OH)_8 \cdot 6\,H_2O$ [2] with TH representing a thionine cation (=3,7-diamino-phenothiazin-5-ium). The composition of a dark red solid is $(SAF)_6Mn_2(C_4N_2S_2)_5(OH)_2 \cdot 11\,H_2O$ with SAF representing tolusafranine cation (=3,7-diamino-2,8-dimethyl-5-phenylphenazinium cation). Incorporation of the thionine cation into the thiolate complex increases the photoreactivity of the dye, while the opposite is true for tolusafranine cation [2].

Insoluble complexes with ligand 1 or ligand 9 of composition $[Mn_2(C_2H_4S_2)_3O(OH)_3]_n$ or $[Mn_2(C_7H_6S_2)_4O(OH)(H_2O)_8]_n$, respectively, were synthesized by the reaction of aqueous $KMnO_4$ with an ethanolic solution of the ligand (1:2 mole ratio). Magnetic measurements, IR, ESR, and mass spectra indicate that the materials are polynuclear compounds with oxygen and disulfide bridges between the Mn atoms [3].

References:

[1] Kaneko, M.; Araki, H.; Yamada, A. (Polym. Prepr. Am. Chem. Soc. Div. Polym. Chem. **20** [1979] 1053/4).
[2] Kaneko, M.; Yamada, A. (Makromol. Chem. **182** [1981] 105/12).
[3] Kaneko, M.; Ishehara, W.; Yamada, A. (Makromol. Chem. **182** [1981] 89/99).

35.1.6 With Mercapto Sulfonic Acids

No.	ligand	formula
1	$HSCH_2CH_2SO_3H$	$C_2H_6O_3S_2$
2	$HSCH_2$–CH(SH)–CH_2SO_3H	$C_3H_8O_3S_3$
3	$HSCH_2$–CH(SH)–CH_2–O–$CH_2CH_2SO_3H$	$C_5H_{12}O_4S_3$
4	$HSCH_2$–CH(SH)–CH_2–S–$CH_2CH_2SO_3H$	$C_5H_{12}O_3S_4$
5	$HSCH_2$–CH(SH)–CH_2–$S(O_2)$–$CH_2CH_2SO_3H$	$C_5H_{12}O_5S_4$

Complexes of manganese(II) with ligand 1 have a stability comparable to that of manganese(II) complexes with 2-aminoethanesulfonic acid (taurine) [1].

The stability constants of the complexes of manganese(II) with the sodium salts of ligands 2 through 5 in basic solution, I=0.1M (KNO_3), were determined by the pH method under hydrogen. The values of $\log K_1$ and $\log \beta_2$ are: 16.10 for $Mn(C_3H_5NaO_3S_3)$ and 21.10 for

$Mn(C_3H_5NaO_3S_3)_2^{2-}$; 16.65 for $Mn(C_5H_9NaO_4S_3)$ and 23.25 for $Mn(C_5H_9NaO_4S_3)_2^{2-}$; 16.00 for $Mn(C_5H_9NaO_3S_4)$ and 21.00 for $Mn(C_5H_9NaO_3S_4)_2^{2-}$; 15.70 for $Mn(C_5H_9NaO_5S_4)$ and 20.50 for $Mn(C_5H_9NaO_5S_4)_2^{2-}$ [2].

A brown solution is formed in acidic media and a brown-green solution is formed in ammoniacal solution from the complexation of Mn^{II} by the sodium salt of ligand 2 [3].

References:

[1] Sakurai, H.; Takeshima, S. (Ganryu Aminosan **5** [1982] 255/60 from C.A. **99** [1983] No. 195397).

[2] Ryabushko, O. P.; Pilipenko, A. T.; Krivokhizhina, L. A.; Emchenko, N. L. (Ukr. Khim. Zh. **39** [1973] 1051/3; Soviet Progr. Chem. **39** No. 10 [1973] 76/8).

[3] Vol'f, A. L. (Zh. Vses. Khim. Obshchestva **5** [1960] 232; C.A. **1960** 18162).

35.1.7 With Mercapto Carboxylic Acids or Their Esters

Remark. Mn^{II} complexes with mercapto amino carboxylic acids are treated in "Manganese" D 4, 1985, pp. 290 and 292, together with other amino carboxylic acids.

No.	ligand	formula
1	$HSCH_2COOH$	$C_2H_4O_2S$
2	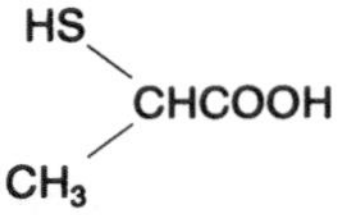	$C_3H_6O_2S$
3	$HSCH_2CH_2COOH$	$C_3H_6O_2S$
4	HS–CHCOOH CH_2COOH	$C_4H_6O_4S$
5	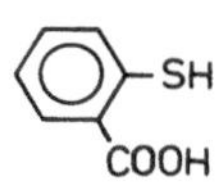	$C_7H_6O_2S$
6	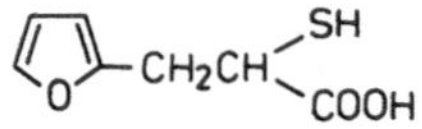	$C_7H_8O_3S$

No.	ligand		formula
7	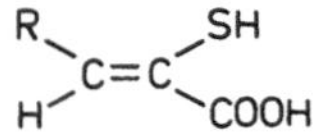	R = C_6H_5	$C_9H_8O_2S$
8		R = 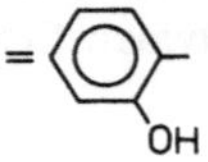	$C_9H_8O_3S$
9		R = 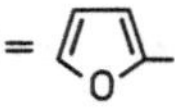	$C_7H_6O_3S$
10		R = 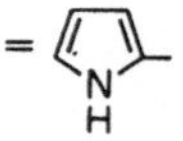	$C_7H_7NO_2S$
11		R = 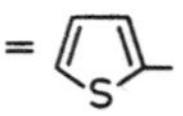	$C_7H_6O_2S_2$
12	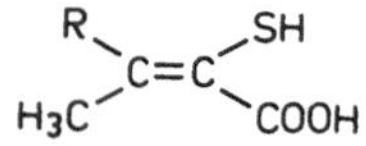	R = C_6H_5	$C_{10}H_{10}O_2S$
13		R = 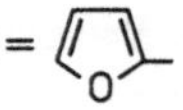	$C_8H_8O_3S$
14		R = 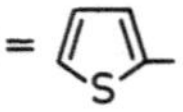	$C_8H_8O_2S_2$
15	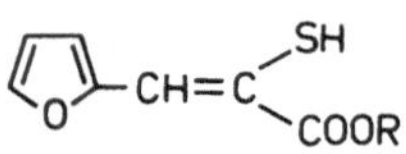	R = CH_3	$C_8H_8O_3S$
16		R = C_2H_5	$C_9H_{10}O_3S$

35.1.7.1 Manganese(II) Complexes in Solution

Stability constants of Mn^{II} complexes with mercapto carboxylic acids in aqueous or aqueous-organic medium (v/v organic solvent given) were determined by potentiometric pH titrations [1 to 5, 7, 8]. The titrations were performed in a nitrogen atmosphere by [1 to 3, 7, 9].

ligand	t in °C	medium	I	log K_1	log K_2	Ref.
1 (= H_2L)	room temp.	aqueous	0.1(KCl)	4.38	3.18	[1]
3[a] (= H_2L)	25	50% ethanol	0.1($KClO_4$)	5.70	—	[2]
	35	50% ethanol	0.1($KClO_4$)	6.30	—	[2]
4 (= H_3L)	35	aqueous	0.2($NaClO_4$)	4.92	—	[8]
	45	aqueous	0.2($NaClO_4$)	4.87	—	[8]
	55	aqueous	0.2($NaClO_4$)	4.82	—	[8]
5 (= H_2L)	17	50% ethanol	0.05($NaClO_4$)	5.07	—	[4]
	30	45% ethanol	0.15($NaClO_4$)	5.04	4.05	[3]
	40	45% ethanol	0.15($NaClO_4$)	5.15	4.18	[3]
	50	45% ethanol	0.15($NaClO_4$)	5.28	4.28	[3]
6 (= H_2L)	25	10% ethanol	0.1 KNO_3	3.68	2.90	[7]
9 (= H_2L)	30	10% ethanol	0.1 KNO_3	5.07	3.83	[7]
11 (= H_2L)	25	50% dioxane	0.1($NaClO_4$)	6.26	4.15	[5]

[a] In [2] the formula of ligand 3 is given, but its designation is α-mercapto carboxylic acid.

For the formation of the complex with ligand 4, according to $Mn^{2+} + C_4H_6O_4S \rightleftharpoons Mn(C_4H_4O_4S) + 2H^+$, the equilibrium constant, log K = 5.91, was found at 31°C and I = 0.1 M ($NaClO_4$) by potentiometric titration (pH method). At pH > 5.8 the neutral complex dissociates. The equilibrium constant of the reaction $Mn(C_4H_4O_4S) \rightleftharpoons Mn(C_4H_3O_4S)^- + H^+$ is log K = -7.48 ± 0.24 at 31°C, I = 0.1 M ($NaClO_4$) [9]. The stability constant, log $\beta_2 = 3.44$, of the protonated complex with ligand 8, formed by $Mn^{2+} + 2C_9H_7O_3S^- \rightleftharpoons Mn(C_9H_7O_3S)_2$, was determined by distribution between two phases at ambient temperature [10].

Complex formation reactions of manganese with mercapto carboxylic acids have found use in analytical chemistry. Thus, Mn^{II} forms a yellow-brown complex in aqueous ammonia, while other metals form colorless complexes [11]. Yellow, yellow-green, or green color reactions of Mn^{II} were observed with ligands 2 [12], 8 [13], and 9 [14] in aqueous ammonia, with ligands 8 [13], 9 [14], and 15, 16 [15] in aqueous sodium acetate, and with ligands 12, 13, and 14 in isoamyl alcohol [16]. Reddish color reactions were observed in aqueous ammonia with ligands 9, 10, and 11 [17], and in aqueous pyridine with ligands 12, 13, and 14 [16]. The limit of dilution for all the reactions is between $1:10^5$ and $1:10^6$ [12 to 17].

Ligand 3 has been used as a masking agent in the $EDTA^{4-}$ determination of Mn^{II} in the presence of other metal ions [18]. Mn^{III} can be determined amperometrically by the reaction with ligand 5 [19]. Similarly, using coulometric titration with Mn^{III}, ligands 1, 4, and 5 can be determined in glacial acetic acid [20]. The 1:2 complex of Mn^{II} with ligand 7 is quantitatively extracted into isoamyl alcohol at pH 6.3 to 9.4. The extract shows maximal absorbance at 625 nm ($\varepsilon = 7.3 \times 10^3$ L·mol^{-1}·cm^{-1}). The procedure can be used for the selective determination of Mn [6].

A complex of Mn^{III} with ligand 1, formed in acid media, acts as a free radical initiator in the polymerization of methyl methacrylate [21].

References:

[1] Leussing, D. L. (J. Am. Chem. Soc. **80** [1958] 4180/3).
[2] Srivastava, P. C. (Thermochim. Acta **55** [1982] 125/33).

[3] Ramamani, R.; Shanmuganathan, S. (Current Sci. [India] **37** [1968] 39/40).
[4] Reddy, M. V.; Bhattacharya, P. K. (J. Inorg. Nucl. Chem. **32** [1970] 2321/4).
[5] Wagner, J.; Vitali, P.; Schoun, J.; Giroux, E. (Can. J. Chem. **55** [1977] 4028/36).
[6] Izquierdo, A.; Prat, M. D.; Garriga, N.; Allegria, J. M. (Analyst [London] **111** [1986] 309/11).
[7] Izquierdo, A.; Garcia Puignou, L.; Guasch, J. (Polyhedron **5** [1986] 1253/7).
[8] Reddy, M. V.; Bhattacharya, P. K. (J. Indian Chem. Soc. **46** [1969] 1058/60).
[9] Mohanty, N. K.; Patnaik, R. K. (J. Indian Chem. Soc. **57** [1980] 224/5).
[10] Izquierdo, A.; Giné, M.; Compañó, R. (J. Inorg. Nucl. Chem. **43** [1981] 617/8).

[11] Přibil, R.; Vaselý, V. (Talanta **8** [1961] 880/4).
[12] Buscarons, F.; Casassas, E. (Anales Real Soc. Espan. Fis. Quim. [Madrid] B **55** [1959] 655/62).
[13] Izquierdo, A.; Giné, M. (Anales Quim. **74** [1978] 53/7).
[14] Izquierdo, A.; Calmet, J. (Quim. Anal. **28** [1974] 148/52).
[15] Izquierdo, A.; Guasch, J.; Callao, M. P. (Quim. Anal. [Barcelona] **2** [1983] 174/86).
[16] Izquierdo, A.; Garcia Puignou, L. (Anales Quim. B **79** [1983] 254/9).
[17] Izquierdo, A.; Guasch, J.; Miguel, J. (Anales Quim. B **80** [1984] 382/9).
[18] Yamaguchi, K.; Ueno, K. (Talanta **10** [1963] 1195/7).
[19] Bessarabova, I. M.; Songina, O. A.; Zakharova, L. V. (Khim. Khim. Tekhnol. [Alma Ata] No. 22 [1977] 41/6; C.A. **92** [1980] No. 121204).
[20] Pastor, T. J.; Vajgand, V. J.; Antonijevic, V. V. (Glasnik Hem. Drustva Beograd **44** [1979] 651/6; C.A. **92** [1980] No. 226164).

[21] Samal, R. K.; Suryanarayan, G. V.; Panda, G.; Das, D. P.; Nayak, M. C. (J. Appl. Polym. Sci. **26** [1981] 41/8).

35.1.7.2 Isolated Manganese(II) Compounds

$Mn[Sb(C_2H_2O_2S)_2]_2$ was prepared by the addition of an aqueous solution of $Mn(NO_3)_2$ to a hot aqueous solution of the hydrogen antimony salt of ligand 1, $Sb(C_2H_2O_2S)(C_2H_3O_2S)$, in a mole ratio of 1:2. The light pink precipitate, which formed slowly, was isolated, washed with hot water, hot ethanol, and dried in vacuum at 50°C. The magnetic moment of the solid is μ_{eff} = 5.88 μ_B. The IR spectrum of $C_4H_5O_4S_2Sb$ indicates the presence of an Sb–S bond. The IR spectrum of the complex suggests that the carboxy group has a bridging function. The complex melts >280°C. It is insoluble in common polar and nonpolar solvents. On the basis of its properties a polymeric structure is assumed [1].

$[Mn(C_3H_4O_2S)(H_2O)_2]\cdot 2H_2O$ was prepared by the reaction of equimolar amounts of an aqueous solution of $MnCl_2$ with that of ligand 2. The pink precipitate, formed after adjusting the resulting solution to pH 6.5 and concentrating the mixture, was isolated, washed with water, ethanol, acetone, petroleum ether, and dried at 80°C. The magnetic moment of the solid increases from μ_{eff} = 4.29 μ_B at 82.0 K to μ_{eff} = 4.48 μ_B at 295.5 K. The susceptibility data obey the Curie-Weiss law with the Weiss constant $\Theta = -8$ K. The magnetic data are consistent with a square-planar Mn^{II} complex with partial spin pairing. The IR spectrum of the complex in a KBr disk shows prominent bands at 1690, 1405, and 725 cm^{-1}, which were assigned to ν_{as}(COO), ν_s(COO), and ν(C–S) vibrations, respectively. Shifts with respect to the corresponding free ligand bands indicate coordination of the ligand dianion to the metal via the sulfur atom and one oxygen atom of the carboxylate group, which is unidentate. The complex is soluble in water and insoluble in common organic solvents [2].

Li[Mn($C_4H_3O_4S$)]·3H_2O and M[Mn($C_4H_3O_4S$)]·2H_2O (M = Na, K). The complexes were prepared under nitrogen by the combination of equimolar amounts of an aqueous solution of $Mn(CH_3COO)_2 \cdot 4H_2O$ and the appropriate alkali salt of ligand 4. The reaction mixture was heated at 60°C for 15 min. The pale blue precipitate that formed was collected, washed with water and acetone, and dried under vacuum for 6 h. The magnetic moments of the solids are: μ_{eff} = 5.08, 5.04, 5.08 μ_B for M = Li, Na, or K, at 90 K and μ_{eff} = 6.23, 5.95, and 6.06 μ_B for M = Li, Na, or K, at 295 K. The Weiss constants are Θ = −80, −70, and −63 K for M = Li, Na, or K. The IR spectra of the complexes in Nujol mull show characteristic absorptions in the regions 3380 to 3300, 1550 to 1545, 1410, and 640 to 635 cm^{-1}, which were assigned to the $\nu(OH)_{H_2O}$, $\nu_{as}(COO)$, $\nu_s(COO)$, and $\nu(CS)$ vibrations, respectively. The electronic absorption spectrum shows bands at approximately 16500, 17400, 21000, 23000, 23500, 26300, and 28000 cm^{-1}. The bands arise from the transitions of the octahedral high-spin Mn^{II}: $^6A_{1g} \to {}^4T_{1g}(^4G)$, $\to {}^4T_{2g}(^4G)$, $\to {}^4A_{1g}(^4G)$, $\to {}^4E_g(^4G)$, $\to {}^4T_{2g}(^4D)$, $\to {}^4E_g(^4D)$. Several intense absorption bands at high frequencies arise from charge transfer and intra-ligand absorptions. The physical properties of the complexes indicate coordination of the carboxylate and thiolate groups, both of which bridge in a two-dimensional polymeric structure [3].

[Mn($C_4H_4O_4S$)(H_2O)$_2$] was prepared under nitrogen by the reaction of equimolar amounts of an aqueous solution of $Mn(CH_3COO)_2 \cdot 4H_2O$ and ligand 4. Upon addition of ethanol to the reaction mixture, a light blue precipitate formed. The product was collected, washed with absolute ethanol, acetone, and dried under vacuum for 6 h. The magnetic moment of the solid increases from μ_{eff} = 5.18 μ_B at 90 K to μ_{eff} = 5.90 μ_B at 295 K, with a Weiss constant of Θ = −38 K. The IR spectrum of the complex in Nujol mull shows prominent bands at 3300, 1550, 1400, and 622 cm^{-1}, which were assigned to $\nu(OH)_{H_2O}$, $\nu_{as}(COO)$, $\nu_s(COO)$, and $\nu(CS)$ vibrations, respectively. The electronic absorption spectrum of the solid, in diffuse reflectance, shows maxima at 16000, 17500, 20900, 23700, 26500, and 27900 cm^{-1}. The bands arise from the transitions of octahedral Mn^{II}: $^6A_{1g} \to {}^4T_{1g}(^4G)$, $\to {}^4T_{2g}(^4G)$, $\to {}^4A_{1g}(^4G)$, $\to {}^4T_{2g}(^4D)$, $\to {}^4E_g(^4D)$, respectively.

The physical properties of the complex suggest that the octahedral Mn^{II} atom is coordinated to the dinegative bidentate ligand via bridging carboxylate and thiolate groups in a one-dimensional polymeric structure [3].

References:

[1] Sharma, H. K.; Lata, S.; Dubey, S. N.; Puri, D. M. (Indian J. Chem. A **20** [1981] 1031/3).
[2] Srivastava, S. K.; Pandeya, K. B.; Nigam, H. L. (Indian J. Chem. **13** [1975] 498/500).
[3] Larkworthy, L. F.; Sattari, D. (J. Inorg. Nucl. Chem. **42** [1980] 551/9).

35.1.7.3 Isolated Manganese(III) Compound

[P(C_6H_5)$_4$][Mn($C_7H_4O_2S$)$_2$($C_3H_4N_2$)]·CH_2Cl_2 ($C_3H_4N_2$ = 1H-imidazole). The complex was prepared by the reaction of [Mn(acac)$_3$], triethylamine, ligand 5, and 1H-imidazole (mole ratio 1:1:2:2) in CH_2Cl_2 at −78°C. Addition of $[P(C_6H_5)_4]Br$ to the intensely red reaction mixture yielded red-black crystals of the complex in 25% yield. The single crystal X-ray diffraction structure of the complex has been determined at −155°C. The crystal data are: monoclinic system, space group $P2_1/n\text{-}C^5_{2h}$, No. 14 (standard setting $P2_1/c$) with a = 21.898(11), b = 13.933(6), c = 14.065(6) Å, β = 99.12(2)°; Z = 4. The structure was solved up to R = 6.53%. In the structure the Mn atom is five-coordinate, square-pyramidal with the imidazole nitrogen atom in the apical position and *cis* thiolate sulfur and *cis* carboxylate oxygen atoms in the

Formulas of ligands are tabulated on p. 50

basal positions (see **Fig. 10**). The mean bond distances are: Mn–S 2.284(4) Å, Mn–O 1.940(8) Å. The Mn–N distance is 2.176(6) Å. Mean bond angles are: N–Mn–S 104.82°, N–Mn–O 96.09°, S–Mn–O 159.07°. The DMSO solution magnetic moment is $\mu_{eff} = 5.11\ \mu_B$. The electronic absorption spectrum of the complex (in CH_3CN with excess imidazole) shows bands at 415 (sh, $\varepsilon = 2940$), 450 ($\varepsilon = 2760$), 500 ($\varepsilon = 2500$) nm. The visible absorption bands have been assigned to thiolate sulfur-to-Mn charge transfers. The band at 500 nm is similar to a 515 nm (2460) band of acid phosphatase, and supports O,N,S coordination of a five-coordinate Mn^{III} atom in the enzyme also.

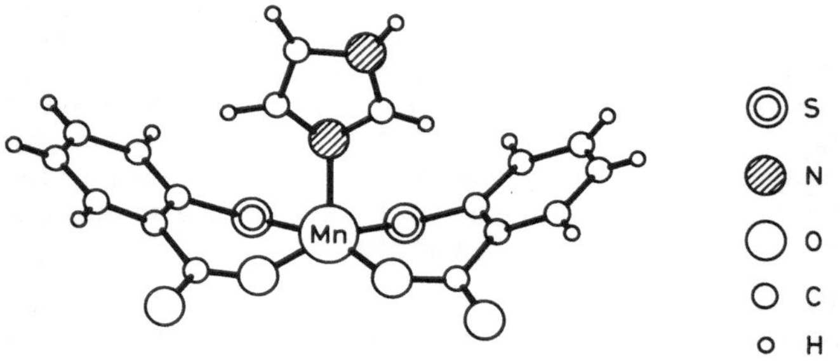

Fig. 10. Structure of the complex with thiosalicylic acid and 1H-imidazole, $[Mn^{III}(C_7H_4O_2S)_2(C_3H_4N_2)]^-$.

Reference:

Bashkin, J. S.; Huffman, J. C.; Christou, G. (J. Am. Chem. Soc. **108** [1986] 5038/9).

35.1.8 With Mercaptoacetamides $HSCH_2C(=O){-}NHR$

ligand	R	formula
1	C_6H_5	C_8H_9NOS
2	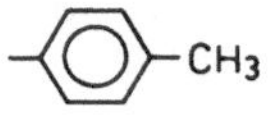	$C_9H_{11}NOS$
3	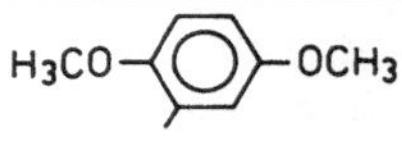	$C_{10}H_{13}NO_3S$
4	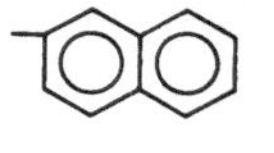	$C_{12}H_{11}NOS$

Complex formation of manganese(II) with ligands 1 to 4 in 75 vol% dioxane-water was studied by potentiometric pH-titration at 30°C. Stability constants are: log $K_1 \approx 6.3$ for ligand 1, log $K_1 \approx 6.4$ for ligands 2 and 3, corrected to $I = 0$ [1], and log $K_1 = 4.4$, log $K_2 = 4.4$ for ligand 4, at $I = 0.1(NaClO_4)$ [2].

References:

[1] Martin, D. F. (J. Am. Chem. Soc. **83** [1961] 1076/8).
[2] Burger, K.; Korecz, L.; Toth, A. (Magy. Kem. Folyoirat **73** [1967] 455/8; C.A. **68** [1968] No. 16591).

35.1.9 With Thiodiketones or Mercapto Ketones

R–C(O)–CH_2–C(S)–R′ ligand 1 with R = R′ = C_6H_5 (= $C_{15}H_{12}OS$)
ligand 2 with R = CF_3, R′ = C_6H_5 (= $C_{10}H_7F_3OS$)
ligand 3 with R = CF_3, R′ = 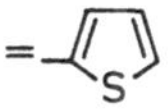 (= $C_8H_5F_3OS_2$)

CH_3–C(S)–CH_2–C(S)–CH_3 ligand 4 (= $C_5H_8S_2$)

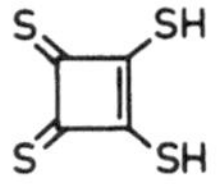

ligand 5 (= $C_4H_2S_4$)

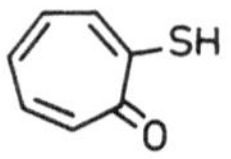

ligand 6 (= C_7H_6OS)

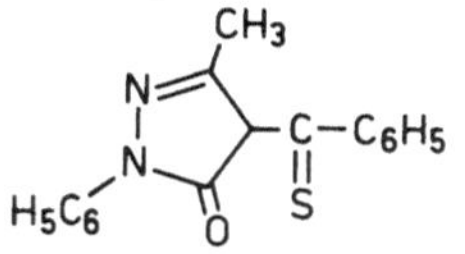

ligand 7 (= $C_{17}H_{14}N_2OS$)

Remark. On reaction of a manganese(II) salt with 2,4-pentanedithione or 1-phenyl-1,3-butanedithione in the air, ligand oxidation occurs with formation of the resonance-stabilized five-membered 1,2-dithiolium cation, containing an S–S bond. The 1,2-dithiolium cation (I) does not coordinate to the manganese atom, but forms halomanganate salts in the presence of halide ions. For example in the case of ligand 4 in the presence of chloride ions the 3,5-dimethyl-1,2-dithiolium tetrachloromanganate(II), $(C_5H_7S_2)_2MnCl_4$, is formed. Reduction of this compound with a strong reducing agent, such as $NaBH_4$, leads then to the desired Mn^{II} 2,4-pentanedithionate (II).

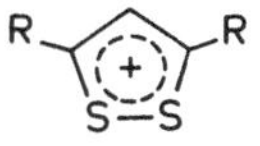

I

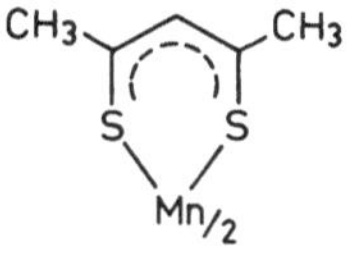

II

Because of the noncoordination of the S-containing ligands in the dithiolium salts, these are not treated here. The compound $(C_5H_7S_2)_2MnCl_4$ (from [1]) was already described in "Mangan" C 5, 1978, p. 209. Other references concerning this compound are [2 to 4, 8, 9]. References concerning dithiolium salts derived from 1-phenyl-1,3-butanedithione are [5, 6, 8]. A general reference concerning 1,2-dithiolium salts is [7].

References:

[1] Furuhashi, A.; Watanuki, K.; Ouchi, A. (Bull. Chem. Soc. Japan **41** [1968] 110/4).
[2] Furuhashi, A.; Takeuchi, T.; Ouchi, A. (Bull. Chem. Soc. Japan **41** [1968] 2049/54).
[3] Heath, G. A.; Martin, R. L.; Stewart, I. M. (Australian J. Chem. **22** [1969] 83/95).
[4] Takahashi, Y.; Nakatani, M.; Ouchi, A. (Nippon Kagaku Zasshi **91** [1970] 636/9; C.A. **73** [1970] No. 94182).
[5] Takahashi, Y.; Nakatani, M.; Ouchi, A. (Bull. Chem. Soc. Japan **42** [1969] 274).
[6] Ouchi, A.; Eguchi, H.; Takeuchi, T.; Furuhashi, A. (Bull. Chem. Soc. Japan **42** [1969] 2259/63).
[7] Lockyer, T. N.; Martin, R. L. (Progr. Inorg. Chem. **27** [1980] 223/324, 227/54).

[8] Ouchi, A.; Takahashi, Y.; Nakatani, M.; Furuhashi, A. (Sci. Papers Coll. Gen. Educ. Univ. Tokyo **19** [1969] 53/70; C.A. **72** [1970] No. 38365).
[9] Huang, J.; Huang, J. (Gaodeng Xuexiao Huaxue Xuebao **6** [1985] 521/6; C.A. **103** [1985] No. 226225).

35.1.9.1 Complexes in Solution

Stability constants of binary manganese(II) complexes with deprotonated ligands 1 and 5 in dioxane-water medium were determined potentiometrically (pH method) at 30°C [1, 2]: log K_1 = 7.43, log K_2 = 7.31, I = 0 corr. in 75% v/v dioxane-water for ligand 1 [1] (previous results in [9]) and log K_1 = 5.80, log K_2 = 4.85 for I = 0.1 M ($NaClO_4$) in 50% v/v dioxane-water for ligand 5 [2]. The stability constant of the ternary complex of Mn^{II} with ligand 6 and bipyridine, $[Mn(C_7H_5OS)bpy]^+$, is log $\beta_{1,1}$ = 5.70, and of the ternary complex with ligand 6 and phenanthroline, $[Mn(C_7H_5OS)phen]^+$, is log $\beta_{1,1}$ = 5.29, both at the above conditions [2].

Manganese(II) complexes with thio-β-diketones have found use in analytical chemistry. The formation of a stable 1:2 complex of Mn^{II} with ligand 2 allows the extraction of the metal from aqueous solution with $CHCl_3$ [3]. Mn^{II} was extracted at pH 8.0 to 8.5 with ligand 3 in CCl_4 in the presence of butanol, and the absorbance of the organic phase was measured at 450 nm ($\varepsilon = 3.81 \times 10^3$ L·mol^{-1}·cm^{-1}). The sensitivity was 0.014 μg·mL^{-1}·[cm^2]$^{-1}$ [4, 5]. Mn^{II} reacts with ligand 3 and 1,10-phenanthroline (phen) to form the $[Mn(C_8H_4F_3OS_2)_2phen]$ complex [6, 7]; for determination of manganese in the xylene extract of the complex by spectrophotometry at 375 nm ($\varepsilon = 3.58 \times 10^4$ L·mol^{-1}·cm^{-1}) see [7], or in the $CHCl_3$ extract by atomic absorption spectrophotometry see [6]. The 2,4-pentanedithionate, $[Mn(C_5H_7S_2)_2]$, was extracted into CCl_4 to give an orange-brown solution with an absorption maximum at 380 nm. The R_f value of the complex on silica gel TLC plates developed with CH_3CCl_3 is 0.57 [8]. Mn^{II} can be extracted as the bis-chelate, $Mn(C_{17}H_{13}N_2OS)_2$, from aqueous solution (containing tartrate as supporting ligand) with a $CHCl_3$ solution of ligand 7. The reaction was studied by atomic absorption spectrometry, and the extraction parameters, $pH_{1/2}$ = 6.2 and log K_{ex} = −7.6, were determined [10].

References:

[1] Uhlemann, E.; Thomas, P.; Gotthard, G.; Arnold, K. (Z. Anorg. Allgem. Chem. **364** [1969] 153/60).
[2] Sharma, J. L.; Srivastava, J. N.; Garg, B. S.; Singh, R. P. (Current Sci. [India] **46** [1977] 103/4).
[3] Rao, G. N.; Chouhan, V. S. (Food Farming Agric. **9** [1977] 19 from C.A. **89** [1978] No. 122418).
[4] Khopkar, S. M.; Solanke, K. R. (Chem. Era **10** [1974] 12/6 from C.A. **83** [1975] No. 157425).
[5] Solanke, K. R.; Khopkar, S. M. (Z. Anal. Chem. **275** [1975] 286).
[6] Deguchi, M.; Okumura, I.; Yamaguchi, K. (Kenkyu Hokoku Hiroshima Daigaku Kogakubu **30** [1981] 27/30 from C.A. **96** [1982] No. 79084).
[7] Deguchi, M.; Hayakawa, S. (Bunseki Kagaku **31** [1982] 612/5 from C.A. **98** [1983] No. 10795).
[8] Honjo, T.; Otaki, T. (Z. Anal. Chem. **300** [1980] 413).
[9] Uhlemann, E.; Suchan, W. W. (Z. Anorg. Allgem. Chem. **342** [1966] 41/5).
[10] Uhlemann, E.; Maack, B.; Raab, M. (Anal. Chim. Acta **116** [1980] 153/60).

35.1.9.2 Isolated Compounds

[$Mn^{II}(C_8H_4F_3OS_2)_2$]. The complex with ligand 3, prepared by solvent extraction from aqueous solution, can be purified by vacuum sublimation. The temperature in the sublimation recrystallization zone was 145 to 204°C. The brown complex sublimed almost quantitatively without thermal decomposition. The procedure is suitable for the purification of the metal [1].

[$Mn^{II}(C_5H_7S_2)_2$]. The complex with 2,4-pentanedithione was prepared by the two-electron reduction of 3,5-dimethyl-1,2-dithiolium tetrachloromanganate(II) with $NaBH_4$ (1:5 mole ratio) in the presence of an Mn^{II} salt in aqueous solution, adjusted to pH 2 to 4 while passing N_2 gas through the solution [2].

[$Mn^{II}(C_{15}H_{11}OS)_2py_2$] and **[$Mn^{II}(C_8H_4F_3OS_2)_2py_2$].** For preparation of the mixed ligand complexes with pyridine, a solution of manganese(II) acetate tetrahydrate in oxygen-free water, containing 3 drops of concentrated sulfuric acid, was added at 15°C to a pyridine solution of ligand 1 or 3. The green precipitates were filtered off and washed with water. The crude adducts were recrystallized from cold pyridine-water [3, 4]. They were dried over $CaCl_2$ [3] or with pyridine vapor [4]. The mixed ligand complex with ligand 3 and 4-methylpyridine (= γ-pic), [$Mn^{II}(C_8H_4F_3OS_2)_2(\gamma\text{-pic})_2$], was prepared in an analogous manner [4]. [$Mn(C_{15}H_{11}OS)_2py_2$] melts at 102 to 105°C. Its magnetic moment is $\mu_{eff} = 6.09\ \mu_B$ at ambient temperature [3]. Molecular weights of [$Mn(C_8H_4F_3OS_2)_2py_2$] and [$Mn(C_8H_4F_3OS_2)_2(\gamma\text{-pic})_2$] are 687 and 715, respectively. Their magnetic moments are 5.99 μ_B at 290 K and 6.06 μ_B at 288 K, respectively. Far-IR bands of [$Mn(C_8H_4F_3OS_2)_2(\gamma\text{-pic})_2$] were assigned as follows ($\tilde{\nu}$ in cm^{-1}): 717, 708 to ν(CS); 552 to ν(Mn–O); 299 to ν(Mn–S); and a band <200 to ν(Mn–N). The properties suggest that the compounds are octahedral containing high-spin manganese(II) [4]. Isotropic ^1H NMR shifts of [$Mn(C_8H_4F_3OS_2)_2(\gamma\text{-pic})_2$] have been measured at 70°C. Relative isotropic shifts are 10.00 for the α-H, 4.82 for the β-H, and 2.44 for the γ-CH_3 proton. The β/γ ratio is 1.98. The β/γ ratio for [$Mn(C_8H_4F_3OS_2)_2py_2$] is −1.90. The values of the β/γ ratio suggest that the isotropic shifts are contact shifts [5]. The pyridine adducts are stable in the air and are readily soluble in polar and nonpolar solvents [3].

[$As(C_6H_5)_4]_2[Mn^{II}_2(C_4S_4)_3$] precipitated on addition of a suspension of $MnSO_4$ (0.50 mmol) to a suspension of $[As(C_6H_5)_4]_2(C_4S_4)$ (0.25 mmol), both in methanol, in a Schlenk tube under N_2. The reaction mixture was stirred for 24 h. The dark brown precipitate was separated, washed repeatedly with water, methanol, and dichloromethane, then dried in vacuum. The yield was 80 to 90%. The IR spectrum of the complex in Nujol shows characteristic intense absorption bands at 1246(sh), 1238, 1220, 1192, 788, 781, and 237 cm^{-1}. The first four bands were assigned to ν(CCS) vibrations. Those at 788 and 781 cm^{-1} were attributed to ν(CS) vibrations, and the band at 237 cm^{-1} to the ν(Mn–S) vibration. A resonance-stabilized chain structure (shown below) with a tetrahedral environment of the Mn^{II} ion was assumed [7].

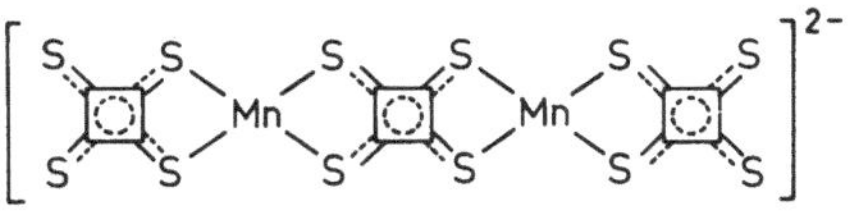

[$Mn^{III}(C_{15}H_{11}OS)_3$]. The manganese(III) complex with ligand 1 was formed by the reaction of [$Mn(acac)_3$] with the ligand in aqueous solution containing triethylamine. The crude product could not be recrystallized, because the solution decomposed rapidly to give the disulfide. The compound is insoluble in water [6].

Formulas of ligands are tabulated on p. 55

References:

[1] Honjo, T.; Shima, S. (Bull. Chem. Soc. Japan **57** [1984] 293/4).
[2] Honjo, T.; Otaki, T. (Z. Anal. Chem. **300** [1980] 413).
[3] Uhlemann, E.; Thomas, P. (Z. Naturforsch. **23b** [1968] 275/7).
[4] Maitani, T.; Kobayashi, T.; Saito, Y.; Chikuma, M.; Araishi, K.; Tanaka, H. (J. Inorg. Nucl. Chem. **41** [1979] 1689/95).
[5] Maitani, T.; Chikuma, M.; Araishi, K.; Tanaka, H. (J. Inorg. Nucl. Chem. **41** [1979] 1697/702).
[6] Cox, M.; Darken, J. (Coord. Chem. Rev. **7** [1971/72] 29/58, 51).
[7] Grenz, R.; Götzfried, F.; Nagel, U.; Beck, W. (Chem. Ber. **119** [1986] 1217/31).

35.1.10 With Heterocyclic Thiols or Thiones

35.1.10.1 With Derivatives of Imidazole

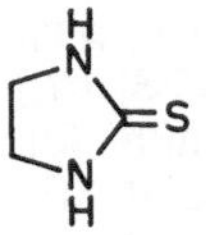

ligand 1 (= $C_3H_6N_2S$)

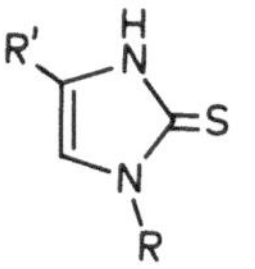

ligands 2 to 5

ligand	R	R'	formula
2	H	H	$C_3H_4N_2S$
3	CH_3	H	$C_4H_6N_2S$
4	H	$CH_2CH_2NH_2$	$C_5H_9N_3S$
5	H	$CH_2CH(NH_2)COOH$	$C_6H_9N_3O_2S$

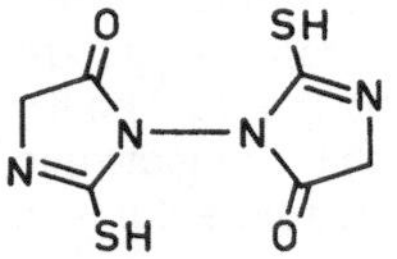

ligand 6 (= $C_6H_6N_4O_2S_2$)

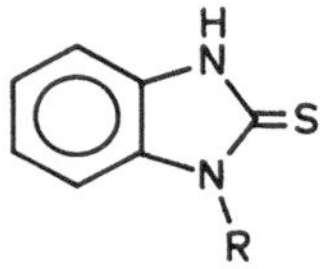

ligand 7 with R = H (= $C_7H_6N_2S$)
ligand 8 with R = CH=CH$_2$ (= $C_9H_8N_2S$)

Complexes in Solution. Stability constants of manganese(II) complexes with ligands 1 to 4 and 7 were determined at 25°C by pH titration in aqueous-organic medium:

ligand	log K_1	log K_2	medium	I in mol/L	Ref.
2	4.78 ± 0.14	3.85 ± 0.07	aqueous	0.1 ($NaClO_4$)	[1]
3	5.20 ± 0.05	4.14 ± 0.10	aqueous	0.1 ($NaClO_4$)	[1]
4	5.35 ± 0.09	4.27 ± 0.20	aqueous	0.1 ($NaClO_4$)	[2]
5	4.07 ± 0.11	4.47 ± 0.09	aqueous	0.1 ($NaClO_4$)	[2]
7	5.00	—	50% ethanol	0.05 ($NaClO_4$)	[3]
7	4.35	—	50% ethanol	0.1 ($NaClO_4$)	[3]
7	4.05	—	50% ethanol	0.15 ($NaClO_4$)	[3]

[$Mn^{II}(C_6H_4N_4O_2S_2)(H_2O)_2$] was prepared by reaction of equimolar amounts of the sodium salt of ligand 6 with an Mn^{II} salt in aqueous solution. A gray precipitate formed immediately and was digested on a water bath for one hour. The product was isolated, washed with water several times, then with ethanol, and dried in vacuum. The magnetic moment of the solid is $\mu_{eff} = 5.39\ \mu_B$. In the IR spectrum, shifts of the free ligand ν(CO) band (at 1730 cm^{-1}) and ν(CS) bands (at ~1400 and 720 cm^{-1}) to lower wavenumbers indicate coordination of the oxygen and the thiolate sulfur to Mn^{II}. A polymeric structure with seven-membered chelate rings and delocalized charge in tetrahedral coordination units is consistent with the IR data [4].

$[Mn^{II}(C_9H_8N_2S)_2Cl_2]$ was prepared by the reaction of ethanolic solutions of $MnCl_2$ and ligand 8. The reaction mixture was heated at 70 to 80°C for 5 to 8 h. The precipitate was isolated, washed with ethanol, and dried at 60 to 70°C under vacuum (3 to 5 Torr) to give a crystalline powder in 25 to 47% yield. In the IR spectrum, the absorption bands of the NH bond (3440 to 3150 cm^{-1}), of the thioamide group (1512 cm^{-1}), and those of the vinyl group are retained. This suggests coordination of the neutral ligand via the sulfur atom. The compound dissolves in organic solvents such as alcohol, acetone, and chloroform, but does not dissolve in water. In acid and alkaline media it decomposes [5].

$[Mn^{II}(C_3H_6N_2S)_2(NCS)_2]$ was prepared by slow concentration of an aqueous solution containing $MnCl_2$, NH_4SCN, and ligand 1 in the mole ratio 5:10:1 at room temperature. The compound was obtained as big, almost colorless triclinic prisms. An X-ray determination of the crystal geometry yielded the space group $P\bar{1}$-C_i^1 (No. 2) and the lattice constants a = 5.67, b = 8.63, c = 8.72 Å, α = 111.6°, β = 96.1°, γ = 107.7°; Z = 1. The calculated density is D_{calc} = 1.70 g/cm^3, the experimental density is D_{exp} = 1.65 g/cm^3. The compound is isostructural with the corresponding complexes of Co^{II} and Ni^{II} [6].

References:

[1] Sakurai, H.; Takeshima, S. (Transition Metal Chem. [Weinheim] **2** [1977] 103/6).
[2] Takeshima, S.; Sakurai, H. (Inorg. Chim. Acta **66** [1982] 119/24).
[3] Zaidi, S. A. A.; Islam, V.; Siddiqi, K. S. (Indian J. Chem. A **16** [1978] 265/7).
[4] Satpathy, K. C.; Mahana, T. D.; Mishra, H. P. (J. Indian Chem. Soc. **57** [1980] 1234/5).
[5] Skvortsova, G. G.; Trzhtsinskaya, B. V.; Abramova, N. D.; Teterina, L. F.; Voronov, V. K.; Sherstyannikova, L. V. (Zh. Obshch. Khim. **46** [1976] 1305/9; J. Gen. Chem. [USSR] **46** [1976] 1286/90).
[6] Nardelli, M.; Chierici, I. (Gazz. Chim. Ital. **88** [1958] 248/54).

35.1.10.2 With Derivatives of Pyrazole or Oxazole

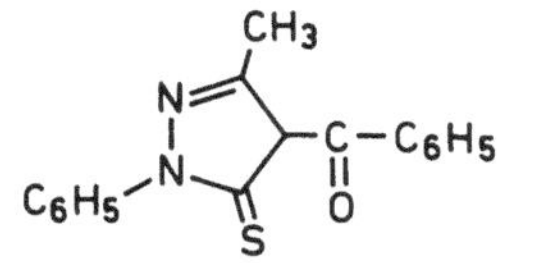

ligand 1 (= $C_{17}H_{14}N_2OS$)

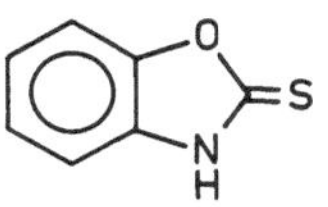

ligand 2 (= C_7H_5NOS)

Manganese(II) forms a complex with ligand 1 in aqueous solution. This is extracted quantitatively by chloroform from solutions of pH 5 to 7. The influence of tartrate as a supporting ligand was calculated and the extraction parameter, log K_{ex} = −8.2, was determined [1].

$[Mn^{II}(C_7H_5NOS)_2X_2]\cdot nH_2O$. Complexes with X = Cl for n = 0 or 2 and X = Br or I for n = 1 were prepared by reaction of MnX_2 salts with a small excess of the molten ligand 2 (m.p. 195 to 196°C). They were purified by repeated washing with ethanol and ether. All the complexes are nonelectrolytes in nitromethane. Color, decomposition point (t_{dec}), magnetic moments at room temperature, bands observed in the diffuse reflectance spectra, and ligand field parameters (both in cm^{-1}) are summarized on p. 60.

complex	$[MnL_2Cl_2]$	$[MnL_2Cl_2]\cdot 2H_2O$	$[MnL_2Br_2]\cdot H_2O$	$[MnL_2I_2]\cdot H_2O$
color	white	pink-white	pale yellow	dark yellow
t_{dec} in °C	201 to 206	190 to 195	198 to 203	145 to 150
μ_{eff} in μ_B	5.4	5.6	5.8	5.7
electronic spectra; transitions from $^6A_{1g}$				
$\rightarrow {}^4T_{1g}(^4G)$	21980	21970	22470	20530
$\rightarrow {}^4E_g, {}^4A_{1g}(^4G)$	26740	26880	26670	26310
$\rightarrow {}^4E_g(^4D)$	31250	31750	31050	31000
$\rightarrow {}^4T_{1g}(^4P)$	37060	36300	36700	36360
$\rightarrow {}^4T_{1g}(^4F)$	43480	43860	42920	42735
B	644	696	626	670
C	4060	3984	4082	3922
10 Dq	7730	7100	6870	7580

The 10 Dq values are in the order Cl > Br. The formulation, $[MnL_2(H_2O)I]I$, that could justify the surprisingly high 10 Dq value for the iodo complex was rejected, because the complex behaves as a nonelectrolyte in nitromethane, and water was not found to be coordinated. The IR spectra of the hydrates reveal ν(OH) and δ(HOH) bands, confirming the presence of water of crystallization, but wagging, twisting, and rocking vibrational modes of coordinated water molecules have not been found in the expected ranges.

Comparison of the IR spectra of the complexes (KBr disks or Nujol mulls) with the spectrum of the free ligand shows a strong red shift of the ν(NH) band and slight blue shifts of the other bands except for the ν_s(COC) band (at 820 cm^{-1}) which is unchanged (see below):

complex	$[MnL_2Cl_2]$	$[MnL_2Cl_2]\cdot 2H_2O$	$[MnL_2Br_2]\cdot H_2O$	$[MnL_2I_2]\cdot H_2O$	ligand
ν(NH)	3180	3160	3190	3240	3250
thioamide I [a]	1510	1512	1510	1512	1505
ν(CS) + δ(NCS)	1285	1290	1285	1289	1282
thioamide II [b]	1248	1248	1245	1250	1245
ν_{as}(COC)	1097	1102	1097	1103	1095
ν(Mn–X)	230	229	204	145	—
ν(Mn–N)	219, 212	218, 211	217, 213	216, 206	—

[a] Vibration mode having major contributions from δ(NH) + δ(CH) + ν(C=N). – [b] Band due to ν(C⋯N) + δ(NH) + δ(CH) + ν(C=S).

The thioamide III and IV bands at 1010 and 745 cm^{-1}, respectively, having their main contribution from ν(C=S), do not show a red shift, thus excluding the presence of a metal-sulfur bond. The data suggest that the complexes possess a tetrahedral geometry around the high-spin Mn^{II} atom with coordination of the ligand via the NH group. For additional bands in the far-IR spectra (not assigned) see the paper [2].

References:

[1] Uhlemann, E.; Maack, B.; Raab, M. (Anal. Chim. Acta **116** [1980] 153/60).
[2] Preti, C.; Tosi, G. (J. Coord. Chem. **8** [1978] 15/25).

35.1.10.3 With Derivatives of Thiazole

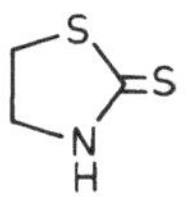

ligand 1 (= $C_3H_5NS_2$)

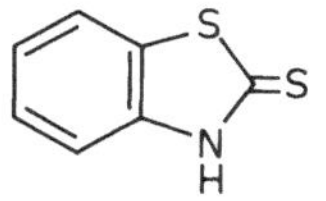

ligand 2 (= $C_7H_5NS_2$)

The complex with ligand 1, **$Mn(C_3H_4NS_2)_2 \cdot H_2O$**, has been prepared by two methods: In the first method, a basic aqueous solution of an Mn^{II} salt (chloride, nitrate, or acetate), maintained at a pH lower than that necessary to precipitate $Mn(OH)_2$, was treated with an aqueous ethanol solution of the ligand (mole ratio 1:2). In the second method the Mn^{II} salt was dissolved in ethanol and added to a solution of the ligand in an ethanol-DMSO mixture. The white precipitate was washed with ethanol and dried over P_4O_{10}. The solid complex decomposes at >350°C. The IR spectrum of the complex in KBr disk or Nujol mull shows bands at 1565, 1250, and 950 cm^{-1}, assigned to the thioamide I, II, and III vibration modes. Bands of ν(CS) were observed at 750 and 665 cm^{-1}, of ν(Mn–S) at 305 and 264 cm^{-1}, and of ν(Mn–N) at 219 and 211 cm^{-1}. Additional bands (not assigned) were found at 430, 284, and 98 cm^{-1}. The spectrum does not display ν(NH) or δ(NH) bands. In the spectrum of the free ligand, bands of thioamide I, II, III were observed at 1490, 1245, and 990 cm^{-1}, of ν(CS) at 690 and 650 cm^{-1}. The solid state electronic spectrum of the complex shows maxima at 28735, 24200, and 20830 cm^{-1}, which were assigned to the high-spin tetrahedral Mn^{II} transitions: $^6A_1 \rightarrow {}^4E_g$ (D), $\rightarrow {}^4E_g, {}^4A_{1g}$ (G), $\rightarrow {}^4T_{1g}$ (^{4}G), respectively. The ligand field parameters derived from the spectral results are: Dq = 567 cm^{-1}, B = 647 cm^{-1} and C = 3546 cm^{-1}. The physical properties of the complex indicate a structure with a uninegative bidentate ligand chelated to the metal via the exocyclic sulfur and the ring nitrogen atoms [1].

The iodo complex, **$[Mn(C_3H_5NS_2)_4I_2]$**, and hydrates of composition **$Mn(C_3H_5NS_2)_nX_2 \cdot xH_2O$** with n = 1, 2, or 4 and x = 1 or 2 (see the table below) were prepared by the reaction of MnX_2 salts with a small excess of the molten ligand 1. The complexes were purified by means of repeated washings with toluene. Decomposition points of the white (X = Cl, ½ SO_4, CH_3COO), light yellow (X = Br, ClO_4), or orange-yellow (X = I) complexes are shown below together with absorption maxima observed in the electronic spectra of the solid compounds and ligand field parameters (in cm^{-1}) derived from the spectral results (L = $C_3H_5NS_2$):

complex	t_{dec} in °C	ν_{max} in cm^{-1} for transitions $^6A_1 \rightarrow$ $^4T_{1g}(^4G)$	$^4E_g, {}^4A_{1g}(^4G)$	$^4E_g(^4D)$	$^4T_{1g}(^4P)$	$^4T_{1g}(^4F)$	B	C	Dq
$[MnL_4Cl_2] \cdot 4H_2O$	88 to 90	19050	24100	29240	33330	39850	734	3352	657
$[MnL_4Br_2] \cdot 4H_2O$	75 to 77	19610	23420	28570	31150	38900	736	3212	536
$[MnL_4I_2]$	78 to 80	19800	22885	28250	30030	38610	766	3045	540
$[MnL_2SO_4] \cdot H_2O$	108 to 110	20000	25125	28985	35335	39850	551	3923	678
$[MnL_4](BF_4)_2 \cdot 2H_2O$	68 to 70	20000	24270	29150	32450	40815	697	3460	712
$[MnL_4](ClO_4)_2 \cdot 2H_2O$	60 to 62	19800	24450	28400	34480	39370	564	3762	725
$[MnLCH_3COO_2] \cdot H_2O$	110 to 112	19610	24820	28985	34840	40480	595	3774	787

The ambient-temperature magnetic moments of the solid compounds (μ_{eff} = 6.0 μ_B for X = I, 5.9 μ_B for X = Cl, Br, ½SO_4, 5.8 μ_B for X = ClO_4, BF_4, and 5.5 μ_B for X = CH_3COO) are in the range expected for a high-spin Mn^{II} complex. Characteristic bands observed in the IR spectra of the complexes in KBr disk or Nujol mulls were assigned as follows:

compound	ν(NH)	thio-amide I	ν(C=S)+ ν(NCS)	ν(CS) [a)]	ν(M–N)	ν(M–X)	other bands
ligand (in $CHCl_3$)	3400	1490	1288	690, 650	—	—	445, 434, 290, 94
$[MnL_4Cl_2]\cdot4H_2O$	3050	1520	1300	700, 660	213	229	434, 291
$[MnL_2Br_2]\cdot4H_2O$	3160	1500	1300	690, 660	215	210	433, 290
$[MnL_4I_2]$	3160	1500	1300	700, 652	213	160	434, 288w
$[MnL_2SO_4]\cdot H_2O$	3150	1520	1300	700, 658	211	—	432, 420sh, 288, [1140, 1100, 1055, 1002, 620, 585] [b)]
$[MnL_4](BF_4)_2\cdot2H_2O$	3140	1515	1300	700, 655	212	—	440, 288, [1050, 525] [c)]
$[MnL_4](ClO_4)_2\cdot2H_2O$	3150	1500	1300	700, 650	211	—	436, 292, [1085, 540] [d)]
$[MnL(CH_3COO)_2]\cdot H_2O$	3200	1520	1300	700, 660	210	—	440, 434, 287sh, [1640, 1570, 1420, 1395] [b)], 630 (OCO)

[a)] Symmetric and antisymmetric. – [b)] Bands for a bridged SO_4 group. – [c)] Bands for an ionic BF_4 group. – [d)] Bands for an ionic ClO_4 group. – [e)] ν_s(OCO) and ν(COO) for bridging and terminal acetato groups.

The properties of the complexes indicate tetrahedral or octahedral structures with the neutral monodentate ligands coordinating to the metal via the nitrogen atom of the ring [2].

$Mn(C_7H_4NS_2)_2$ was obtained on reaction of hydrated manganese(II) sulfate with the sodium salt of 2-mercaptobenzothiazole (ligand 2) in aqueous solution (mole ratio 1:2). The deep yellow precipitate was washed several times with water, then ethanol, and was dried in vacuum over P_4O_{10}. The magnetic moment of the solid compound increases from μ_{eff} = 4.47 μ_B at 77 K to 4.58 μ_B at 294.5 K. The low moment is indicative of the metal-metal interaction in a polymeric structure. No bands resulting from d-d transitions were found in the diffuse reflectance spectrum. IR absorption bands (not assigned) are tabulated in the paper together with bands of the corresponding complexes of Co^{II}, Ni^{II}, and Cu^{II}. The IR spectrum indicates a square-planar coordination of manganese with N–C–S bridges between the metal atoms, since some strong bands in the 300 to 200 cm^{-1} region might result from Mn–N vibrations. The complex is insoluble in both polar and nonpolar solvents, i.e., dimethyl sulfoxide, dimethylformamide, acetone, ethanol, benzene, toluene, and carbon tetrachloride. The properties are consistent with a polymeric structure [3].

$Mn(C_7H_5NS_2)_2SO_4$ was prepared by the reaction of ethanol solutions of ligand 2 and $MnSO_4$ (mole ratio 2:1). The reaction mixture was refluxed for 1 h and concentrated to one half its volume. Upon addition of carbon tetrachloride and cooling, a precipitate separated out. This was washed well with ethanol, finally with diethyl ether, and dried in vacuum. The magnetic moment of the solid is μ_{eff} = 5.85 μ_B at room temperature. The IR spectrum of the complex in KBr shows the bands characteristic of ν(SH) at 2500 cm^{-1}, ν(CN) at 1290 cm^{-1}, and ν(CS) at 620 cm^{-1}. The shifts of the ligand bands upon coordination of the Mn^{II} suggest that the neutral bidentate ligand is bound to the metal via the ring sulfur and mercapto sulfur atoms. The complex shows some fungitoxicity, although less than the free ligand [4].

References:

[1] Preti, C.; Tosi, G. (Can. J. Chem. **54** [1976] 1558/62).
[2] Preti, C.; Tosi, G. (Australian J. Chem. **29** [1976] 543/9).
[3] Pruchnik, F.; Chwolka, D. (Roczniki Chem. **51** [1977] 653/8; C.A. **87** [1977] No. 94705).
[4] Srivastava, S. K.; Gupta, A.; Verman, A. (Egypt. J. Chem. **26** [1983] 173/6; C.A. **100** [1984] No. 167041).

35.1.10.4 With Derivatives of 1,2,4-Triazole or 1,3,4-Oxadiazole

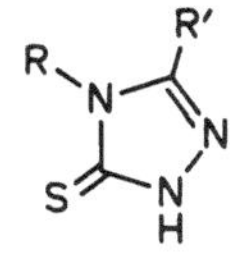

ligand 1 with R = NH_2, R′ = 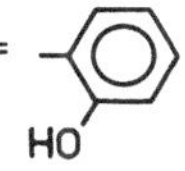 (= $C_8H_8N_4OS$)

ligand 2 with R = H, R′ = 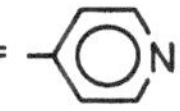 (= $C_7H_6N_4S$)

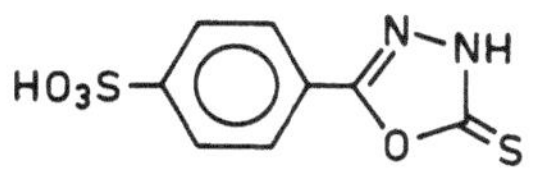

ligand 3 (= $C_8H_6N_2O_4S_2$)

$[Mn^{II}(C_8H_8N_4OS)_2Cl_2]$ was prepared by the reaction of an ethanol solution of $MnCl_2$ with a hot ethanol solution of ligand 1 (mole ratio 1:2). The reaction mixture was refluxed for 1 to 2 h on a water bath to give a yellow precipitate. The product was isolated, washed with hot ethanol, and dried. The magnetic moment of the solid is $\mu_{eff} = 5.65\ \mu_B$ at 300 K. In the IR spectrum of the Mn^{II}, Co^{II}, Ni^{II}, and Cu^{II} complexes (in KBr), the ν(NH) band shows a negative shift of ~100 to 3000 cm^{-1}. While the thioamide I, II, and III bands have not changed, a bathachromic shift (10 to 25 cm^{-1}) or split occurred for the thioamide IV band (at 775 cm^{-1} for the free ligand). The changes indicate chelating by the neutral bidentate ligand through the thioketone sulfur atom and the N atom of the amino group to the Mn atom in an octahedral complex. Far-IR bands were located as follows: ν(Mn–N) at 400 to 440 cm^{-1}, ν(Mn–S) at ~350 cm^{-1}, and ν(Mn–Cl) at 290 cm^{-1}. The electronic spectrum of the complex in Nujol mull shows characteristic octahedral high-spin Mn^{II} bands. Analysis of the spectrum yielded the ligand field parameters: 10Dq = 10600 cm^{-1}, B = 597 cm^{-1}, C = 2627 cm^{-1}, and β = 0.62 [1].

$[Mn^{II}(C_7H_6N_4S)(H_2O)_3]Cl_2$ was prepared by the reaction of an aqueous solution of $MnCl_2 \cdot 4H_2O$ with a refluxing ethanolic solution of ligand 2 (mole ratio 1:3). The reaction mixture was refluxed for 5 h and excess pyridine was then added. The resulting solution was filtered and the filtrate evaporated to a small volume. The solution was refluxed for an additional 2 h and then evaporated nearly to dryness. The light yellow precipitate that formed was isolated, washed with cold ethanol, ethyl methyl ketone, and dried at 120°C. The complex is a 1:2 electrolyte in dimethylformamide. The magnetic moment of the solid is $\mu_{eff} = 5.8\ \mu_B$ at ambient temperature. The IR spectrum of the complex shows the prominent absorptions: ν(H_2O) at 3400 cm^{-1}; ν(NH) + ν(CH) at 3120, 3080 cm^{-1}; δ(H_2O) at 1630 cm^{-1}; thioamide IV at 820 cm^{-1} (free ligand at 850 cm^{-1}); ν(Mn–O) at 470 and 430; and ν(Mn–S) at <400 cm^{-1}. The electronic spectrum of the complex shows strong maxima at 28100 and 29200 cm^{-1}. The spectral data suggest that the coordination geometry around the Mn^{II} atom is tetrahedral with three coordinated water molecules and one neutral unidentate sulfur-donor ligand. The complexes are soluble in dimethylformamide. This suggests that they are monomeric [2].

$Na[Mn^{II}(C_8H_5N_2O_4S_2)(H_2O)_3](CH_3COO)_2$. The microcrystalline light pink complex with ligand 3 was prepared by refluxing equimolar amounts of $Mn(CH_3COO)_2 \cdot 2H_2O$ and the sodium salt of the ligand in water for 1 to 2 h. The mixture was then evaporated to dryness, washed with ethanol, and dried in air. The deliquescent compound decomposes above 300°C. Its magnetic moment, $\mu_{eff}=5.8\ \mu_B$, results from susceptibility measurements by the Cahn-Faraday method at room temperature. In the IR spectrum of the complex (in Nujol) a broad band at $\sim 3000\ cm^{-1}$ is ascribed to $\nu(NH)+\nu(OH)$ vibrations. The lowering of the band from 3500 to $3300\ cm^{-1}$ indicates coordination of the NH and OH (of H_2O) groups to the Mn atom. No changes are observed for the bands of the C=S, SO_3^-, and COO^- groups. The photoacoustic spectrum of the powdered complex (recorded in the range between ~13000 and 25000 cm^{-1}) shows bands at 15380, 16810, and 18180 cm^{-1}, which were assigned to the high-spin tetrahedral Mn^{II} transitions $^6A_1\to{}^4T_1(G)$, $\to{}^4T_2(^4G)$, $\to{}^4E(G)$. The UV-visible spectrum of the complex in aqueous solution exhibits a charge-transfer band at 22220 cm^{-1} and two intraligand bands ($\pi\to\pi^*$) at 33330 and 50000 cm^{-1} (free ligand at 33330 and 40816 cm^{-1}) [5]. The ESR spectra of the polycrystalline compound at −150°C and at room temperature show one isotropic signal at g = 2.018. The physical properties indicate a tetrahedral complex with coordination of the unidentate ligand by the N atom in the 3-position and of three H_2O molecules. The complex shows fungitoxic activity [4].

$[Mn^{III}(C_8H_8N_4OS)_2(CH_3COO)_2]CH_3COO$. The acetato complex was obtained by the reaction of ligand 1 and $Mn(CH_3COO)_3\cdot 2H_2O$ (mole ratio 2:1) in ethanol. The reaction mixture was refluxed for 2 h, then cooled to yield green crystals. The product was isolated, washed thoroughly with ethanol, and dried. The yield is ~60%. The magnetic moment of the complex is $\mu_{eff}=4.75\ \mu_B$ at 300 K. The IR spectra of Cr^{III}, Mn^{III}, Fe^{III}, and CO^{III} complexes show shifts of the $\nu(NH)$ and thioamide IV bands on complexation, similar to those described for $[Mn^{II}(C_8H_8N_4OS)_2Cl_2]$ (on p. 63). The $\nu(OH)$ band (at 3200 cm^{-1}) is unchanged. The $\nu(Mn–S)$ band appears in the 350 to 400 cm^{-1} region. The IR data are consistent with a neutral bidentate ligand chelated to the octahedral Mn^{III} atom via the thioketone sulfur atom and the NH_2 group [3].

References:

[1] Sengupta, S. K.; Sahni, S. K.; Kapoor, R. N. (Indian J. Chem. A **20** [1981] 692/4).
[2] Singh, B.; Tiwari, H. N. (J. Indian Chem. Soc. **60** [1983] 526/8).
[3] Sengupta, S. K.; Sahni, S. K.; Kapoor, R. N. (Indian J. Chem. A **19** [1980] 703/5).
[4] Mishra, L. (Syn. React. Inorg. Metal-Org. Chem. **16** [1986] 831/9).
[5] Tiwari, L. B.; Mishra, L.; Thakur, S. N. (Spectrochim. Acta A **44** [1988] 327/9).

35.1.10.5 With Derivatives of Pyridine

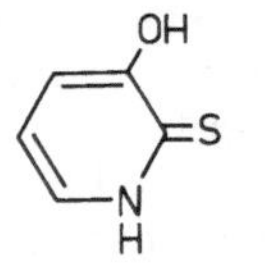

ligand 1 (= C_5H_5NOS)

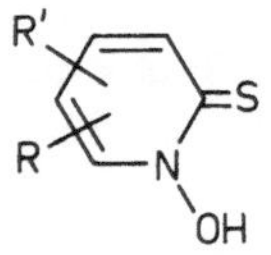

ligands 2 to 8 (= HL)

ligand	R	R'	formula
2	H	H	C_5H_5NOS
3	3-CH_3	H	C_6H_7NOS
4	4-CH_3	H	C_6H_7NOS
5	5-CH_3	H	C_6H_7NOS
6	6-CH_3	H	C_6H_7NOS
7	4-CH_3	6-CH_3	C_7H_9NOS
8	5-Br	H	C_5H_4BrNOS

The stability constant of the 1:1 complex of manganese(II) with ligand 1 in 50% aqueous dioxane at 25°C is log $K_1=5.60$ (ionic strength not given) [1].

Manganese(III) complexes of composition $[MnL_3]$ with HL = ligands 2 to 8 precipitated on addition of excess aqueous NaL solution to an aqueous $KMnO_4$ solution at pH 4.6. The reaction mixture was heated to boiling and, after cooling, the complexes were extracted into $CHCl_3$. The dark green solution of the complex with ligand 2, $[Mn(C_5H_4NOS)_3]$, shows maxima in the electronic absorption spectrum at 696 ($\varepsilon = 477\ L \cdot mol^{-1} \cdot cm^{-1}$), 316 (9400) and 253 nm (31000). The solution is very unstable toward temperature and light. The solutions of $[MnL_3]$ complexes with the derivatives of ligand 2 are more stable. The solution of the complex with ligand 8 is stable for 1 to 2 d. Thin-layer chromatography (TLC) of the complexes with ligands 2 to 5 and 8 on Kieselgel-60 plates, developed with various solvents, yielded the R_F values listed below [2, 3]:

R_F values of $[MnL_3]$ complexes with ligand	2	3	4	5	8
90% CH_2Cl_2 + 10% THF	0.13[a)]	0.18[a)]	0.14[a)]	0.29[a)]	0.60
90% $CHCl_3$ + 10% THF	0.07[a)]	0.11[a)]	0.07[a)]	0.18[a)]	0.59
90% $CHCl_3$ + 10% methyl isobutyl ketone	0.01[a)]	0.01[a)]	0.01[a)]	0.02[a)]	0.41

[a)] The solution decomposed.

The R_F value of the complex with ligand 8, developed with $CHCl_3$, is 0.25 (tailing). Complexes with ligands 6 and 7 decomposed during TLC [3].

$[Mn^{II}(C_5H_4NOS)_2] \cdot H_2O$ was prepared by the reaction of an Mn^{II} salt with the sodium salt of ligand 2 in water. The yellow-green complex precipitated immediately, was washed successively with water, methanol, and ether, and dried in vacuum over P_4O_{10} [4, 5]. The optimum pH range for precipitation of the complex from aqueous solution is 5 to 8 [2]. The melting point of the solid is 255°C. The complex is monomeric and a nonelectrolyte in dimethylformamide and dimethylacetamide. The magnetic moment of the solid is $\mu_{eff} = 5.72\ \mu_B$ at 22°C. The IR spectrum of the complex in KBr shows the prominent absorptions: ν(NO) at 1080 and ν(CS) at 1150 cm^{-1} (1112 and 1142 cm^{-1}, respectively, for the free ligand). The electronic absorption spectrum of the complex in DMF solution shows maxima at 698 ($\varepsilon = 56$), 336 (660), 299 (17800) nm. Of these, the band at 698 nm was assigned to the $^6A_{1g} \rightarrow {}^4T_{1g}(G)$ transition. The spectral properties are consistent with a high-spin Mn^{II} complex having a tetrahedral geometry with two uninegative bidentate ligands chelated via the thioketone sulfur and the negative oxygen atoms. The complex is moderately soluble in dimethylformamide and dimethyl sulfoxide [4]. $[Mn^{II}(C_5H_4NOS)_2] \cdot H_2O$ shows antibacterial and antifungal activity [5].

References:

[1] Kushwaha, V.; Katyal, M.; Singh, R. P. (Talanta **21** [1974] 763/9).
[2] Steinbrech, B.; Koenig, K.-H. (Z. Anal. Chem. **316** [1983] 465/71).
[3] Steinbrech, B.; Koenig, K.-H. (Z. Anal. Chem. **316** [1983] 689/95).
[4] Robinson, M. A. (J. Inorg. Nucl. Chem. **26** [1964] 1277/81).
[5] Olin Mathieson Chemical Corp. (Brit. 761171 [1956]).

35.1.10.6 With 8-Quinolinethiol and Its Derivatives

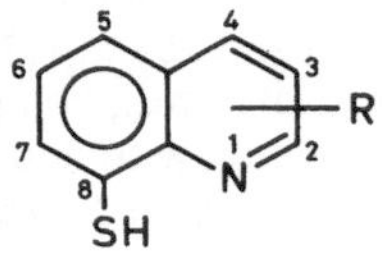

(=HL)

ligand	R	formula	ligand	R	formula
1	H	C_9H_7NS	17	6-OCH_3	$C_{10}H_9NOS$
2	5-F	C_9H_6FNS	18	4-SCH_3	$C_{10}H_9NS_2$
3	6-F	C_9H_6FNS	19	5-SCH_3	$C_{10}H_9NS_2$
4	7-F	C_9H_6FNS	20	7-SCH_3	$C_{10}H_9NS_2$
5	5-Cl	C_9H_6ClNS	21	5-SC_5H_{11}	$C_{14}H_{17}NS_2$
6	6-Cl	C_9H_6ClNS	22	5-$SC_{10}H_{21}$	$C_{19}H_{27}NS_2$
7	7-Cl	C_9H_6ClNS	23	2-CH_3	$C_{10}H_9NS$
8	3-Br	C_9H_6BrNS	24	4-CH_3	$C_{10}H_9NS$
9	5-Br	C_9H_6BrNS	25	6-CH_3	$C_{10}H_9NS$
10	6-Br	C_9H_6BrNS	26	7-CH_3	$C_{10}H_9NS$
11	7-Br	C_9H_6BrNS	27	2-$CH(CH_3)_2$	$C_{12}H_{13}NS$
12	5-I	C_9H_6INS	28	4-C_6H_5	$C_{15}H_{11}NS$
13	7-I	C_9H_6INS	29	2-CH_3, 6-CH_3	$C_{11}H_{11}NS$
14	5-SO_3Na	$C_9H_6NNaO_3S_2$	30	2-CH_3, 7-CH_3	$C_{11}H_{11}NS$
15	4-OCH_3	$C_{10}H_9NOS$	31	2-CH_3-5-SO_3H	$C_{10}H_9NO_3S_2$
16	5-OCH_3	$C_{10}H_9NOS$	32	2-CH_3, 4-CH_3, 6-CH_3	$C_{12}H_{13}NS$

Survey

8-Quinolinethiol (thioxine), the sulfur analog of 8-quinolinol (oxine), behaves as an S,N-donor ligand to form five-membered chelate rings with metal ions. These compounds are more stable than the metal complexes with 8-quinolinol, indicating that the metal-sulfur bond is more covalent than the metal-oxygen bond. The reason for this is not only the lower electronegativity of sulfur, but also the greater conjugation of the bonds in the chelate rings, due to the presence of the donor π-bonding between the metal and sulfur atoms.

Usually, $Mn^{II}L^+$ and $Mn^{II}L_2$ complexes are formed by manganese(II) with ligand anions in aqueous or mixed aqueous-organic solution. The existence of $Mn^{II}L_3^-$ complexes in DMF and DMSO has also been established. The complexes isolated were all of the $Mn^{II}L_2$ or $Mn^{II}L_2 \cdot n\,H_2O$ type. With the exception of complexes with sulfonated ligands 14 and 31, these compounds are insoluble in water, and readily extractable in chloroform, benzene, or carbon tetrachloride. The existence of $Mn^{II}L_2 \cdot 2HL$ species (with two additional molecules of protonated ligand) was usually established in chloroform extracts. Extraction with chloroform is employed in the determination of stability in the water-chloroform system, in the analytical determination of manganese in small amounts, and in its separation from other metal ions. The studies of the redox chemistry of Mn^{II} complexes with 8-quinolinethiol provide insights into the redox chemistry of manganese metalloenzymes.

35.1.10.6.1 Complexes in Solution

The existence of the air-sensitive species $\mathbf{Mn^{II}L^+}$, $\mathbf{Mn^{II}L_2}$, and $\mathbf{Mn^{II}L_3^-}$ has been established by potentiometric, polarographic, and spectrophotometric studies [1 to 5]. Stability constants of the various complexes with ligands 1 and 14, determined under air-free conditions, are summarized below:

HL	t in °C	medium	I in mol/L	method[a)]	log K_1	log β_2	log β_3	Ref.
1	25	DMF	0.1 (TEAP)[b)]	pol.	—	21.6	27.8	[1]
1	25	DMSO	0.1 (TEAP)[b)]	pol., pot.	—	16.8	22.2	[2, 3]
1	27	50% aqueous dioxane	0.1 ($NaClO_4$)	pot.	6.74	—	—	[4]
14	20	aqueous	0.1 ($NaClO_4$)	pot.	5.2	9.8	—	[5, 6]

[a)] pol. = polarographic, pot. = potentiometric. – [b)] Tetraethylammonium perchlorate.

Thermodynamic parameters for the formation of $Mn(C_9H_6NS)^+$, determined calorimetrically in 50 vol% aqueous dioxane at 25°C and I = 0.1 mol/L ($NaClO_4$), are: $\Delta G = -38.5$ and $\Delta H = -14.6$ kJ/mol, and $\Delta S = 79.5\ J\cdot mol^{-1}\cdot K^{-1}$. Comparison with the heat of chelation for the corresponding Mn^{II} complex with 8-quinolinol shows that the metal-sulfur bond is stronger than the metal-oxygen one [7]. Thermodynamic parameters for the formation of $Mn(C_9H_6NS)_2$ and $Mn(C_9H_6NS)_3^-$ in DMSO at 25°C and I = 0.1 mol/L (TEAP) are: $\Delta G = -95.8$ and -127 kJ/mol, $\Delta H = -65.4$ and -106 kJ/mol, and $\Delta S = 103$ and $71\ J\cdot mol^{-1}\cdot K^{-1}$, respectively [3]. The complex with ligand 14 is less stable than the complex with 8-hydroxy-5-quinolinesulfonic acid. On comparison with the 8-mercapto-5-quinolinesulfonato complexes of other metals a stability series was found, which correlates with the stability series of the sulfanes [5].

The existence of $\mathbf{MnL_2\cdot 2HL}$ species (with two additional molecules of the protonated ligand) in chloroform was established by solvent extraction of weakly acidic or buffered aqueous solutions containing Mn^{2+} and the ligand. Extractibility of the complexes is dependent on their solubility, i.e., it increases with increasing polarity of the solvent. For extracting into chloroform the optimal aqueous phase pH lies between 7 and 12. The composition of the complexes was determined spectrophotometrically, using the method of molar ratio. The chloroform extracts of the $MnL_2\cdot 2HL$ complexes show absorption maxima between 400 and 428 nm with an ε value between 15000 and 21000 $L\cdot mol^{-1}\cdot cm^{-1}$ (detailed information about the absorption spectra is listed under the isolated complexes, see pp. 70/2). The stability constant, β', for the formation in chloroform-water, according to the reaction $Mn^{2+} + 2L^- + 2HL \rightleftharpoons \overline{MnL_2\cdot 2HL}$, was determined spectrophotometrically by the distribution method as $\beta' = \beta\cdot R = \overline{S}/L$, where a bar over a symbol signifies the organic phase, β is the stability constant of the complex in water, R is the distribution constant between aqueous and organic solution, L is the solubility product, and $\overline{S}$ is the solubility in chloroform (details about solubility and solubility product are listed under the isolated complexes, see pp. 70/2). The values of log β' at 20°C and I = 0.1 mol/L ($NaClO_4$) are shown below:

ligand	1	7	11	15	17	18	19	24	25	26	28
log β'	28.1	31.3	33.8	25.8	27.6	32.6	33.5	30.0	32.9	34.9	42.2
Ref.	[8, 9]	[10]	[10]	[11]	[12]	[13]	[14, 15]	[16]	[16]	[16]	[17]

In general, the introduction of one substituent group into the molecule of 8-quinolinethiol leads to an increase in the stability of the complex. The complexes are more stable than the corresponding complexes with 8-quinolinol [8].

The complexes with sulfonated ligands 14 and 31 are not extractable into inert organic solvents. In the presence of tetraphenylarsonium cations in aqueous phase, the complex with ligand 31 may be extracted into chloroform as the **$[As(C_6H_5)_4]_2[Mn^{II}(C_{10}H_7NO_3S_2)_2]$** complex (absorption maximum at 415 nm with $\varepsilon = 14300$ L·mol^{-1}·cm^{-1}) [18].

Several electrochemical studies in DMF established the existence of different manganese(II) complexes with ligand 1 in this solvent, and complicated electrode reactions [19, 20]. The polarogram of $Mn^{II}(C_9H_6NS)_2 \cdot 2C_9H_7NS$ at a dropping mercury electrode in DMF with 0.1M $[N(C_4H_9)_4]ClO_4$ as supporting electrolyte shows waves with $E_{1/2} = -0.80$, -1.64, -2.18, and -2.55 V. The first wave may be caused by the discharge of the Mn^{III} complex, arising as a result of the partial oxidation of $Mn(C_9H_6NS)_2 \cdot 2C_9H_7NS$ in DMF. The second irreversible wave probably corresponds to the reduction of the Mn^{II} complex to Mn^0 with the simultaneous elimination of the ligand at -2.55 V. A small wave at -2.18 V may be caused by the reduction of the ligand molecule [19]. The redox reactions of $Mn^{II}(C_9H_6NS)_2$ and $Mn^{II}(C_9H_6NS)_3^-$ in DMF were examined by polarography, cyclic voltammetry, and controlled potential electrolysis. These complexes give a one-electron oxidation wave at a glassy carbon electrode and a two-electron reduction wave at a dropping mercury electrode. The oxidation and reduction mechanisms for both complexes are discussed [20]. The electrochemical behavior of $Mn(C_9H_6NS)_2$ and the corresponding complexes of Fe^{II}, V^{IV}, Zn^{II}, and Co^{II} were studied in aqueous ethanol at 25°C by classical and voltamperic oscillopolarography. The complexes cause the catalytic liberation of hydrogen at a mercury electrode with a wave in the range -1.5 to -1.7 V vs. SCE. The catalytic effect, given by the maximum limiting current, increases in the series of complexes $Co^{II} > V^{IV} > Fe^{II} > Mn^{II}$, which is in agreement with the stability order [21].

Manganese(III) complexes with 8-quinolinethiols are observed in solution, as a result of the partial oxidation of Mn^{II} complexes during the electrochemical studies. The reaction of $Mn^{III}(C_9H_6NS)_3$ (preparation not given) with NO in toluene yields the adduct **$Mn(NO)(C_9H_6NS)_2$**. The EPR parameters of the nitrosyl complex, recorded during the reaction in toluene, are: $g_0 = 2.012 \pm 0.002$, $A_0(Mn^{55}) = 81.2 \times 10^{-4}$ cm^{-1} at 235 K, and $g_\parallel = 1.993 \pm 0.002$, $g_\perp = 2.022 \pm 0.002$, $A_\parallel = 155.2 \times 10^{-4}$, $A_\perp = 44.0 \times 10^{-4}$ cm^{-1} at 125 K [22].

References:

[1] Nakabayashi, Y.; Masuda, Y.; Sekido, E. (J. Electroanal. Chem. Interfacial Electrochem. **176** [1984] 243/57).
[2] Nakabayashi, Y.; Masuda, Y.; Sekido, E. (Electrochim. Acta **30** [1985] 347/51).
[3] Nakabayashi, Y.; Masuda, Y. (Electrochim. Acta **32** [1987] 307/9).
[4] Corsini, A.; Fernando, Q.; Freiser, H. (Anal. Chem. **35** [1963] 1424/8).
[5] Ashaks, Ya. V.; Bankovskii, Yu. A.; Deme, A. M. (Zh. Analit. Khim. **31** [1976] 865/72; J. Anal. Chem. [USSR] **31** [1976] 706/11).
[6] Ashaks, Ya. V.; Bankovskii, Yu. A. (Latvijas PSR Zinatnu Akad. Vestis Kim. Ser. **1968** 122; C.A. **69** [1968] No. 70636).
[7] Gutnikov, G.; Freiser, H. (Anal. Chem. **40** [1968] 39/44).
[8] Bankovskii, Yu. A.; Chera, L. M.; Ievin'sh, A. F. (Zh. Analit. Khim. **23** [1968] 1284/90; J. Anal. Chem. [USSR] **23** [1968] 1134/9).
[9] Chera, L. M.; Bankovskii, Yu. A.; Zaruma, D. E.; Abolinya, M. Ya. (Latvijas PSR Zinatnu Akad. Vestis Kim. Ser. **1969** 297/303; C.A. **71** [1969] No. 90676).
[10] Bankovskii, Yu. A.; Zaruma, D. E.; Krasovska, M. E.; Tsirule, Ya. A.; Miezere, R. S. (Latvijas PSR Zinatnu Akad. Vestis Kim. Ser. **1974** 629/30; C.A. **82** [1975] No. 90687).

[11] Zaruma, D. E.; Krasovska, M. E.; Bankovskii, Yu. A.; Grike, M. O. (Latvijas PSR Zinatnu Akad. Vestis Kim. Ser. **1976** 734/5; C.A. **86** [1977] No. 111687).
[12] Bankovskii, Yu. A.; Zaruma, D. E.; Brusilovskii, P. I.; Ignatovich, L. G. (Latvijas PSR Zinatnu Akad. Vestis Kim. Ser. **1975** 569/76; C.A. **84** [1976] No. 50519).
[13] Leeis, Ya. E.; Bruvere, A. Ya.; Bankovskii, Yu. A. (Latvijas PSR Zinatnu Akad. Vestis Kim. Ser. **1982** 19/25; C.A. **96** [1982] No. 149991).
[14] Leeis, Ya. E.; Bankovskii, Yu. A. (Latvijas PSR Zinatnu Akad. Vestis Kim. Ser. **1976** 102/3; C.A. **84** [1976] No. 141446).
[15] Bankovskii, Yu. A.; Leeis, Ya. E. (Latvijas PSR Zinatnu Akad. Vestis Kim. Ser. **1979** 265/76; C.A. **91** [1979] No. 167606).
[16] Sturis, A. P.; Sturis, A. K.; Bankovskii, Yu. A. (Latvijas PSR Zinatnu Akad. Vestis Kim. Ser. **1976** 623/4; C.A. **86** [1977] No. 22494).
[17] Ashaks, Ya.; Yansons, G.; Dobrovodsky, I.; Zikmund, M.; Bankovskii, Yu. A. (Chem. Zvesti **36** [1982] 73/89; C.A. **96** [1982] No. 206284).
[18] Bankovskii, Yu. A.; Ashaks, Ya. V.; Zakharova, I. A.; Deme, A. M.; Leeis, Ya. E.; Brusilovskii, P. I.; Abolinya, M. Ya. (Latvijas PSR Zinatnu Akad. Vestis Kim. Ser. **1978** 387/410; C.A. **89** [1978] No. 225289).
[19] Toropova, V. F.; Budnikov, G. K.; Zhiyangulova, F. G. (Zh. Obshch. Khim. **47** [1977] 1148/52; J. Gen. Chem. [USSR] **47** [1977] 1054/7).
[20] Nakabayashi, Y.; Masuda, Y.; Sekido, E. (J. Electroanal. Chem. **176** [1984] 243/57).

[21] Budnikov, G. K.; Medyantseva, E. P.; Frolova, V. P.; Toropova, V. F. (Zh. Obshch. Khim. **45** [1975] 2361/4; J. Gen. Chem. [USSR] **45** [1975] 2319/21).
[22] Yordanov, N. D.; Iliev, V.; Shopov, D.; Jezierksi, A.; Jeżowska-Trzebiatowska, B. (Inorg. Chim. Acta **60** [1982] 9/15).

35.1.10.6.2 Isolated Compounds

$Mn^{II}(C_9H_6NS)_2$. The air-sensitive complex was prepared by addition of the sodium salt of 8-quinolinethiol (ligand 1) in 50% excess to an aqueous buffered manganese(II) salt solution at 65 to 75°C and pH 4.6. After 10 min, the yellow-brown product was collected, washed with hot (80°C) filtrate adjusted to the desired pH, then with hot water, and dried over silica gel. All operations were carried out under nitrogen [1]. The complex was also precipitated, according to [3], from hydrochloric acid by the method described for $Ni(C_9H_6NS)_2$ and $Pd(C_9H_6NS)_2$ in [4]. Precipitation of a brown amorphous complex from alkaline and ammoniacal solutions containing sodium tartrate was reported in [2]. The effect of acidity on the precipitation of various bivalent metal complexes has been studied. The minimum pH value required for favorable precipitation of 8-quinolinethiol complexes increases in the order $Cd^{II} < Ni^{II} < Co^{II} < Zn^{II} < Cu^{II} < Pb^{II} < Mn^{II}$ [1].

The IR spectra of the complex, recorded in the 4000 to 400 cm^{-1} range (KBr disks of Nujol mulls), are given in [5, 6]. The spectra show two metal-sensitive bands at 994 and 674 cm^{-1}. Relationships between the frequency of these bands and the atomic weight of the central metal for various divalent metal thioxinates have been found [5]. The band at 994 cm^{-1} is assigned to a ν(C–S) stretching vibration [7]. Bands at 974 and 655 cm^{-1} may arise from the disulfane, $(C_9H_6NS)_2$, which probably exists in a mixture with the $Mn(C_9H_6NS)_2$ complex [5]. The ν(SH) band at (2520 ± 5) cm^{-1} in the spectrum of the free ligand is absent in the spectrum of the complex, indicating the absence of the SH groups, i.e., the coordination of the ligand to Mn

 Formulas of ligands are tabulated on p. 66

through the sulfur atoms [5, 7]. Depending on the nature of the solvent, the appearance of various ligand forms (shown below) may be expected in solution:

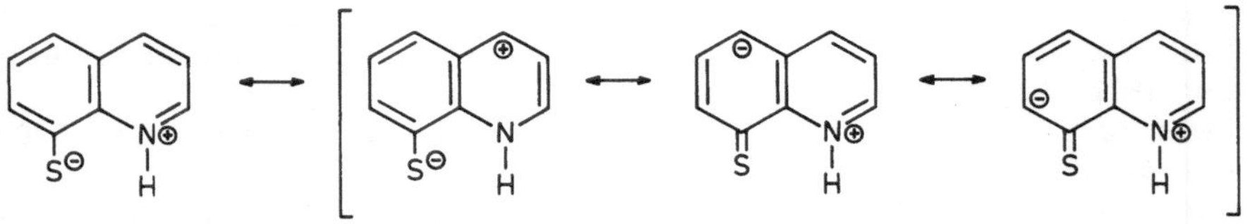

a) b) c) d)

The IR spectra of several metal complexes with ligand 1 were correlated with their electronic structure. The zwitterionic forms b) and c) were suggested for the complexes of transition metals, and the thiol form a) for the complexes of nontransition metals [6]. The X-ray photoelectronic spectra could not generally confirm this hypothesis. The spectrum of solid $Mn(C_9H_6NS)_2$ allows no clear distinction between different canonical structures [8].

The mass spectrum of the complex at 70 eV ionizing voltage and 250°C source temperature shows m/e values (with relative intensity in parentheses) at 375 (22%) for the parent ion $Mn(C_9H_6NS)_2^+$, at 215 (33%) for $Mn(C_9H_6NS)^+$, at 171 (10%) for $Mn(C_8H_6N)^+$, at 161 (100%) for $C_9H_7NS^+$, and at 160 (40%) for $C_9H_6NS^+$. To examine the stabilities of various divalent metal complexes with 8-quinolinethiol under electron impact, the intensity ratio of a fragment ion (ML^+) to a molecular ion (ML_2^+) was calculated, giving the stability order Cd > Zn > Ni > Pd ~ Co > Mn > Cu > Hg > Pb. Definite correlations between these stabilities (intensity ratios) and the charge-radius ratio, e/r, and the ionization potential of the central metal atoms were found [3]. The electronic absorption spectrum of $Mn(C_9H_6NS)_2$ in pyridine shows a maximum at 410 nm ($\varepsilon = 7970\ L \cdot mol^{-1} \cdot cm^{-1}$) [9]. The extraction of the complex into chloroform [9 to 12], benzene, and carbon tetrachloride [9] leads to the formation of the $Mn(C_9H_6NS)_2 \cdot 2C_9H_7NS$ complex with absorption maxima between 412 and 419 nm (ε between 16700 and 15900 $L \cdot mol^{-1} \cdot cm^{-1}$). This band is assigned to a n-π^* transition in the quinoline nucleus due to the shifts of electronic density as a result of the formation of the Mn–N and Mn–S bonds [11]. The similar shape of the absorption spectra and the ca. doubled intensity for $Mn(C_9H_6NS)_2 \cdot 2C_9H_7NS$ indicate that the two additional ligand molecules are bonded to manganese in a way similar to those in $Mn(C_9H_6NS)_2$. A strong absorption at 250 nm ($\varepsilon = 34000\ L \cdot mol^{-1} \cdot cm^{-1}$) in chloroform [11, 12] is assigned to the $\pi \rightarrow \pi^*$ ring transitions [11]. These spectral properties were exploited to determine small quantities of manganese [12].

The complex is insoluble in water and soluble in chloroform, benzene, and carbon tetrachloride [12 to 15]. The solubility product of $Mn(C_9H_6NS)_2$ at 20°C in aqueous solution at pH 6.5 and I = 0.25 mol/L (NaCl) is $-\log K_{s0} = 15.94$, and the solubility in chloroform is $S = 1.1 \times 10^{-3}$ mol/L. A comparison with other metal chelates of 8-quinolinethiol shows that the solubility decreases in the order Mn > V > Pb > Ni ~ Fe > Co ~ Zn > Bi [13]. The solubility product, $-\log K_{s0} = 30.3$, was reported in [14]. The complex is stable in the pH range 6 to 10. In the presence of H_2S it decomposes in the pH range 1 to 3 [16]. Potentiometric titrations of $Mn(C_9H_6NS)_2$ in acetone or acetonitrile with 0.1M $HClO_4$ show basic properties of this complex of limited stability, according to the equation $Mn(C_9H_6NS)_2 + 4HClO_4 \rightarrow Mn(ClO_4)_2 + 2[C_9H_8NS]ClO_4$ [13].

Complex formation is employed in the analytical determination of manganese in small amounts and for separation from other metal ions: by extraction with $CHCl_3$, using spectrophotometry [12], substoichiometric method of isotope dilution [17], or radiometric methods [18, 19], and also by use of coprecipitation [20], electrotitrimetric methods [21], thin-layer chromatography [22], and solubility products [13]. Thin-layer chromatography of the complex on Kiesselgel-60 with development by THF, CH_2Cl_2, and $CHCl_3$ yields R_f values 0.66, 0.27, and 0.22, respectively [22].

$Mn^{II}(C_9H_5ClNS)_2$. The deep brown microcrystalline compound was obtained from a manganese(II) salt and ligand 7 by precipitation from 30 to 50% (v/v) aqueous ethanol at pH 8 [26]. Its solubility in chloroform at 20°C is $-\log S = 3.46$, and the solubility product is $-\log K_{s0} = 34.8$ [27]. Extraction into $CHCl_3$ led to the formation of $Mn(C_9H_5ClNS)_2 \cdot 2C_9H_6ClNS$, which shows an absorption maximum at 411 nm ($\varepsilon = 16850$ $L \cdot mol^{-1} \cdot cm^{-1}$) [26].

$Mn^{II}(C_9H_5BrNS)_2$ (with ligand 8 or 9). The complex with ligand 9 was prepared by the addition of an excess of buffered (pH 2 to 3) aqueous solution of a manganese(II) salt to an ethanolic solution of the ligand. The deep brown crystals, precipitating from 50 to 70% (v/v) aqueous ethanol at pH 8, were washed with ethanol and dried in air. The complex with ligand 8 was obtained from hot aqueous ethanolic solution (50 vol%) with an excess of the ligand. The possibility of polymeric structures for both compounds was discussed. The complexes are slightly soluble in chloroform and benzene, and insoluble in ether [28]. The chloroform solubility of the complex with ligand 9 is $-\log S = 2.74$ (S in mol/L) [14].

As was shown by spectrophotometric analysis, the composition of the complex with ligand 8 in benzene corresponds to that of the crystalline compound with an absorption maximum at 415 nm ($\varepsilon \approx 9500$ $L \cdot mol^{-1} \cdot cm^{-1}$). In organic solvents in the presence of an excess of the ligand or ethylenediamine, the complex with ligand 9 forms $Mn(C_9H_5BrNS)_2 \cdot 2C_9H_6BrNS$ or $Mn(C_9H_5BrNS)_2 \cdot en$, respectively. Electronic spectra of the former complex in chloroform or benzene show an absorption maximum at 425 nm ($\varepsilon = 20200$ $L \cdot mol^{-1} \cdot cm^{-1}$) or 430 nm ($\varepsilon = 18200$ $L \cdot mol^{-1} \cdot cm^{-1}$). This band is assigned to the n-π^* transition. The complex with ligand 8 is less stable than the complex with ligand 9 [28].

$Mn^{II}(C_9H_5NNaO_3S_2)_2 \cdot 3.5H_2O$ was obtained by reacting the sodium salt of ligand 14 (3.3 g) with $MnCl_2 \cdot 4H_2O$ (0.87 g) in aqueous solution containing CH_3COOH (pH 6). The slightly brown crystalline compound, precipitating after addition of ethanol, was washed with ethanol and ether [23]. The electronic spectrum of the compound in methanolic solution shows absorption maxima at 252, 271, and 407 nm with $\varepsilon = 32600$, 36800, and 11400 $L \cdot mol^{-1} \cdot cm^{-1}$, respectively. The complex is soluble in water and is not extracted by organic solvents [24]. It hydrolyzes in water solution [23].

$Mn^{II}(C_{10}H_8NOS)_2 \cdot 0.5H_2O$ was precipitated from aqueous ethanolic solutions (50 to 80 vol%) containing a manganese(II) salt and an excess of ligand 17. In chloroform the complex forms a deep brown solution of composition $Mn(C_{10}H_8NOS)_2 \cdot 2C_{10}H_9NOS$, which shows an absorption maximum at 401 nm ($\varepsilon = 17040$ $L \cdot mol^{-1} \cdot cm^{-1}$) [25].

Other $Mn^{II}L_2$ Compounds. The complexes with ligands 2 [14, 29], 3 [30], 4, 11, 13 [31, 32], 5 [14, 30], 10 [30], 12 [14, 29, 30], 15 [33, 34], 16 [33, 35], 18, 20 [36], 19 [37], 21 [37, 39], 22 [38], 23 to 27 [40 to 44], 28 [45], 29, 30, 32 [40], and 31 [24] have been precipitated, but not isolated. The compounds are insoluble in water, but are readily soluble in organic solvents, with the exception of $Mn(C_{10}H_8NO_3S_2)_2$, a complex with the sulfonated ligand 31 [24], see p. 68. The complexes were extracted into chloroform and the solubility ($-\log S$), the solubility product ($-\log K_{s0}$), and the electronic spectra of the dissolved $MnL_2 \cdot 2HL$ complexes (λ_{max} in nm, ε in $L \cdot mol^{-1} \cdot cm^{-1}$) determined. For comparison, data from the isolated compounds with HL = ligand 1, 7, or 9 are also included in the table:

Formulas of ligands are tabulated on p. 66

ligand[a]	$-\log S$[b]	$-\log K_{s0}$	λ_{max}	ε_{max}	Ref.
1	2.96	30.3	412	16700	[9, 14]
2	2.68	—	418	14640	[14, 29]
4	—	—	410	15400	[31]
5	2.82	—	—	—	[14]
7	3.46	34.8	411	16850	[27, 31]
9	2.74	—	425	20200	[14, 28]
11	3.46	37.3	415	19100	[27, 31]
12	2.60	—	428	21250	[14, 29]
13	—	—	420	19250	[31]
15	3.47	29.3	400	21000	[34]
16	—	—	426	15300	[35]
18	2.48	—	416	20000	[36]
19	—	—	425	20000	[37]
21	—	—	428	20140	[37]
24	1.74	31.7	410	20000	[41, 44]
25	2.28	35.2	413	15500	[42, 44]
26	2.42	37.3	415	16300	[43, 44]
28	<0.3	—	420	18750	[45]

[a] Formulas of ligands are tabulated on p. 66. – [b] Solubility S in mol/L.

Usually, the extractability of $Mn^{II}L_2$ complexes decreases with increasing acidic properties of the mercapto group. With few exceptions, the chloroform solubility of the complex with the unsubstituted ligand 1 is higher than the solubility of the compounds with substituted 8-quinolinethiols. The significantly higher solubility of the $Mn(C_{15}H_{10}NS)_2$ complex with ligand 28 was explained by the increased hydrophobicity of the ligand [45]. The solubilities are compared to analogous complex compounds of other metals (Tl^{I}, Pb^{II}, Ni^{II}, Co^{III}, Zn^{II}, Fe^{III}, In^{III}, Hg^{II}, Bi^{III}) [14, 34, 36, 44, 45]. Relative solubilities in chloroform and carbon tetrachloride were determined for various metal(II) complexes with ligands 1, 2, 3, 5, 6, 9, 10, and 12. The solubility of complexes with 5-substituted halogen ligands is smaller than that of the complex with ligand 1 and decreases in the ligand order 2>5>9>12. Decreasing solubility of the complexes with bromo-substituted ligands in the order 9>8>10 correlates with the order of decreasing pK_{LH} value [30]. The complexes with ligands 8 to 10 were extracted from basic aqueous solution with CCl_4, $CHCl_3$, $C_5H_{11}OH$, and diethyl ether. The solubility of the complexes is greater in strongly polar solvents [46].

Comparison of ε values with analogous complexes of other metals with the compositions $M^{II}L_2$, $M^{III}L_3$, or $M^{IV}L_4$ shows that the color intensity is proportional to the number of ligand molecules bonded to the metal atoms. This band is assigned to the n-π^* transition of the quinoline nucleus. Coordinate bonding between manganese and two additional ligand molecules was assumed for the $MnL_2 \cdot 2HL$ complexes, i.e., coordination number eight was suggested [9, 11].

The coprecipitation of complexes with ligands 23 to 26, $Mn(C_{10}H_8NS)_2$, or with ligand 30, $Mn(C_{11}H_{10}NS)_2$, with the corresponding disulfanes was investigated. Co-crystallization is the mechanism for coprecipitation of the complexes having ligands 24 to 26, while adsorption is responsible for the coprecipitation of those complexes having ligands 23 and 30 [47].

References:

[1] Sekido, E.; Fujiwara, I.; Masuda, Y. (Talanta **19** [1972] 479/87).
[2] Bankovskii, Yu. A.; Ievin'sh, A. F.; Luksha, E. A. (Zh. Analit. Khim. **14** [1959] 222/6; J. Anal. Chem. [USSR] **14** [1959] 237/41).
[3] Kidani, Y.; Naga, S.; Koike, H. (Chem. Pharm. Bull. **23** [1975] 1652/6).
[4] Dalziel, I. A. W.; Kealey, D. (Analyst **89** [1964] 411).
[5] Mido, Y.; Sekido, E. (Bull. Chem. Soc. Japan **44** [1971] 2130/4).
[6] Bankovskii, Yu. A.; Zuika, I. V.; Sturis, A. P.; Leeis, Ya. E. (Latvijas PSR Zinatnu Akad. Vestis Kim. Ser. **1978** No. 3, pp. 278/89; C.A. **89** [1978] No. 145999).
[7] Akimov, V. K.; Busev, A. I.; Zaitsev, B. E.; Bragina, S. I. (Zh. Obshch. Khim. **40** [1970] 1331/4; J. Gen. Chem. [USSR] **40** [1970] 1320/2).
[8] Nefedov, V. I.; Salyn', Ya. V.; Zakharova, I. A.; Bankovskii, Yu. A. (Koord. Khim. **1** [1975] 1545/51; Soviet J. Coord. Chem. **1** [1975] 1278/83).
[9] Bankovskii, Yu. A.; Ievin'sh, A. F.; Buka, M. R.; Luksha, E. A. (Zh. Neorgan. Khim. **8** [1963] 110/7; Russ. J. Inorg. Chem. **8** [1963] 56/60).
[10] Kharkover, M. Z.; Barkovskii, V. F. (Latvijas PSR Zinatnu Akad. Vestis Kim. Ser. **1966** No. 2, pp. 167/72; C.A. **65** [1966] 9807).

[11] Bankovskii, Yu. A.; Ievin'sh, A. F.; Zaruma, D. E.; Luksha, E. A. (Latvijas PSR Zinatnu Akad. Vestis Kim. Ser. **1966** No. 4, pp. 418/30; C.A. **66** [1967] No. 120375).
[12] Bankovskii, Yu. A.; Ievin'sh, A. F.; Luksha, E. A. (Zh. Analit. Khim. **14** [1959] 222/6; J. Anal. Chem. [USSR] **14** [1959] 237/41).
[13] Kharkover, M. Z.; Barkovskii, V. F.; Vdovina, V. M.; Gurova, L. P. (Zh. Analit. Khim. **25** [1970] 30/3; J. Anal. Chem. [USSR] **25** [1970] 22/4).
[14] Chera, L. M.; Bankovskii, Yu. A.; Veveris, O. E.; Ievin'sh, A. F. (Latvijas PSR Zinatnu Akad. Vestis Kim. Ser. **1968** No. 2, pp. 247/9).
[15] Akimov, V. K.; Busev, A. I.; Emel'yanova, I. A. (Zh. Obshch. Khim. **40** [1970] 2706/8; J. Gen. Chem. [USSR] **40** [1970] 2698/700).
[16] Bankovskii, Yu. A.; Ievin'sh, A. F.; Liepinya, Z. E. (Zh. Analit. Khim. **15** [1960] 4/9; J. Anal. Chem. [USSR] **15** [1960] 1/6).
[17] Grosheva, E. I. (Tezisy Dokl. 3rd Konf. Molodykh Nauchn. Rab. Inst. Neorgan. Khim., Riga 1973 [1974], pp. 25/7; C.A. **87** [1977] No. 126592).
[18] Mikulski, J.; Stroński, J. (Nukleonika **7** [1962] 769/73).
[19] Veveris, O. E.; Bankovskii, Yu. A.; Pelekis, L. L.; Aynbinder, N. G.; Shekhtmeyster, L. A. (J. Radioanal. Chem. **9** [1971] 47/53).
[20] Vircavs, M. B.; Bankovskii, Yu. A.; Veveris, O. E. (Latvijas PSR Zinatnu Akad. Vestis Kim. Ser. **1980** No. 3, pp. 344/8; C.A. **93** [1980] No. 87841).

[21] Suprunovich, V. I.; Torokhtei, L. P.; Zherebtsova, E. I. (Vopr. Khim. Khim. Tekhnol. No. 53 [1978] 133/8; C.A. **91** [1979] No. 48884).
[22] Schneeweis, G.; König, K.-H. (Z. Anal. Chem. **316** [1983] 16/22).
[23] Mezharaup, G. P.; Deme, A. M.; Ashaks, Ya. V.; Bankovskii, Yu. A. (Latvijas PSR Zinatnu Akad. Vestis Kim. Ser. **1976** No. 2, pp. 131/7; C.A. **85** [1976] No. 55909).
[24] Bankovskii, Yu. A.; Ashaks, Ya. V.; Zakharova, I. A.; Deme, A. M.; Leeis, Yu. E.; Brusilovskii, P. I.; Abolinya, M. Ya. (Latvijas PSR Zinatnu Akad. Vestis Kim. Ser. **1978** No. 4, pp. 387/410; C.A. **89** [1978] No. 225289).
[25] Bankovskii, Yu. A.; Zaruma, D.; Brusilovskii, P. I.; Ignatovich, L. G. (Latvijas PSR Zinatnu Akad. Vestis Kim. Ser. **1975** No. 5, pp. 569/76; C.A. **84** [1976] No. 50519).
[26] Bankovskii, Yu. A.; Krasovska, M. E.; Chera, L. M.; Leeis, Ya. E.; Zaruma, D. E.; Miezere, R. S. (Latvijas PSR Zinatnu Akad. Vestis Kim. Ser. **1975** No. 2, pp. 169/79; C.A. **83** [1975] No. 104043).

[27] Bankovskii, Yu. A.; Zaruma, D. E.; Krasovska, M. E.; Tsirule, Ya. A.; Miezere, R. S. (Latvijas PSR Zinatnu Akad. Vestis Kim. Ser. **1974** No. 5, pp. 629/30; C.A. **82** [1975] No. 90687).
[28] Bankovskii, Yu. A.; Zaruma, D. E.; Krasovska, M. E.; Ievin'sh, A. F.; Labrence, I. K. (Latvijas PSR Zinatnu Akad. Vestis Kim. Ser. **1965** 662/88; C.A. **65** [1966] 683).
[29] Bankovskii, Yu. A.; Buka, M. R.; Zaruma, D. E.; Abolinya, A. Yu.; Ievin'sh, A. F.; Krasovska, M. E. (Latvijas PSR Zinatnu Akad. Vestis Kim. Ser. **1969** No. 3, pp. 265/76; C.A. **71** [1969] No. 91248).
[30] Bankovskii, Yu. A.; Zaruma, D. E.; Mezharaup, G. P.; Ievin'sh, A. F. (Latvijas PSR Zinatnu Akad. Vestis Kim. Ser. **1966** No. 1, pp. 3/17; C.A. **65** [1966] 9804).

[31] Bankovskii, Yu. A.; Krasovska, M. E.; Chera, L. M.; Leeis, Ya. E.; Zaruma, D. E.; Miezere, R. S. (Latvijas PSR Zinatnu Akad. Vestis Kim. Ser. **1975** No. 2, pp. 169/79; C.A. **83** [1975] No. 104043).
[32] Bankovskii, Yu. A.; Krasovskaya, M. E.; Chera, L. M.; Leeis, Ya. E.; Lazdina, R. S. (Latvijas PSR Zinatnu Akad. Vestis Kim. Ser. **1971** No. 2, pp. 742/4; C.A. **76** [1972] No. 104485).
[33] Zaruma, D. E.; Krasovska, M. E.; Brusilovskii, P. I. (Org. Reagenty Anal. Khim. Tezisy Dokl. 4th Vses Konf., Kiev 1976, Vol. 1, pp. 100/2; C.A. **88** [1978] No. 44408).
[34] Zaruma, D. E.; Krasovska, M. E.; Bankovskii, Yu. A.; Grike, M. O. (Latvijas PSR Zinatnu Akad. Vestis Kim. Ser. **1976** No. 6, pp. 734/5; C.A. **86** [1977] No. 148069).
[35] Krasovska, M. E.; Zaruma, D. E.; Bankovskii, Yu. A. (Latvijas PSR Zinatnu Akad. Vestis Kim. Ser. **1977** No. 1, pp. 101/2; C.A. **86** [1977] No. 149851).
[36] Leeis, Ya. E.; Bruvere, A.; Bankovskii, Yu. A. (Latvijas PSR Zinatnu Akad. Vestis Kim. Ser. **1982** No. 1, pp. 19/25; C.A. **96** [1982] No. 149991).
[37] Bankovskii, Yu. A.; Leeis, Ya. E. (Latvijas PSR Zinatnu Akad. Vestis Kim. Ser. **1979** No. 3, pp. 265/76; C.A. **91** [1979] No. 167606).
[38] Leeis, Ya. E.; Bankovskii, Yu. A. (Latvijas PSR Zinatnu Akad. Vestis Kim. Ser. **1976** No. 3, pp. 359/60; C.A. **85** [1976] No. 136397).
[39] Leeis, Ya. E.; Bankovskii, Yu. A. (Latvijas PSR Zinatnu Akad. Vestis Kim. Ser. **1976** No. 3, pp. 352/3; C.A. **85** [1976] No. 136396).
[40] Sturis, A. P.; Bankovskii, Yu. A.; Sturis, A. K. (Tezisy Dokl. 12th Vses. Chugaevskoe Soveshch. Khim. Kompleksn. Soedin., Novosibirsk 1975, Vol. 3, pp. 423/4; C.A. **85** [1976] No. 201400).

[41] Tsirule, Ya. A.; Sturis, A. P.; Bankovskii, Yu. A.; Ievin'sh, A. F. (Latvijas PSR Zinatnu Akad. Vestis Kim. Ser. **1967** No. 4, pp. 407/8; C.A. **68** [1968] No. 83912).
[42] Sturis, A. P.; Bankovskii, Yu. A. (Latvijas PSR Zinatnu Akad. Vestis Kim. Ser. **1976** No. 1, pp. 100/1; C.A. **84** [1976] No. 186854).
[43] Bankovskii, Yu. A.; Sturis, A. K.; Sturis, A. P. (Latvijas PSR Zinatnu Akad. Vestis Kim. Ser. **1970** No. 6, pp. 727/8; C.A. **74** [1971] No. 57836).
[44] Sturis, A. P.; Sturis, A. K.; Bankovskii, Yu. A. (Latvijas PSR Zinatnu Akad. Vestis Kim. Ser. **1976** No. 6, pp. 623/4; C.A. **86** [1977] No. 22494).
[45] Ashaks, Ya. V.; Yansons, G.; Dobrovodsky, J.; Zikmund, M.; Bankovskii, Yu. A. (Chem. Zvesti **36** [1982] 73/89; C.A. **96** [1982] No. 206284).
[46] Bankovskii, Yu. A.; Lobanova, E. F. (Zh. Analit. Khim. **14** [1959] 523/8; J. Anal. Chem. [USSR] **14** [1959] 569/74).
[47] Vircavs, M. V.; Sturis, A. P. (Latvijas PSR Zinatnu Akad. Vestis Kim. Ser. **1978** No. 5, pp. 515/20; C.A. **90** [1979] No. 58665).

35.1.10.7 With Derivatives of Quinoxaline or Pyrazino[2,3-g]quinoxaline

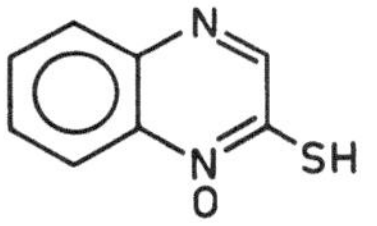

ligand 1 (= $C_8H_6N_2OS$)

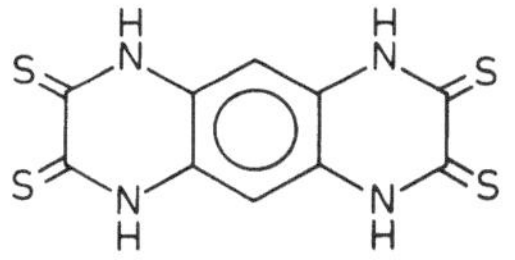

ligand 2 (= $C_{10}H_6N_4S_4$)

The complex **$Mn^{II}(C_8H_5N_2OS)_2$** precipitates on reaction of $MnSO_4$ with the sodium salt of ligand 1 (mole ratio 1:2) in aqueous or aqueous alcohol solution. The compound shows antimicrobial and antifungal activity [1, 2].

An orange dichroic polarizer, consisting of polyvinyl alcohol stained with an oligomeric **Mn^{II} chelate** of ligand 2, was prepared by the reaction of a hot aqueous solution of Mn^{II} acetate and polyvinyl alcohol containing ammonia with a hot aqueous solution of ligand 2 and polyvinyl alcohol. The oligomeric chelate is assumed to consist of chains of 2 to 10 repeating molecular units of Mn^{II} chelated with the bis-bidentate bridging ligand. The preparation of polarizers which also contain oligomeric Ni chelates with ligand 2 is described, as well [3].

References:

[1] Douglass, M. L. to Colgate-Palmolive Co. (U.S. 3733323 [1969/73]).
[2] Douglass, M. L. to Colgate-Palmolive Co. (U.S. 3971725 [1974/76]).
[3] Bloom, S. M. to Polaroid Corp. (U.S. 4133775 [1974/79]).

35.1.10.8 With Derivatives of Pyrimidine or Purine

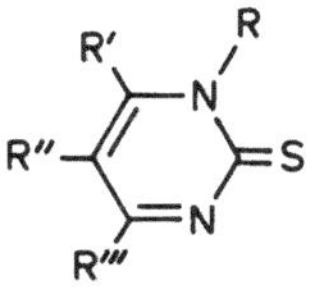

ligands 1 to 4

ligand	R	R′	R″	R‴	formula
1	H	H	H	H	$C_4H_4N_2S$
2	CH_3	H	H	H	$C_5H_6N_2S$
3	H	CH_3	H	CH_3	$C_6H_8N_2S$
4	CH_3	CH_3	H	CH_3	$C_7H_{10}N_2S$

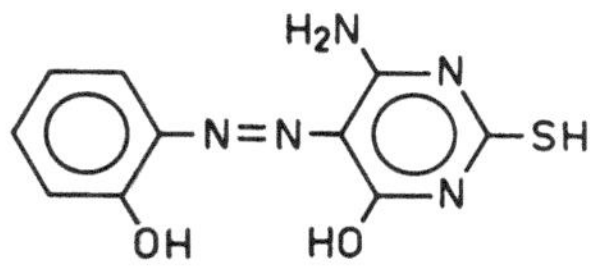

ligand 5 (= $C_{10}H_9N_5O_2S$)

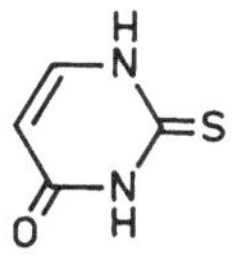

ligand 6 (= $C_4H_4N_2OS$)

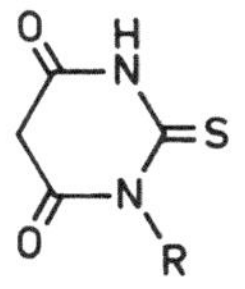

ligands 7 to 12

ligand	R	formula
7	H	$C_4H_4N_2O_2S$
8	CH_3	$C_5H_6N_2O_2S$
9	C_6H_5	$C_{10}H_8N_2O_2S$

ligand	R	formula
10	$C_6H_4CH_3$-2	$C_{11}H_{10}N_2O_2S$
11	$C_6H_4CH_3$-3	$C_{11}H_{10}N_2O_2S$
12	$C_6H_4CH_3$-4	$C_{11}H_{10}N_2O_2S$

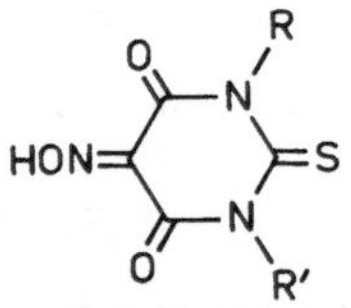
ligands 13 to 21

ligand	R	R′	formula
13	H	H	$C_4H_3N_3O_3S$
14	H	C_6H_5	$C_{10}H_7N_3O_3S$
15	H	$C_6H_4CH_3$-2	$C_{11}H_9N_3O_3S$
16	H	$C_6H_4CH_3$-3	$C_{11}H_9N_3O_3S$
17	H	$C_6H_4CH_3$-4	$C_{11}H_9N_3O_3S$
18	C_6H_5	C_6H_5	$C_{16}H_{11}N_3O_3S$
19	$C_6H_4CH_3$-2	$C_6H_4CH_3$-2	$C_{18}H_{15}N_3O_3S$
20	$C_6H_4CH_3$-3	$C_6H_4CH_3$-3	$C_{18}H_{15}N_3O_3S$
21	$C_6H_4CH_3$-4	$C_6H_4CH_3$-4	$C_{18}H_{15}N_3O_3S$

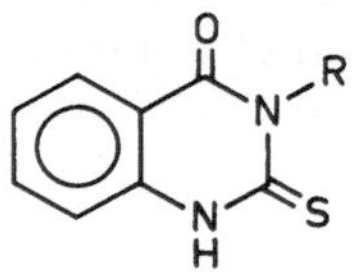
ligand 22 with R = CH_3 (= $C_9H_8N_2OS$)
ligand 23 with R = C_6H_5 (= $C_{14}H_{10}N_2OS$)

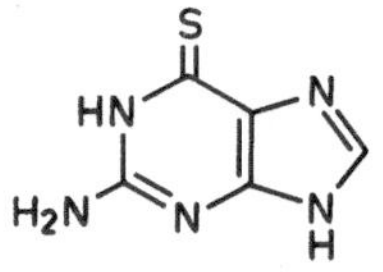
ligand 24 (= $C_5H_5N_4S$)

35.1.10.8.1 Complexes in Solution

Stability constants of 1:1 and 1:2 complexes of manganese(II) with deprotonated ligands 7 through 24 in aqueous or dioxane-water medium were determined by potentiometric pH titrations [1 to 8]. Bubbling of presaturated N_2 through the titration solution is reported by [1, 2, 7].

ligand	t in °C	I in mol/L[a)]	vol% dioxane	log K_1	log K_2	Ref.
7	31	0 (corr.)	0	5.11	3.75	[5]
8	31	0 (corr.)	0	5.23	3.87	[5]
9	31	0 (corr.)	0	5.50	4.14	[5]
10	31	0 (corr.)	0	5.35	3.99	[5]
11	31	0 (corr.)	0	5.44	4.08	[5]
12	31	0 (corr.)	0	5.62	4.26	[5]
13	30	0.1 ($NaClO_4$)	50	2.93	—	[1]
14	31	0 (corr.)	80	5.90	—	[4]
15	31	0 (corr.)	80	5.88	—	[4]
16	31	0 (corr.)	80	5.92	—	[4]
17	31	0 (corr.)	80	6.08	—	[4]
18	30	0.1 ($NaClO_4$)	75	3.09	—	[1]
				3.05[b)]	—	[2]
				2.90[c)]	—	[2]

ligand	t in °C	I in mol/L[a)]	vol% dioxane	log K_1	log K_2	Ref.
19	30	0.1($NaClO_4$)	75	2.52[b)]	1.95[b)]	[2]
				2.70[c)]	—	[2]
20	30	0.1($NaClO_4$)	75	2.39[b)]	1.99[b)]	[2]
				2.82[c)]	—	[2]
21	30	0.1($NaClO_4$)	75	2.91[b)]	2.17[b)]	[2,3]
				3.04[c)]	—	[2]
22	27	0.1($NaClO_4$)	75	4.75	3.79	[6]
23	27	0.1	75	2.90	—	[7]
24	25	0.01	50	4.3	—	[8]

[a)] I = 0 (corr.) means corrected to zero ionic strength by use of an empirical formula. – [b)] Values obtained by the weighed least-squares method. – [c)] Values obtained by the correction term method.

Additional values of log K_1 and log K_2 at 40, 45, and 50°C are reported in [3], at 18 and 42°C in [4, 5]. Comparison of log K_1 values of Mn^{II} complexes with thiobarbituric acid (ligand 7) and thiovioluric acid (ligand 13) with those of their N-substituted derivatives shows decreasing log K_1 values in the order 4-$CH_3C_6H_4$ > C_6H_5 > 3-$CH_3C_6H_4$ > 2-$CH_3C_6H_4$ > H for monosubstituted thiobarbituric acid [5], 4-$CH_3C_6H_4$ > 3-$CH_3C_6H_4$ > C_6H_5 > 2-$CH_3C_6H_4$ for monosubstituted thiovioluric acid [4], and C_6H_5 > 4-$CH_3C_6H_4$ > 3-$CH_3C_6H_4$ > 2-$CH_3C_6H_4$ for disubstituted thiovioluric acid [1 to 3]. The order of log K values is essentially the same as that of the pK_a values of the corresponding ligands. Substituent effects decrease in the order: monosubstituted thiobarbituric acid > monosubstituted thiovioluric acid > disubstituted thiovioluric acid [1 to 5].

Thermodynamic formation parameters for manganese(II) complexes with the cited ligands are given below (conditions as above; ΔH in kcal/mol, ΔS in $cal \cdot mol^{-1} \cdot K^{-1}$):

ligand	$-\Delta H_1$	$-\Delta S_1$	$-\Delta H_{\beta 2}$	$-\Delta S_{\beta 2}$	Ref.
7	9.72	8.92	16.99	15.76	[5]
8	10.02	9.14	17.38	15.71	[5]
9	13.68	19.90	21.32	26.18	[5]
10	11.52	13.51	18.96	19.82	[5]
11	12.99	14.67	20.54	20.92	[5]
12	14.50	22.10	22.25	28.25	[5]
14	12.13	13.02	—	—	[4]
15	12.63	14.73	—	—	[4]
16	13.22	16.52	—	—	[4]
17	14.18	18.94	—	—	[4]
21	—	—	16.38	30.3	[3]

The stepwise complexation reactions are found to be predominantly enthalpy driven processes [5].

Multidentate ligand 5 forms an orange solution when it is added to a basic solution containing Mn^{2+} ions. The limit of detection is 1 ppm [9].

Formulas of ligands are tabulated on pp. 75/6

References:

[1] Chawla, R. S.; Singh, R. P. (Microchem. J. **18** [1973] 646/51).
[2] Mathur, P.; Goel, D. P.; Singh, R. P. (Indian J. Chem. A **16** [1978] 890/2).
[3] Mathur, P.; Goel, D. P.; Singh, R. P. (J. Indian Chem. Soc. **55** [1978] 879/81).
[4] Singh, B. R.; Ghosh, R. (Thermochim. Acta **53** [1982] 257/62).
[5] Singh, B. R.; Jain, R. K.; Jain, M. K.; Ghosh, R. (Thermochim. Acta **78** [1984] 175/80).
[6] Raheja, D. H.; Laxmeshwar, N. B. (J. Indian Chem. Soc. **61** [1984] 507/9).
[7] Raheja, D. H.; Fulwadhva, U. P.; Laxmeshwar, N. B. (J. Indian Chem. Soc. **61** [1984] 584/5).
[8] Bag, S. P.; Fernando, Q.; Freiser, H. (Arch. Biochem. Biophys. **106** [1964] 379/80).
[9] Tarin, P.; Blanco, M. (Anales Quim. B **78** [1982] 257/9).

35.1.10.8.2 Isolated Compounds

[$Mn^{II}L_2$] complexes with HL = ligands 18, 19, and 21 precipitated on reaction of saturated aqueous solutions of $MnCl_2$ or $Mn(NO_3)_2$ with concentrated aqueous solutions of the ammonium salt of the appropriate ligand. The precipitates were washed with water [1 to 3], and the complex with ligand 18 was recrystallized from acetone. The slate-colored crystalline compounds are stable at ordinary temperatures [1 to 3]. [$Mn(C_{16}H_{10}N_3O_3S)_2$] melts with decomposition at 295°C [1]; the complexes with ligand 19 or 21, [$Mn(C_{18}H_{14}N_3O_3S)_2$], melt at 243°C [2] and 254°C [3], respectively.

They are only slightly soluble in water, but are soluble in organic solvents, such as acetone, alcohol, or ether. The blue solutions in acetone show absorption maxima at 498, 457, or 437 nm in the electronic spectra. In aqueous solution or aqueous acetone some dissociation into Mn^{2+} ions and the ligands was observed [2, 3]. Coordination by the ligands (in the nitroso-enolic form) through the N atom of the NO group and the O atom of the adjacent CO group is assumed [1 to 3].

[$Mn^{II}(C_4H_3N_2O_2S)Cl$]. The complex with ligand 7 was prepared by the reaction of aqueous solutions of $MnCl_2$ and an equimolar amount of the ligand at pH 4 to 7. The reaction mixture was stirred for a few hours at room temperature and then kept overnight. The yellowish pink precipitate that formed was isolated, washed with water, and dried in an oven at 60°C. The yield of the product was 48%. The magnetic moment of the solid is $\mu_{eff} = 2.31\ \mu_B$ at 21°C (from susceptibility measurements by the Faraday method). The IR spectrum of the complex shows prominent absorptions at the following wavenumbers (in cm^{-1}; free-ligand bands in parentheses): ν(NH) 3060, 3050 (3200, 3060); ν(CS) 1205 (1204); ν(CO) 1557 (1630); δ(NH) + ν(CN) 1522, 1511 (1520); ν(ring) 1448, 1395, 1320 (1462, 1400, 1324); ν(Mn–Cl) 330, 290; ν(Mn–O) 270; ν(Mn–N) 220. The reflectance spectrum shows maxima at 17860, 26315, 31250, and 36360 cm^{-1}, which were assigned to the low-spin Mn^{II} transitions: $^2T_2 \rightarrow {}^4T_1$, charge transfer, $^2T_2 \rightarrow {}^2A_1$, $^2T_2 \rightarrow {}^2E$. The physical properties indicate a low-spin octahedral Mn^{II} complex in a polymeric structure. The IR spectral results suggest coordination of the ligand via oxygen and nitrogen atoms with no binding via the sulfur atom. Chlorine atoms are assumed to be in the axial positions. The complex melts at 240°C with decomposition. It is insoluble in common organic solvents [4].

[$Mn^{II}L_2X_2$] (X = Cl, Br). The complexes with ligand 1 were prepared by reaction of MnX_2 with the ligand (mole ratio 1:1.7) in a small volume of hot, dried, and deaerated ethanol under nitrogen. The mixture was heated until the reaction was complete. The product was filtered off under nitrogen, washed with ethanol and ether, and dried in vacuum over P_4O_{10} [5]. The complexes with ligands 2 [6], 3 [7], and 4 [8] were prepared from stoichiometric amounts of hydrated Mn salt in ethanol and the ligands in ethanol, acetone, or propanol [6 to 8]. The mixtures were heated to 60°C for 10 to 15 min [7, 8]; the complexes either precipitated immediately or after cooling for several days [6 to 8]. They were washed with acetone [6, 8] or ethanol [7], then with ether, and dried in vacuum [6 to 8] at 80°C [8]. The complexes with ligand 6 were obtained by refluxing (24 to 48 h) the appropriate metal salt with 75% of the stoichiometric amount of thiouracil in ethyl acetate containing a small quantity of 2,2-dimethoxypropane. The reaction mixture was protected from atmospheric moisture, as the products are somewhat hygroscopic. They were dried in vacuum at 100°C [9].

Far-IR absorptions of the yellow complexes were assigned as follows:

ligand	complex	IR bands (in cm^{-1}) ν(Mn–X)	ν(Mn–L)	Ref.
1	$[Mn(C_4H_4N_2S)_2Cl_2]$	228, 210	a)	[5]
	$[Mn(C_4H_4N_2S)_2Br_2]$	180, 166	231	[5]
2	$[Mn(C_5H_6N_2S)_2Cl_2]$	256	a)	[6]
	$[Mn(C_5H_6N_2S)_2Br_2]$	213	233	[6]
3	$[Mn(C_6H_8N_2S)_2Cl_2]$	234	—	[7]
	$[Mn(C_6H_8N_2S)_2Br_2]$	—	—	[7]
4	$[Mn(C_7H_{10}N_2S)_2Cl_2]$	263, 254	229	[8]
	$[Mn(C_7H_{10}N_2S)_2Br_2]$	—	225	[8]
6	$[Mn(C_4H_4N_2OS)_2Cl_2]$	—	227 b)	[9]
	$[Mn(C_4H_4N_2OS)_2Br_2]$	—	224 b)	[9]

a) Hidden under strong, broad ν(Mn–X) bands. – b) Measured between 200 and 400 cm^{-1}. Other strong bands at 339 (for X = Cl) and at 335 cm^{-1} (for X = Br) were not assigned.

X-band ESR spectra of the complexes with ligands 1 to 4 show fine structure due to zero-field splittings in the monomeric pseudooctahedral complexes [5 to 8]. For the bromo complex with ligand 1, $[Mn(C_4H_4N_2S)_2Br_2]$, a numerical analysis of the ESR spectrum gave $D = 0.2\ cm^{-1}$ and $\lambda(= E/D) = 0.25$ [5]. The X-band ESR spectra of the complexes with ligand 6, however, show a broad signal at $g = 2$ with no evidence of fine structure, thus suggesting a polymeric structure with bridging halide ions for these compounds. Bridging halide ions for the latter complexes are also indicated by the lack of ν(Mn–X) bands above 200 cm^{-1} [9].

$[Mn^{II}(C_7H_{10}N_2S)_3]X_2$ (X = ClO_4, I). The complexes were prepared by the reaction of a solution of $Mn(ClO_4)_2$ or freshly prepared MnI_2 (1 mmol in 10 mL ethanol) with a solution of ligand 4 (3 mmol in 60 mL of hot propanol). The yellow solids precipitated almost immediately or after cooling for a few days. They were washed with ethanol, acetone, and ether, then dried in vacuum. X-ray powder photographs showed the following series to be isomorphous: a) $[M(C_7H_{10}N_2S)_3](ClO_4)_2$ with M = Mn^{II}, Co^{II}, Cu^{II}, and Zn; b) $[M(C_7H_{10}N_2S)_3]I_2$ with M = Mn, Co, and Zn. The X-band ESR spectrum of the powdered perchlorate was analyzed to yield the zero-field splitting parameters $D = 0.055\ cm^{-1}$ and $\lambda(= E/D) = 0.05$. The low values are consistent with an octahedral Mn^{II} complex with a *fac* arrangement of the neutral bidentate ligand molecules, chelated via the thioketone sulfur and ring imine nitrogen atoms. The far-IR spectra of the complexes in Nujol mull show the prominent bands: ν(Mn–L) at 233, 211 cm^{-1} and ligand bands at 413, 336, 312, 290 cm^{-1} for X = ClO_4; ν(Mn–L) at 220, 210 cm^{-1} for X = I; and ligand bands at 401, 331, 314, 292 cm^{-1} [10].

$[Mn^{II}(C_5H_6N_2S)_4](ClO_4)_2 \cdot 2CH_3COCH_3$ was prepared by the reaction of ligand 2 and hydrated manganese(II) perchlorate (mole ratio 2:1) in acetone or ethanol solution. A yellow solid

 Formulas of ligands are tabulated on pp. 75/6

formed on cooling the reaction mixture for several days. The product was isolated, washed with acetone and ether, and dried in air. The far-IR spectrum of the complex shows a characteristic band at 218 cm^{-1} for ν(Mn–N). The X-band ESR spectrum of the microcrystalline complex at room temperature shows a strong band in the g = 2 region, weaker lines at lower field, and a pronounced shoulder at higher field. Analysis of the spectrum of the Mn^{II} complex doped into the isomorphous Zn^{II} complex yielded the zero-field splitting parameters D = 0.047 cm^{-1} and λ = 0.3. Based on the spectral results and the observation (by X-ray powder diffraction) that the Mn^{II} complex is isomorphous with the tetrahedral Zn^{II} and Co^{II} complexes, it is suggested that the Mn^{II} complex has a tetrahedral geometry with four Mn–N bonds [6].

References:

[1] Gambhir, I. R.; Singh, R. P. (Proc. Indian Acad. Sci. A **23** [1946] 330/4).
[2] Singh, R. P. (J. Indian Chem. Soc. **36** [1959] 198/200).
[3] Singh, R. P. (J. Indian Chem. Soc. **37** [1960] 569/72).
[4] Siddiqi, K. S.; Khan, P.; Khan, S.; Zaidi, S. A. A. (Syn. React. Inorg. Metal-Org. Chem. **12** [1982] 681/94).
[5] Abbot, J.; Goodgame, D. M. L.; Jeeves, I. (J. Chem. Soc. Dalton Trans. **1978** 880/4).
[6] Goodgame, D. M. L.; Leach, G. A. (J. Chem. Soc. Dalton Trans. **1978** 1705/9).
[7] Goodgame, D. M. L.; Jeeves, I.; Leach, G. A. (Inorg. Chim. Acta **38** [1980] 247/51).
[8] Goodgame, D. M. L.; Leach, G. A. (Inorg. Chim. Acta **32** [1979] 69/73).
[9] Goodgame, D. M. L.; Leach, G. A. (Inorg. Chim. Acta **37** [1979] L505/L506).
[10] Goodgame, D. M. L.; Leach, G. A. (Inorg. Chim. Acta **25** [1977] L127/L129).

35.2 Complexes with Sulfanes or Related Compounds

35.2.1 With Diorganylsulfanes

Remark. Complexes with ligands of the type R–S–R′, derived from pyrimidinylazo compounds (= R′) with R = CH_3 or C_2H_5, have already been described in "Manganese" D 5, 1987, p. 309. For complexes with Schiff bases containing RS groups, see "Manganese" D 6, 1989, pp. 8, 16, 82, 129/30, and 268/9.

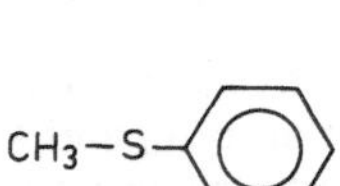

ligand 1 (= C_7H_8S)

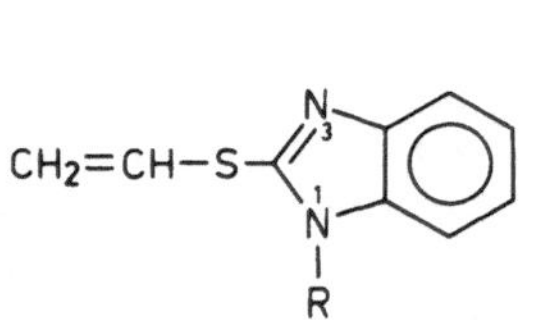

ligand 2 with R = H (= $C_9H_8N_2S$)
ligand 3 with R = CH=CH$_2$ (= $C_{11}H_{10}N_2S$)

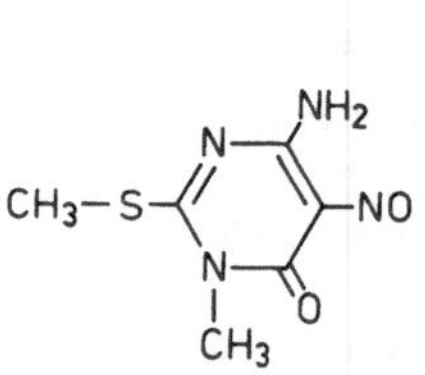

ligand 4 (= $C_6H_8N_4O_2S$)

Manganese(II) Complexes in Solution. Stability constants of $Mn(C_6H_7N_4O_2S)^+$ and $Mn(C_6H_7N_4O_2S)_2$ complexes with the deprotonated tautomeric form of ligand 4 have been determined at ionic strengths I = 0.01 to 0.1 M (KCl) and temperatures of 25, 30, 40, and 50°C. The constants are practically unaffected by both ionic strength and temperature in the range studied. Values at 25°C and I = 0.1 M are log K_1 = 3.55, log K_2 = 4.02. Enthalpy changes of ΔH_1 = −19.2 and ΔH_2 = −12.8 kJ/mol have been calculated, but incorrect values for entropy changes (ΔS_1 = 132.0 and ΔS_2 = 120.0 J·mol^{-1}·K^{-1}) reported. The stability constants at I = 0, log K_1 =

3.57 and log $K_2 = 4.08$, and changes of free energy, $\Delta G_1 = 20.38$ and $\Delta G_2 = 23.29$ kJ/mol, have been calculated at 25°C. It is supposed that coordination of the ligand occurs via the oxygen atoms of the nitroso and oxo groups [1].

[$Mn(NO)_3C_7H_8S$], characterized only by spectral data, was prepared by adding a threefold molar excess of ligand 1 to a pentane solution of [$Mn(NO)_3THF$], as described on p. 6 for the corresponding complex with acetonitrile, [$Mn(NO)_3C_2H_3N$]. The NMR spectrum reveals the shift parameter $\delta(^{55}Mn) = -1130$ ppm (relative to saturated aqueous $KMnO_4$ solution). The ^{55}Mn shielding, caused by the coordinated thioether, is compared with that of other ligands coordinated by O, N, Se, Te, C, or P atoms [2].

$Mn^{II}(C_9H_8N_2S)_2Cl_2$ and **$Mn^{II}(C_{11}H_{10}N_2S)_2Cl_2$** complexes were prepared by heating manganese(II) chloride and ligand 2 or 3, respectively, in alcohol for 5 to 8 h at 70 to 80°C. The precipitates were washed with alcohol and dried at 60 to 70°C under reduced pressure (3 to 5 Torr). A quantum mechanical calculation of the distribution of π-electron density implied that the ligands should coordinate through the nitrogen atom in the 3-position of the heterocycle. No IR shifts due to the complex formation are revealed in the stretching frequencies of the heterocycle in the 1600 to 1400 cm^{-1} range. A small shift of ν(NH) in the $Mn(C_9H_8N_2S)_2Cl_2$ spectrum may be related to the localization of the metal on the N(3) nitrogen atom. CS vibrations do not appear. The monomeric complexes are stable in air. They are soluble in alcohol, acetone, chloroform, but insoluble in water. They decompose in acid or alkaline aqueous media [3].

References:

[1] López-Garzón, R.; Martinez-Garzón, A. M.; Gutierrez-Valero, M. D.; Domingo-Garcia, M. (Thermochim. Acta **108** [1986] 181/7).
[2] Rehder, D.; Ihmels, K.; Wenke, D.; Oltmanns, P. (Inorg. Chim. Acta **100** [1985] L11/L12).
[3] Skvortsova, G. G.; Trzhtsinskaya, B. V.; Abramova, N. D.; Teterina, L. F.; Voronov, V. K.; Sherstyannikova, L. V. (Zh. Obshch. Khim. **46** [1976] 1305/9; Russ. J. Gen. Chem. **46** [1976] 1286/90).

35.2.2 With Sulfanediylbisalkanols or -phenols

ligand HOR–S–ROH ...	1	2	3	4
ROH	CH_2CH_2OH	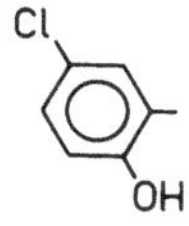	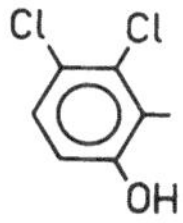	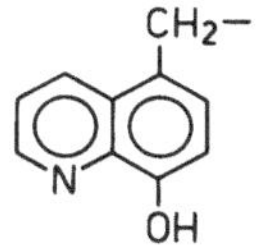
formula	$C_4H_{10}O_2S$	$C_{12}H_8Cl_2O_2S$ (= H_2L)	$C_{12}H_6Cl_4O_2S$ (= H_2L)	$C_{20}H_{16}N_2O_2S$ (= H_2L)

Manganese(II) Complexes in Solution. The stability constant, log $K_1 = -0.22 \pm 0.06$, for the complex with ligand 1, $Mn(C_4H_{10}O_2S)^{2+}$, was determined spectrophotometrically in aqueous solution at 25°C and I = 1 M ($NaClO_4$) by comparison with the spectrum of the corresponding Cu^{II} complex [1]. The stability constants of MnL and MnL_2 complexes with ligands 2 and 3 were determined pH-potentiometrically in a 3:1 ethanol-water mixture (v/v) at 25°C and I = 1 M ($NaClO_4$) under nitrogen: log $K_1 = 5.98 \pm 0.05$ for $Mn(C_{12}H_6Cl_2O_2S)$, log $K_2 \approx 4.5$ for $Mn(C_{12}H_6Cl_2O_2S)_2^{2-}$, log $K_1 = 5.32 \pm 0.05$ for $Mn(C_{12}H_4Cl_4O_2S)$, and log $K_2 = 4.02 \pm 0.04$ for $Mn(C_{12}H_4Cl_4O_2S)_2^{2-}$ [2].

$[Mn^{II}(C_{20}H_{14}N_2O_2S)]_n$. The coordination polymer was prepared by the reaction of ligand 4 (5 mmol) with the same amount of manganese(II) acetate (each dissolved in 50 mL dimethylformamide). After addition of sodium acetate (0.5 g), the mixture was stirred and refluxed for 2 h. The precipitate was washed with dimethylformamide, then with water, and air-dried. The brown complex does not melt below 360°C. The magnetic moment, $\mu_{eff} = 4.83\ \mu_B$ at room temperature, is somewhat lower than the expected spin-only value. The IR spectrum of the complex (in KBr) reveals no ν(OH) vibration mode. The in-plane δ(OH) vibration of the free ligand at 1430 cm^{-1} is shifted to higher frequency, indicating the formation of a manganese-oxygen bond. The ν(C=N) vibration of the free ligand at 1600 cm^{-1} is shifted towards lower frequency due to coordination of manganese by the imine nitrogen atom. A weak band around 3200 cm^{-1} probably indicates the presence of absorbed water molecules. The TGA data reveal an appreciable weight loss up to 280°C, which may be due to the presence of water strongly absorbed by the polymer. The complex is insoluble in common organic solvents [3].

$[Mn^{II}(C_4H_{10}O_2S)Cl_2]$ was prepared by the reaction of an equimolar amount of ligand 1 and $MnCl_2$ in absolute ethanol. The mixture was refluxed for 2 h and the solution concentrated by distillation. Addition of pentane to the reaction mixture gave a flesh-colored precipitate, which was recrystallized from 2-propanol (containing a small amount of ethanol) by the addition of chloroform. The compound melts at 192°C with decomposition. The magnetic moment of the solid is $\mu_{eff} = 6.07\ \mu_B$ at 20°C. The IR spectrum of the complex in hexachlorobutadiene shows ν(OH) bands at 3430 and 3315 cm^{-1}; the electronic spectrum of the complex in Nujol shows bands at 34800 and 41600 cm^{-1}. The complex behaves as a nonelectrolyte in ethanol. The physical properties are consistent with a pentacoordinate complex of D_{3h} symmetry with the tridentate neutral ligand binding to the metal via the two oxygen and the sulfur atoms. The complex is fairly stable in air. It picks up water only after being exposed to air for several hours [4].

References:

[1] Sigel, H.; Rheinberger, V. M.; Fischer, B. E. (Inorg. Chem. **18** [1979] 3334/9).
[2] Fogg, A. G.; Gray, A.; Thornburn Burns, D. (Anal. Chim. Acta **51** [1970] 265/70).
[3] Patel, R. D.; Patel, H. S.; Patel, S. R. (Eur. Polym. J. **23** [1987] 229/32).
[4] Sen, B.; Johnson, D. A. (J. Inorg. Nucl. Chem. **34** [1972] 609/18).

35.2.3 With Sulfanylacetic Acids or 2-Ethylsulfanylbenzoic Acid

Remark. Complexes with S-methylcysteine or methionine and related compounds have been described in "Manganese" D 4, 1985, pp. 291/5, together with complexes of other amino carboxylic acids containing sulfur. Formation constants of complexes with a Schiff base derived from the sulfur-containing amino carboxylic acid $H_2NC_6H_4{-}S{-}CH_2COOH$ and salicylaldehyde are reported in "Manganese" D 6, 1988, p. 35.

ligand 1 to 11 $R{-}S{-}CH_2COOH$

ligand	R	formula
1	C_2H_5	$C_4H_8O_2S$
2	C_3H_7	$C_5H_{10}O_2S$
3	$(CH_3)_2CH$	$C_5H_{10}O_2S$
4	$C_6H_5CH_2$	$C_9H_{10}O_2S$

ligand	R	formula
5	C_6H_5	$C_8H_8O_2S$
6	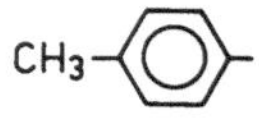	$C_9H_{10}O_2S$

ligand	R	formula
7	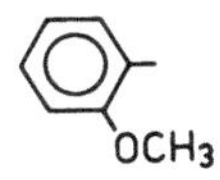	$C_9H_{10}O_3S$
8	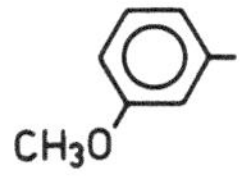	$C_9H_{10}O_3S$
9	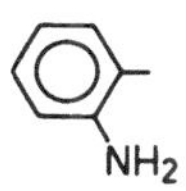	$C_8H_9NO_2S$
10	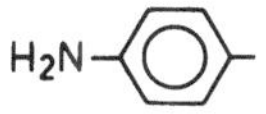	$C_8H_9NO_2S$
11	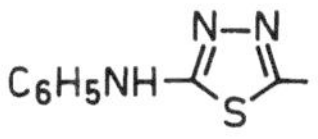	$C_{10}H_9N_3O_2S_2$

ligand 12 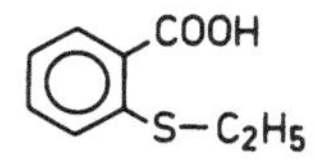 (= $C_9H_{10}O_2S$)

Complexes in Solution. The stability constants of $\mathbf{Mn^{II}L^+}$ complexes with HL = ligands 5, 7, 8, and 10 were determined potentiometrically in aqueous solution at ionic strength I = 0.1 M (KNO_3) and 25°C:

complex	$Mn(C_8H_7O_2S)^+$	$Mn(C_9H_9O_3S)^+$	$Mn(C_9H_9O_3S)^+$	$Mn(C_8H_8NO_2S)^+$
ligand	5	7	8	10
log K_1	0.72	0.51	0.59	3.27*)
Ref.	[1]	[1]	[1]	[2]

*) At I = 0.05 M under nitrogen.

Measurements carried out under nitrogen in 50% dioxane-water (v/v) at 25°C and I = 0.1 M ($NaClO_4$) reveal the following constants:

ligand	complex	log K_1	Ref.
1	$Mn(C_4H_7O_2S)^+$	1.85	[3]
2	$Mn(C_5H_9O_2S)^+$	1.83	[4]
3	$Mn(C_5H_9O_2S)^+$	1.85	[4]
4	$Mn(C_9H_9O_2S)^+$	1.83	[4]
5	$Mn(C_8H_7O_2S)^+$	1.70	[4]
6	$Mn(C_9H_9O_2S)^+$	1.79	[4]

The values are in the order expected from the basicities of the carboxylate groups, i.e., there is no, or only a weak, coordination of the sulfur atom. H NMR spectra of ligand 1, with increasing amounts of the paramagnetic ions Cu^{2+} and Mn^{2+}, confirm the results: with Mn^{2+} the spectra suggest that interaction with the sulfur atom is weak and the simple carboxylate complex probably dominates, whereas in the case of Cu^{2+} the sulfur is strongly coordinated [3, 4].

$\mathbf{Mn^{II}(C_8H_8NO_2S)_2}$ was prepared by the reaction of an aqueous solution of manganese(II) chloride with the potassium salt of ligand 9. The reaction mixture was filtered and the filtrate cooled overnight, whereupon the product separated. The magnetic moment of the solid is $\mu_{eff} = 5.65\ \mu_B$ at 20°C. The IR spectrum of the complex in hydrocarbon mull shows nearly the same vibration modes as the potassium salt of the ligand. Only a weak coordination of the ligand anion to Mn^{II} is suggested, whereas the ligand seems to be tridentate in the corresponding complexes of Ni^{II}, Co^{II}, and Cu^{II}. The compound is soluble in water [5].

[$Mn^{II}(C_{10}H_8N_3O_2S_2)_2(H_2O)_2$] and [$Mn^{II}(C_9H_9O_2S)_2(H_2O)_2$] complexes were prepared by refluxing stoichiometric amounts of manganese(II) choride and ligand 11 or 12, respectively, in ethanol for several hours [6, 7]. The brownish green [$Mn(C_9H_9O_2S)_2(H_2O)_2$], separating on cooling the solution, was washed with water, ethanol, ether, and finally air-dried at 80°C. Its magnetic moment at 308 K is 5.84 μ_B [7]. For [$Mn(C_{10}H_8N_3O_2S_2)(H_2O)_2$] a magnetic moment of 5.88 μ_B was found. The IR data indicate that the nitrogen and sulfur of the thiadiazole ring and the nitrogen of the NH group of the ligand are not taking part in coordination. The position of the C=O absorption band is also unchanged. However, a negative shift of ~50 cm^{-1} in the exocyclic C–S and the complete disappearance of the OH bands in the complex indicates that the exocyclic sulfur is involved in coordination together with the oxygen atom of the monodentate carboxylate anion [6]. Similar results were obtained from the IR spectrum of [$Mn(C_9H_9O_2S)_2(H_2O)_2$]. Coordination through oxygen and sulfur was supported by ν(Mn–O) and ν(Mn–S) bands in the ranges 625 to 620 and 340 to 325 cm^{-1}, respectively. Coordination of water molecules was evident by the appearance of a broad hump at 3450 cm^{-1} [7] and bands in the region 800 to 650 cm^{-1} [6]. Both complexes are nonelectrolytes. The electronic spectrum of [$Mn(C_{10}H_8N_3O_2S_2)_2(H_2O)_2$] shows two d-d bands, one at ~25000 cm^{-1}, another at 16205 cm^{-1}, and a charge transfer band at 36000 cm^{-1}. The d-d bands may be assigned to the $^6A_{1g} \rightarrow {}^4A_{1g}$(G), ^4E_g(G), and $^6A_{1g} \rightarrow {}^4T_{1g}$(G) transitions. On this basis a distorted octahedral geometry is proposed. The complex melts >250°C. The loss of water at relatively high temperatures (150 to 250°C) also indicates coordination of the water molecules [6].

[$Mn^{II}(C_9H_9O_2S)_2py_2$] and [$Mn^{II}(C_9H_9O_2S)_2(C_6H_7N)_2$]. The mixed-ligand complexes were obtained by refluxing the ethanolic solution of $MnCl_2$ and ligand 12 with pyridine or 2-methylpyridine (= C_6H_7N), respectively, in a mole ratio of 1:2:2 at pH 8 for 6 h. The precipitates were washed and dried as described above for the complex [$Mn(C_9H_9O_2S)_2(H_2O)_2$]. The magnetic moments of the solids are μ_{eff} = 5.84 and 5.80 μ_B, respectively. Coordination of nitrogen from pyridine or 2-methylpyridine is confirmed by the appearance of a band in the 280 to 230 cm^{-1} range. The electronic spectra of the mixed-ligand complexes exhibit three bands of low extinction coefficient in the 16500 to 17000, 26400 to 27150, and 28800 to 29000 cm^{-1} regions, due to the transitions $^6A_{1g} \rightarrow {}^4T_{1g}(G)(\nu_1)$, $\rightarrow {}^4T_{2g}(G)(\nu_2)$, $\rightarrow {}^4E_g(G)(\nu_3)$, respectively. The values of 10 Dq, B, and β (not given in detail) are within the range of values for octahedral Mn^{II} complexes [7].

References:

[1] Ford, G. J.; Gans, P.; Pettit, L. D.; Sherrington, C. (J. Chem. Soc. Dalton Trans. **1972** 1763/5).
[2] Dubey, K. P.; Puri, M. K. (Rev. Chim. Minerale **12** [1975] 255/8).
[3] Sigel, H.; Griesser, R.; Prijs, B.; McCormick, D. B.; Joiner, M. G. (Arch. Biochem. Biophys. **130** [1969] 514/20).
[4] Griesser, R.; Hayes, M. G.; McCormick, D. B.; Prijs, B.; Sigel, H. (Arch. Biochem. Biophys. **144** [1971] 628/35).
[5] Takahashi, Y.; Omura, T.; Nakatani, M.; Ouchi, A. (J. Inorg. Nucl. Chem. **35** [1973] 650/2).
[6] Srivastava, R. S.; Yadav, L. D. S.; Khare, R. K.; Srivastava, A. K. (Indian J. Chem. A **20** [1981] 516/8).
[7] Sharma, T.; Sharma, S.; Kumar, M. (J. Indian Chem. Soc. **64** [1987] 695/6).

35.2.4 With Sulfanediyl Dicarboxylic Acids

Remark. Stability constants of complexes with a derivative of iminodiacetic acid, CH_3SCH_2-$CH_2N(CH_2COOH)_2$, have been tabulated in "Manganese" D 5, 1987, p. 9, together with the stabilities of other iminodicarboxylate complexes.

ligand HOOCR–S–RCOOH	1	2	3	4
R	CH_2	C_2H_4	$CH(CH_3)$	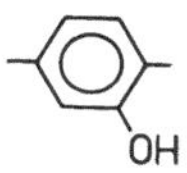
formula	$C_4H_6O_4S$	$C_6H_{10}O_4S$	$C_6H_{10}O_4S$	$C_{14}H_{10}O_6S$

Complexes in Solution. Stability constants of **Mn^{II}L** complexes with H_2L = ligands 1 to 3 have been determined pH-potentiometrically (mostly under nitrogen) in aqueous solution at 25°C and different ionic strengths:

ligand	complex	I in mol/L	log K_1	Ref.
1	$Mn(C_4H_4O_4S)$	0.1 ($NaClO_4$)	1.7	[1]
1		0.1 ($NaClO_4$)	1.72	[2]
1		0.1 ($NaClO_4$)	2.92	[3]
1		→0	2.83	[3]
2	$Mn(C_6H_8O_4S)$	0.1 ($NaClO_4$)	0.5	[4]
2		0.1 (KNO_3)	1.77	[5]
2		0.1 ($NaClO_4$)	2.72	[3]
2		0.05 (KNO_3)	3.30	[6]
2		→0	2.35	[3]
3	$Mn(C_6H_8O_4S)$	0.1 (KNO_3)	2.1	[5]

Additional data for ionic strengths 0.2 and 0.3 M and temperatures of 35 and 45°C are given for the complexes with ligands 1 and 2 in [3]. For the complex with ligand 4, $Mn(C_{14}H_6O_6S)^{2-}$, a stability constant of log K_1 = 7.3 has been determined potentiometrically in 60:40 (v/v) ethanol-water at 25°C, I = 0.1 M (KNO_3); log K_2 = 3.10 was obtained by calculation [7]. According to [8], the protonated MnH_2L complex, $Mn(C_{14}H_8O_6S)$, is formed on reaction of Mn^{2+} with ligand 4 in 50% aqueous ethanol. Formation of $Mn(C_{14}H_6O_6S)^{2-}$ was assumed in the reaction with the di- or tetrasodium salt of ligand 4. The additional formation of an $Mn_2^{II}L$ complex at pH > 2 is discussed. The formation constants, log K_1 = 7.3 and log K_2 (?) = 3.1, were determined at I = 0.1 M ($NaClO_4$) and 25°C [8]. Thermodynamic parameters have been determined for the 1:1 complexes with ligand 1, 2, or 4 in aqueous solution at I = 0.1 M ($NaClO_4$):

ligand	complex	t in °C	ΔG in kJ/mol	ΔH in kJ/mol	ΔS in J·mol^{-1}·K^{-1}	Ref.
1	$Mn(C_4H_4O_4S)$	25	−16.7	9.2	24.7	[3]
2	$Mn(C_6H_8O_4S)$	25	−15.5	−8.4	23.4	[3]
4	$Mn(C_{14}H_6O_6S)^{2-}$	35	−40.1	−78.9	−113.1	[9]

Conductometric titrations confirming the complex formation were carried out with ligand 4 in 50% aqueous ethanol [8].

[Mn^{II}($C_4H_4O_4S$)(H_2O)]. The complex was prepared by the reaction of equimolar amounts of hydrated manganese(II) perchlorate and the disodium salt of ligand 1 in aqueous ammonium acetate at pH 6. The pale pink crystals, slowly formed overnight, were washed with water, ethanol, diethyl ether, and air-dried. Interplanar spacings and relative intensities of X-ray powder diagram lines are given in the paper. The magnetic moment of the solid compound is 5.93 μ_B at 25°C. The IR spectrum of the complex shows the prominent bands: ν(OH) at 3000, ν_{as}(COO) at 1600, ν_s(COO) at 1396, and ν(CS) at 693 cm^{-1}. An octahedral polymeric structure, where Mn^{II} is coordinated by the dinegative tridentate ligand, bridging carboxylato groups, and a water molecule, was suggested. Coordination to the sulfur atom is supported by the fact that the ν(C–S) vibration mode is shifted by 33 or 35 cm^{-1} to lower wavenumbers, compared to free thiodiacetic acid and its disodium salt. The thermal analysis data indicate that the coordinated water is lost at 240°C, while the ligand leaves at about 300°C. A solubility product of 2.30 × 10^{-4} mol^2/L^2 was calculated using the stability constant at I = 0.1 M ($NaClO_4$) and 25°C, see p. 85 [2]. The solubility in 0.1 M $NaClO_4$ solution is 2.79 × 10^{-2} mol/L [10].

$Mn^{II}(C_{14}H_8O_6S)\cdot4H_2O$ was prepared by the reaction of an ethanol solution of ligand 4 with an equimolar amount of an Mn^{II} salt in water. The pH of the reaction mixture was raised to 4.2 by the addition of NaOH. Heating the reaction mixture on a steam bath gave a light pink precipitate. The product was washed with water, ethanol, acetone, and air-dried. The magnetic moment of the solid is 4.49 μ_B at 300 K. The IR spectrum of the complex in KBr shows the prominent bands: ν_{as}(COO) at 1550, ν_s(COO) at 1625, and ν(Mn–O) at 420 cm^{-1}; ν_{as}(COO) and ν_s(COO) vibration modes of the free ligand were observed at 1660 and 1600 cm^{-1}, respectively. The electronic absorption spectrum of the complex in a mull shows maxima at 26320, 31250, and 35700 cm^{-1}, which were assigned to the transitions $^6A_{1g}\rightarrow{}^4A_{1g}(G)$, $^4E_g(G)$, $\rightarrow{}^4E_g(D)$, $\rightarrow{}^4A_{2g}(F)$, respectively. A ligand field parameter B = 800 cm^{-1} was calculated from the data. The physical properties of the compound suggest that it has a polymeric structure with bridging carboxylato groups and metal-metal interactions in the solid state. The complex melts with decomposition at 250°C. It is insoluble in water and common organic solvents [11].

References:

[1] Suzuki, K.; Yamasaki, K. (J. Inorg. Nucl. Chem. **28** [1966] 473/80).
[2] Podlaha, J.; Podlahová, J. (Inorg. Chim. Acta **4** [1970] 521/5).
[3] Dubey, S. N.; Beweja, R. K.; Puri, D. M. (Indian J. Chem. A **22** [1983] 450/1).
[4] Suzuki, K.; Karaki, C.; Mori, S.; Yamasaki, K. (J. Inorg. Nucl. Chem. **30** [1968] 167/70).
[5] Laing, D. K.; Pettit, L. D. (J. Chem. Soc. Dalton Trans. **1975** 2297/301).
[6] Dubey, K. B.; Puri, M. K. (Rev. Chim. Minerale **12** [1975] 255/8).
[7] Kumar, A. N.; Srivastava, P. C.; Nigam, H. L. (Indian J. Chem. **9** [1971] 488/9).
[8] Capitán, F.; Salinas, F.; Martínez-Vidal, J. L.; Pino-Osuna, J. L. (Rev. Soc. Quim. Mex. **28** [1984] 89/92; C.A. **103** [1985] No. 93791).
[9] Srivastava, P. C. (Thermochim. Acta **55** [1982] 125/33).
[10] Podlaha, J.; Podlahová, J. (Inorg. Chim. Acta **4** [1970] 549/53).

[11] Srivastava, P. C.; Pandeya, K. B.; Nigam, H. L. (J. Inorg. Nucl. Chem. **35** [1973] 3613/7).

35.2.5 With Bis(sulfanylacetic Acids) $HOOCCH_2$–S–R–S–CH_2COOH

ligand	1	2	3	4	5	6
R	CH_2	C_2H_4	C_4H_8	C_2H_4–S–C_2H_4	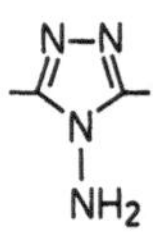	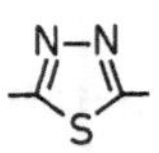
formula ...	$C_5H_8O_4S_2$	$C_6H_{10}O_4S_2$	$C_8H_{14}O_4S_2$	$C_8H_{14}O_4S_3$	$C_6H_6N_4O_4S_2$	$C_6H_6N_2O_4S_3$

Manganese(II) Complexes in Solution. Stability constants of the complexes with ligands 2 and 4 in aqueous solution have been determined pH-potentiometrically under nitrogen at 25°C and $I=0.1M$ ($NaClO_4$): log $K_1=1.04$ for $Mn(C_6H_8O_4S_2)$ [1], log $K_1=1.7$ for $Mn(C_8H_{12}O_4S_3)$ [2]; log $K_1=1.47$ at $I=0.2M$ ($NaClO_4$) for the $Mn(C_8H_{12}O_4S_3)$ complex [3]. Stability constants of the complexes with ligand 1, $Mn(C_5H_6O_4S_2)$ and $Mn(C_5H_6O_4S_2)_2^{2-}$, have been determined pH-potentiometrically in aqueous media at 20, 30, and 40°C and $I=0.1M$ ($NaClO_4$). Values of log K_1 and log K_2, and the overall changes of free energy, enthalpy, and entropy evaluated at 30°C are: log $K_1=2.99$, log $K_2=2.71$, $\Delta G=-7.92$ kcal/mol, $\Delta H=-17.36$ kcal/mol, $\Delta S=-31.27$ cal · $mol^{-1} \cdot K^{-1}$ [4]. The same formation constant value, log $K^{Mn}_{MnHL}=0.7$, was reported for both of the complexes with monoprotonated ligand 2, $Mn(C_6H_9O_4S_2)^+$ [1], or ligand 4, $Mn(C_8H_{13}O_4S_3)^+$ [2].

$Mn^{II}(C_6H_8O_4S_2)$ was prepared by the reaction of equimolar amounts of $MnSO_4$ and the sodium salt of ligand 2 in aqueous solution (0.2 M). The pale pink precipitate was washed with water, ethanol, and ether, and dried at room temperature. Interplanar distances and relative intensities of X-ray powder diagram lines are given in the paper. The magnetic moment of the solid is $\mu_{eff}=5.92\ \mu_B$ at 25°C. The IR spectrum of the complex in KBr shows prominent bands of ν_{as}(COO) at 1588, ν_s(COO) at 1395, and ν(CS) at 676 cm^{-1}, which are shifted to lower wavenumbers in comparison to the free ligand or its disodium salt. The thermal analysis indicates that the complex decomposes in the range 330 to 370°C. A polymeric structure is assumed for the solid compound, the carboxyl groups of the dinegative tetradentate ligand bridging the Mn^{II} atoms. A solubility product of 3.80×10^{-4} mol^2/L^2 in water was calculated by using the stability constant at 25°C and $I=0.1M$ ($NaClO_4$) cited above [1].

$[Mn^{II}(C_8H_{12}O_4S_2)(H_2O)_2]$ was prepared by the reaction of an aqueous solution of an Mn^{II} salt with an aqueous-ethanol (70%) solution of an equimolar amount of ligand 3. Addition of CH_3COONa to the reaction mixture gave a precipitate. The product was washed with water, ethanol, and dried at 80°C. The magnetic moment of the solid is 6.09 μ_B at ambient temperature. The IR spectrum of the complex in KBr disk shows the prominent bands (vibration modes of the free ligand in parentheses): ν(OH) at 3150 (3050 br), ν(COO) at 1565 (1695, 1710), ν_s(COO) at 1370 (1425), ν(CS) at 700 (730), $\delta(H_2O)$ at 670, and ν(Mn–S) at 370 cm^{-1}. The thermal analysis shows the loss of coordinated water at 140°C and the decomposition of the ligand at 170°C. The physical properties of the complex suggest that the structure consists of an octahedral complex with the tetradentate dinegative ligand coordinated to Mn^{II} via the two oxygen and two sulfur donors plus the coordinated water molecules. The complex is insoluble in water and common organic solvents [5].

$Mn^{II}(C_8H_{12}O_4S_3) \cdot 2H_2O$ was obtained from equimolar amounts of $MnSO_4$ and the disodium salt of ligand 4 in aqueous solution. The pale pink precipitate was washed with water, ethanol, diethyl ether, and air-dried at room temperature. Interplanar distances and relative intensities of the X-ray powder diagram lines are given in the paper. The magnetic moment of the solid compound is 5.97 μ_B at 25°C. Prominent IR bands (in cm^{-1}) of the complex in KBr and the free ligand (given in parentheses) are ν(OH) at 3440 (3100), ν_{as}(COO) at 1596 (1700, 1730), ν_s(COO) at 1410 (1406), and 705 (716). The thermal analysis indicates that the complex loses one water molecule at 70°C, the second one at 160°C, while decomposition of the ligand occurs at around 280°C. The physical properties suggest that the dinegative ligand is behaving in a tetradentate fashion with binding to the metal via two sulfur and two oxygen groups. A solubility product of 6.24×10^{-5} mol^2/L^2 at 25°C was calculated, using the stability constant at $I=0.1M$ ($NaClO_4$) cited above. The solubility in 0.1M $NaClO_4$ solution is 9.76×10^{-3} mol/L [2].

$[Mn^{II}(C_6H_4N_2O_4S_3)(H_2O)]$ and **$[Mn^{II}(C_6H_4N_4O_4S_2)(H_2O)_2] \cdot H_2O$** complexes with ligand 6 or 5, respectively, were obtained by refluxing hydrated manganese(II) acetate (0.01 mol), dissolved in 30 to 40 mL methanol, with a hot methanol solution of the corresponding ligand (0.01 mol) on a steam bath for about 15 min. The complexes either separated out immediately or on cooling the mixture to room temperature. The pinkish white compounds were washed with methanol and ether, and dried in air. The magnetic moments, $\mu_{eff} = 5.84$ and $6.08\ \mu_B$, are in the range of high-spin octahedral complexes. The IR spectra of the complexes are discussed. Negative shifts of $\nu_{as}(COO)$ and positive shifts of $\nu_s(COO)$, compared to the free ligand, indicate that both carboxylate oxygens are involved in bonding. It appears that one carboxylate oxygen is bonded to water present in the complexes, while the second one is bonded to the metal atom. Bonding of the NH_2 group (of ligand 5) and the ring sulfur atom (of ligand 6) is suggested, due to the observed negative shifts of the $\nu(NH)$ and $\nu(C–S)$ vibration modes. Coordination of the sulfanyl sulfur atoms is also suggested. The ligands are assumed to have a tetradentate behavior; ligand 5 is probably bridging two metal atoms [6].

References:

[1] Podlaha, J.; Podlahová, J. (Inorg. Chim. Acta **5** [1971] 413/9).
[2] Podlaha, J.; Podlahová, J. (Inorg. Chim. Acta **5** [1971] 420/4).
[3] Kotek, J.; Klierová, H.; Doležal, J. (J. Electroanal. Chem. **31** [1971] 451/62).
[4] Saxena, R. S.; Parikh, R. (J. Chem. Soc. Pak. **6** [1984] 207/10).
[5] Kaul, B. B.; Kapahi, A.; Pandeya, K. B. (J. Inorg. Nucl. Chem. **39** [1977] 1719/21).
[6] Keshari, B.N.; Mishra, L. K. (J. Indian Chem. Soc. **57** [1980] 279/81).

35.2.6 With Sulfanediyl Tri- or Tetracarboxylic Acids

Remark. Stability constants of complexes with sulfur-containing amine-N-polycarboxylic acids have been tabulated in "Manganese" D 5, 1987, pp. 22/3, together with those of other amine-N-polycarboxylic acids.

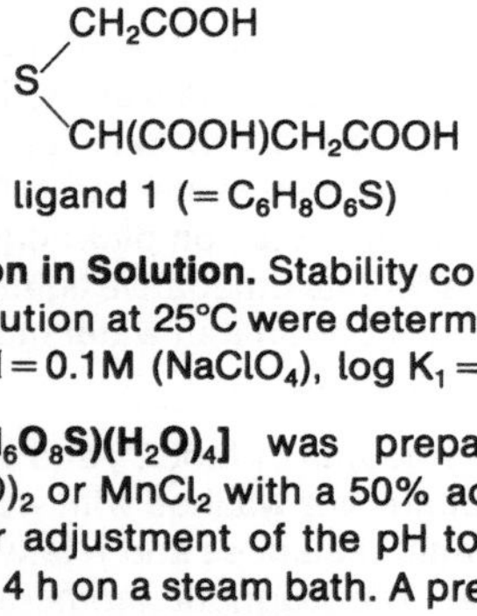

ligand 1 (= $C_6H_8O_6S$)

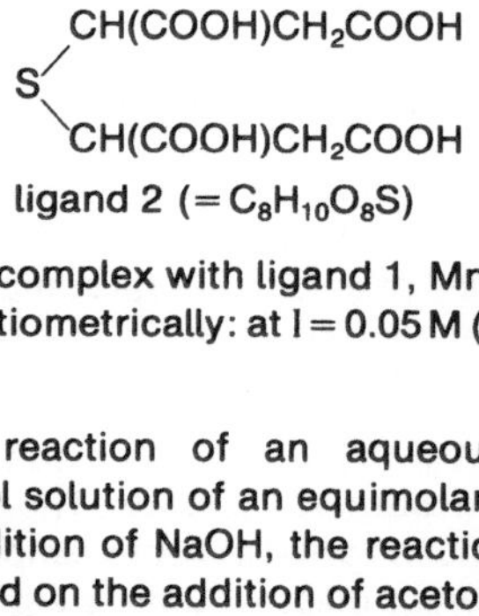

ligand 2 (= $C_8H_{10}O_8S$)

Formation in Solution. Stability constants of the complex with ligand 1, $Mn^{II}(C_6H_5O_6S)^-$, in aqueous solution at 25°C were determined pH-potentiometrically: at I = 0.05 M (KNO_3), log K_1 = 3.55 [1]; at I = 0.1 M ($NaClO_4$), log K_1 = 2.11 [2].

[$Mn^{II}(C_8H_6O_8S)(H_2O)_4$] was prepared by the reaction of an aqueous solution of $Mn(CH_3COO)_2$ or $MnCl_2$ with a 50% aqueous ethanol solution of an equimolar amount of the ligand. After adjustment of the pH to 5 to 6 by addition of NaOH, the reaction mixture was refluxed for 4 h on a steam bath. A precipitate formed on the addition of acetone. The grayish product was isolated, washed with ethanol, acetone, diethyl ether, and dried in vacuum. The magnetic moment of the solid is $5.48\ \mu_B$ at ambient temperature. The IR spectrum shows the prominent bands (vibration modes of the free ligand in parentheses): $\nu(H_2O)$ at 3350, $\nu_{as}(COO)$ at 1560 (1690), $\nu_s(COO)$ at 1375 (1455, 1300), and $\nu(CS)$ at 725, 600 (725, 660) cm^{-1}. The electronic spectrum of the solid in diffuse reflectance shows maxima at 17850, 20400, and 26380 cm^{-1}, assigned to the octahedral Mn^{II} transitions: $^6A_{1g}(F) \rightarrow {}^4T_{1g}(G), \rightarrow {}^4T_{2g}(G), \rightarrow {}^4A_{1g}(G)$ and a charge-transfer band at 40810 cm^{-1}. The ligand field parameters derived from the data are: Dq = 818 cm^{-1}, B = 743 cm^{-1}, $\beta = 0.774$. The physical properties suggest that the solid contains a polymeric structure with the ligand coordinated to the metal via bridging carboxylate groups. The thermal analysis shows that four H_2O molecules are lost from the complex before decomposition at 200°C. The complex is slightly soluble in water and insoluble in common organic solvents [3].

References:

[1] Dubey, K. P.; Puri, M. K. (Rev. Chim. Minerale **12** [1975] 2155/9).
[2] Casassas, E.; Arias-Leon, J. J.; Garcia-Montelongo, F. (J. Chim. Phys. **74** [1977] 324/8).
[3] Tiwari, S. K.; Rathore, D. P. S. (Natl. Acad. Sci. Letters [India] **2** [1979] 293/6).

35.2.7 With Tetrakis(sulfanylalkanoic Acids)

$$\text{HOOCH}_2\text{C–S} \qquad \text{S–CH}_2\text{COOH}$$
$$\diagdown\text{CH–CH}\diagup$$
$$\text{HOOCH}_2\text{C–S} \qquad \text{S–CH}_2\text{COOH}$$

$$\text{HOOC(CH}_3\text{)HC–S} \qquad \text{S–CH(CH}_3\text{)COOH}$$
$$\diagdown\text{CH(CH}_2)_3\text{CH}\diagup$$
$$\text{HOOC(CH}_3\text{)HC–S} \qquad \text{S–CH(CH}_3\text{)COOH}$$

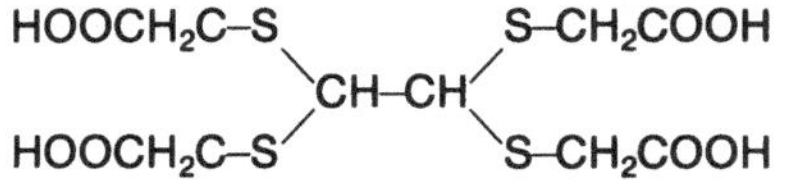

ligand 1 ($= C_{10}H_{14}O_8S_4$)   ligand 2 ($= C_{17}H_{28}O_8S_4$)

Manganese(II) Complexes in Solution. Stability constants of the 1:1 complexes with ligand 1 have been determined pH-potentiometrically at 25°C and $I = 0.1M$ (NaClO$_4$): log $K_1 = 2.32$ for Mn(C$_{10}$H$_{10}$O$_8$S$_4$)$^{2-}$, log $K^{Mn}_{MnHL} = 6.41$ for Mn(C$_{10}$H$_{11}$O$_8$S$_4$)$^-$, and log $K^{Mn}_{MnH_2L} = 9.48$ for Mn(C$_{10}$H$_{12}$O$_8$S$_4$). A formation constant for an Mn$_2$L complex could not be determined [1], whereas a value of log $K = 1.15$ was determined polarographically by [2] for the Mn$_2$(C$_{10}$H$_{10}$O$_8$S$_4$) chelate complex. The structure of the different species was discussed in [1].

Mn^{II}(C$_{10}$H$_{12}$O$_8$S$_4$)·6H$_2$O and **Mn$^{II}_2$(C$_{10}$H$_{10}$O$_8$S$_4$)·12H$_2$O.** The complex Mn(C$_{10}$H$_{12}$O$_8$S$_4$)·6H$_2$O was prepared under nitrogen by slow crystallization (1 d at 25°C) of an aqueous solution containing equivalent amounts of Mn(ClO$_4$)$_2$ and the disodium salt of ligand 1. The complex Mn$_2$(C$_{10}$H$_{10}$O$_8$S$_4$)·12H$_2$O was obtained by mixing the metal perchlorate with the tetrasodium salt of the ligand in the appropriate ratio. This mixture was stored for 3 d at 5°C. The crystals were dried at room temperature. For interplanar distances and relative intensities of X-ray powder diagram lines, see the paper. The magnetic moments at 20°C are $\mu_{eff} = 5.96$ and 5.87 μ_B, respectively. Prominent IR bands (in cm^{-1}) observed in the IR spectrum of the complexes in KBr are shown below together with those of the di- or tetrasodium salts of the ligand:

compound	ν(H$_2$O)	ν_{as}(COOH)	ν_s(COOH)	ν_{as}(COO)	ν_s(COO)	ν(CS)
Na$_2$H$_2$L·2H$_2$O	3430	1700	1253	1570	1392, 1387	670
Na$_4$L·2H$_2$O	3430	—	—	1610	1408	685
Mn^{II}(C$_{10}$H$_{12}$O$_8$S$_4$)·6H$_2$O	3440	1710	1252	1620, 1570	1405	695
Mn$^{II}_2$(C$_{10}$H$_{10}$O$_8$S$_4$)·12H$_2$O	3485	—	—	1598, 1588	1400	719

The electronic spectra of the solid complexes in diffuse reflectance show maxima at 32300 and 44400 cm^{-1} or 33300 and 44400 cm^{-1}, respectively. The absorption spectrum of the Mn(C$_{10}$H$_{10}$O$_8$S$_4$)$^{2-}$ ion in aqueous solution shows maxima at 33900 (log $\varepsilon = 1$) and 45500 cm^{-1} (log $\varepsilon = 3.46$). The first band is assigned to a ligand-to-metal charge transfer (S → Mn) and the second to an intraligand charge transfer. The physical properties suggest that the ligand is behaving in a tetradentate dinegative fashion, binding the manganese via two sulfur and two carboxylate oxygen atoms. In the case of Mn^{II}(C$_{10}$H$_{12}$O$_8$S$_4$)·6H$_2$O coordination around the metal probably is completed by two water molecules in *cis* positions. As shown by DTA, the complex loses three water molecules at 160°C. The second three molecules are lost at 185°C. Decomposition occurs at 320°C. The coordination sphere of the metal in Mn$_2$(C$_{10}$H$_{10}$O$_8$S$_4$)· 12H$_2$O is probably completed by intermolecular bridges, formed by the ligand through carbonyl oxygens. The thermal analysis shows that ten water molecules are lost at 60°C, while the last two water molecules are lost at 110°C; decomposition occurs at 240 to 280°C. The solubility of the complexes in water is 0.93 and 12.3 g/100 g H$_2$O, respectively [3].

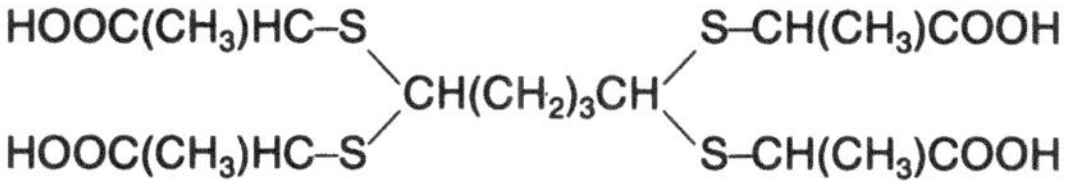

$[Mn^{II}_2(C_{17}H_{24}O_8S_4)(H_2O)_2]\cdot 2H_2O$ was prepared by addition of the stoichiometric amount of $MnCl_2\cdot 4H_2O$ dissolved in water to the solution of ligand 2 in ethanol. The mixture was adjusted to pH 6 with concentrated sodium hydroxide solution and refluxed over a steam bath for 30 min. The grayish precipitate was washed with 1:1 water-ethanol mixture, and with acetone, and dried in vacuum over anhydrous $CaCl_2$. The magnetic moment at 298 K is 5.26 μ_B. The IR spectra of the manganese complex and similar compounds of Ni^{II}, Co^{II}, and Cu^{II} reveal negative shifts of the $\nu_{as}(COO)$, $\nu_s(COO)$, and $\nu(CS)$ vibration modes, compared to the spectrum of the free ligand. Coordination of the ligand through the sulfur atoms and the oxygen atoms of the carboxylato groups was suggested. The strong band observed near 3400 cm^{-1} indicates the presence of water molecules in the complexes. The band observed at ~900 cm^{-1} was assigned to coordinated water molecules. The electronic spectrum of the manganese(II) complex consists of three main bands at 16400, 22220, and 25000 cm^{-1}, due to the transitions $^6A_{1g}\rightarrow{}^4T_{1g}$, $\rightarrow{}^4T_{2g}$, $\rightarrow{}^4A_{1g}$ in an octahedral Mn^{II} environment. A charge-transfer band was observed at 38460 cm^{-1}. The ligand field parameters Dq = 751.6 cm^{-1}, B = 683.3 cm^{-1}, E_p = 10248.5 cm^{-1}, and β = 0.795 have been calculated. The value of β indicates a low degree of covalency. The thermogravimetric analysis indicates that $[Mn_2(C_{17}H_{24}O_8S_4)(H_2O)_2]$ forms at 110°C, loss of the coordinated water molecules occurs at 180 to 200°C, and decomposition of the complex at 285°C. $[Mn_2(C_{17}H_{24}O_8S_4)(H_2O)_2]\cdot 2H_2O$ is insoluble in water and common organic solvents [4].

References:

[1] Petráš, P.; Podlahová, J.; Podlaha, J. (Collection Czech. Chem. Commun. **38** [1973] 3221/7).
[2] Kotek, J.; Klierowá, H.; Doležal, J. (J. Electroanal. Chem. **31** [1971] 451/62).
[3] Petráš, P.; Podlaha, J. (Inorg. Chim. Acta **6** [1972] 253/8).
[4] Tiwari, S. K.; Rathore, H. P. S. (J. Chem. Res. S **1985** 374; J. Chem. Res. M **1985** 3960/74).

35.2.8 With Disulfanediyl Compounds

For correlation purposes it should be pointed out that several complexes of manganese with ligands containing –S–S– bridges have already been described in the "Manganese" D series: complexes with homocystine in "Manganese" D 4, 1985, p. 295, with triazene derivatives in "Manganese" D 5, 1987, p. 319, with Schiff base ligands in "Manganese" D 6, 1988, pp. 137/9 and 192.

The R–S–S–R ligands presented in this section are the following:

ligand	1	2	3	4	5
R	CH_2COOH	C_2H_4COOH	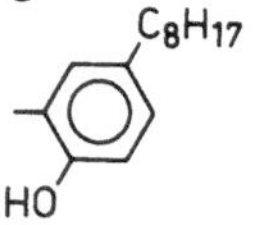	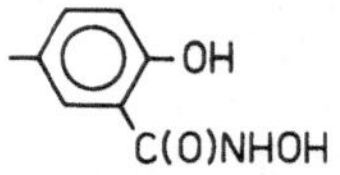	$C(O)C_6H_5$
formula	$C_4H_6O_4S_2$ (= H_2L)	$C_6H_{10}O_4S_2$ (= H_2L)	$C_{28}H_{42}O_2S_2$	$C_{12}H_{12}N_2O_6S_2$ (= H_4L)	$C_{14}H_{10}O_2S_2$

Complexes in Solution. The stability constant in aqueous solution, log K_1 = 1.7, of the chelate complex with ligand 1, $Mn^{II}(C_4H_4O_4S_2)$, has been determined pH-potentiometrically at 25°C and I = 0.1 M ($NaClO_4$) [1]. The formation of the complexes with ligand 2, $Mn^{II}(C_6H_8O_4S_2)$ and $Mn^{II}(C_6H_8O_4S_2)_2^{2-}$, in aqueous solution in the range of pH 3.3 to 5.3 has been demonstrated by potentiometric and conductometric measurements. The stability constants were determined pH-potentiometrically at 20, 30, and 40°C, and I = 0.1 M ($NaClO_4$). Values of log K_1 = 3.24

and log $K_2 = 3.00$ and the overall changes $\Delta G = -8.64(\pm 0.07)$ kcal/mol, $\Delta H = -14.62$ (± 0.20) kcal/mol, and $\Delta S = -19.73(\pm 0.28)$ $cal \cdot mol^{-1} \cdot K^{-1}$ have been evaluated at 30°C [2].

Formation constants of complexes with ligand 4, log $K_1 = 7.4$ for $Mn^{II}(C_{12}H_8N_2O_6S_2)^{2-}$ and log $K^{Mn}_{MnH_2L} = 3.8$ for $Mn^{II}(C_{12}H_{10}N_2O_6S_2)$, were determined pH-potentiometrically at 20 ± 0.05°C and I = 0.1 M (KNO_3). Plots of the conditional stability constants against pH reveal maxima at pH ~11 and ~9.4, respectively [3]. Formation of an $Mn^{II}_2(C_{12}H_8N_2O_6S_2)$ complex at mole ratio Mn : ligand = 2 : 1 was shown by conductometric and potentiometric measurements. The possible structure of the different species in solution was discussed [4]. A brown precipitate, formed on reaction of ligand 4 with Mn^{2+} ions in aqueous solution, was found to be extractable (between pH 7 to 9) into a benzene solution of trioctylmethylammonium chloride [5].

$Mn^{II}(C_{28}H_{41}O_2S_2)_2$ was obtained from stoichiometric amounts of $MnCl_2$ and ligand 3 in alcohol or aqueous alcohol at pH 7 to 8 [6, 7]. The antiwear efficiencies of the compound and of similar metal complexes in an ester lubricant was found to decrease in the order: Cu > Ni > Co > Mn > Fe > Ba. The length of the alkyl substituent has little effect on the additive performance [7]. The ability of the complexes, as lubricating oil additives, to inhibit the oxidation of esters of pentaerythritol and C_5–C_9 monocarboxylic acids has been investigated [6].

$[Mn^{II}(C_{14}H_{10}O_2S_2)Cl_2]$ was prepared by refluxing equimolar quantities of anhydrous $MnCl_2$ with ligand 5 in dry benzene. The white complex was recovered by adding petroleum ether to the solution. It melts at 133°C. A magnetic moment of 5.94 μ_B was found at room temperature. The IR spectrum of the complex in Nujol reveals ν(CO) bands at 1610 and 1580 cm^{-1}, which have been shifted to lower wavenumbers, compared to the free ligand bands at 1750 and 1670 cm^{-1}. A ν(Mn–O) vibration mode was observed at 410 cm^{-1}. A tetrahedral stereochemistry is suggested, in which Mn^{II} is coordinated by the carbonyl oxygen atoms. The complex is soluble in common organic solvents. In nitrobenzene it behaves as a nonelectrolyte. The molecular weight of 394.7, determined in benzene, reveals a monomeric species. As shown by TGA the complex decomposes between 140 and 150°C with formation of benzoyl chloride and manganese(II) sulfide [8].

References:

[1] Suzuki, K.; Karaki, C.; Mori, S.; Yamasaki, K. (J. Inorg. Nucl. Chem. **30** [1968] 167/70).

[2] Saxena, R. S.; Gupta, A. (J. Indian Chem. Soc. **61** [1984] 210/2).

[3] Martinez-Vidal, J. L.; Gonzales-Parra, J.; Salinas, F. (Bull. Soc. Chim. Belges **94** [1985] 155/6).

[4] Martinez-Vidal, J. L.; Gonzales-Parra, J.; Salinas, F. (Anales Quim. B **81** [1985] 338/41).

[5] Salinas, F.; Martinez-Vidal, J. L.; Gonzales-Parra, J. (Proc. Indian Acad. Sci. Chem. Sci. **95** [1985] 265/74).

[6] Kovtun, G. A.; Zhukovskaya, G. B.; Berenblyum, A. S.; Moiseev, I. I. (Neftekhimiya **22** [1982] 501/3; C.A. **97** [1982] No. 219212).

[7] Kovtun, G. A.; Zhukovskaya, G. B. (Neftepererab. Neftekhim. [Moscow] **1986** No. 7, pp. 8/9; C.A. **105** [1986] No. 175514).

[8] Puri, J. K.; Miller, J. M. (Inorg. Chim. Acta **97** [1985] 179/82).

35.2.9 With Polysulfides

Remark. The chemistry of metal-polysulfido complexes, containing one or more polysulfide anions S_n^{2-} with $n \geqq 2$, has become of ever-increasing interest in recent years; see for example [1]. The coordinated S_n^{2-} ligands exhibit an especially rich structural chemistry and an unusual and characteristic reactivity. For this reason the polysulfide complexes of manganese are treated in this volume. The sulfides MnS_2 and MnS_3 have been described in "Mangan" C 6, 1976, pp. 35/9.

$[P(C_6H_5)_4]_2[(S_5)Mn^{II}(S_6)]$. The complex was prepared by oxidation of $[P(C_6H_5)_4]_2[Mn(C_6H_5S)_4]$, see p. 25, with dibenzyltrisulfane. (In polar media the RS^--catalyzed dissociation of aliphatic trisulfides results in the generation of S_n° fragments (n = 2 to 6), that in the presence of suitable reducing agents form the anionic polysulfide ligands. The $C_6H_5S^-$ ligands of the $[Mn(C_6H_5S)_4]^{2-}$ complex are readily oxidized by the S_n° fragments or by elemental sulfur.) To a solution of 1 g (0.95 mmol) of $[P(C_6H_5)_4]_2[Mn(C_6H_5S)_4]$ in 30 mL acetonitrile was added 2.6 g (9.3 mmol) of $(C_6H_5CH_2)_2S_3$ under anaerobic conditions. The color of the solution immediately changed from yellow to red. After stirring for 10 min, diethyl ether was added until nucleation was apparent. A yellow precipitate (which did not contain manganese) was separated after 10 min, and more ether was added to the filtered solution. The orange-red crystals formed were isolated after 2 to 3 days. Single crystals were obtained by slow diffusion of diethyl ether into the acetonitrile solution of the complex [2].

The X-ray structure determination showed the crystals to be monoclinic with the lattice parameters: a = 23.266(5), b = 20.390(3), c = 23.894(5) Å, α = 90.00°, β = 118.03(1)°, γ = 90.00°; space group $P2_1/n-C_{2h}^5$ (No. 14); Z = 8. The structure was refined to R = 0.121 (C_6H_5 as rigid molecules); atom coordinates deposited by the use of 6495 reflections (>3σ). **Fig. 11** shows the two independent anions in the asymmetric unit which also contains the four $[P(C_6H_5)_4]^+$ cations. The MnS_5 unit displays the chair conformation, the MnS_6 unit the crown conformation. Mn^{II} is tetrahedrally coordinated. A crystallographic disorder was found and was interpreted as the consequence of the site occupations according to the composition $[(MnS_{11})_{0.85}(MnS_{10})_{0.15}]$. Intramolecular bond distances and angles in the $[(S_5)Mn(S_6)]^{2-}$ complex are given in the paper. The Mn(1)–S distances range from 2.416(7) to 2.451(7) Å, the Mn(2)–S distances from 2.393(6) to 2.446(6). The calculated density D = 1.40 g/cm³ corresponds to the value D = 1.42(2), determined by flotation in a CCl_4-C_5H_{12} mixture. The magnetic moment of the complex is $\mu_{eff} = 5.65\ \mu_B$. The electronic spectrum of the complex in dimethylformamide solution shows a band at 610 nm characteristic of MS_n^{2-} compounds (M = Zn^{II}, Cd^{II}, Mn^{II}; n ≧ 5). The transition is very likely due to the $\cdot S_3^-$ radical anion obtained by ligand dissociation. In acetonitrile solution this transition does not appear, but intraligand and ligand → Mn transitions do occur between 410 and 390 nm. The complex reacts readily with CS_2 to give the corresponding complex with carbonoperthioic acid, $[P(C_6H_5)_4]_2[Mn^{II}(CS_4)(S_6)]$, described on p. 210 [2].

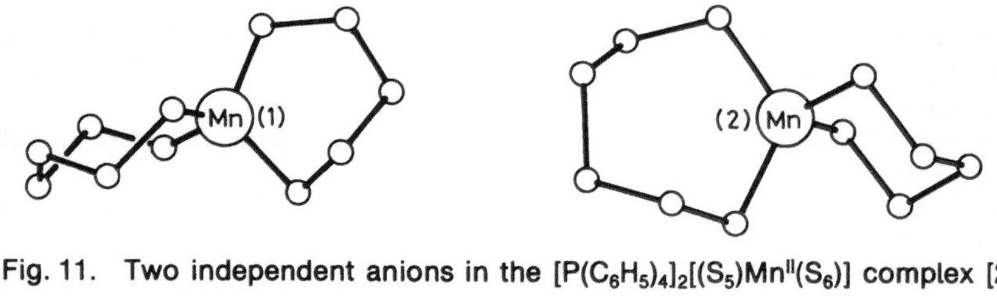

Fig. 11. Two independent anions in the $[P(C_6H_5)_4]_2[(S_5)Mn^{II}(S_6)]$ complex [2].

$[Mn^{II}(C_9H_{21}N_3)(S_4)(H_2O)]$. The pale yellow complex with 1,4,7-trimethyl-1,4,7-triazacyclononane (= $C_9H_{21}N_3$) was obtained on reaction of the dinuclear manganese(III) complex $[(C_9H_{21}N_3)_2Mn_2(\mu\text{-}O)(\mu\text{-}CH_3COO)_2]^{2+}$ with sulfide anions: A solution of 1 g $Mn(CH_3COO)_3 \cdot 2H_2O$ and 1 g of the triazacyclononane derivative in 50 mL of methanol was stirred at room temperature for 30 min under argon, then 2 mL of a 40% aqueous solution of $(NH_4)_2S$ was added. The red-brown solution immediately became pale yellow. The complex precipitated after reducing the volume of the solution to 15 mL (under reduced pressure) and standing for 24 h at 0°C. Crystals for X-ray measurements were grown from ethanol solution at room temperature. They are orthorhombic with a=8.321(2), b=15.128(4), c=26.37(1) Å, space group Pbca–D_{2h}^{15} (No. 61); Z=8. Refinements were made to R=0.058 and R_w=0.041, using anisotropic thermal parameters for all nonhydrogen atoms. The manganese(II) ion is surrounded by three nitrogen atoms of the triazacyclononane macrocycle, two sulfur atoms of a bidentate S_4 ligand, and one oxygen atom of a coordinated water molecule. The distorted octahedral geometry is shown in **Fig. 12a**. The whole molecule with selected distances is shown in **Fig. 12b**. The five-membered MnS_4 ring is strongly puckered, it adopts either a λ or δ conformation; the three five-membered chelate rings formed by coordination of the cyclic amine also adopt either (λλλ) or (δδδ) conformations. The Mn–S bonds are only slightly smaller than the sum of the ionic radii, indicating little covalent character of the Mn–S bond. The calculated density of the complex is D=1.491 g/cm³. The magnetic moment of the pale yellow complex, μ_{eff}=5.82(4) μ_B, is independent of temperature between 300 and 100 K [3].

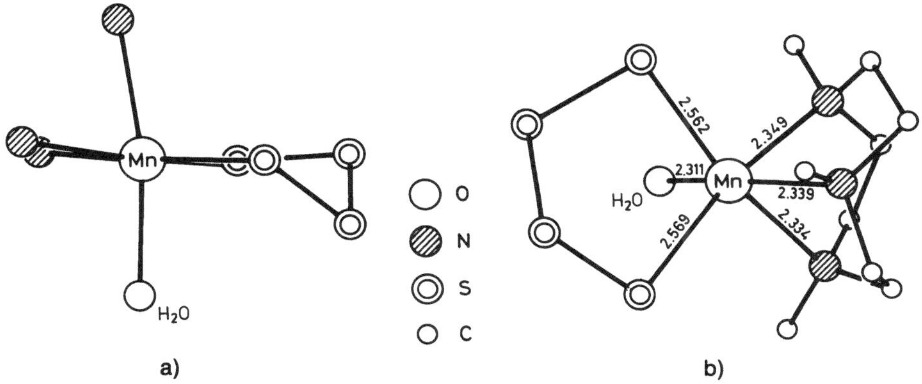

Fig. 12. a) View of the $MnON_3S_4$ core and b) molecular structure of $[Mn(C_9H_{21}N_3)(S_4)(H_2O)]$ with selected distances in Å [3].

References:

[1] Müller, A.; Diemann, E. (Advan. Inorg. Chem. Radiochem. **31** [1987] 89/122).
[2] Coucouvanis, D.; Patil, P. R.; Kanatzidis, M. G.; Detering, B.; Baenziger, N. C. (Inorg. Chem. **24** [1985] 24/39).
[3] Wieghardt, K.; Bossek, U.; Nuber, B.; Weiss, J. (Inorg. Chim. Acta **126** [1987] 39/43).

35.3 Complexes with Sulfoxides

General Reference:

Davies, J. A., The Coordination Chemistry of Sulfoxides with Transition Metals in: Advan. Inorg. Chem. Radiochem. **24** [1981] 115/87, 160.

General. Interest in the coordination chemistry of sulfoxide complexes has developed as a result of the excellent solvent action of the sulfoxides and their use in the solvent extraction of metals during refining processes. Manganese complexes containing coordinated sulfoxides, R_2SO, have been described in great detail especially those of the compositions $[Mn^{II}(R_2SO)_6]X_2$

and $[Mn^{II}(R_2SO)_4X_2]$. Complexes of the type MnL_3X_2 or MnL_2X_2 are formed with bidentate sulfoxides RS(O)RS(O)R.

For their preparation it is often sufficient to mix a hydrated manganese salt with the sulfoxide but in many cases an appropriate solvent (e.g., methanol or acetone) is used as medium for the reaction. Infrared studies and magnetic susceptibility measurements show that most of the complexes are high-spin with O-bonded sulfoxide ligands. Coordination to the metal generally leads to a decrease in the frequency of the band assigned as ν(SO). The existence of $[Mn(C_2H_6OS)_6]^{2+}$ ions in dimethyl sulfoxide solutions of $Mn(ClO_4)_2$ was shown by various physical data (see p. 95) but no stability constant has been reported.

Only one manganese(III) compound, a complex with dimethyl sulfoxide, $[Mn(C_2H_6OS)_6]$-$(ClO_4)_3$, is known. For preparation and properties, see p. 99.

35.3.1 With Dimethyl Sulfoxide $(CH_3)_2SO$ (= C_2H_6OS = DMSO)

35.3.1.1 $[Mn^{II}(C_2H_6OS)_6]X_2$ and $[Mn^{II}(C_2H_6OS)_6]MnX_4$ Compounds

$[Mn(C_2H_6OS)_6](NO_3)_2$. The complex was prepared by the reaction of hydrated manganese(II) nitrate with neat dimethyl sulfoxide. The solution was heated under reduced pressure to drive the water off, then cooled to room temperature for crystallization [1, 2]. Successive recrystallization from DMSO gave pale golden crystals of the pure product at 25°C [3]. Bands in the IR spectrum of the complex at 1357 and 835 cm^{-1} for ionic NO_3^- reveal the presence of noncoordinated nitrate ions. The diffuse reflectance spectrum shows the ionic NO_3^- $n \rightarrow \pi^*$ band at 315 nm. The electronic absorption spectrum of the complex in DMSO solution shows the free nitrate $n \rightarrow \pi^*$ band at 312.5 nm ($\varepsilon = 5.5$) [3]. On heating, $Mn(NO_3)_2 \cdot 3C_2H_6OS$ is formed at 50°C [2].

$[Mn(C_2H_6OS)_6][Mn(NO_3)_4]$. The pale yellow complex, also formulated as $Mn(NO_3)_2 \cdot 3C_2H_6OS$ [5], was prepared by crystallizing $[Mn(C_2H_6OS)_6](NO_3)_2$ from DMSO at 50°C [3]. Bands in the IR spectrum at 1475, 1280, 1030, 820, and 811 cm^{-1} indicate that coordinated NO_3^- is present in the complex. It seems therefore that the trisolvate is more correctly formulated as containing the tetranitrato anion [4]. The diffuse reflectance spectrum of the solid shows the nitrate $n \rightarrow \pi^*$ band as an inflection at 320 nm [3]. Heating of the complex under TGA in vacuum yields free DMSO and the intermediate $Mn(NO_3)_2 \cdot C_2H_6OS$ [5]. At temperatures >156°C decomposition was observed [2]. Above 200°C the DMSO is oxidized by the nitrate ion [5].

$[Mn(C_2H_6OS)_6][MnCl_4]$. The tan to pale yellow complex, also formulated as $MnCl_2 \cdot 3C_2H_6OS$, was prepared by the addition of anhydrous $MnCl_2$ to excess DMSO so as to form a saturated solution at 60 to 70°C. After addition of an equal volume of ethanol and standing at room temperature for several hours, crystals formed. The product was isolated, washed with 1:4 ethanol-diethyl ether in the absence of moisture. The complex decomposed immediately in water [2, 6]. The complex can also be synthesized by the reaction of hydrated $MnCl_2$ with neat DMSO [1, 7]. Subsequent removal of the excess DMSO in vacuum gave the product [7]. Alternatively, the reaction mixture can be heated to drive off water and then cooled to yield crystals which were washed with acetone and dried in vacuum over silica gel [1]. TGA of the solid in vacuum shows the formation of two intermediates upon heating the complex: $MnCl_2 \cdot 2C_2H_6OS$ and $MnCl_2 \cdot 0.5C_2H_6OS$ [8].

Magnetic measurements (temperature not given) reveal a magnetic moment $\mu_{eff} = 6.02\ \mu_B$ [32]. The IR spectrum of the complex in KBr shows a band of the ν(SO) vibration mode at 950 cm^{-1} which is at lower wavenumber than for the free ligand [9]. The far-IR spectrum of the complex in Nujol mull shows bands at 417, 388, 345, 317, and 295 cm^{-1}. The band at 417 cm^{-1}

was assigned to the ν(Mn–O) mode while the band at 295 cm^{-1} was assigned to a mode of $MnCl_4^{2-}$ [10]. A band at 1004 cm^{-1} in the nitromethane solution spectrum of the complex has also been assigned to ν(SO) [6, 11]. The ESCA spectrum of the solid complex shows lines at 165.1 and 530.9 eV which are assigned to $S(2p_{3/2})$ and O(1s) ionizations of the DMSO. The spectrum is consistent with an oxygen-bonded sulfoxide ligand [12]. The electronic absorption spectrum of the complex in nitromethane shows absorption maxima at 442 nm (ε = 3.10), 428 nm (ε = 3.91), 371 nm (ε = 12.35) [6].

$[Mn(C_2H_6OS)_6](ClO_4)_2$. The pale pink complex was prepared by the reaction of hydrated manganese(II) perchlorate with excess DMSO either neat [7] or in acetone solution [13]. Better crystals are obtained if the salt is present in slight excess. The complex was recrystallized from acetone containing a small amount of diethyl ether to reduce its solubility [13]. The crystals melt at 210 to 212°C [13] and explode under thermogravimetric analysis at 200 to 235°C [5]. The complex was also isolated from a neat DMSO solution of hydrated manganese(II) perchlorate similar to $[Mn(C_2H_6OS)_6](NO_3)_2$. The crystals were washed with acetone and dried in vacuum over silica gel [1, 2].

The IR spectrum of the complex in KBr disks, Nujol mulls, or nitromethane solution indicates that the ligand is bonded to the metal via the oxygen donor. The ν(SO) mode of the complex (at 1001 cm^{-1} [13] or at 955 cm^{-1} [9, 14]) is lower in comparison to the free (uncomplexed) ligand. The infrared spectra of $[M(C_2H_6OS)_6](ClO_4)_2$ and $[M(C_2H_6OS)_6](ClO_4)_3$ complexes with M = Zn, Mn, Ni, Co or Ga, Sc, Fe, respectively, were discussed in [33]. Assignments of ligand and metal-ligand vibrations were based on the band shifts induced by ^{18}O-labeling, by deuteration of the methyl groups and by the effects of metal ion substitution. Bands of $[Mn(C_2H_6OS)_6]^{2+}$ (in cm^{-1}) were assigned as follows: 998, 945, 940 to $\varrho(CH_3)$ + ν(SO), 716, 679 to ν(C–S), 623 to $\nu(ClO_4)$, 418 to ν(Mn–O), 325 to δ(CSO), 317 to ν(CSO), 218, 170 to δ(O–Mn–O) + δ(O–Mn–S). Similar band positions for ν(CS) were reported in [13, 14], of ν(Mn–O) in [15, 16]. A simple Mn–O=S< bent vibrational model has been used to estimate the force constants, F_{Mn-O} = 1.45 mdyn/Å for ν(M–O) 418 cm^{-1} and F_{SO} = 6.79 mdyn/Å for ν(SO) 955 cm^{-1} [17].

The electronic spectrum of the complex in DMSO shows maxima at 19600, 23584, 25906, 25188, 27932, 29154, and 31400 cm^{-1}. The bands were assigned to the transitions $^6A_{1g} \rightarrow {}^4T_{1g}$, $\rightarrow {}^4T_{2g}$, $\rightarrow {}^4E_g$, $\rightarrow {}^4A_{1g} \rightarrow {}^4T_{2g}$, $\rightarrow {}^4E_g$, $\rightarrow {}^4T_{1g}$. The ligand field parameters derived from the analysis of the spectrum are Dq = 710 cm^{-1}, B = 780 cm^{-1} [18]. The electronic spectrum has also been interpreted with the aid of MCD measurements. The spin-forbidden transitions from the ground state $^6A_{1g}$ to the following levels are assigned to the observed bands: $\rightarrow {}^4T_{1g}(G)$ 512 nm, $\rightarrow {}^4T_{2g}(G)$ 425 nm, $\rightarrow {}^4T_{2g}(D)$ 358 nm, $\rightarrow {}^4T_{1g}(P)$ 314 nm, and from ground state $^2I \rightarrow {}^2T_{2g}$ 382 nm, $\rightarrow {}^2T_{1g}$ 335 nm. The ligand field splitting 10 Dq = 7.100 cm^{-1}, is found to be lower for the $[Mn(C_2H_6OS)_6]^{2+}$ complex than that for $[Mn(H_2O)_6]^{2+}$ with 10 Dq = 7800 cm^{-1} [19].

The ESR spectrum of the complex shows a normal six-line pattern at room temperature [20]. Variation of the line width with temperature is interpreted in terms of a model in which DMSO fluctuations about the manganese(II) ion modulate the ligand field and relax the electron spin through spin-orbit interaction [21]. The same conclusion is drawn from consideration of dielectric correlation times [22]. NMR line broadening and contact shift measurements indicate that DMSO ligand exchange is rapid between 21 and 80°C and that downfield contact shifts of the DMSO proton resonances suggest a ligand π to metal t_{2g} unpaired spin density transfer. The value of the electron spin-nuclear spin coupling constant, calculated from the contact shift, is $A_n = (2.0 \pm 0.5) \times 10^5$ Hz [23]. The proton NMR relaxation study of the DMSO solutions of $Mn(ClO_4)_2$ between 298 and 413 K over a wide frequency range indicates six-coordinate solvation for the metal ion. At 25°C the ligand exchange rate constant is 10^8 s^{-1} and the activation energy is 14 ± 2 kJ/mol [24]. The measurements of the concentration dependence of the molar conductance of $Mn(ClO_4)_2$ in DMSO and DMSO-benzene mixtures at 25°C yield first-step association constants of the cation and anion, log K = 1.28 and 1.86 for the two solvent systems, respectively [25].

The paper chromatographic separation of $[Mn(C_2H_6OS)_6]^{2+}$ is achieved by ascending on Whatman No. 1 with ethyl alcohol ($R_f = 0.22$), n-butyl alcohol ($R_f = 0.03$), water ($R_f = 0.47$), dimethylformamide ($R_f = 0.55$), and a 2:1:1 water-acetone-ethyl alcohol mixture ($R_f = 0.40$) [30]. The thin-layer chromatographic separation of $[Mn(C_2H_6OS)_6]^{2+}$ on Silicagel-G layers reveals $R_f = 0.25$, 0.30, 0.15, or 0.22 if methyl, ethyl, propyl, or butyl alcohol, respectively, was used as solvent. Additional R_f values are given for water (0.50), acetone (0.20), the 2:1:1 water-acetone-ethyl alcohol mixture (0.27), and dimethylformamide [31].

$[Mn(C_2H_6OS)_6]Br_2$ and **$[Mn(C_2H_6OS)_6]I_2$**. The faint brown bromide and the rose-colored iodide were prepared by reaction of the hydrated manganese(II) salts and neat DMSO with subsequent heating to drive off the water [1, 2]. After cooling the reaction mixture to room temperature crystals form. The products were washed with acetone and dried in vacuum over silica gel. The complex loses dimethyl sulfoxide upon TGA in vacuum with the formation of the intermediates $[Mn(C_2H_6OS)_6][MnX_4]$ (X = Br or I) and $MnBr_2 \cdot 2C_2H_6OS$ [8].

$[Mn(C_2H_6OS)_6](HSO_4)_2$ was prepared by reaction of MnS with the mixed nonaqueous system DMSO–SO_2 and characterized by elemental analysis and by IR spectroscopy [34].

$[Mn(C_2H_6OS)_6]S_2O_7$ was prepared by the oxidation of manganese metal with SO_2 in DMSO at ambient temperature. Addition of diethyl ether to the reaction mixture yields the product. Heating of the solid at 120°C liberates the coordinated DMSO while at 400°C $MnSO_4$ is formed. The IR spectrum was investigated [26]. For preparation see also [35].

$[Mn(C_2H_6OS)_6](BF_4)_2$. The pink-flesh-colored complex was prepared by reaction of $[Mn(H_2O)_6](BF_4)_2$ with excess DMSO in acetone. Removal of the solvent and excess ligand in vacuum yielded crystals which were washed with benzene and dried over P_4O_{10} [27]. An alternative preparation uses the electrochemical oxidation of manganese metal in the presence of 40% aqueous HBF_4 and DMSO (6 V, 50 mA, 7 h) at ambient temperature [28].

$[Mn(C_2H_6OS)_6][C(CN)_3]_2$ was prepared by addition of CCl_4 or benzene to a solution of anhydrous $Mn[C(CN)_3]_2$ in DMSO at ambient temperature. The light brown precipitate was washed with CCl_4 and dried in vacuum. The solid melts at 72 to 73°C. The IR spectrum of the complex shows the anion band, ν(CN), at 2175 cm^{-1}. A molar conductivity of 78.32 $cm^2 \cdot \Omega^{-1} \cdot mol^{-1}$ was found for a 2.57×10^{-4} M solution in DMSO [29].

References:

[1] Schläfer, H. L.; Schaffernicht, W. (Angew. Chem. **72** [1960] 618/26).
[2] Schläfer, H. L.; Opitz, H. P. (Z. Anorg. Allgem. Chem. **313** [1961] 178/86).
[3] Addison, C. C.; Sutton, D. (J. Chem. Soc. A **1966** 1524/8).
[4] Moniz, W. B.; Poranski, C. F.; Venezky, D. L. (U.S. Clearinghouse Fed. Sci. Tech. Inform. AD 663552 [1967] 1/43; C.A. **69** [1968] No. 15455).
[5] Glavaš, M.; Škerlak, T. (Glasnik Hem. Tehnol. Bosne Hercegovine **15** [1967] 41/6; C.A. **69** [1968] No. 64250).
[6] Meek, D.; Straub, D. K.; Drago, R. S. (J. Am. Chem. Soc. **82** [1960] 6013/6).
[7] Cotton, F. A.; Francis, R. (J. Am. Chem. Soc. **82** [1960] 2986/91).
[8] Glavaš, M.; Škerlak, T. (Glasnik Hem. Tehnol. Bosne Hercegovine **15** [1967] 31/40; C.A. **69** [1968] No. 48796).
[9] Cotton, F. A.; Francis, R., Horrocks, W. D., Jr. (J. Phys. Chem. **64** [1960] 1534/6).
[10] Johnson, B. F. G.; Walton, R. A. (Spectrochim. Acta **22** [1966] 1853/8).

[11] Drago, R. S.; Meek, D. (J. Phys. Chem. **65** [1961] 1446/7).
[12] Su, Chang-Cheng; Faller, J. W. (Inorg. Chem. **13** [1974] 1734/6).

[13] Selbin, J.; Bull, W. E.; Holmes, L. H., Jr. (J. Inorg. Nucl. Chem. **16** [1961] 219/24).
[14] Currier, W. F.; Weber, J. H. (Inorg. Chem. **6** [1967] 1539/43).
[15] Berney, C. V.; Weber, J. H. (Inorg. Chem. **7** [1968] 283/7).
[16] Adams, D. M.; Trumble, W. R. (Inorg. Chem. **15** [1976] 1968/73).
[17] James, B. R.; Morris, R. H. (Spectrochim. Acta A **34** [1978] 577/82).
[18] Schläfer, H. L.; Opitz, H. P. (Z. Elektrochem. **65** [1961] 372/5).
[19] Larcher, D.; Gabriel, M. (J. Phys. [Paris] **36** [1975] 447/9).
[20] Lohmann, W.; Fowler, C. F.; Perkins, W. H.; Sanders, J. L. (Nature **209** [1966] 908/9).

[21] Garrett, B. B.; Morgan, L. O. (J. Chem. Phys. **44** [1966] 890/7).
[22] Stockhausen, M. (Z. Naturforsch. **29a** [1974] 1767/70).
[23] Vigee, G. S.; Ng, P. (J. Inorg. Nucl. Chem. **33** [1971] 2477/89).
[24] von Goldammer, E.; Kreysch, W. (Ber. Bunsenges. Physik. Chem. **82** [1978] 463/8).
[25] Libuś, W.; Grzybkowski, W.; Pastewski, R. (J. Chem. Soc. Faraday Trans. I **77** [1981] 147/56).
[26] Harrison, W. D.; Gill, J. B.; Goodall, D. C. (J. Chem. Soc. Dalton Trans. **1979** 847/50).
[27] Kůtek, F. (Collection Czech. Chem. Commun. **33** [1968] 1930/2).
[28] Habeeb, J. J.; Said, F. F.; Tuck, D. G. (J. Chem. Soc. Dalton Trans. **1981** 118/20).
[29] Köhler, H. (Z. Anorg. Allgem. Chem. **336** [1965] 245/51).
[30] Sharma, S. D.; Misra, S. (J. Liquid Chromatog. **8** [1985] 1731/8).

[31] Sharma, S. D.; Misra, S. (J. Liquid Chromatog. **8** [1985] 2991/8).
[32] Haberditzl, W.; Friebe, R.; Havemann, R. (Z. Physik. Chem. [Leipzig] **228** [1965] 73/80, 77).
[33] Griffiths, G.; Thornton, D. A. (J. Mol. Struct. **52** [1979] 39/45).
[34] Harrison, W. D.; Gill, J. B.; Goodall, D. C. (J. Chem. Soc. Chem. Commun. **1988** 728/9).
[35] Jeffreys, B.; Gill, J. B.; Goodall, D. C. (J. Chem. Soc. Dalton Trans. **1985** 99/100).

35.3.1.2 $[Mn^{II}(C_2H_6OS)_4X_2]$ Compounds

A complex of composition $[Mn(C_2H_6OS)_4I_2]$ was obtained on reaction of hydrated manganese(II) iodide with neat DMSO. Pink crystals were formed by removal of the excess DMSO under vacuum [1].

$[Mn(C_2H_6OS)_4(NCS)_2]$ was prepared by the reaction of a methanolic solution of $Mn(NCS)_2$ with DMSO in a 1:6 molar ratio. After standing for 24 h in a vacuum dessicator the reaction mixture yielded pink crystals, which were washed with diethyl ether and acetone. Refractive indices are $n_\alpha=1.542$, $n_\beta=1.568$, $n_\gamma=1.708$. The IR spectrum of the complex in hydrocarbon mull shows NCS^- bands ν(CN) at 2065 and 2080 cm^{-1}, ν(CS) at 765 cm^{-1}, ν(NCS) at 478 cm^{-1} in addition to DMSO bands ν(SO) at 1010 cm^{-1} and ν(CS) at 720 cm^{-1}. The spectral results indicate an octahedral complex with N-coordinated NCS and O-bonded DMSO. The complex is fairly stable in air. It dissolves readily in water, acetone, and methanol, but not in ether or benzene [2]. The complex is involved in the ion exchange separations with Dowex macroporous MSA-1 resin [3].

$[Mn(C_2H_6OS)_4(NCSe)_2]$ has been prepared by the action of DMSO on acetone solutions of $Mn(NCSe)_2$. The compound is isostructural with corresponding complexes of Zn^{II}, Ni^{II}, Co^{II}, Fe^{II}. X-ray diffraction data are given in the paper. IR spectral results indicate that the $NCSe^-$ ions are coordinated in *trans* positions. The ν(CN) and ν(CSe) vibration modes were observed at 2075 and 610 cm^{-1}, respectively [4].

References:

[1] Cotton, F. A.; Francis, R. (J. Am. Chem. Soc. **82** [1960] 2986/91).
[2] Tsintsadze, G. V. (Zh. Neorgan. Khim. **16** [1971] 1160/2; Russ. J. Inorg. Chem. **16** [1971] 614/6).
[3] Qureshi, M.; Tandon, S. N.; Varshney, R. G. (J. Indian Chem. Soc. **58** [1981] 290/2).
[4] Skopenko, V. V.; Tsintsadze, G. V.; Brusilovets, A. I. (Ukr. Khim. Zh. **36** [1970] 474/6; Soviet Progr. Chem. **36** No. 5 [1970] 51/2; C.A. **73** [1970] No. 62129).

35.3.1.3 Other Manganese(II) Compounds

A white solid of composition $\mathbf{Mn(ClO_4)_2 \cdot 3C_2H_6OS \cdot 4H_2O}$ was obtained on reaction of $Mn(ClO_4)_2 \cdot 6H_2O$ with 30 to 40% aqueous DMSO. The IR spectrum shows the ν(SO) band at 954 cm^{-1} [1]. It was assumed that all the dimethyl sulfoxide is coordinated to the metal [2].

The pale yellow $\mathbf{MnBr_2 \cdot 2C_2H_6OS}$ was prepared by the reaction of hydrated manganese(II) bromide with neat DMSO at ambient temperature. Removal of the excess DMSO in vacuum yields the product [1]. The far-IR spectrum shows a ν(Mn–O) band at 417, further bands at 388, 340, and 311 cm^{-1} [3]. The pale pink $\mathbf{MnBr_2 \cdot 3C_2H_6OS \cdot 6H_2O}$ was prepared by the reaction of hydrated manganese(II) bromide with 30 to 40% aqueous DMSO [1]. A ν(SO) band was observed at 952 cm^{-1} [2].

$\mathbf{MnC_2O_4 \cdot C_2H_6OS \cdot H_2O}$ was obtained from $MnC_2O_4 \cdot 3H_2O$ and excess DMSO in ethanol. The suspension was stirred and refluxed for 4 to 6 d, then filtered and washed with alcohol and ether. A polymeric structure with octahedral units bridged by oxalate ions is suggested [7].

$\mathbf{Mn(CN)_2 \cdot C_2H_6OS}$ was prepared by the passage of NH_4CN in an N_2 stream through a solution of anhydrous $MnCl_2$ in DMSO. Addition of ethanol to the reaction mixture yielded a light brown precipitate. The product was dried under vacuum or air at 80°C. The IR spectrum of the complex in a liquid petroleum mull shows characteristic bands of ν(CN) at 2150 cm^{-1}, ν(SO) at 1015 cm^{-1} and further bands at 1322, 960, 720, 625, 500, 435 cm^{-1}. The results suggest that the structure is polymeric with bridging CN^- and oxygen-bonded dimethyl sulfoxide. The interplanar distances and relative intensities from the X-ray powder pattern are given [4].

$\mathbf{Mn[N(CN)_2]_2 \cdot 2C_2H_6OS}$ was prepared by addition of CCl_4 or C_6H_6 to a solution of anhydrous $Mn[N(CN)_2]_2$ in DMSO. The colorless crystals, washed with CCl_4 and dried in vacuum, melt at 143°C. The IR spectrum in hexachlorobutadiene mull shows the ν(CN) band at 2240 and 2193 cm^{-1}. A polymeric structure with bridging $N(CN)_2^-$ groups is suggested [5].

The polymeric solid $\mathbf{Mn(C_2H_6OS)_2Hg(SCN)_2(OCN)_2}$ is formed by the reaction of excess DMSO and $MnHg(SCN)_2(OCN)_2$. The IR spectrum of the complex shows the characteristic ν(CN) bands at 2242, 2145, and 2122 cm^{-1}. An octahedrally coordinated Mn^{II} in a polymeric structure with bridging anions and oxygen-bonded dimethyl sulfoxide is proposed [6].

References:

[1] Cotton, F. A.; Francis, R. (J. Am. Chem. Soc. **82** [1960] 2986/91).
[2] Cotton, F. A.; Francis, R.; Horrocks, W. D., Jr. (J. Phys. Chem. **64** [1960] 1534/6).
[3] Johnson, B. F. G.; Walton, R. A. (Spectrochim. Acta **22** [1966] 1853/8).
[4] Kuntyi, O. I.; Mikalevich, K. N.; Semenishin, D. I. (Koord. Khim. **5** [1979] 685/8; Soviet J. Coord. Chem. **5** [1979] 539/42).
[5] Köhler, H. (Z. Anorg. Allgem. Chem. **336** [1965] 245/51).
[6] Ojha, T. N.; Sharma, S. B. (Chem. Era **19** [1983] 34/7).
[7] Pandey, B. D.; Rupainwar, C. D. (Current Sci. [India] **49** [1980] 336/9).

35.3.1.4 Mixed Ligand Compounds

Preparation and properties of DMSO adducts of β-diketonato complexes, described in [1], are reported in "Mangan" D 1, 1979, pp. 84, 110, 114, and 118. The thermal stability of $[Mn^{II}(DMSO)_2L_2]$ complexes (L^- = anion of acetylacetone or fluorinated diketones) was determined [2].

Manganese(II) perchlorate in DMSO reacts rapidly with bidentate ligands to give mixed ligand complexes. At 292 K in 0.2 M $NaClO_4$ the second-order rate constant for the reaction with phenanthroline $[Mn(C_2H_6OS)_6]^{2+}$ + phen ⇄ $[Mn(C_2H_6OS)_4phen]^{2+}$ is log k = 3.89. The corresponding value for the reaction with 8-quinolinol is 2.30. The reaction with 2,2'-bipyridine is too fast to measure. The reaction between a large excess of $[Mn(C_2H_6OS)_6]^{2+}$ and 2,2',2''-terpyridine (= terpy) is complicated; an initial very rapid reaction with log k = 3.21 at 292 K probably involves the formation of a binuclear intermediate. The following slow step is associated with final chelate ring closure to form $[Mn(C_2H_6OS)_3\,terpy]^{2+}$ with the first-order constant, $k = (2.58 \pm 0.46) \times 10^{-3}\,s^{-1}$ [3, 4]. The rate of dissociation of $[Mn(C_2H_6OS)_4L']^{2+}$ depends less strongly on the chelate ring. For L' = phen the first-order rate constant is $k = 33.7 \pm 1.7\,s^{-1}$, while for L' = terpy $k = 46.1 \pm 0.7\,s^{-1}$ [3]. The equilibrium constants for L' binding, derived from the rate data, are log K = 2.6 and 5.0 for phen and terpy, respectively [4].

References:

[1] Matsushita, T.; Masuda, I.; Shono, T. (Technol. Rept. Osaka Univ. **24** [1974] 345/53; C.A. **81** [1974] No. 144920).

[2] Mazurenko, E. A.; Volkov, S. V.; Bublik, Zh. W. (Probl. Khim. Primen. Beta Diketonatov Metal. Mater. 4th Vses. Semin., Kiev 1978 [1982], pp. 57/61 from C.A. **98** [1983] No. 64624).

[3] Buck, D. M. W.; Moore, P. (J. Chem. Soc. Chem. Commun. **1974** 60/1).

[4] Buck, D. M. W.; Moore, P. (J. Chem. Soc. Dalton Trans. **1976** 638/42).

35.3.1.5 Manganese(III) Compound

$[Mn(C_2H_6OS)_6](ClO_4)_3$. The complex was prepared by dissolution of $Mn(CH_3COO)_3 \cdot 2H_2O$ in the minimum amount of DMSO. The addition of $HClO_4$ (mole ratio 1 Mn : 3 $HClO_4$) in methanol at 0°C yielded violet crystals, which were washed with diethyl ether and dried in vacuum over P_4O_{10} [1]. The compound can be further purified by recrystallization from $HClO_4$-DMSO [2]. The magnetic moment of the solid is 5.13 μ_B at 298 K. The IR spectrum of the complex in Nujol mull shows bands of $\nu_3(ClO_4^-)$ and $\nu_4(ClO_4^-)$ at 1100 and 625 cm^{-1}, respectively, and a ν(SO) band at 915 cm^{-1}. The reflectance spectrum of the solid shows a broad band at 20000 cm^{-1}. The band can be assigned to the high-spin d^4 octahedral transition $^5E_g \rightarrow {}^5T_{2g}$. The electronic absorption spectrum of the complex in DMSO shows a broad band at 19800 cm^{-1} (ε = 248) [1]. More detailed analysis of the solution spectrum allows a resolution of the broad band into three components at 20000, 16670, and 14810 cm^{-1}. The spectrum can be interpreted on the basis of a Jahn-Teller tetragonal distortion of the octahedral complex. The three bands are assigned to the transitions: $^5E_{1g} \rightarrow {}^5B_{1g}$ (20000 cm^{-1}), $^5B_{2g} \rightarrow {}^5B_{1g}$ (16670 cm^{-1}), $^5A_{1g} \rightarrow {}^5B_{1g}$ (14810 cm^{-1}). The ligand field parameters are: Dq = 1667 cm^{-1}, Ds = 2590 cm^{-1}, and Dt = 850 cm^{-1} [2]. An alternative assignment of the bands at 19500 and 17900 cm^{-1} has been made to $^5E_g \rightarrow {}^5T_{2g}$ and $^5E_g \rightarrow {}^3T_{2g}$, respectively, with oscillator strengths 1.4×10^{-3} and 1.2×10^{-4} [3].

The complex is a 1:3 electrolyte in CH_3CN solution. The molar conductance of the complex in CH_3CN (10^{-3} M) is 361 $cm^2 \cdot \Omega^{-1} \cdot mol^{-1}$ [1]. Solutions of $[Mn(C_2H_6OS)_6](ClO_4)_3$ in DMSO are unstable and slowly decompose to form Mn^{II}. The initial step in the process is the oxidation of DMSO by Mn^{III} by inner-sphere electron transfer, with $k = 1.294 \times 10^{-4}\,s^{-1}$ at 40°C. The rate of formation of the DMSO cation radical is inhibited by the presence of Mn^{II} in H_2SO_4 and $HClO_4$ [4]. The free radical initiates polymerization of acrylonitrile [5].

References:

[1] Prabhakaran, C. P.; Patel, C. C. (J. Inorg. Nucl. Chem. **30** [1968] 867/9).
[2] Prabhakaran, C. P.; Patel, C. C. (J. Inorg. Nucl. Chem. **34** [1972] 2371/4).
[3] Fredrick, F. C.; Johnson, K. E. (J. Inorg. Nucl. Chem. **43** [1981] 1483/7).
[4] Devi, N. Ganga; Mahadevan, V. (Chem. Commun. **1970** 797/9).
[5] Devi, N. Ganga; Mahadevan, V. (J. Polym. Sci. Polym. Chem. Ed. **11** [1973] 1553/64).

35.3.2 With Dipropyl or Dibutyl Sulfoxide R_2SO

ligand 1 with $R=C_3H_7$ ($=C_6H_{14}OS$) ligand 2 with $R=C_4H_9$ ($=C_8H_{18}OS$)

$MnCl_2$ in 1 to 9M HCl solution is not extracted with dibutyl sulfoxide in benzene [1].

$[Mn^{II}(C_6H_{14}OS)_6](ClO_4)_2$ and $[Mn^{II}(C_8H_{18}OS)_6](ClO_4)_2$. The complexes were prepared by the reaction of methanol solutions of manganese(II) perchlorate and the ligand (mole ratio 1:6). Prior to combination both solutions were dehydrated by treatment with 2,2-dimethoxypropane for 25 h at 40 or 50°C, respectively. The final reaction mixture with dipropyl sulfoxide was stirred for 30 to 60 min at 40°C. The solution was concentrated and anhydrous diethyl ether added. The white crystals were dried in vacuum over $CaCl_2$ for 10 to 15 h. The reaction mixture with dibutyl sulfoxide was heated at 50°C for 30 to 60 min and the solvent evaporated to yield a heavy oil. Crystals were obtained by allowing a CH_2Cl_2 solution of the oil to evaporate slowly. Recrystallization of the complex from methanol was accomplished in the presence of added 2,2-dimethoxypropane and ligand at 50°C. After 30 to 60 min the solution was evaporated and the residue was washed with diethyl ether. $[Mn(C_6H_{14}OS)_6](ClO_4)_2$ melts at 109 to 111.5°C. The magnetic moment of the solid is 6.16 μ_B at 16°C. The IR spectrum of the complex in Nujol mull shows the ν(SO) band at 982 cm^{-1}, while the same band is seen at 983 cm^{-1} for nitromethane solution. The white crystals of $[Mn(C_8H_{18}OS)_6](ClO_4)_2$ melt at 53 to 56°C. The IR spectrum of the complex in Nujol mull shows the ν(SO) band at 972 cm^{-1}, while the same band is seen at 985 cm^{-1} for CH_3NO_2 solution [2].

$[Mn^{II}(C_5HF_6O_2)_2(C_8H_{18}OS)_2]$. The gas chromatographic behavior of the mixed ligand complex in cyclohexane (obtained by synergistic extraction of Mn^{II}, with hexafluoroacetylacetone and DMSO in cyclohexane) was studied. No thermal decomposition is noted and the detection limit of the metal is 110 nanograms [3].

References:

[1] Shanker, R.; Venkateswarla, K. S. (J. Inorg. Nucl. Chem. **32** [1970] 2369/81, 2372).
[2] Currier, W. F.; Weber, J. H. (Inorg. Chem. **6** [1967] 1539/43).
[3] O'Brien, T. P.; O'Laughlin, J. W. (Talanta **23** [1976] 805/10).

35.3.3 With Dicyclohexyl Sulfoxide $(C_6H_{11})_2SO$ ($=C_{12}H_{22}OS$)

$[Mn^{II}(C_{12}H_{22}OS)_n(NO_3)_2]$. The nitrato complex $[Mn(C_{12}H_{22}OS)_4(NO_3)_2]$ was prepared by shaking a saturated aqueous solution of manganese(II) nitrate with a saturated hexane solution of the ligand for 10 to 15 min at ambient temperature. The crystals formed were dried over P_4O_{10} in vacuum for several days. The complex is a nonelectrolyte in dry acetone. The IR spectrum of the complex in liquid paraffin and benzene shows ν(SO) at 40 cm^{-1} lower than the free sulfoxide, while the unidentate NO_3^- bands are seen at 1470, 1300, 1020, 820, and 750 cm^{-1}. Molecular weight measurements indicate that the complex is partially dissociated in benzene solution [1].

The complex $[Mn(C_{12}H_{22}OS)_2(NO_3)_2]$ was obtained on reprecipitation of $[Mn(C_{12}H_{22}OS)_4(NO_3)_2]$ from benzene with hexane, three times. The complex is a nonelectrolyte in dry acetone, and monomeric undissociated in benzene solution. The dipole moment, 6.48 D, of the complex in benzene is consistent with a *cis* octahedral complex. The IR spectrum of the complex in liquid paraffin or benzene shows the ν(SO) bands at 40 and 90 cm^{-1} lower than that of free sulfoxide, while bidentate NO_3^- bands are found at 1495, 1295, 1020, 820, and 750 cm^{-1} [1].

$MnCl_2 \cdot 2C_{12}H_{22}OS$ was prepared by the extraction of manganese(II) chloride from aqueous solution by the neat liquid ligand [2]. Gas chromatography has been used to measure the activity coefficients, partial molar changes in enthalpy, entropy, and free energy of dissolution of vapors of pentane, heptane, benzene, cyclohexane, carbon tetrachloride, toluene, and m-xylene in the pure ligand and in the complex of $MnCl_2$ and the ligand [2, 3].

References:

[1] Kolosnitsyn, V. S.; Murinov, Yu. I.; Nikitin, Yu. E. (Zh. Neorgan. Khim. **25** [1980] 2194/7; Russ. J. Inorg. Chem. **25** [1980] 1216/8).
[2] Kolosnitsyn, V. S. (Issled. Obl. Khim. Vysokomol. Soedin. Neftekhim. **1977** 41 from C.A. **92** [1980] No. 170102).
[3] Murinov, Yu. I.; Kolyadina, O. A.; Kolosnitsyn, V. S.; Nikitin, Yu. E. (Zh. Fiz. Khim. **51** [1977] 527/8; Russ. J. Phys. Chem. **51** [1977] 313; Deposited Doc VINITI 3702-76 [1976] 1/11; C.A. **87** [1977] No. 12443).

35.3.4 With Other Dialkyl Sulfoxides or Petroleum Sulfoxide

Manganese(II) chloride is extracted from aqueous 8 to 10 M HCl solution by methyl 3-ethylheptyl sulfoxide ($=C_{10}H_{22}OS=L_1$), methyl 4,8-dimethylnonyl sulfoxide ($=C_{12}H_{26}OS=L_2$), or methyl pentadecyl sulfoxide ($=C_{16}H_{34}OS=L_3$) dissolved in p-xylene at ambient temperature. The order of the efficiency of metal extraction for the sulfoxide is $L_2 > L_1 > L_3$ [1].

The yellow-green **$[Mn^{II}(C_{14}H_{30}OS)_2Cl_2]$** was obtained from $MnCl_2$ and a saturated solution of diheptyl sulfoxide in hexane or benzene [3]. **$[Mn^{II}(PSO)_4(NCS)_2]$** was formed in decane by the extraction of $Mn(NCS)_2$ from aqueous thiocyanate solution with PSO (petroleum sulfoxide) in decane. The instability constant for the complex is 1.4 mol/L and the equilibrium constant for the extraction process ~100 [2].

References:

[1] McDowell, W. J.; Harmon, H. D. (J. Inorg. Nucl. Chem. **33** [1971] 3107/17).
[2] Gorbanev, A. I.; Tsvetkova, Z. W.; Fomin, G. S.; Kalmykova, R. V. (Zh. Obshch. Khim. **45** [1975] 2266/9; J. Gen. Chem. **45** [1975] 2226/9).
[3] Nikitin, Yu. E.; Kolosnitsyn, V. S.; Murinov, Yu. I.; Baranovskaya, E. M. (Neftekhimiya **16** [1976] 299/303).

35.3.5 With Dibenzyl Sulfoxide $(C_6H_5CH_2)_2SO$ ($=C_{14}H_{14}OS$)

$[Mn^{II}(C_{14}H_{14}OS)_6](ClO_4)_2$ was prepared by the reaction of hydrated manganese(II) perchlorate and a slight excess above the stoichiometric amount of ligand in *tert*-butyl alcohol.

After warming the reaction mixture for a moment, the solution was cooled to room temperature, whereupon crystals formed. The white crystals were washed with anhydrous ether and dried in vacuum over $Mg(ClO_4)_2$ for 15 h. The product is obtained in yields of 60 to 80%. The solid melts at 139 to 145°C. The magnetic moment of the solid is 6.00 μ_B at room temperature. The IR spectrum of the Nujol mull of the complex shows ν(SO) vibrations at 986 and 970 cm^{-1}.

$[Mn^{II}(C_{14}H_{14}OS)_4(C_3H_6O)_2](ClO_4)_2$. The mixed ligand complex with acetone was prepared by the reaction of hot acetone solutions of hydrated manganese(II) perchlorate and the ligand. The reaction mixture was stirred for 10 min, cooled to room temperature, and refrigerated for 15 h to yield white crystals. The product was isolated and dried at ambient temperature in vacuum over $Mg(ClO_4)_2$ for 15 h. The solid melts at 139 to 145°C under decomposition. The magnetic moment of the solid is 6.02 μ_B at room temperature. The IR spectrum of the complex in Nujol mull shows characteristic bands of ν(SO) at 976 and 987 cm^{-1}, ν(CO) at 1683 cm^{-1}, δ(CO) at 556 and 535 cm^{-1}, δ(CCC) at 381 cm^{-1}, δ(CSO) at 476 and 339 cm^{-1}, ν(Mn–O) at 200 cm^{-1}.

Reference:

Weber, J. H. (Inorg. Chem. **8** [1969] 2813/5).

35.3.6 With Diphenyl Sulfoxide $(C_6H_5)_2SO$ ($=C_{12}H_{10}OS$)

$[Mn^{II}(C_{12}H_{10}OS)_6](ClO_4)_2$ was prepared by the reaction of acetone solutions of hydrated manganese(II) perchlorate and a small excess of the ligand. The reaction mixture was allowed to evaporate slowly in air. The light yellow crystals were washed with petroleum ether and recrystallized from acetone-ethanol. The solid melts at 151 to 160°C [1].

An alternative preparation of the complex involves the reaction of dehydrated methanol solutions of manganese(II) perchlorate and the ligand in 1:6 molar ratio. After keeping the reaction mixture at 40°C for 30 to 60 min, it was concentrated and diethyl ether was added to give white crystals. The product, which was dried in vacuum over $CaCl_2$, melts at 169 to 171°C. The magnetic moment of the solid is ~5.98 μ_B at 290 K [2] and 5.56 μ_B at 298 K [3]. The IR spectrum of the complex in a KBr disk shows characteristic bands for the ligand at 1152, 1065, 1015, 1000, and 910 cm^{-1}, for ClO_4^- at 1080 and 613 cm^{-1}, and for ν(SO) at 983 cm^{-1} [4]. Other workers report ligand bands at 1030, 1015, 760, and 745 cm^{-1}, bands of ClO_4^- at 1100 and 625 cm^{-1}, of ν(SO) at 998 cm^{-1}, ν(CS) at 697 and 685 cm^{-1}, ν(Mn–O) at 405 cm^{-1}, and ν(SO) at 480 cm^{-1} [3]. The Nujol mull and nitromethane solution spectra show ν(SO) at 990 cm^{-1} and ν(CS) at 720 cm^{-1} [2].

$[Mn^{II}(C_{12}H_{10}OS)_6](BF_4)_2$. Crystals of the white complex slowly form from the reaction mixture of hydrated manganese(II) tetrafluoroborate and a small excess of the stoichiometric amount required of the ligand in acetone. The compound melts at 187 to 190°C [1]. The IR spectrum of the complex in a KBr disk shows the characteristic bands of the ligand at 1160, 1088, 1065, 1039, 1023, 1007, and 920 cm^{-1} and a $\nu_3(BF_4^-)$ band at 1052 cm^{-1}, as well as ν(SO) at 991 cm^{-1} [4].

References:

[1] van Leeuwen, P. W. N. M.; Groeneveld, W. L. (Recl. Trav. Chim. **85** [1966] 1173/6).
[2] Currier, W. F.; Weber, J. H. (Inorg. Chem. **6** [1967] 1539/43).
[3] Gopalakrishnan, J.; Patel, C. C. (Inorg. Chim. Acta **1** [1967] 165/8).
[4] van Leeuwen, P. W. N. M. (Recl. Trav. Chim. **86** [1967] 201/8).

35.3.7 With Methyl Phenyl Sulfoxide $CH_3S(O)C_6H_5$ (= C_7H_8OS)

$[Mn^{II}(C_7H_8OS)_6](ClO_4)_2$ was prepared by the reaction of anhydrous methanolic solutions of $Mn(ClO_4)_2$ and the ligand at 40°C for 30 to 60 min. Both solutions were dehydrated prior to combination by reaction with 2,2-dimethoxypropane. After the reaction mixture was concentrated, diethyl ether was added. The white crystals were dried in vacuum over $CaCl_2$ for 10 to 15 h. The complex melts at 150.5 to 152°C. The magnetic moment is 6.13 μ_B at 18°C. The infrared spectrum of the complex in Nujol mull shows the characteristic ν(SO) bands at 991 cm^{-1}, $\delta(CH_3)$ at 950 cm^{-1}, ν(CS) at 720 cm^{-1}, while the nitromethane solution spectrum shows ν(SO) at 990 cm^{-1}, $\delta(CH_3)$ at 960 cm^{-1}, ν(CS) at 720 cm^{-1}.

Reference:

Currier, W. F.; Weber, J. H. (Inorg. Chem. **6** [1967] 1539/43).

35.3.8 With Tetrahydrothiophene or Tetrahydrothiopyran 1-Oxides

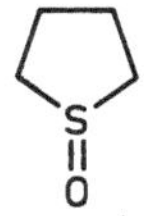
(= C_4H_8OS)

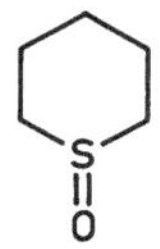
(= $C_5H_{10}OS$)

$[Mn^{II}(C_4H_8OS)_6](ClO_4)_2$ and **$[Mn^{II}(C_5H_{10}OS)_6](ClO_4)_2$** complexes were prepared by the reaction of hydrated manganese(II) perchlorate and the ligands (mole ratio 1:6) in acetone. After the reaction mixture was dehydrated by reaction with ethyl orthoformate, crystals were isolated and washed with acetone. All operations were carried out in a dry box [1].

The IR spectrum of the complex with tetrahydrothiophene 1-oxide has been studied extensively. For Nujol mulls the $\nu_3(ClO_4^-)$ and $\nu_4(ClO_4^-)$ bands are seen at 1085 and 622 cm^{-1}, respectively, while the ν(SO) is at 965 cm^{-1} [1]. The far-IR spectrum shows prominent ligand bands at 540, 334, and 260 cm^{-1}, as well as ν(Mn–O) at 315 cm^{-1} [2]. Other workers report ν(SO) at 974 and 948 cm^{-1} and ν(Mn–O) at 388 cm^{-1} [3]. Additional far-IR bands are assigned as follows: δ(ring) 544 cm^{-1}, δ(ring fold) 277 cm^{-1}, δ(SO) 341 and 325 cm^{-1}, δ(Mn–O) 211 cm^{-1}. The force constants, F_{Mn-O} = 1.055 mdyn/Å and F_{O-Mn-O} = 0.283 mdyn/Å, were calculated from the observed frequencies, assuming a model for $[Mn(C_4H_8OS)_6]^{2+}$ with S_6 symmetry. In addition a simple bent Mn–O–S vibrational model has been used to estimate force constants. For ν(S–O) observed at 974 cm^{-1} F_{SO} = 7.22 mdyn/Å was evaluated while for ν(MnO) observed at 388 cm^{-1} follows F_{MO} = 1.22 mdyn/Å [4]. The interplanar spacings and relative intensities from X-ray powder patterns are given [1].

The IR spectrum of the complex with tetrahydrothiopyran 1-oxide in Nujol mull shows ligand bands at 1148, 1018, 960, and 889 cm^{-1}, ClO_4^- bands at 1093 and 626 cm^{-1}, and ν(SO) at 978 cm^{-1} [1].

References:

[1] van Leeuwen, P. W. N. M.; Groeneveld, W. L. (Recl. Trav. Chim. **86** [1967] 721/30).
[2] Reedyk, J.; van Leeuwen, P. W. N. M.; Groeneveld, W. L. (Recl. Trav. Chim. **87** [1968] 1073/8).
[3] Berney, C. V.; Weber, J. H. (Inorg. Chim. Acta **5** [1971] 375/80).
[4] James, B. R.; Morris, R. H. (Spectrochim. Acta A **34** [1978] 577/82).

35.3.9 With 1,3-Dithiane 1-Oxide or Its 2-Phenyl Derivative

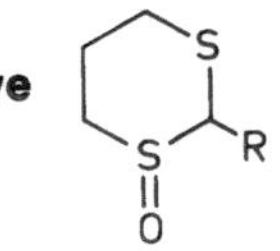

ligand 1 with R = H (= $C_4H_8OS_2$)
ligand 2 with R = C_6H_5 (= $C_{10}H_{12}OS_2$)

$[Mn^{II}(C_4H_8OS_2)_6]X_2$ and **$[Mn^{II}(C_{10}H_{12}OS_2)_6]X_2$** complexes with X = ClO_4 or BF_4 were prepared by reaction of the appropriate MnX_2 salt (obtained by reaction of the hydrated compound with ethyl orthoformate) with the ligand in ethanol solution (mole ratio 1:6). Upon standing the reaction mixture yields white crystals, which were washed with diethyl ether and dried in a vacuum at room temperature. Melting points (m.p.) and characteristic IR bands of the compounds in Nujol mulls (in cm^{-1}) are shown below:

compound	m.p. in °C	ν(SO)	ν(Mn–O)
$[Mn(C_4H_8OS_2)_6](ClO_4)_2$	177 to 178	994, 976	412
$[Mn(C_4H_8OS_2)_6](BF_4)_2$	171 to 172	994, 977	413
$[Mn(C_{10}H_{12}OS_2)_6](ClO_4)_2$	174 to 175	988	396
$[Mn(C_{10}H_{12}OS_2)_6](BF_4)_2$	183 to 185	987	398

The absence of splitting of the ClO_4^- and BF_4^- bands indicates that the anions are not coordinated.

Reference:

Driessen-Fleur, A. H. M.; Groeneveld, W. L. (Inorg. Chim. Acta **7** [1973] 139/43).

35.3.10 With 1,4-Dithiane 1-Oxide

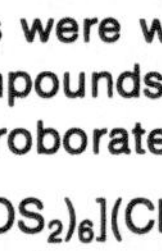

(= $C_4H_8OS_2$)

$[Mn^{II}(C_4H_8OS_2)_6]X_2$ (X = ClO_4, I, BF_4). For preparation, the respective manganese(II) salt was dehydrated with ethyl orthoformate in ethanol and an excess of the ligand (mole ratio 1:8) in ethanol was added. Upon standing, the reaction mixture yields white crystals of the perchlorate and tetrafluoroborate [1]. The yellow iodide separates on addition of diethyl ether to the reaction mixture [2]. The complexes were washed twice with dry diethyl ether and dried in vacuum. Melting points of the compounds are 246 to 248°C (perchlorate), 128 to 130°C (iodide), and 228 to 233°C (tetrafluoroborate).

The crystal structure of $[Mn(C_4H_8OS_2)_6](ClO_4)_2$ has been determined by single crystal X-ray methods. The crystallographic parameters of the triclinic solid are: a = 11.201(5), b = 11.331(5), c = 11.536(5) Å, α = 68.22(2)°, β = 62.68(2)°, γ = 60.18(2)°; Z = 1, space group $P\bar{1}$-C_i^1 (No. 2). The coordination sphere of the slightly distorted octahedral complex, see **Fig. 13**, contains 6 oxygen-bonded sulfoxide ligands with average distances between sulfur and oxygen of 1.51 Å and an Mn–O distance of 2.16 Å. The ligands have the chair conformation with the oxygen of the SO group in the axial position. Atomic coordinates are given in the paper [3].

The IR spectrum of the complexes (Nujol mulls) reveal bands of the SO stretching vibration at 988 cm^{-1} (X = ClO_4, BF_4) and 977 cm^{-1} (X = I). The shift to lower wavenumbers, in comparison to the free ligand where ν(SO) = 1019 cm^{-1}, indicates that the ligand coordinates via the oxygen atom. For additional ligand bands observed between 1200 and 250 cm^{-1}, see the paper. Coupling of internal ligand vibrations with Mn–O vibrations in the range 480 to 380 cm^{-1} is assumed. Bands of ionic ClO_4^- were found at 1092, 1088, and 618 cm^{-1}, of ionic BF_4^- at 1055, 1035, and 517 cm^{-1}. The electronic absorption spectrum of the iodide shows maxima at 18400 and 24900 cm^{-1} (charge transfer). The conductance of a 10^{-3}M solution in nitromethane indicates that the iodide does not dissociate completely in this solvent. Association of the I^- ions with the ligand molecules is likely to occur [2].

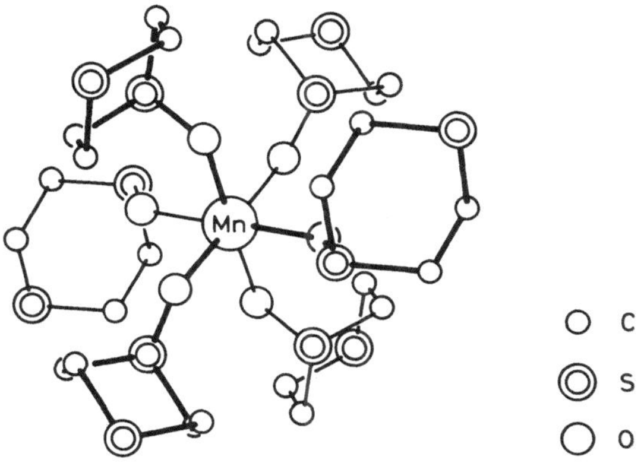

Fig. 13. Arrangement of the ligands about Mn in $[Mn(C_4H_8OS_2)_6](ClO_4)_2$ [3].

$[Mn^{II}(C_4H_8OS_2)_3X_2]$ (X = NO_3, Br). The compounds were prepared from manganese(II) nitrate or bromide and the ligand by a procedure similar to that for the $[Mn(C_4H_8OS_2)_6]X_2$ complexes, but using a mole ratio of 1:6. Addition of diethyl ether to the reaction mixture yields white crystals which were washed with diethyl ether and dried in vacuum [3]. The nitrato complex melts at 133 to 135°C, the bromo complex at 164 to 166°C. The IR spectra of the solids show bands of ν(SO) at 984 and 985 cm^{-1}, respectively, of coordinated NO_3^- at 1015 to 1030 cm^{-1}, and other prominent ligand bands at 414 and 397 cm^{-1} (X = NO_3) or at 416 and 409 cm^{-1} (X = Br). The electronic absorption spectrum of the solid nitrato complex shows maxima at 19400, 23400, and 24300 cm^{-1}, of the bromo complex at 19200, 21500, 22500, and 23000 cm^{-1}. Both compounds are nonelectrolytes in nitromethane [2].

$[Mn^{II}(C_4H_8OS_2)_2(NCS)_2]$ was obtained from manganese(II) thiocyanate and the ligand in ethanol (mole ratio 1:6) as described above. It melts at 173 to 175°C. The IR spectrum of the complex shows the prominent bands: ν(SO) at 906 cm^{-1}, ν(CN) at 2090 cm^{-1}, and ν(CS) at 780 cm^{-1}, both assigned to bridging NCS^-, δ(NCS) at 472 and 465 cm^{-1}, and other ligand bands at 419 and 395 cm^{-1}. The electronic absorption spectrum of the solid shows maxima at 20000, 23000, and 23300 cm^{-1}. The complex is a nonelectrolyte in nitromethane [2].

$[Mn^{II}(C_4H_8OS_2)Cl_2]$ precipitates from an ethanolic solution of manganese(II) chloride (dehydrated with ethyl orthoformate) and the ligand (mole ratio 1:6) after addition of diethyl ether. The white crystals melt at >300°C. The IR spectrum of the solid shows the prominent bands: ν(SO) at 995 cm^{-1}, ν(Mn–Cl) at 306 and 290 cm^{-1}, and other ligand bands at 419 and 412 cm^{-1}. The electronic absorption spectrum of the solid complex shows maxima at 18100, 21900, and 23800 cm^{-1}. The complex is a nonelectrolyte in nitromethane [2].

References:

[1] Reedijik, J.; Fleur, A. H. M.; Groeneveld, W. L. (Recl. Trav. Chim. **88** [1969] 1115/31).
[2] Fleur, A. H. M.; Groeneveld, W. L. (Recl. Trav. Chim. **91** [1972] 317/30).
[3] Spek, A. L. (Cryst. Struct. Commun. **2** [1973] 331/4).

35.3.11 With 1,4-Oxathiane 4-Oxide

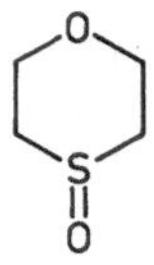

(= $C_4H_8O_2S$)

$[Mn^{II}(C_4H_8O_2S)_6](ClO_4)_2$ was prepared by the reaction of ethanolic, methanolic, or acetone solutions of hydrated manganese(II) perchlorate and excess ligand (mole ratio ~1:10). The reaction mixture was cooled to yield white crystals which were recrystallized from methanol and dried in vacuum at 65°C. The IR spectrum of the complex in a KBr disk shows a strong ν(SO) band at 958 and 970 cm^{-1}. This frequency is lower than that of the free ligand, indicating metal bonding to the sulfoxide oxygen atom.

Reference:

Edwards, J. O.; Goetsch, R. J.; Stritar, J. A. (Inorg. Chim. Acta **1** [1967] 360/4).

35.3.12 With 2-Alkylsulfinylpyridine 1-Oxides

ligand 1 with R = CH_3 (= $C_6H_7NO_2S$)
ligand 2 with R = C_2H_5 (= $C_7H_9NO_2S$)

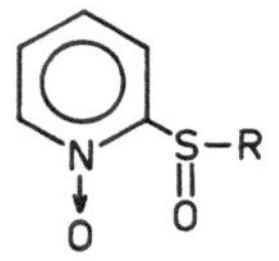

$[Mn^{II}(C_6H_7NO_2S)_2(H_2O)_2](ClO_4)_2$ and **$[Mn^{II}(C_7H_9NO_2S)_2(H_2O)_2](ClO_4)_2$** complexes were prepared by reaction of methanolic solutions of hydrated manganese(II) perchlorate and 2,2-dimethoxypropane with the respective ligand in the molar ratio 1:4. The reaction mixture was evaporated with gentle heating. Multiple trituration of the residue with anhydrous diethyl ether under a nitrogen atmosphere yielded a yellow solid which was stored in vacuum at 80°C. The complexes are extremely hygroscopic [1, 2]. The IR spectra of $[Mn(C_6H_7NO_2S)_2(H_2O)_2](ClO_4)_2$ (in KBr or Nujol) show bands of ν(NO) at 1225 cm^{-1}, ionic $\nu(ClO_4^-)$ at 1120 and 623 cm^{-1}, ν(SO) at 990 cm^{-1}, δ(NO) at 839 cm^{-1}, δ(CH) at 775 cm^{-1}, $\nu(Mn{-}O_S)$ at 531 cm^{-1}, $\nu(Mn{-}O_N)$ at 422 cm^{-1} [1]. The spectra of $[Mn(C_7H_9NO_2S)_2(H_2O)_2](ClO_4)_2$ reveal bands of ν(NO) at 1220 cm^{-1}, ionic $\nu(ClO_4^-)$ at 1110 and 628 cm^{-1}, $\nu(Mn{-}O_N)$ at 328 cm^{-1}, $\nu(Mn{-}O_S)$ at 422 cm^{-1}, δ(NO) at 841 cm^{-1}, and δ(CH) at 773 cm^{-1} [2]. The electronic absorption spectrum of the complex with ligand 1 in Nujol mull shows d-d bands at 18010 and 21550 cm^{-1}, charge transfer (metal → ligand) bands at 27780 and 31060 cm^{-1}, and intraligand bands $n \rightarrow \pi^*$ at 37170 and $\pi \rightarrow \pi^*$ at 40160 cm^{-1} [1]. The electronic absorption spectrum of the complex with ligand 2 shows maxima at 11980, 18380, 26670, 38020, and 40490 cm^{-1}.

The physical properties of the complex suggest an octahedral structure with neutral bidentate ligands coordinated to Mn^{II} via the N-oxygen and S-oxygen atoms. The molar conductivities in acetonitrile, Λ = 222 and 218 $cm^2 \cdot \Omega^{-1} \cdot mol^{-1}$, respectively, are consistent with 1:2 electrolytes [1, 2].

References:

[1] Chartier, L. H.; Kohrman, R. E.; West, D. X. (J. Inorg. Nucl. Chem. **41** [1979] 657/61).
[2] Chartier, L. H.; Kohrman, R. E.; West, D. X. (J. Inorg. Nucl. Chem. **41** [1979] 663/6).

35.3.13 With Other Monosulfinyl Compounds

No.	ligand	formula
1	$HOCH_2CH_2S(O)CH_2CH_2OH$	$C_4H_{10}O_3S$
2	$C_6H_5S(O)CH_2COOH$	$C_8H_8O_3S$
3	$C_6H_5CH_2S(O)CH_2COOH$	$C_9H_{10}O_3S$
4	$C_6H_5NS(O)$	C_6H_5NOS

Complexes in Solution. Stability constants for the formation of $Mn^{II}L^+$ complexes with HL = ligand 2 or 3 in 50% dioxane-water (I = 0.1 M) at 25°C are log K_1 = 1.93 and 1.67, respectively [1].

$Mn^{II}(C_4H_{10}O_3S)_2(ClO_4)_2$. The complex was prepared by the reaction of hydrated manganese(II) perchlorate with ligand 1 (mole ratio 1:2) in ethanol solution. After removal of the solvent at reduced pressure the product was dried in vacuum. The pale rose complex was recrystallized from ethanol-diethyl ether, washed with the solvent and diethyl ether, and then dried in a vacuum desiccator over P_4O_{10}. The compound melts between 160 and 162°C. Interlayer distances and relative intensities from X-ray powder pattern are given in the paper. A magnetic moment of 5.98 μ_B was observed at room temperature. The IR spectra of the complex in KBr or Nujol show the prominent bands: ν(OH) at 3350 cm^{-1}, ν(SO) at 984 cm^{-1}, ionic ν(ClO_4) at 1125 and 624 cm^{-1}. The spectrum supports the assignment of a structure with a tridentate neutral ligand coordinated to the octahedral Mn^{II} via the two OH groups and the oxygen of the sulfoxide group. The complex is a 1:2 electrolyte in acetonitrile and nitromethane. The solution molecular weight of the complex in CH_3CN indicates that the complex is partially dissociated at lower dilution [2].

$[Mn^{II}(C_6H_5NOS)_2Cl_2]$. For preparation a solution of $MnCl_2 \cdot 4H_2O$ (4 mmol) in ethanol (10 mL) was slowly added with stirring to a solution of ligand 4 (16 mmol) in ethanol (10 mL). On stirring the reaction mixture for 1 h a pale pink solid separated out, which was centrifuged, washed a few times with ethanol and finally with ether. The vacuum-dried sample did not melt up to 300°C. The complex was found to be paramagnetic and exhibits a magnetic moment of 5.8 μ_B. On the basis of IR and electronic spectral data a tetrahedral geometry of the complex is assumed with N-sulfinyl aniline acting as monodentate ligand and bonded through its nitrogen atom. The solubility of the complex in most of the organic solvents indicates its monomeric nature [3].

References:

[1] Griesser, R.; Hayes, M. G.; McCormick, D. B.; Prijs, B.; Sigel, H. (Arch. Biochem. Biophys. **144** [1971] 628/35).
[2] Giesbrecht, E.; Lakatos Osorio, V. K. (J. Inorg. Nucl. Chem. **37** [1975] 1409/15).
[3] Arulsamy, K. S.; Ashok, R. F. N.; Agarwala, U. C. (Indian J. Chem. A **23** [1984] 21/5).

35.3.14 With Disulfinyl Compounds

No.	ligand	formula
1	$CH_3S(O)CH_2CH_2S(O)CH_3$	$C_4H_{10}O_2S_2$
2	$C_2H_5S(O)CH_2CH_2S(O)C_2H_5$	$C_6H_{14}O_2S_2$
3	$C_3H_7S(O)CH_2CH_2S(O)C_3H_7$	$C_8H_{18}O_2S_2$
4	$CH_3S(O)CH_2CH_2CH_2S(O)CH_3$	$C_5H_{12}O_2S_2$
5	$HOOCCH_2S(O)CH_2CH_2S(O)CH_2COOH$	$C_6H_{10}O_6S_2$ (= H_2L)

[$Mn^{II}L_3$](ClO_4)$_2$. The complex with ligand 1, $Mn(C_4H_{10}O_2S_2)_3(ClO_4)_2$, was prepared by the reaction of a 2,2-dimethoxypropane solution of hydrated manganese(II) perchlorate with an acetonitrile solution of excess disulfinyl compound. The reaction mixture was refluxed for 2 h then cooled. Crystallization was induced by the addition of CH_2Cl_2. The white solid, isolated in 80% yield, melts at 227°C [1]. An alternative preparation of the complex involves the reaction of a methanolic solution of hydrated manganese(II) perchlorate and the ligand (mole ratio 1:3). The crystals were washed with warm methanol, then diethyl ether, and dried in vacuum over P_4O_{10} at 80°C [2]. The complexes with ligand 2 or 4, $[Mn(C_6H_{14}O_2S_2)_3](ClO_4)_2$ and $[Mn(C_5H_{12}O_2S_2)_3](ClO_4)_2$, were prepared in a similar way with ethanol as solvent. The reaction mixture was cooled and sometimes reduced in volume to induce precipitation [3]. The magnetic moment of $[Mn(C_4H_{10}O_2S_2)_3](ClO_4)_2$ is 6.07 μ_B at 300 K [2], the complexes with ligand 2 or 4 exhibit magnetic moments of 5.85 and 5.80 μ_B, respectively [3]. The IR spectra of the complexes in Nujol show bands at 1100 and 622 cm^{-1} indicative for ionic perchlorate groups and a ν(SO) band at ~1000 cm^{-1} [3] or at 988 cm^{-1} [1]. A band at 480 cm^{-1} (or at 422 cm^{-1} in the case of the complex with ligand 4) was assigned to ν(Mn–O) [3]. The complex with ligand 1 behaves as 1:2 electrolyte in N,N-dimethylacetamide [1].

[$Mn^{II}(C_4H_{10}O_2S_2)_2(NO_3)_2$]. The complex was prepared by the reaction of a solution of hydrated manganese(II) nitrate and 2,2-dimethoxypropane in isopropyl alcohol and a $CHCl_3$ solution of excess ligand 1. After heating the reaction mixture to reflux for 2 h, it was concentrated to give a white solid in 90% yield. It melts at 198°C. The IR spectrum shows a ν(SO) band at 1003 cm^{-1} while bands of the coordinated NO_3^- groups are found in the region 1250 to 1500 cm^{-1}. The complex is a 1:1 electrolyte in N,N-dimethylacetamide. Dissociation of the coordinated NO_3^- increases with dilution [1].

[$Mn^{II}(C_4H_{10}O_2S_2)(H_2O)_2Cl_2$]. The chloro complex was prepared by the reaction of $MnCl_2$ in tributyl phosphate with excess ligand 1 in CH_2Cl_2. The white crystalline precipitate was dried in vacuum at 120°C for 5 h. The product, obtained in 85% yield, melts at 208°C. The IR spectrum shows a ν(SO) band at 1000 cm^{-1}, about 15 cm^{-1} lower than for the free ligand. The complex is a nonelectrolyte in N,N-dimethylacetamide [1].

[$Mn^{II}(C_8H_{18}O_2S_2)NO_3$]$NO_3$. The nitrato complex was prepared by the reaction of hydrated manganese(II) nitrate and ligand 3 (mole ratio 1:2) in methanol. The reaction mixture was refluxed for 2 h and the solvent removed by distillation. Addition of hexane to the wet residue and heating under reflux gave a white powder. The product was isolated, washed with $CHCl_3$, and kept in vacuum over silica gel. The complex which was obtained in 70% yield, melts at 247°C with decomposition. The complex is a 1:1 electrolyte in DMF. The IR spectrum in KBr disk shows bands of ν(SO) at 978 cm^{-1}, ν(Mn–O) at 445 cm^{-1}, ν(NO_3) of the ionic group at 1387, 830, 720 cm^{-1}, and the coordinated one at 1495, 1290, 1030, 800, and 740 cm^{-1} [4].

$Mn^{II}(C_6H_8O_3S_2)\cdot 2H_2O$ was prepared as follows: 5 mmol of the *meso* form (α form) of ligand 5 were dissolved in 50 mL warm water and diluted with 150 mL ethanol. A solution of 5 mmol of hydrated manganese(II) perchlorate in 30 mL ethanol was added slowly to the solution of the acid. Formation of a precipitate began after a few minutes and was completed by standing 24 h at room temperature and 24 h in a refrigerator. The complex was washed with ethanol and dried in vacuum over anhydrous $CaCl_2$. It is practically insoluble in the usual solvents, including water. The similarities in the patterns of $M(C_6H_8O_6S_2)\cdot 2H_2O$ complexes, where M = Zn, Mn, Ni or Co, indicate that they are isomorphous. Selected IR vibration modes (in cm^{-1}) of the Mn^{II} complex and the disodium salt of the ligand (given in parentheses) are: $\nu(OH)$ 3370, 3220 (3450); $\delta(H_2O)$ and $\nu_{as}(COO^-)$ 1680, 1615 (1610); $\nu_s(COO^-)$ ~1395, 1375 (1420, 1380); $\nu(SO)$ 1007 (1028). The negative shift of $\nu(SO)$ evidences that the SO groups are coordinated to manganese. The complex exhibits a magnetic moment of 6.0 μ_B at 292 K. A pseudooctahedral ligand field is assumed, as shown for the corresponding Ni^{II} or Co^{II} compounds. The complex can be heated several hours at ca. 70°C in vacuum over P_4O_{10} without any change in composition and can be heated until 250°C without melting. Decomposition occurs above 150°C as evidenced by the darkening of the sample [5].

References:

[1] DuPreez, J. G. H.; Steyn, W. J. A.; Basson, A. J. (J. South African Chem. Inst. [2] **21** [1968] 8/17).
[2] Madan, S. K.; Hull, C. M.; Herman, L. J. (Inorg. Chem. **7** [1968] 491/5).
[3] Zipp, A. P.; Madan, S. K. (Inorg. Chim. Acta **22** [1977] 49/53).
[4] Filgueiras, C. A. L.; Marques, E. V. (Transition Metal Chem. [Weinheim] **10** [1985] 241/3).
[5] Lakatos Osorio, V. K.; Giesbrecht, E. (Anais Acad. Brasil. Cienc. **52** [1980] 695/701).

35.4 Complexes with Arenesulfinic Acids or N,N-Dimethylmethanesulfinamide

ligand 1	$C_6H_5SO_2H$	$(=C_6H_6O_2S)$
ligand 2	$4\text{-}CH_3C_6H_4SO_2H$	$(=C_7H_8O_2S)$
ligand 4	$CH_3S(O)N(CH_3)_2$	$(=C_3H_9NOS)$

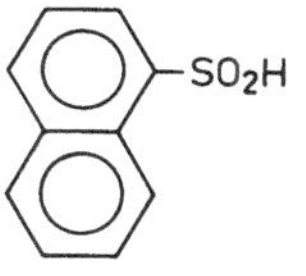

ligand 3 $(=C_{10}H_8O_2S)$

$[Mn^{II}(RSO_2)_2(H_2O)_2]$ complexes with ligand 1 [1], 2 [2], or 3 [3] were prepared by reaction of hydrated manganese(II) chloride with the sodium salt of the appropriate ligand (mole ratio 1:2). The compounds were isolated after several hours, washed with water followed by THF, and dried in vacuum. The magnetic moment of 5.93 μ_B for $[Mn(C_6H_5O_2S)_2(H_2O)_2]$ at 293 K and of 5.23 μ_B for $[Mn(C_{10}H_7O_2S)_2(H_2O)_2]$ (temperature not given) show that the compounds are hexacoordinate high-spin complexes [1, 3]. IR spectral data suggest that the sulfinato ligands act as bidentate chelating agents, coordinating through the two oxygen atoms. Vibration modes of $\nu_{as}(SO_2)$ and $\nu_s(SO_2)$ were observed at 988 and 952 cm^{-1}, respectively [1, 2]. They are shifted to lower wavenumbers in comparison to the bands of the free anions and are typical for O,O'-coordinated complexes [1 to 3]. The spectra also reveal bands in the regions 3500 to 3200, 1673 to 1615, and 870 to 858 cm^{-1}, characteristic of coordinated water [1, 3]. The absorption of $[Mn(C_6H_5O_2S)_2(H_2O)_2]$ in the 3500 to 3200 cm^{-1} region remains, after refluxing the complex with 2,2-dimethoxypropane [1]. The visible spectrum of $[Mn(C_{10}H_7O_2S)_2(H_2O)_2]$ exhibits a shoulder at 25000 cm^{-1}, which may be assigned to the transition $^6A_{1g} \rightarrow {}^4E_g$ (G) in an octahedral environment [3]. It cannot be established conclusively from the spectra whether the ligands are linked intramolecularly or intermolecularly, as shown by the structures a) and b) on p. 110, but the insolubility of the complexes in organic solvents and even in water points to a polymeric intermolecular linkage [2, 4]:

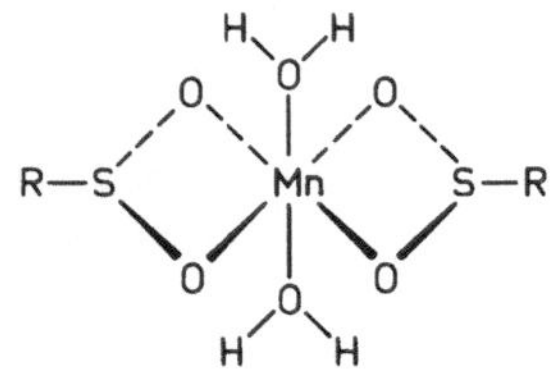

a)

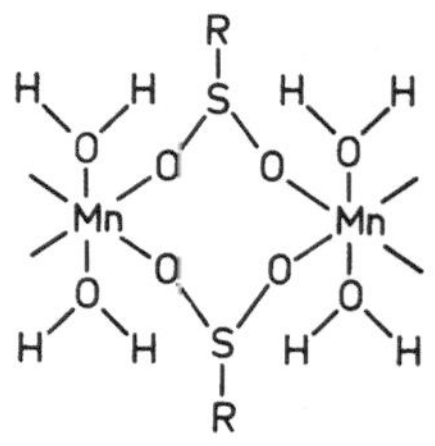

b)

The X-ray powder photographs of $[Mn(C_6H_5O_2S)_2(H_2O)_2]$ and the corresponding complexes of Ni^{II}, Co^{II}, and Fe^{II} show that they are isomorphous [1].

The complexes are stable in air and insoluble in H_2O and organic solvents [1 to 3]. The solubility of $[Mn(C_{10}H_7O_2S)_2(H_2O)_2]$ in DMSO may be due to adduct formation. Analysis of the TGA curve (heating rate 5°C/min) indicates that the elimination of water takes place in a single step at temperatures >120°C, which is characteristic of coordinated water molecules. The anhydrous salt is fairly stable up to 280°C and decomposes to the metal oxide in a single step at temperatures >280°C [3].

$[Mn(C_7H_7O_2S)en_2]X$ ($X = C_7H_7O_2S$, $B(C_6H_5)_4$) and **$[Mn(C_7H_7O_2S)_2en_2]$**. Reaction of ethylenediamine with $[Mn(C_7H_7O_2S)_2(H_2O)_2]$ in ethanol at room temperature yields the ionic compound $[Mn(C_7H_7O_2S)en_2]C_7H_7O_2S$ (I), which exhibits conductivity in ethanol. The cation can be characterized in the form of $[Mn(C_7H_7O_2S)en_2][B(C_6H_5)_4]$ (II). On repeated recrystallization from ethanol with tetrahydrofuran the metastable compound $[Mn(C_7H_7O_2S)en_2]C_7H_7O_2S$ is converted irreversibly into the nonpolar isomer $[Mn(C_7H_7O_2S)_2en_2]$ (III), which is insoluble in all organic solvents. The coordinated $C_7H_7O_2S^-$ group is bonded to the Mn^{2+} ion via both oxygen atoms. The IR spectrum of I additionally contains SO_2 stretching vibrations of the sulfinate anion, observed in the spectrum of the sodium salt $C_7H_7O_2SNa$ (IV):

compound	I	II	III	IV
$\nu_s(SO_2)$ in cm^{-1}	979	970	977	979
$\nu_{as}(SO_2)$ in cm^{-1}	1027, 1009	1000	1012	1027

Thus, ethylenediamine is assumed to be bidentate in I and II, but monodentate in III [5].

$[Mn(C_6H_5O_2S)_2py_2]$ and **$[Mn(C_7H_7O_2S)_2py_2]$** complexes were precipitated from the solutions obtained by stirring the corresponding diaqua complexes, $[Mn(RSO_2)_2(H_2O)_2]$, with pyridine for 3 h at 15°C. Nearly quantitative precipitation occurs, if the solution is then gently stirred and layered with THF. The crystals were washed with THF, then with ether, and dried in vacuum. $[Mn(C_7H_7O_2S)_2py_2]$ melts at 269°C and reveals a magnetic moment of 5.42 μ_B at 293 K. The $\nu_{as}(SO_2)$ and $\nu_s(SO_2)$ bands observed indicate the compounds to be pseudooctahedral O,O′-coordinated complexes. No linkage or stereo isomers could be detected. A polymeric structure with the pyridine molecules in *trans* positions was assumed, similar to that of the $[Mn(RSO_2)_2(H_2O)_2]$ complexes which are formed in the reversible reaction with water. The $[Mn(RSO_2)_2py_2]$ complexes are soluble in methanol or hot pyridine, but insoluble in ether, acetone, or hydrocarbons [6].

$[Mn(C_7H_7O_2S)_2bpy]$. Two isomers of this complex type are known: The complex (A), where the sulfinato groups are assumed to be O,O′-coordinated, was prepared by the reaction of $[Mn(C_7H_7O_2S)_2(H_2O)_2]$ (4.8 mmol suspended in 50 mL H_2O) with 2,2′-bipyridine (10 mmol). The yellow solution, formed after stirring for a longer period at 70°C, was evaporated to dryness in vacuum at 50°C, the residue treated with ether, and dried in vacuum. The yellow complex melts at 136°C [6]. It is identical with the compound prepared earlier in THF solution, see [7].

The S-coordinated sulfinato complex of the same composition (B) was obtained by the reaction of $[Mn(C_7H_7O_2S)_2(H_2O)_2]$ (5 mmol) with 2,2'-bipyridine (11 mmol) in 50 mL pyridine at 50°C for 5 h. The colorless crystals which precipitated from the brown-yellow solution were identified as $[Mn(C_7H_7O_2S)_2py_2]$. After evaporating the filtrate to 15 mL and adding 75 mL THF, the pale yellow $[Mn(C_7H_7O_2S)_2bpy]$ complex (B) precipitated. It was washed with ether and dried in vacuum. The compound melts at 191°C [6].

Magnetic moments at 293 K and bands of the $\nu(SO_2)$ vibration modes (in cm^{-1}) for both isomers are shown below:

$[Mn(C_7H_7O_2S)_2bpy]$	μ_{eff} in μ_B	$\nu_{as}(SO_2)$	$\nu_s(SO_2)$
O,O' isomer (A)	6.15	1014	975
S isomer (B)	5.05	1258, 1168, 1156	1014

The results indicate an octahedral structure for the O,O' isomer and a tetrahedral structure for the S isomer. Both complexes are soluble in H_2O or pyridine and insoluble in ether, acetone, and hydrocarbons [6].

$[Mn^{II}(C_3H_9NOS)_6](ClO_4)_2$. The slightly hygroscopic, white compound was prepared by addition of the ligand 4 (>0.014 mol) to a solution of anhydrous $Mn(ClO_4)_2$ (0.002 mol) in a mixture of methanol (2 mL) and 2,2-dimethyoxypropane (5 mL). After stirring for 10 min, ether was added to complete the precipitation of the complex, which was then washed with anhydrous ether and dried under vacuum for 30 min. Measurements of magnetic susceptibility at 22°C yielded $\mu_{eff}=6.03\ \mu_B$. The IR spectrum of the complex dissolved in N,N-dimethylmethanesulfinamide shows a $\nu(SO)$ absorption band at 996 cm^{-1} (ligand at 1073 cm^{-1}). The existence of only one $\nu(SO)$ band reflects the coordination of the manganese to six oxygen atoms in an octahedral geometry [8].

References:

[1] Dudley, C. W.; Oldham, C. (Inorg. Chim. Acta **2** [1968] 199/201).
[2] Lindner, E.; Vitzthum, G.; Weber, H. (Z. Anorg. Allgem. Chem. **373** [1970] 122/32, 131).
[3] Natarajan, C.; Athappan, P. (Indian J. Chem. A **15** [1977] 1102/3).
[4] Vitzthum, G.; Lindner, E. (Angew. Chem. **83** [1971] 315/27; Angew. Chem. Intern. Ed. Engl. **10** [1971] 315/26, 319).
[5] Lindner, E.; Vitzthum, G. (Angew. Chem. **82** [1970] 322; Angew. Chem. Intern. Ed. Engl. **9** [1970] 308).
[6] Lindner, E.; Lorenz, I.-P.; Vitzthum, G. (Chem. Ber. **106** [1973] 211/9).
[7] Lindner, E.; Vitzthum, G. (Chem. Ber. **102** [1969] 4062/9).
[8] Nykerk, K. M.; Eyman, D. P.; Smith, R. L. (Inorg. Chem. **6** [1967] 2262/4).

35.5 Complexes with Sulfur Dioxide

Only a few manganese(II) complexes with sulfur dioxide have been prepared. An unstable violet compound of composition **$Mn^{II}SO_4 \cdot SO_2$** was extracted with ether from an aqueous solution of manganese(II) sulfate saturated with sulfur dioxide [1]. Formation of **$Mn^{II}(AsF_6)_2 \cdot nSO_2$** or **$Mn^{II}(SbF_6)_2 \cdot nSO_2$** adducts was observed on oxidation of manganese with AsF_5 or SbF_5 in liquid sulfur dioxide. The adhering sulfur dioxide molecules can be removed by heating the product up to 50°C in vacuum ($<10^{-3}$ Torr) [2]. The preparation of a fluoroarsenato complex with two coordinated SO_2 molecules is described below. Complexes of composition **[$Mn^{II}L(SO_2)_nX_2$]** with L = phosphine or phosphine oxide ligands will be described in "Manganese" D 8.

[$Mn^{II}(SO_2)_2(AsF_6)_2$]. For preparation 20 mmol of Mn powder or shavings were placed in a Teflon-sealed glass vessel and 20 mL of dry SO_2 and 0.1 mol of AsF_5 were condensed on it. The mixture was allowed to warm to room temperature with stirring under strict exclusion of moisture until a clear solution resulted. An intermediate blue color was attributed to S_8^{2+} or SO_2^- ions. Solvent SO_2, excess of AsF_5, and some AsF_3 were evaporated at room temperature; the rest of the AsF_3 formed was evaporated 2 h at −20°C in vacuum (oil pump). A quantitative yield of the pure complex remained. The IR spectrum of the compound in Nujol shows bands of $\nu_{as}(SO_2)$ at 1320, $\nu_s(SO_2)$ at 1150, δ(OSO) at 530 cm^{-1}, attributed to oxygen-bonded sulfur dioxide molecules. The ν(AsF) vibration modes at 765, 728, 701 sh, 673, 568, 381 cm^{-1} were assigned to bridging AsF_6 groups bidentately coordinated to Mn. The proposed distorted octahedral coordination sphere at Mn^{II} agrees with the X-ray structure obtained from the corresponding Mg species, and with the electronic spectra obtained from SO_2 solutions of the Ni^{II}, Co^{II}, Fe^{II}, or Cu^{II} complexes. The SO_2 molecules are located in *trans* positions [3].

References:

[1] Kashtanov, L. I.; Gulyanskaya, Ts. A. (Zh. Obshch. Khim. **6** [1936] 227/31; C. A. **1937** 1319).
[2] Dean, P. A. W. (J. Fluorine Chem. **5** [1975] 499/507).
[3] Hoppenheit, R.; Isenberg, W.; Mews, R. (Z. Naturforsch. **37b** [1982] 1116/21).

35.6 Complexes with Sulfonyl Compounds

Remark. An azo complex of Mn^{II} containing a diphenylsulfonyl moiety is described together with other azo complexes in "Manganese" D 5, 1987, pp. 281/2.

35.6.1 With Tetrahydrothiophene 1,1-Dioxide

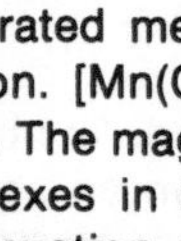

(= $C_4H_8O_2S$ = sulfolane)

[$Mn^{II}(C_4H_8O_2S)_6$][$SbCl_6$]$_2$ and **[$Mn^{II}(C_4H_8O_2S)_6$][MCl_4]$_2$** (M = Al, In). The chloroantimonate was prepared by blending stoichiometric amounts of dehydrated manganese(II) chloride with $SbCl_5 \cdot 2C_4H_8O_2S$ in excess sulfolane in a dry box. When the solids disappeared, dry toluene was added for crystallization. The white complex melts at 238 to 239°C. The white chloroaluminate or chloroindate complexes were prepared in a similar way by dissolving the dehydrated metal halides in an excess of sulfolane at 30°C and adding dry toluene to the solution. [$Mn(C_4H_8O_2S)_6$][$AlCl_4$]$_2$ melts at 155 to 160°C and [$Mn(C_4H_8O_2S)_6$][$InCl_4$]$_2$ at 215 to 217°C. The magnetic moment of [$Mn(C_4H_8O_2S)_6$][$SbCl_6$]$_2$ is μ_{eff} = 5.94 μ_B. The IR spectra of the complexes in Nujol show negative shifts of the ligand's SO_2 stretching frequencies upon complexation, which indicate that the ligand is coordinated to the lone electron pair of one of the sulfone oxygen atoms. Compared with sulfoxide complexes the shift magnitudes are smaller. For shift ranges, see the paper. The spectrum of [$Mn(C_4H_8O_2S)_6$][$InCl_4$]$_2$ shows a vibration mode at 191 cm^{-1}, assigned to ν(Mn–O), and an anion vibration mode at 330 cm^{-1}.

Because of the nonplanarity of the ligand C_4S ring, the symmetry of the complexes must be lower than octahedral. Their approximately octahedral environment and weak coordination were concluded from the electronic spectra of the corresponding $[M(C_4H_8O_2S)_6][SbCl_6]_2$ complexes with $M = Ni^{II}$, Co^{II}, Fe^{II}, Cu^{II}. The complexes are very sensitive to hydrolysis.

$Mn^{II}(C_4H_8O_2S)_{4/3}Cl_2$ was prepared by dissolving dehydrated $MnCl_2$ in an excess of ligand and adding dry toluene to the clear solution. The complex precipitated, was washed with toluene, and dried in vacuum. The rose-colored compound does not melt below 250°C and is very sensitive to hydrolysis. The operations were carried out in a dry box.

Reference:

Reedijk, J.; Vrijhof, P.; Groeneveld, W. L. (Inorg. Chim. Acta **3** [1969] 271/7).

35.6.2 With Other Sulfonyl Compounds

$HC[S(O)_2C_2H_5]_3$

ligand 1 ($=C_7H_{16}O_6S_3$)

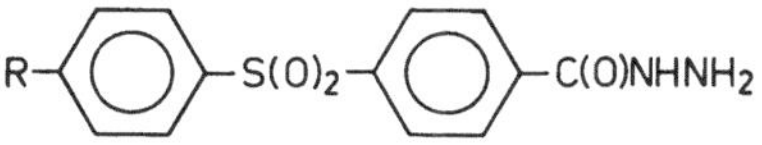

ligand 2 with R=H ($=C_{13}H_{12}N_2O_3S$)
ligand 3 with R=Cl ($=C_{13}H_{11}ClN_2O_3S$)
ligand 4 with R=Br ($=C_{13}H_{11}BrN_2O_3S$)

The aqua complex **$[Mn^{II}(H_2O)_6](C_7H_{15}O_6S_3)_2$** was prepared by refluxing a suspension of ligand 1 and excess of manganese(II) carbonate in water for 15 min. The mixture was filtered and the solution evaporated. The white complex was obtained by recrystallizing the residue from ethanol-ether and drying the compound in vacuum for a short time at room temperature. **$Mn^{II}(C_7H_{15}O_6S_3)_2$** was obtained by drying the aqua complex in high vacuum at 100°C. The ligand acts as a weak acid with the carbon and the sulfur atoms in the same plane and, the anion has C_s symmetry, as indicated by X-ray and IR spectral data. A polymeric octahedral structure is proposed for the anhydrous complex. It is extremely hygroscopic.

The white pyridine adduct **$Mn^{II}(C_7H_{15}O_6S_3)_2 \cdot 2py$** was obtained by dissolving $Mn(C_7H_{15}O_6S_3)_2$ in a small amount of pyridine at 110°C. On addition of ether and cooling the mixture an oily product separated; this crystallized on standing. Yellow-green **$[Mn^{II}phen_3](C_7H_{15}O_6S_3)_2$** was formed by mixing ethanol solutions of phenanthroline and $Mn(C_7H_{15}O_6S_3)_2$ or $[Mn(H_2O)_6]$-$(C_7H_{15}O_6S_3)_2$ and was precipitated by carefully adding ether. $Mn(C_7H_{15}O_6S_3)_2 \cdot 2py$ loses pyridine at room temperature; $[Mn phen_3](C_7H_{15}O_6S_3)_2$ is soluble in acetone or chloroform, and is insoluble in benzene or petroleum ether [1].

$Mn^{II}L_2Cl_2$ complexes with ligands 2 to 4 were prepared by mixing the solution of $MnCl_2 \cdot 4H_2O$ in the minimum quantity of distilled water with the stoichiometric amount of the ligand dissolved in dimethylformamide. The microcrystalline brown complexes precipitated on addition of 20% aqueous ammonia until pH 8 was obtained. The compounds were washed with small amounts of water and dimethylformamide [2]. The electronic spectra of the complexes in cyclohexanol solution show a band in the 328 to 330 nm region ($\varepsilon = 3.4$ to 6.9×10^3 L·mol^{-1}·cm^{-1}), assigned to the $\pi \rightarrow \pi^*$ transition of the C(O)NH group (which appears in the free ligand spectra in the 311 to 317 nm region; $\varepsilon = 1.1$ to 1.3×10^4 L·mol^{-1}·cm^{-1}). A second band at 245 nm, present in the complex and the ligand spectra, is assigned to the benzene ring system. Metal d-d transitions are observed only with the corresponding Co^{II} and Fe^{II} species and are described to an octahedral environment. The ligands are bidentate, bonding to the metals through the oxygen atom of the carbonyl group and the nitrogen atom of the terminal amino group [3]. The complexes are barely soluble in water, alcohol, acetone, slightly soluble in ether, and soluble in cyclohexanol or dimethyl sulfoxide [2].

References:

[1] Beck, W.; Johansen, O.; Fehlhammer, W. P. (Z. Anorg. Allgem. Chem. **361** [1968] 147/56).
[2] Georgescu, G.; Mavrodin, A. (Farmacia [Bucharest] **22** [1974] 493/8; C.A. **84** [1976] No. 83492).
[3] Georgescu, G. (Farmacia [Bucharest] **24** [1976] 187/92; C.A. **86** [1977] No. 105295).

35.7 Complexes with Sulfonic Acids

Remark. Complexes with sulfonic acids, RSO_3H, where R contains additional donor atoms, are described in the section of the special ligand type in the "Manganese" D series; e.g., complexes with pyrocatecholsulfonic acids in "Mangan" D 1, 1979, p. 49, with hydroxyquinolinesulfonic acids in "Manganese" D 3, 1982, p. 184. Complexes with azo compounds or Schiff bases containing SO_3H groups are described in "Manganese" D 5 and D 6, respectively. In the following some information on alkane- or benzenesulfonato complexes are given:

$Mn^{II}(RSO_3)_2 \cdot nH_2O$ and $Mn^{II}[R'(SO_3)_2] \cdot nH_2O$. Alkanesulfonato compounds with n = 2 for R = CH_3, C_3H_7, C_4H_9, C_5H_{11} and n = 4 for R = C_2H_5, and disulfonato compounds with n = 2 for R' = CH_2 or C_2H_4, n = 4 for R' = C_3H_6 or C_4H_8, and n = 5 for R' = C_5H_{10} were prepared by reaction of the appropriate barium sulfonate with $MnSO_4$ in aqueous solution. Pale rose crystals of the hydrates separated from the slowly concentrated filtrate. Dehydration temperatures and stability regions for the anhydrous compounds resulting from TGA and DTA analysis are given in [1]. $Mn(CH_3SO_3)_2 \cdot 4H_2O$ was prepared by neutralizing $MnCO_3$ with the acid in aqueous medium. The anhydrous salt was obtained by dehydrating the hydrate in vacuum at 120°C for 48 h. Compounds with R = 4-$CH_3C_6H_4$ (n = 7) and R = 4-ClC_6H_4 (n = 2) were prepared and dehydrated in a similar way [2]. $Mn(CH_3SO_3)_2$ was also obtained by heating anhydrous $MnCl_2$ with methanesulfonic acid in an oil bath at 160 to 170°C under suction till the reaction mixture was free of Cl^- ions. The resulting insoluble solid was washed with dry ether and dried in vacuum [3]. A benzenedisulfonate of composition $[Mn[C_6H_3(SO_3)_2] \cdot 3.5H_2O$ was obtained from $MnCO_3$ and 1,3-benzenedisulfonic acid after recrystallization from water and drying over $CaCl_2$ [4].

$Mn(CH_3SO_3)_2$ has been assigned an octahedral arrangement with six oxygen atoms around Mn^{II}, due to the multidentate character of the CH_3SO_3 group. The spectral and magnetic studies of the 4-tolyl- and 4-chlorophenylsulfonates indicate that these are also octahedral systems, due to the presence of a similar moiety [2]. $Mn(CH_3SO_3)_2$ exhibits a magnetic moment of 5.46 μ_B at room temperature [2, 3]. The compounds with R = CH_3, 4-$CH_3C_6H_4$, and 4-ClC_6H_4 form stable complexes with pyridine or bipyridyl (see below). The conductivity of manganese 1,3-benzenedisulfonate was measured at 25°C in methanol-water mixtures containing 0 to 100% CH_3OH [4] or dioxane-water mixtures from 0 to 40% dioxane [5]. The ion association in these mixtures was investigated. The kinetics of ion association has been studied by the pressure-jump relaxation technique in anhydrous methanol at 25°C [6].

$[Mn^{II}(RSO_3)_2py_4]$ (R = CH_3, 4-$CH_3C_6H_4$) and **$[Mn^{II}(RSO_3)_2py_2]$** (R = 4-ClC_6H_4). The complexes were obtained by refluxing the anhydrous $Mn(RSO_3)_2$ salts with pyridine for about 10 h and then precipitating with ether. The complexes were repeatedly washed with ether then dried in vacuum [2]. The white methanesulfonato complex, $[Mn(CH_3SO_3)_2py_4]$, has a magnetic moment of 5.63 μ_B at room temperature. The electronic spectrum shows bands at ~13300, 15600, and 20000 cm^{-1}, but no intensities could be measured [3].

[$Mn^{II}(CH_3SO_3)_2bpy_2$] was prepared by shaking a stoichiometric amount of $Mn(CH_3SO_3)_2$ with 2,2'-bipyridine in methanol. The complex precipitated after concentrating the reaction mixture, and was washed with ether. The light pink complex reveals a magnetic moment of 6.19 μ_B at room temperature and bands in the electronic spectrum at ~13800, 15000, 20000, and 28600 cm^{-1} [3].

References:

[1] Charbonnier, F.; Gauthier, J. (Compt. Rend. C **271** [1970] 830/3).
[2] Kumar, S.; Gupta, S. K.; Sharma, S. K. (Thermochim. Acta **71** [1983] 193/7).
[3] Sharma, S. K.; Mahajan, R. K.; Kapila, B.; Kapila, V. P. (Polyhedron **2** [1983] 973/5).
[4] Hallada, C. J.; Atkinson, G. (J. Am. Chem. Soc. **83** [1961] 3759/62).
[5] Atkinson, G.; Hallada, C. J. (J. Am. Chem. Soc. **84** [1962] 721/4).
[6] Marci, G.; Petrucci, S. (Inorg. Chem. **9** [1970] 1009/14).

35.8 Complexes with Sulfonamides or 2-Sulfobenzoic Imide

Introduction

The chelating ability of sulfonamides, R'–SO_2NH–R, arises from the acid hydrogen on the sulfonamide nitrogen and a geometrically suitably placed donor atom on the substituent R. As shown below, chelation is achieved by tautomeric shift of the sulfonamide hydrogen either by isoamide tautomerism (a) or by amine-imine tautomerism (b):

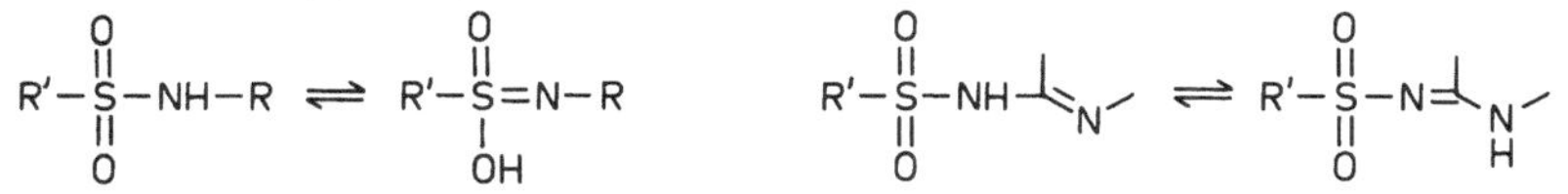

(a) (b)

The coordination to the manganese atom usually occurs through the oxygen atom and the nitrogen atom of the substituent R, thus forming a six-membered chelate ring. In some cases a four-membered chelate ring with the coordination via sulfonamide nitrogen has been postulated.

Generally, 1:1 and 1:2 complexes of manganese(II) with ligand anions or with protonated ligands are formed in mixed aqueous-organic solution. Stabilities were found to follow the Irving-Williams sequence $Cu^{II} > Zn^{II} > Cd^{II} > Ni^{II} > Co^{II} > Mn^{II}$. The type of the isolated complexes depends on the nature of the ligand and the pH value of the reaction medium. In general, reactions of aqueous or organic solutions of the ligand HL and a hydrated Mn^{II} salt at pH 4 to 5.5 yield $Mn(HL)_2X_2$ complexes. Addition of a base to the reaction mixture (pH 7 to 9) results in the deprotonation of the ligand and the formation of MnL_2 complexes. In the case of complexes with sulfanilamides (4-aminobenzenesulfonamide), the addition of acid (pH 1 to 1.4) results in the formation of $(H_2L)_2[MnX_4]$ complexes (with twice-protonated ligands). These three types of complexes are interconvertable by the change of medium pH (see p. 122). In the $(H_2L)_2[MnX_4]$ complexes the twice-protonated sulfonamide ligands are not coordinated to the manganese atom. They are therefore not treated in this volume. References concerning this type of complex are [1 to 6].

It has been found that metal chelates of sulfonamides show more antibacterial and antifungal properties than the ligands themselves with decreasing toxicity in the order $Cu^{II} > Ni^{II} > Zn^{II} > Mn^{II}$.

For the sake of congruity a number of manganese complexes with ligands containing sulfonylamino groups were already described in preceding volumes of the "Manganese" D series. These ligands are listed on pp. 126/7.

References:

[1] Tskitishvili, M. G.; Gogorishvili, P. G.; Machkhoshvili, P. I.; Zhorzholiani, N. B. (Izv. Akad. Nauk Gruz. SSR Ser. Khim. **5** [1979] 295/301; C.A. **93** [1980] No. 18280).
[2] Tskitishvili, M. G.; Shvelashvili, A. E.; Mikadze, I. I.; Chrelashvili, M. V.; Zhorzholiani, N. B. (Izv. Akad. Nauk Gruz. SSR Ser. Khim. **7** [1981] 204/12; C.A. **96** [1982] No. 134774).
[3] Gogorishvili, P. V.; Tskitishvili, M. G.; Kalandarishvili, D. Z. (Issled. Obl. Khim. Kompleksn. Prostykh Soedin. Nekot. Perekhodnykh Redk. Metal. No. 2 [1974] 152/73; C.A. **82** [1974] No. 38083).
[4] Gogorishvili, P. V.; Tskitishvili, M. G. (Issled. Obl. Khim. Kompleksn. Prostykh Soedin. Nekot. Perekhodnykh Redk. Metal. No. 3 [1978] 5/22; C.A. **90** [1979] No. 214469).
[5] Tskitishvili, M. G.; Gogorishvili, P. V.; Chrelashvili, M. V.; Shvelashvili, A. E. (Izv. Akad. Nauk Gruz. SSR Ser. Khim. **5** [1979] 13/19; C.A. **92** [1980] No. 163917).
[6] Gogorishvili, P. V.; Tskitishvili, M. G. (Issled. Obl. Khim. Kompleksn. Prostykh Soedin. Nekot. Perekhodnykh Redk. Metal. No. 2 [1974] 17/29; C.A. **82** [1974] No. 38073).

35.8.1 With Sulfanilamide or Its N-Substituted Derivatives

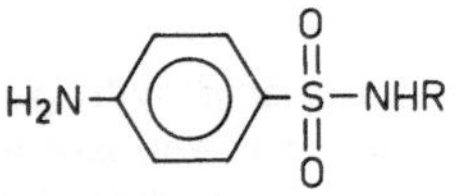

(=HL)

ligand	R	name	formula
1	H	sulfanilamide	$C_6H_8N_2O_2S$
2	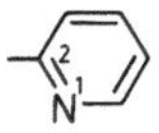	sulfapyridine (=sulfidine)	$C_{11}H_{11}N_3O_2S$
3	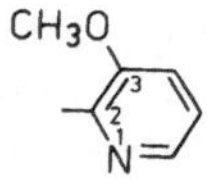	—	$C_{12}H_{13}N_3O_3S$
4	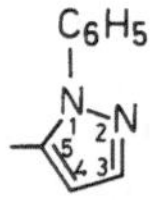	sulfaphenazole	$C_{15}H_{14}N_4O_2S$
5	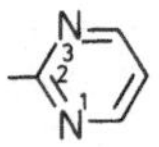	sulfadiazine (=sulfapyrimidine)	$C_{10}H_{10}N_4O_2S$
6	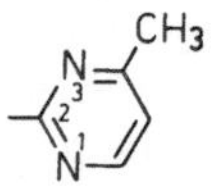	sulfamerazine	$C_{11}H_{12}N_4O_2S$
7	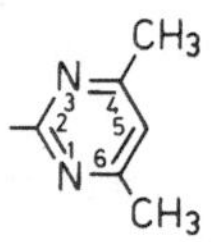	sulfadimidine (=sulfadimezine)	$C_{12}H_{14}N_4O_2S$
8	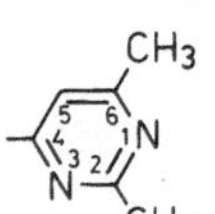	sulfadimetine (=sulfaisodimerazine)	$C_{12}H_{14}N_4O_2S$

ligand	R	name	formula
9	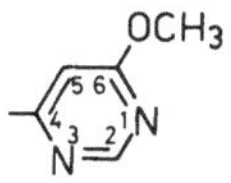	sulfamonomethoxine	$C_{11}H_{12}N_4O_3S$
10	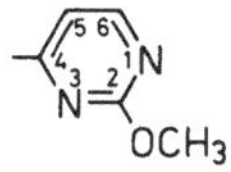	—	$C_{11}H_{12}N_4O_3S$
11	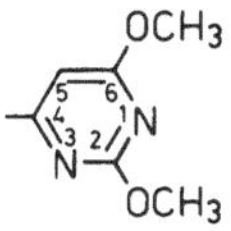	sulfadimethoxine (= sulfadimethoxypyrimidine)	$C_{12}H_{14}N_4O_4S$
12	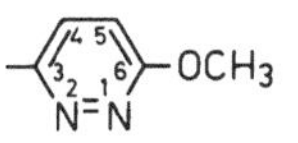	sulfapyridazine (= sulfamethoxypyridazine)	$C_{11}H_{12}N_4O_3S$
13	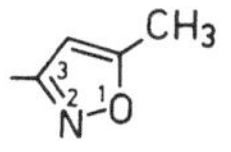	sulfamethoxazole	$C_{10}H_{11}N_3O_3S$
14		sulfathiazole (= norsulfazole)	$C_9H_9N_3O_2S_2$
15	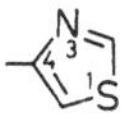	—	$C_9H_9N_3O_2S_2$
16	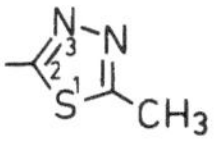	sulfamethizole (= sulfathiadiazole)	$C_9H_{10}N_4O_2S_2$
17	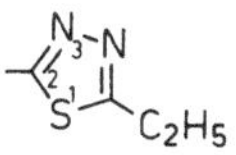	sulfaethidole (= ethazole)	$C_{10}H_{12}N_4O_2S_2$
18	$C(O)CH_3$	albucid	$C_8H_{10}N_2O_3S$
19	$C(O)NHC_4H_9$	bucarban	$C_{11}H_{17}N_3O_3S$
20	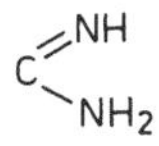	sulfaguanidine	$C_7H_{10}N_4O_2S$

35.8.1.1 Complexes in Solution

Potentiometric pH titrations of solutions containing Mn^{2+} ions and one of the ligands in aqueous-organic solvents at varying temperatures and ionic strengths (0.01, 0.05, 0.1, 0.2 mol/L) reveal the formation of the $\mathbf{Mn^{II}L^+}$ and $\mathbf{Mn^{II}L_2}$ species. Stepwise stability constants, log K_1 and log K_2, are given for I = 0.1 mol/L [1 to 3, 5 to 8]. Ionic strengths were maintained by addition of $NaClO_4$ [1, 2, 7], KNO_3 [3, 8], or KCl [5, 6]. Overall stability constants, log β_2 for $I \rightarrow 0$, were obtained by extrapolation. From the latter values, the total free energy changes $\Delta G°$ (in kcal/mol) were calculated:

ligand	t in °C	medium	log K_1	log K_2	log β_2	$-\Delta G°$	Ref.
2	27	50 vol% ethanol	6.00	3.90	10.80	14.82	[1]
4	27	50 vol% ethanol	4.60	2.20	7.35	10.12	[1]
5	27	50 vol% ethanol	5.04	3.30	9.02	12.41	[2]
	25	60 vol% DMSO	3.51	2.49	—	—	[3]
	30	60 vol% DMSO	3.30	1.90	—	—	[3]
6	27	50 vol% ethanol	5.20	3.34	9.30	12.78	[2]
7	27	50 vol% ethanol	5.46	3.50	9.75	13.40	[2]
	25	50 vol% acetone	2.90 a)	2.46 a)	—	—	[4]
	30	50 vol% methanol	3.75	1.90	6.50	9.07	[5]
	40	50 vol% methanol	2.98	1.80	—	—	[5]
11	27	50 vol% ethanol	5.05	3.50	9.30	12.79	[1]
	25	50 vol% acetone	2.45 a)	2.20 a)	—	—	[4]
12	25	50 vol% acetone	3.75 a)	2.97 a)	—	—	[4]
	27	50 vol% ethanol	3.85	1.70	6.20	8.52	[1]
13	30	50 vol% ethanol	5.00	3.30	11.70	16.33	[6]
	40	50 vol% ethanol	4.50	2.92	—	—	[6]
14	25	50 vol% acetone	2.44 a)	2.09 a)	—	—	[4]
	27	50 vol% ethanol	5.60	3.85	10.40	14.30	[7]
	35	35 vol% ethanol	5.81	2.86	—	—	[8]
	45	35 vol% ethanol	2.92	2.6	—	—	[8]
16	27	50 vol% ethanol	5.70	3.90	10.55	14.51	[7]

a) Ionic strength at the beginning of the titration: 7.5×10^{-3} mol/L.

The stability of metal complexes with ligands 7, 11, 12, and 14, measured at 25, 35, 45, and 55°C in 50% aqueous acetone, decreases in the ligand order 12>7>11>14, and metal order $Mn^{II} > Mg^{II} > Ca^{II} > Sr^{II} > Ba^{II}$ (log K values are not specified in the publication). Thermodynamic parameters for stepwise formation of MnL_2 complexes were calculated from the log K values. Enthalpies in the range −2 to −3 kcal/mol and entropies between 8 and 9 $cal \cdot mol^{-1} \cdot K^{-1}$ at 308 K support the fact that ligands are deprotonated in the complexes [9].

Potentiometric pH titration of solutions containing MnX_2 (X = Cl, Br, or NCS) and ligand 7, 9, 11, 12, 14, 18, or 19 in 50% (v/v) acetone-water at 25°C and pH 3 to 5.5 reveal the formation of **$Mn^{II}(HL)X_2$** and **$Mn^{II}(HL)_2X_2$** complexes, according to the equilibria $MnX_2 + HL \overset{K}{\rightleftharpoons} Mn(HL)X_2$ and $Mn(HL)X_2 + HL \overset{K'}{\rightleftharpoons} Mn(HL)_2X_2$. The stability constants of stepwise formation are tabulated below (ionic strengths at the beginning of the titrations were $\sim 6 \times 10^{-2}$ mol/L; standard deviation for log K: ±0.02 to ±0.07) [11]:

ligand	X = Cl		X = Br		X = NCS	
	log K	log K′	log K	log K′	log K	log K′
7	7.20	6.32	7.97	7.19	8.09	7.28
9	6.57	5.61	7.13	6.68	7.39	6.78
11	6.94	5.81	7.83	7.06	7.95	7.20
12	7.54	6.70	8.07	7.45	8.19	7.82
14	6.78	5.70	7.70	6.86	7.82	6.94
18	5.00	4.63	5.16	5.03	5.64	5.14
19	5.73	5.23	6.34	5.84	6.47	5.99

Deviating values for the chloro complexes with ligands 7 and 14 are reported in an earlier publication [10]. The stabilities are compared to those of analogous Mg^{II} complexes [11].

Potentiometric pH titrations of solutions containing equimolar amounts of Mn^{2+} ions, ligand 5 or 7, and nitrilotriacetic acid (=$C_6H_9NO_6$) reveal the formation of the 1:1:1 mixed ligand complexes $[Mn(C_{10}H_9N_4O_2S)(C_6H_6NO_6)]^{2-}$ and $[Mn(C_{12}H_{13}N_4O_2S)(C_6H_6NO_6)]^{2-}$. The formation constants for the complexes, formed from $Mn(C_6H_6NO_6)^-$ and the deprotonated ligands, were determined in 50% (v/v) aqueous ethanol at 25°C: log K=1.99 for $[Mn(C_{10}H_9N_4O_2S)(C_6H_6NO_6)]^{2-}$ at I=0.1 mol/L (NaCl) [12], and log K=1.24 for $[Mn(C_{12}H_{13}N_4O_2S)(C_6H_6NO_6)]^{2-}$ at I=0.1 mol/L ($NaNO_3$) [13]. Comparison of the log K value, 1.99, with the stability constants of the binary Mn^{II} complexes with ligand 5 (see p. 118) shows that the ternary complex possesses lower stability, due to steric hindrance caused by the bulky nitrilotriacetate anion and to the electrostatic repulsion between the two anions [12]. A six-membered strainless chelate ring, achieved by a tautomeric shift of the sulfonamide hydrogen to the sulfonyl oxygen, was proposed [12, 13].

References:

[1] Lal, K. (Indian J. Chem. A **17** [1979] 313/4).
[2] Lal, K. (Chem. Era **14** [1978] 170/2).
[3] Kumari, Madhu; Narain, Lalit (J. Coord. Chem. **12** [1982] 49/52; J. Electrochem. Soc. India **31** [1982] 149/51).
[4] Tskitishvili, M. G.; Mikadze, I. I.; Zhorzholiani, N. B.; Chrelashvili, M. B. (Izv. Akad. Nauk Gruz.SSR Ser. Khim. **8** [1982] 270/4; C.A. **98** [1983] No. 186577).
[5] Abdel-Gawad, F. W.; El-Guindi, N. M.; Abdel-Hamed, S. M. (Orient. J. Chem. **2** [1986] 96/9; C.A. **105** [1986] No. 179285).
[6] Abdel-Gawad, F. M.; El-Guindi, N. M.; Abdel-Hamed, S. M. (J. Drug Res. **16** [1985] 175/80; C.A. **105** [1986] No. 120658).
[7] Lal, K.; Gupta, S. P. (Acta Ciencia Indica **4** [1978] 242/4).
[8] Kumari, Madhu; Narain, Lalit (Trans. SAEST **15** [1980] 265/9; C.A. **94** [1981] No. 91303).
[9] Tskitishvili, M. G.; Mikadze, I. I.; Chrelashvili, M. V.; Zhorzholiani, N. B.; Kalandarishvili, D. Z. (Izv. Akad. Nauk Gruz.SSR Ser. Khim. **10** [1984] 87/92; C.A. **101** [1984] No. 203194).
[10] Tskitishvili, M. G.; Mikadze, I. I. (Issled. Obl. Khim. Kompleksn. Prostykh Soedin. Nekot. Perekhodnykh Redk. Metal. No. 2 [1974] 190/3; C.A. **82** [1975] No. 65088).

[11] Tskitishvili, M. G.; Mikadze, I. I.; Zhorzholiani, N. B.; Chrelashvili, M. V. (Koord. Khim. **9** [1983] 369/72; C.A. **98** [1983] No. 186599).
[12] Shoukry, M. M.; Khater, M. M.; Shoukry, E. M. (Indian J. Chem. A **25** [1986] 488/9).
[13] Shoukry, M. M.; Shoukry, E. M. (Indian J. Chem. A **27** [1988] 364/5).

35.8.1.2 Isolated Compounds

$Mn^{II}L_2$ complexes with deprotonated ligands 3, 5, 8, 10 to 12, 15, and 17 (see pp. 116 and 117) were obtained by reaction of a manganese(II) salt with the sodium salt of the corresponding ligand in aqueous solution (mole ratio 1:2). The resulting precipitate was washed with hot water and recrystallized from DMSO [1]. The complex with ligand 5 was dried at 80°C. Its solubility in water is 0.0381 g/100 mL H_2O [2]. Melting points and IR data (KBr pellets) of the complexes are given on the following page:

 Formulas of ligands are tabulated on pp. 116/7

ligand	complex	m.p. in °C	$\nu(NH_2)$	$\nu(SO_2)$	Ref.
3	$Mn(C_{12}H_{12}N_3O_3S)_2$	172 to 174	3370	1150	[1]
5	$Mn(C_{10}H_9N_4O_2S)_2$	255 (dec.)	—	—	[2]
8	$Mn(C_{12}H_{13}N_4O_2S)_2$	162 to 164	3365	1145	[1]
10	$Mn(C_{11}H_{11}N_4O_3S)_2$	168 to 170	3375	1148	[1]
11	$Mn(C_{12}H_{13}N_4O_4S)_2$	170 to 171	3360	1147	[1]
12	$Mn(C_{11}H_{11}N_4O_3S)_2$	161 to 163	3360	1147	[1]
15	$Mn(C_9H_8N_3O_2S_2)_2$	170 to 172	3340	1140	[1]
17	$Mn(C_{10}H_{11}N_4O_2S_2)_2$	158 to 160	3345	1140	[1]

$Mn^{II}L_2 \cdot nH_2O$ complexes with deprotonated ligands 7 (n = 3), 11, 12 (n = 2), 14 (n = 3.5), or 17 (n = 10) were prepared by the reaction of an aqueous solution of a manganese(II) salt with the sodium salt of the corresponding ligand (mole ratio 1:2) in aqueous solution [3, 4 to 6] or aqueous acetone (50 vol%) [7, 8]. The preparation usually proceeds in weakly alkaline medium (pH 6 to 8) [4, 5, 8]. The precipitation of the complexes with ligand 7 or 14 from the reaction mixture was effected by addition of a few drops of NH_3 to obtain a pH in the range 8 to 9 [6, 7]. The crystalline precipitate was isolated, washed with water, then with ethanol and ether, and dried in air [4 to 8]. The complex with ligand 14, $Mn(C_9H_8N_3O_2S_2)_2 \cdot 3.5H_2O$, melts at 85°C and decomposes at 215°C [6].

The IR spectra of the complex with ligand 7, $Mn(C_{12}H_{13}N_4O_2S)_2 \cdot 3H_2O$, and the corresponding complexes of Cu^{II}, Co^{II}, Zn^{II}, Cd^{II}, and Ni^{II} have been investigated. The splitting of the $\nu(C{=}N)$ band into two components, compared to the spectrum of the free ligand, indicates the participation of at least one heterocycle in the formation of a nitrogen-metal coordinate bond. The presence of a weak bond between the metal and one oxygen atom of the SO_2 group was supposed on the basis of a shift of the $\nu(SO_2)$ band [9]. The IR spectrum of the complex with ligand 17, $Mn(C_{10}H_{11}N_4O_2S_2)_2 \cdot 10H_2O$, taken from liquid paraffin or perfluorinated hydrocarbon mulls shows the following characteristic absorption bands (free ligand bands in parentheses; $\bar{\nu}$ in cm^{-1}): $\nu_{as}(NH_2)$ at 3468, 3430(3478), $\nu_s(NH_2)$ at 3370(3368), $\delta(NH_2)$ at 1640(1637), and $\nu(SO_2)$ at 1278, 1248, 1142 (1300, 1152). The $\nu(NH)$ band of the sulfonamide group is absent from the spectrum, indicating that the proton is ionized in the uninegative bidentate ligand. The $\nu(NH_2)$ bands are very similar to those of the free ligand, suggesting that the amino group does not bond to the metal. The heterocyclic ring in ligand 17 exhibits a pronounced electron-withdrawing nature, which decreases the electron density on the oxygen atom of the sulfonamide group and thus its ability to coordinate to the metal. A comparison of the UV spectrum of the complex in ethanolic solution with the spectrum of the free ligand and the spectrum of the corresponding complexes with Na and Mg indicates that the metal atom probably replaces the hydrogen of an imino group, but does not form coordinate bonds with the ligand. An ionic compound consisting of an aquated metal ion and an anionic ligand was assumed [5].

$MnL_2 \cdot nH_2O$ complexes are air-stable. They decompose on heating in the temperature range 250 to 400°C. They are only slightly soluble in water and common organic solvents. The complexes are nonelectrolytes in methanol [4, 7 to 9]. The solubility of $MnL_2 \cdot nH_2O$ complexes with ligands 7, 11, 12, and 14 was studied in aqueous solutions containing Na^+, Cl^-, and CH_3COO^- ions at 25°C and pH 6.4 to 8.2. The values of the solubility products are between 10^{-9} and 10^{-8} for complexes with ligands 7, 11, 12, and 14 and $\sim 10^{-6}$ for the complex with ligand 12. The solution is assumed to contain $[Mn(H_2O)_6]^{2+}$ cations and deprotonated ligand anions. The solubilities of analogous metal complexes increase in the following order: $Cu^{II} < Co^{II} < Ni^{II} < Mn^{II} < Mg^{II}$ [10]. Addition of an acid to the complex solution results in the formation of $Mn(HL)_2X_2$ complexes at (pH 4 to 5) or $(H_2L)_2[MnCl_4]$ complexes at pH < 1.95 [4, 7 to 9].

$Mn^{II}(C_{12}H_{13}N_4O_2S)_2py_2 \cdot 3H_2O$ is formed from the complex with ligand 7, $Mn(C_{12}H_{13}N_4O_2S)_2 \cdot 3H_2O$, in warm pyridine. On cooling to room temperature the solution deposits a colorless precipitate which was washed with ethanol and dried in air. The air-stable compound loses all its water on heating at 120°C. It is sparingly soluble in water [7].

$[Mn^{II}(C_{12}H_{13}N_4O_2S)(OH)]_2$. A complex of this probable composition (erroneously formulated as $[Mn(C_{12}H_{13}N_4O_2S)(H_2O)_2]$ in the paper) was obtained by the reaction of a manganese(II) salt with ligand 7 (mole ratio 1:2) in ethanol for 8 h.

The X-ray diffraction patterns of the complex and the corresponding chelates of $M = Ni^{II}$, Cu^{II}, and Zn^{II} with ligand 7 show new peaks, which can be assigned to M–OH (3.93 to 4.04 Å), M–N (2.96 to 3.06 Å), and M–O–S=N (1.987 to 1.995 Å) by use of ASTM cards. The IR spectrum exhibits characteristic absorption bands of ν(OH) at 3480, $\nu_{as}(SO_2)$ at 1315, $\nu_s(SO_2)$ at 1140, ν(Mn–O) at 465, and ν(Mn–N) at 370 cm^{-1}. The visible electronic spectrum in DMF exhibits an absorption maximum at 400 nm. The data indicate a monobasic, bidentate ligand coordinating through one heterocyclic nitrogen atom and the sulfonyl oxygen atom, forming a six-membered chelate ring. A dimeric structure with two bridging oxygen atoms from hydroxy groups is discussed. The complex is a nonelectrolyte [11].

$Mn^{II}(HL)_2X_2$ (X = Cl, Br, NCS). The complexes with ligands 1, 7, 9, 11, 12, 18, or 19 (see the table below) were obtained by heating the 1:2 mixture of MnX_2 and the corresponding ligand in methanol-acetone (1:2) in the pH range 4 to 5.5. The precipitates obtained by evaporation on the water bath were washed with benzene, finally with ether, and dried in air [4, 8, 12]. The complex $Mn(C_9H_9N_3O_2S_2)_2(NCS)_2$ precipitated from a mixture of $Mn(NCS)_2$ in acetone and ligand 14 in dimethylformamide (mole ratio 1:2) on standing in air for several days [6]. The complexes can also be obtained by adjusting solutions of $MnL_2 \cdot nH_2O$ compounds (see p. 120) or of the tetraacido complexes $(H_2L)_2[MnX_4]$ to pH 4 to 5.5 [4, 6, 8, 12]. Colors of several solid complexes and the electrical conductivities (Λ in $cm^2 \cdot \Omega^{-1} \cdot mol^{-1}$) of 10^{-3} M solutions in methanol at 25°C at the given pH are summarized below:

ligand	complex	color	Λ	pH	Ref.
1	$Mn(C_6H_8N_2O_2S)_2Cl_2$	white	—	—	[13]
7	$Mn(C_{12}H_{14}N_4O_2S)_2Cl_2$	flesh	105.1	5.8	[8]
9	$Mn(C_{11}H_{12}N_4O_3S)_2Cl_2$	flesh	92.0	5.57	[12]
11	$Mn(C_{12}H_{14}N_4O_4S)_2Cl_2$	pink	—	—	[8]
12	$Mn(C_{11}H_{12}N_4O_3S)_2Cl_2$	slightly brown	—	—	[4]
14	$Mn(C_9H_9N_3O_2S_2)_2(NCS)_2$	—	99.97	—	[6]
18	$Mn(C_8H_{10}N_2O_3S)_2Cl_2$	white	—	—	[13]
19	$Mn(C_{11}H_{17}N_3O_3S)_2Br_2$	white	43.5	6.3	[12]

The complex with ligand 9, $Mn(C_{11}H_{12}N_4O_3S)_2Cl_2$, melts at 85°C [12]. The complex with ligand 12 decomposes on heating to temperatures between 100 and 200°C [4]. All the compounds are air-stable [4, 6, 8, 12]. They are soluble in organic solvents, such as methanol, ethanol, acetone, and dimethylformamide, and behave as 1:2 electrolytes in methanol. By treatment with NaOH (pH 8 to 9) the complexes can be converted into MnL_2 complexes. The addition of acids to obtain the pH range 1 to 1.5 results in the formation of $(H_2L)_2[MnX_4]$ complexes. An interconversion scheme involving amine-imine tautomerism of the coordinated ligands is assumed [4, 6, 8] and is shown on the following page:

Formulas of ligands are tabulated on pp. 116/7

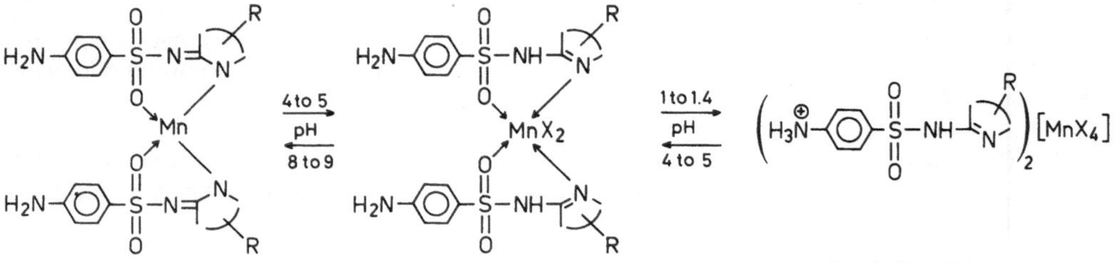

MnL_2 $Mn(HL)_2X_2$ $(H_2L)_2[MnX_4]$

$[Mn^{II}(C_7H_{10}N_4O_2S)Cl_2(H_2O)_2]$. The complex was prepared by the reaction of a manganese(II) salt with ligand 20 (mole ratio 1:2) in ethanolic solution for 8 h. The precipitate was washed with ethanol, then with ether, and dried in vacuum. The IR spectrum exhibits characteristic absorption bands of ν(OH) at 3510, ν(NH) at 3130, $\nu_{as}(SO_2)$ at 1320, $\nu_s(SO_2)$ at 1135, ν(Mn–O) at 470, ν(Mn–N) at 375, and ν(Mn–Cl) at 335 cm^{-1}. The broad band at 3510 cm^{-1} demonstrates the presence of two coordinated water molecules. A considerable negative shift of the ν(NH) band, as compared to the ligand, indicates that the ligand is neutral and coordinated through the nitrogen atom. The negative shift of both the $\nu(SO_2)$ bands and their splitting in two different energy bands suggest that only one O atom is involved in coordination. The band at 335 cm^{-1} and a negligible conductivity of the complex in DMF are in agreement with the coordinative nature of both chlorine atoms. The X-ray diffraction patterns of the compound (and of the corresponding chelates of Ni^{II}, Cu^{II}, and Zn^{II}) with ligand 20 support the IR data. New peaks, which are not present in the spectrum of the ligand, can be assigned to the bond lengths M–Cl (4.14 to 4.09 Å), M–N (3.09 to 3.02 Å), and M–O=S (2.96 to 2.80 Å) by use of ASTM cards. The visible electronic spectrum in DMF exhibits an absorption maximum at 400 nm, suggesting an octahedral structure [11].

$[Mn^{III}(C_{12}H_{13}N_4O_2S)_3]$. A solution of MnO_2 in concentrated aqueous HCl was extracted with ether, and the dark green extract, containing $MnCl_3$, was filtered, and added to a solution of three equivalents of ligand 7 in acetone containing ammonia. The resulting solution (pH 8 to 9) was evaporated on a water bath, and the residue dried in air [7].

References:

[1] Mil'grom, A. E.; Andrianova, L. N.; Pesnya, O. I.; Vladyko, G. V.; Korobchenko, L. V.; Karako, N. I.; Boreko, E. I. (Khim. Farm. Zh. **19** [1985] 163/5; Pharm. Chem. J. [USSR] **19** [1985] 177/9).

[2] Rao, P. Gundu; Ranganatham, P.; Srinivasan, K. K.; Sharada, N. R. (Current Sci. [India] **48** [1979] 99/101).

[3] Ruskin, S. L. (U.S. 2602085 [1952]; C.A. **1953** 8776).

[4] Tskitishvili, M. G.; Gogorishvili, P. V.; Chrelashvili, M. V.; Shvelashvili, A. E. (Izv. Akad. Nauk Gruz.SSR Ser. Khim. **5** [1979] 13/9; C.A. **92** [1980] No. 163917).

[5] Zedelashvili, E. N.; Shvelashvili, A. E.; Gogorishvili, D. A. (Zh. Neorgan. Khim. **25** [1980] 3309/14; Soviet J. Inorg. Chem. **25** [1980] 1812/5; C.A. **94** [1981] No. 57248).

[6] Gogorishvili, P. V.; Tskitishvili, M. G. (Issled. Obl. Khim. Kompleksn. Prostykh Soedin. Nekot. Perekhodnykh Redk. Metal. No. 2 [1974] 17/29; C.A. **82** [1974] No. 38073).

[7] Gogorishvili, P. V.; Tskitishvili, M. G.; Kalandarishvili, D. Z. (Issled. Obl. Khim. Kompleksn. Prostykh Soedin. Nekot. Perekhodnykh Redk. Metal. No. 2 [1974] 152/73; C.A. **82** [1974] No. 38083).

[8] Gogorishvili, P. V.; Tskitishvili, M. G. (Issled. Obl. Khim. Kompleksn. Prostykh Soedin. Nekot. Perekhodnykh Redk. Metal. No. 3 [1978] 5/22; C.A. **90** [1979] No. 214469).

[9] Gogorishvili, P. V.; Tskitishvili, M. G.; Machkhoshvili, R. I.; Kharitonov, Yu. Ya. (Zh. Neorgan. Khim. **20** [1975] 1420/2; Soviet J. Inorg. Chem. **20** [1975] 798/9; C.A. **83** [1975] No. 90119).
[10] Tskitishvili, M. G.; Shvelashvili, A. E.; Mikadze, I. I.; Zhorzholiani, N. B.; Chrelashvili, M. V. (Izv. Akad. Nauk Gruz.SSR Ser. Khim. **7** [1981] 300/4; C.A. **96** [1982] No. 111045).

[11] Abu-El-Wafa, S. M.; El-Ries, M. A.; Aly, F. A.; Issa, R. M. (Egypt. J. Pharm. Sci. **26** [1985] 7/14; C.A. **107** [1987] No. 16689).
[12] Tskitishvili, M. G.; Shvelashivili, A. E.; Mikadze, I. I.; Chrelashvili, M. V.; Zhorzholiani, N. B. (Izv. Akad. Nauk Gruz.SSR Ser. Khim. **7** [1981] 204/12; C.A. **96** [1982] No. 134774).
[13] Tskitishvili, M. G.; Gogorishvili, P. G.; Machkhoshvili, R. I.; Zhorzholiani, N. B. (Izv. Akad. Nauk Gruz.SSR Ser. Khim. **5** [1979] 295/301; C.A. **93** [1980] No. 18280).

35.8.2 With an N,N′-Disubstituted Derivative of Sulfanilamide

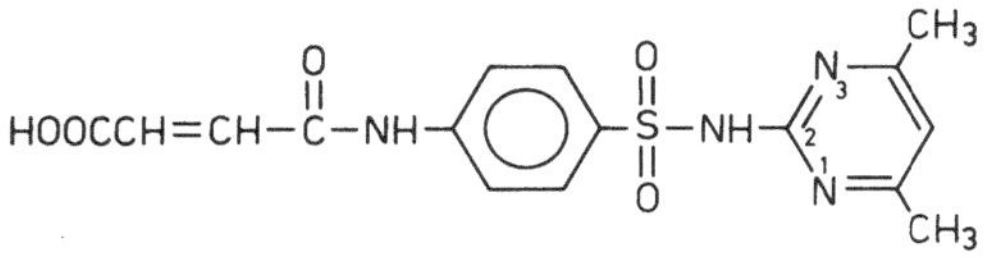

$(=C_{16}H_{16}N_4O_5S = H_2L)$

$Na_2[Mn^{II}(C_{16}H_{14}N_4O_5S)_2]$. An ethanol solution of the monosodium salt of the ligand (1 mol) adjusted to pH 7 was poured into manganese(II) hydroxide (0.5 mol), the mixture brought to 70°C for a few minutes, and filtered. Ethanol (2.5 kg) was added to the filtrate, followed by 10 L anhydrous acetone, and the mixture boiled for several minutes. The compound which separated on cooling was dried in warm air. It is supposed that the mixed salt consists of one manganese(II) atom bridging two sulfonamide nitrogen atoms of the two twice-deprotonated ligand molecules. The complex is soluble in water and shows antimicrobial and fungicidal activity.

Reference:

Vallee, J. P. S. (Brit. 940052 [1960/63]; C.A. **60** [1964] 2844).

35.8.3 With N,N′-Ethanediylbis(2-aminobenzenesulfonamide)

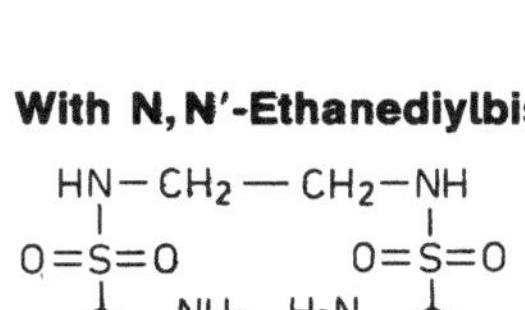

$(=C_{14}H_{18}N_4O_4S_2 = H_2L)$

$K_2[Mn^{II}(C_{14}H_{16}N_4O_4S_2)_2]$. The complex was prepared by addition of an aqueous KOH solution to an ethanolic solution of $MnCl_2 \cdot 4H_2O$ and the ligand (mole ratio 3.6 : 1 : 2) at 65°C. The reaction mixture was refluxed for 20 min, the brown precipitate washed with ethanol and acetone, and finally dried over P_4O_{10} in vacuum. The magnetic moment is $\mu_{eff} = 5.90$ μ_B at room temperature. The IR spectrum of the air-stable crystalline compound (KBr pellets) taken in the 4000 to 250 cm^{-1} region shows the following characteristic absorption bands (free ligand bands in parentheses; $\bar{\nu}$ in cm^{-1}): $\nu_{as}(NH_2)$ at 3440(3458); $\nu_s(NH_2)$ at 3360(3362); $\nu_{as}(SO_2)$ at 1313, 1037(1323), and $\nu_s(SO_2)$ at 1159, 1130, 989, 903(1170, 1162). The changes in the IR spectrum, including the disappearance of the ν(NH) free ligand band, suggest that the twice-deprotonated ligand is coordinated to Mn by the two sulfonamide N atoms and one O atom, the NH_2 groups being uninvolved in coordination. The electronic reflectance spectrum of the solid compound exhibits weak bands at 16530, 24390, and 26950 cm^{-1}, assignable to the electron transitions $^6A_{1g} \rightarrow {}^4T_{1g}(G)$, $\rightarrow {}^4E_g$, $^4A_{1g}(G)$, and $\rightarrow {}^4T_{2g}(D)$ and/or charge transfer, respectively, of the Mn^{II} ion in a distorted octahedral environment.

On heating, the compound does not change up to 280°C. It darkens in the range between 280 and 340°C. It is insoluble in common organic solvents. These properties together with the spectral data suggest a polymeric structure involving bridging L^{2-} units.

Reference:

Perlepes, S. P.; Zafiropoulos, T. F.; Galinos, A. G.; Tsangaris, J. M. (Z. Naturforsch. **38b** [1983] 350/6).

35.8.4 With Derivatives of p-Toluenesulfonamide

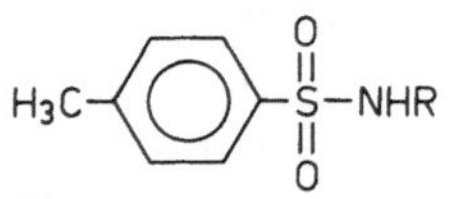
(=HL)

ligand 1 with R = C(O)NHC$_4$H$_9$ (= $C_{12}H_{18}N_2O_3S$)

ligand 2 with R = 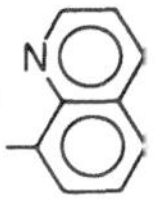 (= $C_{16}H_{14}N_2O_2S$)

Remark. Complexes with p-toluenesulfonamides of amino acids have been described in "Manganese" D 4, 1985, pp. 265 and 289. Hydrazones containing the p-tolylsulfonylamino group were described in "Manganese" D 6, 1988, p. 250. These ligands are listed on pp. 126/7 together with other sulfonamides already described in preceding volumes.

Complexes in Solution. Potentiometric titrations of solutions containing MnX_2 (X = Cl, Br, NCS) and ligand 1 in 50% (v/v) acetone-water at 25°C and pH 3 to 5.5 reveal the formation of 1:1 and 1:2 complexes with the following stability constants [1]:

equilibrium	log K (X = Cl)	log K (X = Br)	log K (X = NCS)
$Mn^{II}X_2 + HL \rightleftharpoons Mn^{II}(HL)X_2$	5.12	5.27	6.30
$Mn^{II}(HL)X_2 + HL \rightleftharpoons Mn^{II}(HL)_2X_2$	5.10	5.19	5.74

The stabilities are compared to those of analogous complexes with derivatives of sulfanilamide, and to those of Mg complexes with ligand 1 [1]. Conductometric titrations and Job's method of continuous variation reveal the formation of a 1:2 complex of manganese(II) and deprotonated form of ligand 1 in 80% aqueous ethanol, also [2].

$Mn^{II}(C_{16}H_{13}N_2O_2S)_2$. The dirty brown complex is quantitatively precipitated on reaction of manganese(II) ion with ligand 2 at pH 5.5 to 9.5. It is unstable towards heat. Coordination of the two N atoms of the ligand to Mn is assumed [4].

$[Mn^{II}(C_{12}H_{17}N_2O_3S)_2(H_2O)_2]$. Metal salt and ligand 1 (mole ratio 1:2) were dissolved separately in minimum quantities of absolute alcohol, and refluxed for 3 h. The dark brown complex precipitated on cooling, and was separated, washed, and dried. It melts at 195°C. The IR spectrum exhibits characteristic absorption bands of ν(CO) around 1085 (absent in the ligand) and ν(M–O) around 660 cm^{-1} (reported for the Mg, Ba, Mn, and Co compounds). The data indicate O,O coordination of the two ligand molecules in an octahedral complex structure with the two water molecules occupying the axial sites [2].

$Mn^{II}(C_{12}H_{18}N_2O_3S)_2Cl_2$. The light brown crystalline compound was prepared by reacting $MnCl_2$ and ligand 1 in the same way as the $Mn^{II}(HL)_2X_2$ complexes described on p. 121. It has the same properties as these compounds [3].

References:

[1] Tskitishvili, M. G.; Mikadze, I. I.; Zhorzholiani, N. B.; Chrelashvili, M. V. (Koord. Khim. **9** [1983] 369/72; C.A. **98** [1983] No. 186599).

[2] Desnavi, A.; Iqbal, S. A. (Orient. J. Chem. **2** [1986] 156/9; C.A. **105** [1986] No. 213991).

[3] Tskitishvili, M. G.; Shvelashivili, A. E.; Mikadze, I. I.; Chrelashvili, M. V.; Zhorzholiani, N. B. (Izv. Akad. Nauk Gruz.SSR Ser. Khim. **7** [1981] 204/12; C.A. **96** [1982] No. 134774).

[4] Agarwal, R. C. (J. Indian Chem. Soc. **52** [1975] 207/9).

35.8.5 With N-Cyclohexyl-2-benzothiazolesulfonamide

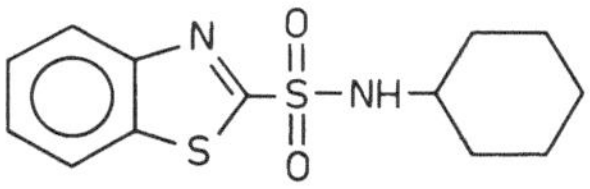

($=C_{13}H_{16}N_2O_2S_2$)

$Mn^{II}(C_{13}H_{16}N_2O_2S_2)_2X_2$ with X = NO_3, Cl, CH_3COO, and $C_6H_2(NO_2)_3O$ were prepared by the reaction of an ethanolic solution of MnX_2 and the ligand in the mole ratio 1:2. The reaction mixture was refluxed for 0.5 h, concentrated to half its volume, and then cooled. The precipitate was collected, washed with ethanol and ether, and dried in vacuum. The magnetic moments of the complexes, resulting from susceptibility measurements at room temperature, are: μ_{eff} = 6.00, 5.90, 5.87, and 5.88 μ_B for X = NO_3, Cl, CH_3COO, and $C_6H_2(NO_2)_3O$, respectively. In the IR spectra shifts to lower wavenumbers were observed for the bands of the sulfonamide group and for the ν(CS) band. A new band appearing around 500 cm^{-1} was assigned to the ν(Mn–N) vibration. The data indicate that the neutral bidentate ligand is bonded to the manganese(II) atom via the ring sulfur atom and the nitrogen atom of the sulfonamide group. For the acetato complexes the ν_{as}(COO) bands were observed in the 1550 to 1530 cm^{-1} region (acetate ion at 1578 cm^{-1}), and the ν_s(COO) bands in the 1450 to 1435 cm^{-1} region (acetate ion at 1425 cm^{-1}). The changes in these modes on complexation indicate that the acetato group acts as a unidentate ligand. The electronic absorption spectra of the solids show four bands: in the regions 19700 to 19800, 23000, 28500 to 29000, and 34000 to 34600 cm^{-1}, which were assigned to the high-spin octahedral Mn^{II} transitions $^6A_{1g} \rightarrow {}^4T_{1g}(G)$, $^6A_{1g} \rightarrow {}^4T_{2g}(G)$, $^6A_{1g} \rightarrow {}^4T_{2g}(D)$, and $^6A_{1g} \rightarrow {}^4T_{1g}(P)$, respectively. All the complexes exhibit significant fungitoxicity [1].

Complexes of the **$Mn^{II}(HL)_2X_2$** type with HL = a ligand named benzothiazolesulfonamide morpholide ($=C_{12}H_{14}N_3S_2O_3$?) and X = NO_3, Cl, CH_3COO, $C_6H_2(NO_2)_3O$ were prepared in a way similar to that described for the $Mn(C_{13}H_{16}N_2O_2S_2)_2X_2$ complexes, and display properties similar to these [2].

References:

[1] Srivastava, S. K.; Verman, A.; Gupta, A. (J. Indian Chem. Soc. **59** [1982] 925/6).

[2] Srivastava, S. K.; Gupta, A.; Verman, A. (Chim. Acta Turc. **11** [1983] 99/106).

35.8.6 With Other Sulfonamides

Manganese complexes with the ligands tabulated below have already been described in the chapters "Complexes with N-Heterocycles" ("Manganese" D 3, 1982), "Complexes with Amino Acids" (Manganese D 4, 1985), "Complexes with Nitrosonaphthols" ("Manganese" D 5, 1987), "Complexes with Azo Compounds" ("Manganese" D 5, 1987), "Complexes with Schiff Bases" ("Manganese" D 6, 1988), and "Complexes with Hydrazones" ("Manganese" D 6, 1988):

ligand			volume, page
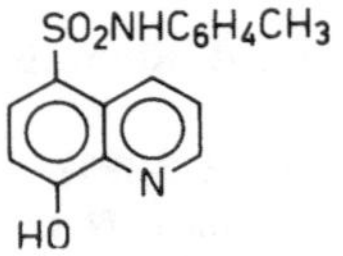		(= $C_{16}H_{14}N_2O_3S$)	"Manganese" D 3, 1982, p. 186
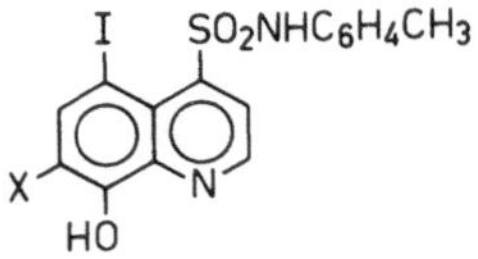	with X = H with X = Cl with X = I	(= $C_{16}H_{13}IN_2O_3S$) (= $C_{16}H_{12}ClIN_2O_3S$) (= $C_{16}H_{12}I_2N_2O_3S$)	"Manganese" D 3, 1982, p. 186
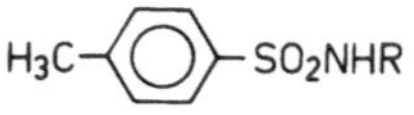	with R = CH_2COOH	(= $C_9H_{11}NO_4S$)	"Manganese" D 4, 1985, p. 265
	with R = $(CH_2)_4CH(NH_2)COOH$	(= $C_{13}H_{20}N_2O_4S$)	"Manganese" D 4, 1985, p. 289
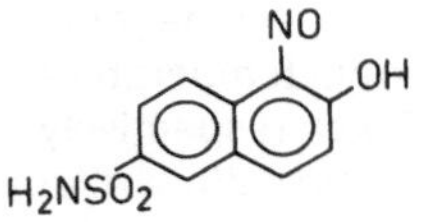		(= $C_{10}H_8N_2O_4S$)	"Manganese" D 5, 1987, p. 254
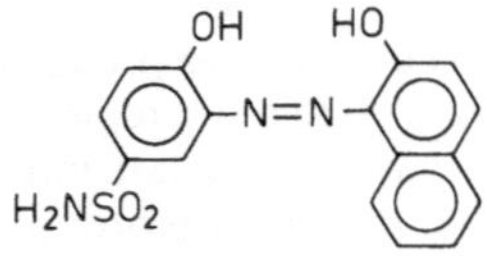		(= $C_{16}H_{13}N_3O_4S$)	"Manganese" D 5, 1987, p. 286
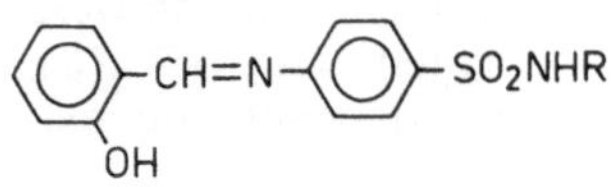	with R = H	(= $C_{13}H_{12}N_2O_3S$)	"Manganese" D 6, 1988, p. 15
	with R = 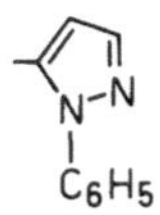	(= $C_{22}H_{18}N_4O_3S$)	
	with R = 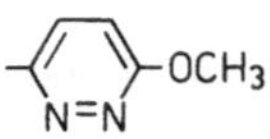	(= $C_{18}H_{16}N_4O_4S$)	

ligand			volume, page
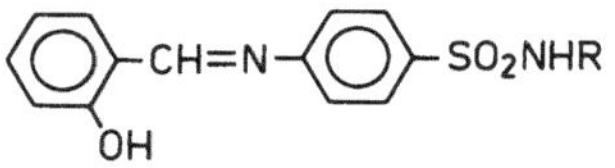	with R = 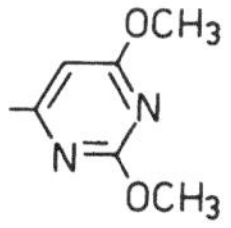	(= $C_{19}H_{18}N_4O_5S$)	"Manganese" D 6, 1988, p. 15
	with R = 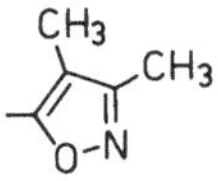	(= $C_{18}H_{17}N_3O_4S$)	
	with R = 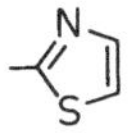	(= $C_{16}H_{13}N_3O_3S_2$)	
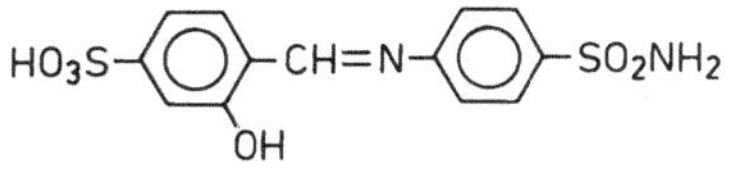		(= $C_{13}H_{12}N_2O_6S_2$)	"Manganese" D 6, 1988, p. 50
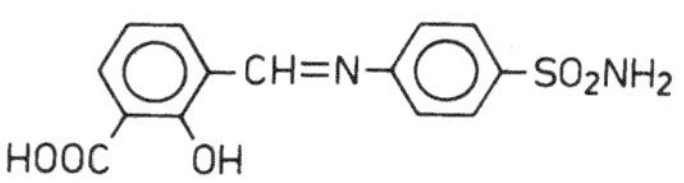		(= $C_{14}H_{12}N_2O_5S$)	"Manganese" D 6, 1988, p. 62
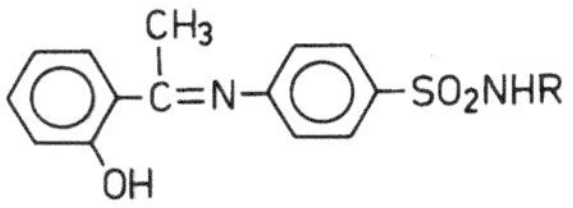	with R = 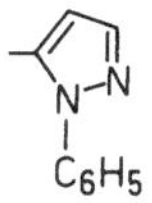	(= $C_{23}H_{20}N_4O_3S$)	"Manganese" D 6, 1988, p. 87
	with R = 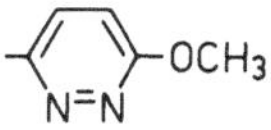	(= $C_{19}H_{18}N_4O_4S$)	
	with R = 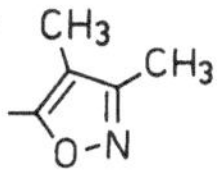	(= $C_{19}H_{19}N_3O_4S$)	
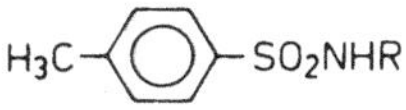	with R = 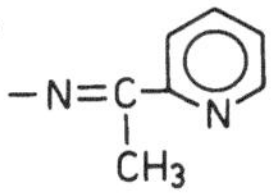	(= $C_{14}H_{15}N_3O_2S$)	"Manganese" D 6, 1988, p. 250
	with R = 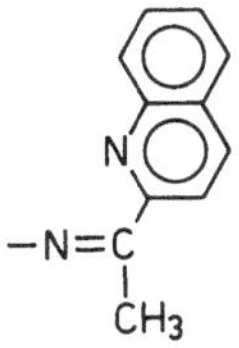	(= $C_{18}H_{17}N_3O_2S$)	
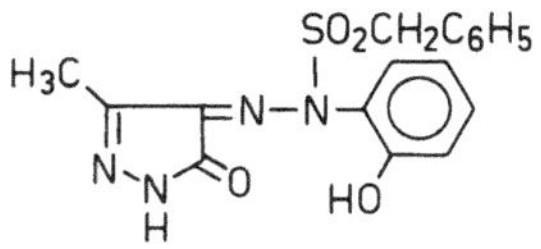		(= $C_{17}H_{16}N_4O_4S$)	"Manganese" D 6, 1988, p. 254

35.8.7 With 2-Sulfobenzoic Imide (= Saccharin)

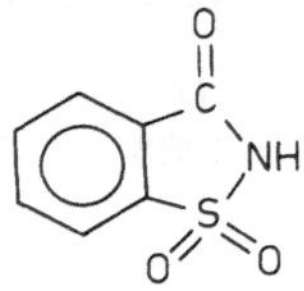

(= $C_7H_5NO_3S$ = HL)

$[Mn^{II}(C_7H_4NO_3S)_2(H_2O)_4]\cdot2H_2O$. The complex was prepared by heating 1.2 g $Mn(CH_3COO)_2\cdot4H_2O$ in 70 mL water with 1.8 g saccharin on a steam bath for 2 h. The solution was filtered and the filtrate kept at room temperature for 2 d, when faintly pink crystals appeared. These were separated, recrystallized from methanol, and washed with acetone [1]. The compound can also be easily obtained by metathesis of sodium saccharinate with $MnCl_2$ in aqueous solution with heating to boiling [2, 4] or by neutralization of an aqueous saccharin solution with $MnCO_3$ [4]. By evaporation of the solution on the water bath, well-formed prismatic crystals were obtained and washed with cold water [2, 4]. In the earlier publications [2, 4 to 6] the compound was erroneously formulated as the tetrahydrate.

A single-crystal diffractometer study reveals a monoclinic lattice, space group $P2_1/c-C^5_{2h}$ (No. 14) [2, 4] with the lattice parameters a = 7.955(2), b = 16.131(6), c = 7.784(1) Å, β = 99.70(3)°; Z = 2. The structure was solved from 2197 independent reflections and refined with anisotropic thermal parameters for all nonhydrogen atoms up to a final reliability of R = 0.028. The measured density is 1.73 g/cm³, the calculated density is 1.78 g/cm³. Atomic coordinates are given in the paper. As shown in **Fig. 14**, the manganese atom has a distorted octahedral geometry with the two N atoms from saccharinate ions in *trans* positions. Four

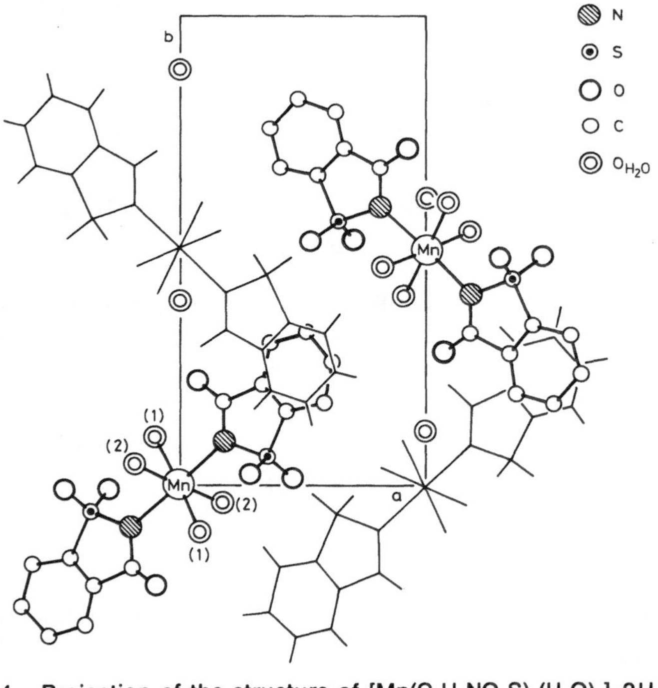

Fig. 14. Projection of the structure of $[Mn(C_7H_4NO_3S)_2(H_2O)_4]\cdot2H_2O$ on the a, b plane [3]. H atoms omitted for clarity.

water molecules are coordinated to the metal, while the remaining two are held in the crystal lattice as water of crystallization. All six water molecules participate in hydrogen bonding with oxygen atoms from saccharinate ions [3]. The saccharinate ions appear to be approximately planar (with the exception of the two oxygens attached to the sulfur) [5, 6].

Selected interatomic distances and angles are given below [3]:

distances in Å		angles in °	
Mn–N	2.281(1)	$O_{H_2O}(1)$–Mn–$O_{H_2O}(2)$	89.5(1)
Mn–$O_{H_2O}(1)$	2.162(1)	$O_{H_2O}(1)$–Mn–N	86.7(1)
Mn–$O_{H_2O}(2)$	2.219(2)	$O_{H_2O}(2)$–Mn–N	86.6(1)
		Mn–N–S	119.6(1)
		Mn–N–C(7)	129.4(1)

The complex is isostructural with analogous saccharinato complexes of Fe^{II}, Cu^{II}, Cd, Ni^{II}, and Zn^{II} [1, 3]. The crystal structure of the complex supposed to be the tetrahydrate was published earlier. While the values of the lattice parameters are in good agreement with those reported in [3], the densities, D_{obs}=1.645 g/cm³ and D_{calc}=1.644 g/cm³, differ significantly [4 to 6].

The IR spectrum of $[Mn(C_7H_4NO_3S)_2(H_2O)_4]\cdot 2H_2O$, recorded from Nujol mulls or KBr disks, shows characteristic absorption bands (in cm^{-1}) which were assigned as follows: five bands in the range between 3550 and 3000 to ν(OH) (of H_2O), bands at 1600 to ν(CO), 1340, 1330 to ν(CN), 1273 to $\nu_{as}(SO_2)$, 1148 to $\nu_s(SO_2)$, 671 to ν(Mn–O_{H_2O}), and 442 to ν(Mn–N). The lowering of the CO and SO_2 stretching frequencies, relative to those of the free ligand, was explained by extensive hydrogen bonding between water molecules and saccharinate oxygen atoms.

The complex is fairly stable toward light or air [1]. The thermal decomposition was studied by differential scanning calorimetry, thermogravimetry, or differential thermogravimetry in air with a heating rate of 2.5°C/min [1] and in flowing air, argon, or nitrogen (flowing rate 50 mL/min) with a heating rate of 10°C/min [7]. It was found that all six water molecules are removed in one step between 97 and 149°C with an enthalpy of dehydration of 49.1 kcal/mol, and that an endothermic solid-solid transition occurs between 167 and 226°C with a transition enthalpy of 33.9 kcal/mol [7]. In the range between 380 and 550°C the anhydrous compound decomposes, leaving MnO_2 as the residue [1, 7]. Melting of $[Mn(C_7H_4NO_3S)_2(H_2O)_4]\cdot 2H_2O$ with decomposition had been observed earlier [2]. The thermal data indicate that there exist superstructures which could be confirmed by weak extra reflections in the X-ray powder patterns [7]. The complex is soluble in hot water, methanol, ethanol, or benzene, only slightly soluble in cold water, and insoluble in acetone or other organic solvents [1, 2]. It is less sweet than saccharin, but leaves a bitter aftertaste in the throat [2].

References:

[1] Haider, S. Z.; Malik, K. M. A. (J. Bangladesh Acad. Sci. **6** [1982] 119/25).
[2] Defournel, M. H. (Bull. Soc. Chim. France [3] **25** [1901] 322/9).
[3] Kamenar, B.; Jovanovski, G. (Cryst. Struct. Commun. **11** [1982] 257/61).
[4] Font-Altaba, M. (Publ. Dept. Crystalogr. Mineral. C.S.I.C. Spain **1** [1954] 133/4; C.A. **1955** 14416).
[5] Font-Altaba, M. (Publ. Dept. Crystalogr. Mineral. C.S.I.C. Spain **2** [1956] 163/77; C.A. **1958** 3457).
[6] Font-Altaba, M. (Structure Reports **20** [1956] 575).
[7] Haider, S. Z.; Malik, K. M. A.; Wadsten, T. (J. Bangladesh Acad. Sci. **7** [1983] 107/10).

35.9 Complexes with Xanthic Acids

35.9.1 With Monoxanthic Acids

RO–C(=S)–SH (=HL)

ligand	R	formula
1	C_2H_5	$C_3H_6OS_2$
2	$HOCH_2CH_2$	$C_3H_6O_2S_2$
3	C_4H_9	$C_5H_{10}OS_2$
4	$i\text{-}C_4H_9$	$C_5H_{10}OS_2$
5	C_6H_5	$C_7H_6OS_2$

ligand	R	formula	
6	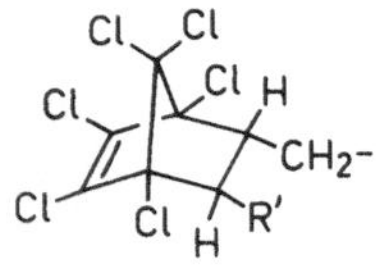	R′ = H	$C_9H_6Cl_6OS_2$
7		R′ = CH_3	$C_{10}H_8Cl_6OS_2$
8	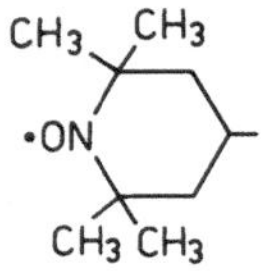		$C_{10}H_{18}NO_2S_2$

35.9.1.1 Manganese(II) Compounds

The complex with ligand 1, **$Mn(C_3H_5OS_2)_2$**, was prepared by the reaction of an aqueous solution of manganese(II) halide with potassium ethyl xanthate (mole ratio 1:2) under nitrogen. The reaction mixture was evaporated to dryness and the residue taken up in acetonitrile. The solution, extremely sensitive to air, was chromatographed on a column of alumina (deactivated with ethyl acetate) and the eluent concentrated by removing the solvent at room temperature under vacuum. Addition of diethyl ether or benzene to the eluent precipitated the yellow complex, which was dried in vacuum [1]. The compound was also produced by electrolysis of manganese in dixanthogen, $[C_2H_5OC(S)S]_2$ under a blanket of argon, but it oxidized immediately in air to the purple-brown Mn^{III} complex (see p. 133) [2]. The yellow-brown $Mn(C_3H_5OS_2)_2$ and the complex with ligand 5, **$Mn(C_7H_5OS_2)_2$**, were precipitated from aqueous Mn^{II} salt solution (pH 7) by addition of the potassium salt of the respective ligand [3]. The complexes with ligand 6, **$Mn(C_9H_5Cl_6OS_2)_2$**, or ligand 7, **$Mn(C_{10}H_7Cl_6OS_2)_2$**, were precipitated from dilute aqueous Mn^{II} salt solutions, slightly acidified with acetic acid. The brown precipitates were washed with water, dried over P_4O_{10}, then dissolved in ethyl acetate at 50°C. Crystals separated from the filtered solution after evaporating and cooling at −18°C. The complexes were washed with diethyl ether [5]. The formation of the complex with ligand 4, **$Mn(C_5H_9OS_2)_2$**, in aqueous solution is reported in [4].

The magnetic moment of $Mn(C_3H_5OS_2)_2$, $\mu_{eff} = 4.8\ \mu_B$ at room temperature, may be reduced by metal-metal interaction due to polymerization, which achieves the hexacoordination at Mn^{II}. The electronic reflectance spectrum shows a shoulder at 15900 cm^{-1}, assigned to the $^6A_{1g} \rightarrow {}^4T_{1g}$ transition of the proposed environment [1].

$Mn(C_3H_5OS_2)_2$ is appreciably soluble in water, soluble in acetonitrile, not soluble in benzene or diethyl ether [1]. $Mn(C_3H_5OS_2)_2$ and $Mn(C_7H_5OS_2)_2$ are practically insoluble in nonpolar solvents [3]. The ESR spectra of these complexes, extracted into toluene from the aqueous reaction mixture, show at 285 K a six-line Mn^{II} pattern with $g = 2.002$ and the isotropic hyperfine coupling constant $A_0 = -87.5 \times 10^{-4}$ and $-87.10^{-4}\ cm^{-1}$, respectively. The extracted species are probably six-coordinated with four coplanar sulfur atoms and two water molecules in apical positions. After several minutes polymerization and a decrease of the hyperfine splittings, probably due to polymerization, was observed [3]. The ESR spectrum of $Mn(C_3H_5OS_2)_2$ in benzene has also been investigated [6].

The polychlorinated complexes, $Mn(C_9H_5Cl_6OS_2)_2$ and $Mn(C_{10}H_7Cl_6OS_2)_2$, were investigated by thin-layer chromatography. R_f values were determined on silica gel GF_{254} and alumina HF_{254} for various eluents and compared with R_f values of Bi, Zn, Cd, Hg, Pb, Ni, Co, Fe, and Cu complexes. At 200°C the complexes pyrolyze quickly and reproducibly to the corresponding alcohol, allowing the chlorinated ligands to be utilized for the determination of ng amounts of metals by electron-capture gas-chromatography [5].

All the complexes are sensitive to air, especially in solution [1, 3, 5].

[$Mn(C_3H_5O_2S_2)_2 \cdot 2H_2O$] was precipitated from aqueous solution containing stoichiometric amounts of manganese(II) chloride and ligand 2. The light yellow complex was washed with methanol and dried in vacuum. The magnetic moment at 30°C is 5.88 μ_B. In the IR spectrum of the complex a weak band in the 3700 to 3680 cm^{-1} region is attributed to coordinated water. Additional water vibration modes were observed at 3500 to 3300 and 1620 to 1605 cm^{-1}. Bands of the free ligand at 1380, 990, 825, and 700 cm^{-1}, assigned to $\nu(CS)$, are shifted to lower wavenumbers upon complexation, which indicates coordination of the sulfur atoms to Mn^{II}. The oxygen atoms are not involved in coordination at Mn^{II}. The complex, when left over concentrated sulfuric acid in vacuum or heated to 130°C does not lose water [7].

$K[Mn(C_3H_5OS_2)_3]$ was obtained from appropriate amounts of $MnSO_4$ and potassium ethyl xanthate in aqueous solution below 5°C. The yellow precipitate was dissolved in a 1:1 ethanol-ether mixture and separated from K_2SO_4 by centrifugation. The organic layer was concentrated below 5°C by a stream of dry nitrogen until yellow crystals appeared. The procedure was repeated three times and the crystals dried under nitrogen in a desiccator. The yellow complex is not stable in air, changes its color reversibly to red at 120°C, and does not melt below 250°C. It is soluble in water, but the aqueous solutions are not stable in acidic medium. The complex is quantitatively extracted from aqueous solution at pH 8.0 to 9.2 into methyl isobutyl ketone. The electronic spectrum of the solid complex dissolved in methyl isobutyl ketone is identical to the spectrum of the methyl isobutyl ketone extract from aqueous solution [8].

Quantitative extraction of Mn^{II} from aqueous solution at pH 6.5 to 9 with methyl isobutyl ketone containing a large excess of xanthic acid was also observed. The organic phase can be separated directly into the flame of the atomic absorption spectrophotometer to determine Mn [9].

$[N(C_2H_5)_4][Mn(C_3H_5OS_2)_3]$ was prepared by the reaction of aqueous solutions of potassium ethyl xanthate, manganese(II) halide, and $[N(C_2H_5)_4]Cl$ in a mole ratio of 3:1:1 under a nitrogen atmosphere. The crude, oily yellow solid solidified after scratching, and was then isolated and dried in vacuum. The product was recrystallized from acetonitrile or dichloromethane with the addition of diethyl ether. The magnetic moment of the solid is 6.0 μ_B at ambient temperature. The electronic reflectance spectrum of the complex shows a shoulder at 14700 cm^{-1} which was assigned to a high-spin Mn^{II} transition $^6A_{1g} \rightarrow {}^4T_{1g}$ [1].

[$Mn(C_3H_5OS_2)_2bpy$] and **[$Mn(C_3H_5OS_2)_2phen$]** complexes were prepared by the reaction of an aqueous solution of potassium ethyl xanthate and a stirred aqueous solution of $[Mn bpy(H_2O)_4]SO_4$ or $[Mn phen(H_2O)_4]SO_4$ in 2:1 mole ratio. The yellow precipitates were washed with 20% ethanol followed by diethyl ether and dried in vacuum [10]. The complexes were also prepared by the reaction of an aqueous solution of potassium ethyl xanthate with an aqueous solution of $MnCl_2$ and the amine in a mole ratio of 2:1:1 or 2:1:2, but the elemental analysis of the complex with bipyridine indicated the presence of approximately 0.9 mol of base [1]. (When the amount of base is increased up to the ratio $C_3H_5OS_2K:MnCl_2:phen =$

2:1:3, the phenanthroline complex, $[Mn phen_3](C_3H_5OS_2)_2$, is formed [11]; see "Manganese" D 3, 1982, p. 246.) The yellow solids were dried in vacuum, and were recrystallized from dichloromethane with addition of diethyl ether. The magnetic moments are 5.6 μ_B for $[Mn(C_3H_5OS_2)_2bpy]$ and 6.0 μ_B for $[Mn(C_3H_5OS_2)_2phen]$ at ambient temperature. The IR spectra of the complexes (in Nujol or KBr) show no exclusive ν(CS) vibration, but bands at 1120, 810, and 620 cm^{-1} were assigned to CS absorptions. Absorptions between 300 and 250 cm^{-1} and at 390 cm^{-1} were assigned to ν(Mn–N) and ν(Mn–S), respectively, indicating the coordination of bipyridine or 1,10-phenanthroline and xanthato groups through the nitrogen and sulfur atoms. The electronic reflectance spectra of the complexes show shoulders at 14700 and 15400 cm^{-1}, respectively. The maxima were assigned to the high-spin Mn^{II} transition $^6A_{1g} \rightarrow {}^4T_{1g}$ [1].

$[Mn(C_{10}H_{17}NO_2S_2)_2phen]$ was prepared by adding the aqueous solution of the potassium salt of ligand 8, a stable radical, to the mixture of manganese(II) halide and 1,10-phenanthroline in a mole ratio of 2:1:1 to 2:1:2. The precipitate was washed with warm water and dried in vacuum. The yellow complex melts at 101 to 102°C. The IR spectrum exhibits the characteristic bands of 1,10-phenanthroline at 1600 and 1480 cm^{-1} and of the xanthate at 1200, 1165, and 1060 cm^{-1}. The ESR spectrum of a polycrystalline sample at room temperature reveals a broad signal, $\Delta H = 245 \pm 10$ G, due to Mn^{II}, instead of a small line caused by the radical ligand. The complex is soluble in ethanol, acetone, chloroform, dimethylformamide and is insoluble in carbontetrachloride. The ethanol solution shows bands at 310, 267, and 233 nm ($\varepsilon = 0.7$ to 1.5 $L \cdot mol^{-1} \cdot cm^{-1}$) [12].

$MnL_2(NO)^+$ and $[MnL_2(NO)Cl]$. The formation of $MnL_2(NO)^+$ species by passing NO gas through toluene solutions of MnL_2 or MnL_3 complexes, followed by argon was reported in [4, 13, 14]. $[MnL_2(NO)Cl]$ complexes with HL = ligand 1, 3, or 5 were prepared by passing NO gas through toluene-water mixtures of MnL_3 complexes (obtained in situ by air oxidation of the MnL_2 complexes) in the presence of Cl^- ions. The complexes were extracted into toluene. After evaporation of the solvent, dark brown solids were obtained. For the ESR studies of the $[MnL_2(NO)Cl]$ complexes, toluene extracts direct from reaction mixtures were used [3]. Analysis of the ESR spectra yielded the parameter $g_0 = 2.011$ and the isotropic hyperfine coupling constant $A_0(^{55}Mn) = 77.1$ G for the ethyl xanthate complex, $Mn(C_3H_5OS_2)_2(NO)^+$, in toluene at 257 K. Values of $g_{\parallel} = 2.003$, $g_{\perp} = 2.015$, $A_{\parallel} = 153.9$ and $A_{\perp} = 39.0$ G were observed in frozen toluene solutions at 77 K [14].

The parameters $g_{\parallel} = 1.994$, $g_{\perp} = 2.023$ and $K = 81.2 \times 10^{-4}$ or 81.3×10^{-4} cm^{-1} ($K = -A_0 + (g_0 - 2.022)P$ with $P \approx 187 \times 10^{-4}$ for ^{55}Mn in $Mn(NO)^{3+}$) have been evaluated for the ethyl and phenyl xanthato complexes, $[Mn(C_3H_5OS_2)_2(NO)Cl]$ and $[Mn(C_7H_5OS_2)_2(NO)Cl]$, at 120 K [3, 15]. Similar data are reported for $MnL_2(NO)^+$ complexes with HL = ligand 3 [16] or ligand 4 [4]. The ESR parameters are consistent with low-spin (S = ½) complexes and linear-coordinated NO groups [3, 15]. For the $[MnL_2(NO)Cl]$ complexes with HL = ligands 1 and 5 a ^{14}N super-hyperfine coupling constant of $A_{\perp} = -5.4 \times 10^{-4}$ cm^{-1} was obtained. The ^{14}N super-hyperfine interaction probably arises from spin-polarization mechanisms involving π orbitals [3]. For changes in the ESR parameters when the Cl^- ion is replaced by thiourea or another sulfur-containing monodentate ligand, see [15].

References:

[1] Holah, G. D.; Murphy, C. N. (Can. J. Chem. **49** [1971] 2726/32).

[2] Casey, A. T.; Vecchio, A. M. (Transition Metal Chem. [Weinheim] **11** [1986] 366/8).

[3] Jezierski, A. (J. Mol. Struct. **115** [1984] 11/4).

[4] Jeżowska-Trzebiatowska, B.; Jezierski, A. (3rd Intern. Symp. Specific Interact. Mol. Ions Proc., Wroclaw-Karpacz 1976, Vol. 1, pp. 257/62; C.A. **88** [1978] No. 112936).

[5] Schuphan, I.; Ballschmitter, K.; Tölg, G. (Z. Anal. Chem. **255** [1971] 116/22).

[6] Yamamoto, D.; Nishimoto, M. (Meiji Daigaku Nogakubu Kenkyu Hokoku No. 66 [1984] 31/7; C.A. **102** [1985] No. 104801).

[7] Kumar, R.; Ansari, M. N.; Jain, M. C. (Indian J. Chem. A **26** [1987] 74/5).

[8] Hayashi, K.; Sasaki, Y.; Tagashira, S.; Yamate, E. (Bunseki Kagaku **27** [1978] 712/6; C.A. **90** [1979] No. 77193).

[9] Aihara, M.; Kiboku, M. (Bunseki Kagaku **23** [1974] 505/10; C.A. **81** [1974] No. 85523).

[10] Malik, W. U.; Bembi, R.; Bhardwaj, V. K. (J. Indian Chem. Soc. **57** [1980] 35/8).

[11] Holah, D. G.; Murphy, C. N. (Inorg. Nucl. Chem. Letters **8** [1972] 1069/72).

[12] Soloshenkin, P. M.; Shvengler, F. A.; Rakitina, E. V. (Dokl. Akad. Nauk Tadzh. SSR **20** [1977] 25/7; C.A. **87** [1977] No. 94710).

[13] Yordanov, N. D.; Iliev, V.; Shopov, D.; Jezierski, A.; Jeżowska-Trzebiatowska, B. (Inorg. Chim. Acta **60** [1982] 9/15).

[14] Garif'yanov, N. S.; Luchina, S. A. (Izv. Akad. Nauk SSSR Ser. Khim. **1969** 471/2; Bull. Acad. Sci. USSR Div. Chem. Sci. **1969** 421/2).

[15] Jezierski, A.; Jeżowska-Trzebiatowska, B. (Nouv. J. Chim. **4** [1980] 599/602).

[16] Jeżowska-Trzebiatowska, B.; Jezierska, J.; Jezierski, A. (Magn. Resonance Relat. Phenom. Proc. 20th Congr. AMPERE, Talinn 1978 [1979], p. 301; C.A. **93** [1980] No. 123135).

35.9.1.2 Manganese(III) Compounds

MnL_3 complexes are formed by air oxidation of the MnL_2 complexes (see p. 130) in solution. An electrochemical synthesis of $Mn(C_3H_5OS_2)_3$ was carried out in an acetone solution (30 mL) of dixanthogen (200 mg) and $(C_2H_5)_4NClO_4$ (30 mg) as current carrier. A manganese metal anode was used in a 2 h run with the initial voltage of 50 V (dc) and 34 mA current. The purple-brown complex separated during electrolysis or was precipitated by adding petroleum ether. It was washed with petroleum ether and dried in vacuum; yield 57%. The magnetic moment at room temperature, $\mu_{eff} = 4.82\ \mu_B$, confirms the high-spin Mn^{III} (d^4) arrangement. The IR spectrum of the complex (in KBr) shows bands assigned to $\nu(CO)$ at 1235, $\nu(CS)$ at 1020, and $\nu(Mn–S)$ at 390 cm^{-1}.

Reference:

Casey, A. T.; Vecchio, A. M. (Transition Metal Chem. [Weinheim] **11** [1986] 366/8.)

35.9.2 With Dixanthic Acids

$$HS-\underset{\underset{S}{\|}}{C}-O-R-O-\underset{\underset{S}{\|}}{C}-SH$$

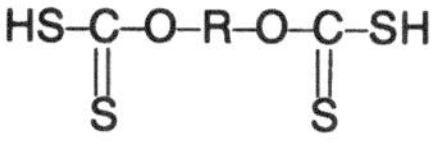

ligand	1	2		3	
R	–CH$_2$CH$_2$–			*cis* form	*trans* form
formula ...	C$_4$H$_6$O$_2$S$_4$	C$_{11}$H$_6$Cl$_6$O$_2$S$_4$		C$_{11}$H$_8$Cl$_6$O$_2$S$_4$	

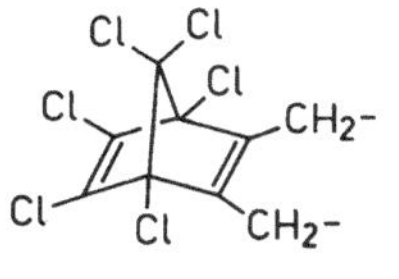

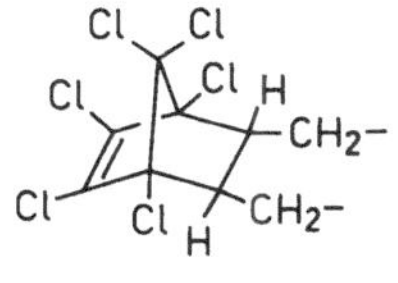

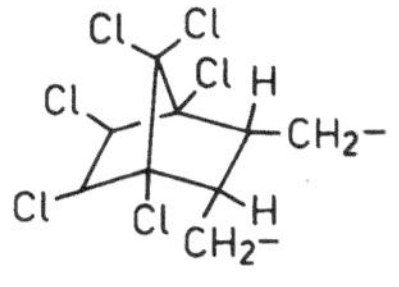

$Mn^{II}(C_{11}H_4Cl_6O_2S_4)$ and $Mn^{II}(C_{11}H_6Cl_6O_2S_4)$. Preparation and properties of the complex with ligand 2 and complexes with the *cis* and *trans* form of ligand 3 are similar to those reported on p. 130 for the complexes $Mn(C_9H_5Cl_6OS_2)_2$ and $Mn(C_{10}H_7Cl_6OS_2)_2$ [1].

$[Mn^{II}(C_4H_4O_2S_4)(H_2O)_2]$ was precipitated from equimolar aqueous solutions of manganese(II) chloride and the potassium salt of ligand 1, purified with methanol and dried in vacuum. The magnetic moment of the yellow complex at 300 K is 5.98 μ_B. The IR spectrum of the complex in KBr indicates the involvement of the sulfur atoms in coordination. The CS vibrations of the free ligand at 1370, 990, 825, and 690 cm^{-1} were shifted to 1340, 860, and 720 cm^{-1} upon complexation. The ν(COC) ligand bands at 1210 and 1120 cm^{-1} are observed at 1210 and 1180 cm^{-1} in the complex spectrum, suggesting that the oxygen atoms are not involved. Additional broad bands in the 3480 to 3380 and 1620 to 1600 cm^{-1} regions are assigned to water molecules, probably coordinated axially at the octahedral Mn^{II} [2].

References:

[1] Schuphan, I.; Ballschmitter, K.; Tölg, G. (Z. Anal. Chem. **255** [1971] 116/22).
[2] Kumar, R.; Ansari, M. N.; Jain, M. C. (J. Indian Chem. Soc. **61** [1987] 12/4).

35.10 Complexes with Dithiocarbamic Acids or N-Heterocyclic N-Carbodithioic Acids

General References:

Thorn, G. D.; Ludwig, R. A.; The Dithiocarbamates and Related Compounds, Elsevier, Amsterdam – New York 1962.
Willemse, J.; Cras, J. A.; Steggerda, J. J.; Keijzers, C. P.; Dithiocarbamates of Transition Group Elements in "Unusual" Oxidation States, Struct. Bonding [Berlin] **28** [1976] 83/126.
Burns, R. P.; McCullough, F. P.; McAuliffe, C. A.; 1,1-Dithiolato Complexes of the Transition Elements, Advan. Inorg. Chem. Radiochem. **23** [1980] 211/80.
Coucouvanis, D.; The Chemistry of the Dithioacid and 1,1-Dithiolate Complexes, Progr. Inorg. Chem. **11** [1970] 233/371.
Coucouvanis, D.; The Chemistry of the Dithioacid and 1,1-Dithiolate Complexes, 1968–1977, Progr. Inorg. Chem. **26** [1979] 301/469.
Eisenberg, R.; Structural Systematics of 1,1- and 1,2-Dithiolato Chelates, Progr. Inorg. Chem. **12** [1970] 295/369.
Hulanicki, A.; Complexation Reactions of Dithiocarbamates, Talanta **14** [1967] 1371/92.
Steggerda, J. J.; Cras, J. A.; Willemse, J.; Reactions of Complexes of Dithiocarbamates and Related Ligands, Recl. J. Royal Neth. Chem. Soc. **100/102** [1981] 41/8.
Bond, A. M.; Martin, R. L.; Electrochemistry and Redox Behaviour of Transition Metal Dithiocarbamates, Coord. Chem. Rev. **54** [1984] 23/98.

General Aspects

In this chapter are described manganese dithiocarbamates together with complexes derived from N-heterocyclic N-carbodithioic acids, because the two complex types are very similar in properties and behavior. On account of these similarities the expression "dithiocarbamates" was used for them both in these general aspects.

Manganese(II) dithiocarbamates are generally prepared from $MnCl_2$ and the sodium dithiocarbamate (NaL) in aqueous or aqueous ethanolic solution. Structures are di- or polynuclear, octahedral, with S bridges for complexes with small substituents on the N atom and tetrahedral for those with bulkier substituents, such as naphthalene. The white or yellow substances are instantaneously oxidized to the dark violet Mn^{III} compounds, when moist or in solution. The various Mn^{II} complexes display large differences in their autoxidizability, and the following order was established (parent amines given): diethylamine > piperidine > thiomorpholine > dimethylamine > pyrrolidine > piperazine. They are also very different in their solubilities. Thus, the piperazinebis(carbodithioate) is practically insoluble, even in hot aqueous solution, while the morpholinecarbodithioate dissolves readily. For other Mn^{II} dithiocarbamates, such as the diethyl derivative, hot precipitation is incomplete and becomes quantitative on cooling. The Mn^{II} compounds easily form stable mixed ligand complexes with, for instance, pyridine or 2,4-pentanedione.

Manganese(III) dithiocarbamates are prepared from Mn^{III} acetate and NaL in methanol or from an Mn^{II} salt and NaL in aqueous solution in the presence of air. They are monomeric, octahedral, and are relatively unstable to light and moisture. Oxidation of $[MnL_3]$ complexes in nonaqueous medium yields Mn^{IV} complexes $[MnL_3]^+$. The $[MnL_3]^+$ cations form dark red salts with large anions, such as ClO_4^-, BF_4^-, or PF_6^-. The Mn^{IV} complexes readily undergo photoreduction to give Mn^{II} species.

In the IR spectra of all the complexes, the ν(CN) band (in the range 1530 to 1430 cm^{-1}) is shifted to higher wavenumbers with respect to that of the free ligands. The position of the band suggests that the bond order of the C⋯N bond is close to two and that the multiple bond character is higher in the complexes than in the free ligands. From the IR data it was concluded that, of the three resonance forms of the dithiocarbamate ion shown below, c) is the main contributor.

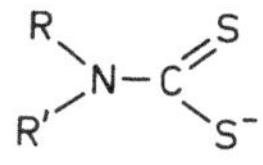

a)

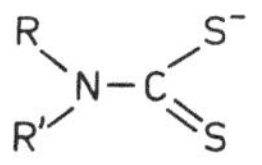

b)

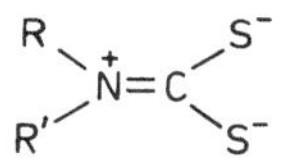

c)

In c) electrons have been delocalized from the nitrogen to the sulfur atoms via the planar π-orbital system, thus increasing the C⋯N double bond character. The extent of delocalization depends on the kind of substituent on the N atom. On chelation delocalization of electrons extends to the Mn atom thus lowering its effective positive charge. This lowering accounts for the stabilization of the high formal oxidation states by the dithiocarbamate ligands.

The easy formation of water insoluble Mn^{II} dithiocarbamates that air oxidize to deeply colored Mn^{III} complexes, soluble in the organic phase, has been applied to separate Mn from various mixtures and to quantitatively determine microamounts of Mn by spectroscopy. The most widely used complex is the diethyldithiocarbamate, and an extensive literature exists on its analytical applications. Several dithiocarbamates, particularly the Mn^{II} ethylenebis(dithiocarbamate), $Mn(C_4H_6N_2S_4)$, usually called Maneb, show fungicidal activity. A number of patents and other publications treat the applications (including agricultural), the properties, and problems arising from the use of this complex and the corresponding (Zn, Mn) complex which shows a higher fungicidal effect.

35.10.1 With Monosubstituted Dithiocarbamic Acids

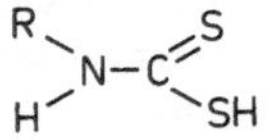

ligand	R	formula	ligand	R	formula
1	CH_3	$C_2H_5NS_2$	9	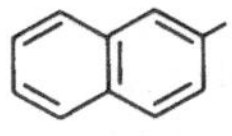	$C_{11}H_9NS_2$
2	C_2H_5	$C_3H_7NS_2$			
3	i-C_3H_7	$C_4H_9NS_2$	10	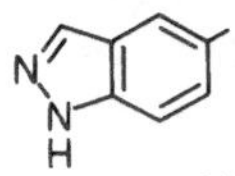	$C_8H_7N_3S_2$
4	C_4H_9	$C_5H_{11}NS_2$			
5	c-C_5H_9	$C_6H_{11}NS_2$	11	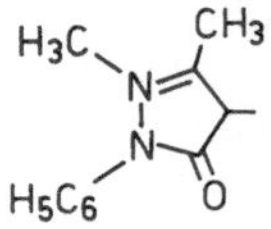	$C_{12}H_{13}N_3OS_2$
6	c-C_6H_{11}	$C_7H_{13}NS_2$			
7	c-C_7H_{13}	$C_8H_{15}NS_2$			
8	C_6H_5	$C_7H_7NS_2$	12	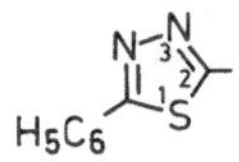	$C_9H_7N_3S_3$

35.10.1.1 Manganese(II) Compounds

MnL_2. The complexes with HL = ligands 1, 2, 3, and 8 were prepared by the reaction of aqueous solutions of hydrated manganese(II) chloride and the sodium salt of the appropriate ligand in the mole ratio 1:2. The yellow solid, which immediately separated, was washed first with water and then with ethanol-diethyl ether and allowed to dry in vacuum. All operations were performed under pure nitrogen with deoxygenated solvents. The complexes with ligands 2, 3, and 8 were recrystallized from ethanol. The magnetic moments at 292 or 293 K are 5.70 for $Mn(C_2H_4NS_2)_2$ and $Mn(C_3H_6NS_2)_2$, 5.75 for $Mn(C_4H_8NS_2)_2$, and 5.65 μ_B for $Mn(C_7H_6NS_2)_2$. On the basis of the magnetic moments an octahedral structure with some degree of polymerism, similar to that of $Mn(C_5H_{10}NS_2)_2$ (see p. 140), was suggested [1]. The ESR spectra of the complexes with ligands 2 and 8 in toluene have been measured at 285 K and analyzed in terms of high-spin ($S = 5/2$) Mn^{II} complexes. The parameter g = 2.002 and the hyperfine coupling constant $A = 88.7 \times 10^{-4}$ cm^{-1} were obtained for both complexes [2].

The compounds are immediately oxidized by air oxygen, when moist or in solution. When dry they are stable in air for some hours to some weeks. The complex with ligand 1 is insoluble in methanol, ethanol, acetone, and dichloromethane. The other complexes are slightly soluble in these solvents. All the complexes are very soluble in pyridine, from which they are precipitated by addition of n-hexane as mono- or bisadducts [1].

The complex with ligand 9, $Mn(C_{11}H_8NS_2)_2$, was prepared by the reaction of an ethanolic solution of $MnCl_2$, CS_2, and 2-naphthylamine. The reaction mixture was stirred for 1 h and then warmed for 15 min. A gray solid formed and was isolated, washed with excess hot ethanol, and dried in vacuum. The product was obtained in 68% yield. The melting point of the solid is 204°C. The magnetic moment is $\mu_{eff} = 5.73$ μ_B at room temperature. The IR spectrum (only ranges given) indicates coordination of the ligand through the S atoms. The electronic absorption spectrum of the complex shows a weak band at 19417 cm^{-1} ($\varepsilon = 75$ $L \cdot mol^{-1} \cdot cm^{-1}$), which can be assigned to the transition $^6A_1 \rightarrow {}^4T_1(G)$. The spectral data are consistent with a tetrahedral geometry for the complex. $Mn(C_{11}H_8NS_2)_2$ is soluble in acetone and is a nonelectrolyte in this solvent [3].

The complex with ligand 10, $Mn(C_8H_6N_3S_2)_2$, was prepared by the reaction of an ethanolic solution of an Mn^{II} salt and $Na(C_8H_6N_3S_2)$ in a mole ratio of 1:2. The brown reaction mixture was digested on a water bath for 30 min. The precipitate was isolated, washed with DMSO, ethanol, diethyl ether, and dried in vacuum. The melting point of the brown solid is 220°C. The magnetic moment is 4.78 μ_B (from susceptibility measurements at room temperature by the Faraday method). Important IR bands (in cm^{-1}) were assigned as follows (bands of $Na(C_8H_6N_3S_2) \cdot 3H_2O$ in parentheses): 1590 (1560) to ν(C=N); 1470 (1460) to ν(C–N); 1020 (1000) to ν(CS). The reflectance spectrum of the complex shows maxima at 18520 and 28990 cm^{-1}, which can be assigned to the high-spin Mn^{II} transitions $^6A_{1g} \rightarrow {}^4T_{1g}(G)$ and $\rightarrow {}^4T_{2g}(D)$, respectively. The complex is insoluble in the usual solvents. The physical properties indicate that the sulfur atoms and the tertiary nitrogen atom are involved in coordination, resulting in the formation of an octahedral ligand-bridged complex [4].

The complex with ligand 11, $Mn(C_{12}H_{12}N_3OS_2)_2$, was prepared by the reaction of an aqueous solution of $MnCl_2$ and an ethanolic solution of the ligand, prepared in situ, in a molar ratio of 1:2. The reaction mixture was stirred and then kept for some time. The precipitate was isolated, washed with ice-cold water, and finally with ethanol. The peach-colored complex was recrystallized from acetone and dried at room temperature in vacuum over P_4O_{10}. The yield of the product was 55%. The complex is monomeric in nitrobenzene solution. It decomposes at 162°C to yield $MnSO_4$. Susceptibility measurements at room temperature yielded a magnetic moment of 3.70 μ_B.

Important IR bands (in cm^{-1}) of the complex in CsI were assigned as follows (bands of $Na(C_{12}H_{12}N_3OS_2)$ in parentheses): 1490 (1485) to ν(CN); 935 (1080) to ν(CS); 375, 365 to ν(Mn–S). The electronic absorption spectrum of the complex in ethanolic solution shows maxima at 24900 and 25300 cm^{-1}, which can be assigned to the high-spin Mn^{II} transitions $^6A_{1g} \rightarrow {}^4E_g$ and $\rightarrow {}^4A_{1g}$, respectively. The spectral data indicate bidentate coordination of the ligand through the S atoms to form a tetrahedral complex. $Mn(C_{12}H_{12}N_3OS_2)_2$ is soluble in the usual organic solvents [5].

$Mn(C_7H_6NS_2)_2 \cdot py$. The mixed ligand complex with ligand 8 and pyridine was prepared by addition of n-hexane to a pyridine solution of $Mn(C_7H_6NS_2)_2$. The magnetic moment of the complex is $\mu_{eff} = 6.15\ \mu_B$ at 293 K [1].

[$MnL_2(NO)Cl$]. The complexes with HL = ligands 2, 4, and 8 were prepared by reaction of air oxidation products of MnL_2 complexes with gaseous NO in the presence of Cl^- ions in water-toluene solution. The compounds were extracted into toluene and, after evaporation of the toluene solvent, they were obtained as dark brown solids [2]. Toluene extracts were subjected to ESR measurements, and the following parameters (given for ^{55}Mn) have been obtained (K and T_{zz} in 10^{-4} cm^{-1}, T in Kelvin; $K = -A_0 + (g_0 - 2.002)P$ with $P \approx 187 \times 10^{-4}$ cm^{-1} for ^{55}Mn; g_0 = isotropic g-factor; A_0 and T_{zz} = isotropic or anisotropic hyperfine coupling constants):

ligand	complex	$g_{\parallel}$	$g_{\perp}$	K	$-T_{zz}$	T	Ref.
2	$[Mn(C_3H_6NS_2)_2(NO)Cl]$	1.993	2.023	82.8	73.7	120	[2]
4	$[Mn(C_5H_{10}NS_2)_2(NO)Cl]$	1.991	2.019	82.4	73.7	130	[6]
8	$[Mn(C_7H_6NS_2)_2(NO)Cl]$	1.993	2.022	82.7	73.6	120	[2]

The spectral parameters are characteristic of low-spin (S = ½) Mn^{II} complexes [2].

References:

[1] Ciampolini, M.; Mangozzi, C.; Orioli, P. (J. Chem. Soc. Dalton Trans. **1975** 2051/4).
[2] Jezierski, A. (J. Mol. Struct. **115** [1984] 11/4).
[3] Siddiqi, K. S.; Shah, M. A. A.; Zaidi, S. A. A. (Indian J. Chem. A **22** [1983] 812/3).
[4] Siddiqi, K. S.; Khan, P.; Singhal, N.; Zaidi, S. A. A. (Bull. Soc. Chim. France **1980** 303/6).

[5] Singh, H. B.; Makeshwari, S.; Srivastava, S.; Rani, V. (Syn. React. Inorg. Metal-Org. Chem. **12** [1982] 659/69).
[6] Jezierski, A.; Jeżowska-Trzebiatowska, B. (Bull. Acad. Polon. Sci. Ser. Sci. Chim. **27** [1979] 481/7).

35.10.1.2 Manganese(III) Compounds

The Mn^{III} complexes with ligands 5, 7, and 12 are formed by air oxidation of Mn^{II} in the presence of excess ligand. Spectral studies show them to be octahedral. The complexes show significant antifungal activity at 1000 ppm, the complexes with ligands 5 and 7 being more active than the corresponding free ligands [1, 2].

$[Mn(C_6H_{10}NS_2)_3]$ and **$[Mn(C_8H_{14}NS_2)_3]$**. The brown complexes with HL = ligands 5 and 7 were prepared by heating an Mn^{II} salt and the sodium salt of the appropriate ligand (mole ratio 1:2) on a water bath for 30 to 45 min. The reaction mixture was cooled, the separated reaction product filtered, washed with water, recrystallized from acetone, and dried over P_4O_{10}. Molecular weight determinations in boiling benzene indicate monomeric compounds. Room temperature susceptibility measurements yielded the magnetic moment $\mu_{eff} = 4.80\ \mu_B$ for both compounds. The IR spectra show a shift of the ν(CN) bands to higher frequencies and one strong ν(CS) band indicating bidentate S,S coordination of the three ligand molecules. This is supported by the appearance of a ν(Mn–S) band in the 400 to 350 cm^{-1} region. Bands (in cm^{-1}) in the electronic spectra were assigned as follows: 15000 to $^5B_{1g} \rightarrow {}^5A_{1g}$, 19500 to 19800 to $^5B_{1g} \rightarrow {}^5B_{2g}$, and 24000 to 24500 to $^5B_{1g} \rightarrow {}^5E_g$ transitions. The complexes are soluble in common organic solvents. They are nonelectrolytes in nitrobenzene [1].

$[Mn(C_9H_6N_3S_3)_3]$. The complex with ligand 12 was prepared by the reaction of an Mn^{II} salt and $Na(C_9H_6N_3S_3)$ (mole ratio 1:3) in 50% aqueous ethanol. The reaction mixture was refluxed for half an hour and then cooled. The complex, which separated, was collected, washed with cold water, and dried over P_4O_{10}. The yield of the dull gray solid was 80 to 85%. A lower yield, 62 to 76%, was obtained by the reaction of 2-amino-1,3,4-thiadiazole, CS_2, NaOH, and an Mn^{II} salt (mole ratio 3:3:3:1) in 50% aqueous ethanol. The reaction mixture was vigorously stirred for 2 h below 5°C and finally refluxed for half an hour. The melting point of the solid is >250°C. The complex is a nonelectrolyte in dioxane. The magnetic moment of the solid is $\mu_{eff} = 4.95\ \mu_B$ at 25°C. A comparison of the IR spectrum of ligand 12 with that of the complex reveals that in the complex the ν(SH) band at 2500 cm^{-1} is absent, and the ν(C=N) band is shifted to a lower frequency. This suggests that the nitrogen atom in the number three position of the thiadiazole ring and one of the S atoms of the dithiocarboxylic group are involved in chelation. Bands in the electronic spectrum at 20000 and 33300 cm^{-1} were assigned to the $^5E_g \rightarrow {}^5T_{2g}$ transition and to $n \rightarrow \pi^*$ charge transfer, respectively [2].

$[Mn(C_7H_{12}NS_2)_3]$ and **$[Mn(C_7H_6NS_2)_3]$** were prepared by the reaction of an Mn^{III} salt with the sodium salt of ligand 6 or 8 in aqueous solution. TGA studies (heating rate 6°C/min) show decomposition of the complexes in the temperature range 120 to 300°C. The final decomposition to Mn_3O_4 and $MnSO_4$ proceeds through the intermediate formation of the isothiocyanate and the sulfide [3].

References:

[1] Pandey, O. P.; Sengupta, S. K.; Tripathi, S. C. (Rev. Roumaine Chim. **32** [1987] 145/50).
[2] Singh, H.; Yadav, L. D. S.; Mishra, S. B. S. (J. Inorg. Nucl. Chem. **43** [1981] 1701/4).
[3] Muthuswamy, S.; Venkappayya, D. (J. Indian Chem. Soc. **64** [1987] 571/3).

35.10.2 With Dialkyl- or Diaryldithiocarbamic Acids

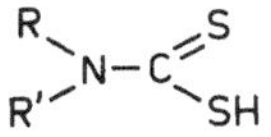

ligand	R	R′	formula	ligand	R	R′	formula
1	CH_3	CH_3	$C_3H_7NS_2$	9	c-C_6H_{11}	c-C_6H_{11}	$C_{13}H_{23}NS_2$
2	C_2H_5	C_2H_5	$C_5H_{11}NS_2$	10	C_6H_5	C_6H_5	$C_{13}H_{11}NS_2$
3	C_3H_7	C_3H_7	$C_7H_{15}NS_2$	11	$C_6H_5CH_2$	$C_6H_5CH_2$	$C_{15}H_{15}NS_2$
4	i-C_3H_7	i-C_3H_7	$C_7H_{15}NS_2$	12	CH_3	c-C_6H_{11}	$C_8H_{15}NS_2$
5	C_4H_9	C_4H_9	$C_9H_{19}NS_2$	13	i-C_3H_7	c-C_6H_{11}	$C_{10}H_{19}NS_2$
6	i-C_4H_9	i-C_4H_9	$C_9H_{19}NS_2$	14	CH_3	C_6H_5	$C_8H_9NS_2$
7	C_5H_{11}	C_5H_{11}	$C_{11}H_{23}NS_2$	15	C_2H_5	C_6H_5	$C_9H_{11}NS_2$
8	i-C_5H_{11}	i-C_5H_{11}	$C_{11}H_{23}NS_2$	16	C_2H_5	3-$CH_3C_6H_4$	$C_{10}H_{13}NS_2$

35.10.2.1 Manganese(II) Compounds

MnL_2. The complexes with HL = ligands 1, 2, 4, and 14 were prepared by the reaction of aqueous solutions of $MnCl_2 \cdot 4H_2O$ and the sodium salt of the appropriate ligand (mole ratio 1:2) [1, 2]. The complex with ligand 11, $Mn(C_{15}H_{14}NS_2)_2$, was prepared in an analogous manner [22]. The yellow solids, which immediately separated out, were isolated, washed first with water and then with ethanol-diethyl ether (2:1 v/v), and allowed to dry in vacuum. The complexes with ligands 2 and 14 were recrystallized from hot ethanol. All operations were performed under nitrogen with deoxygenated solvents [1, 2]. The complexes with HL = ligands 1, 2, 3, and 6 were prepared by the reaction of 50% aqueous ethanol solutions of hydrated manganese(II) acetate and the sodium salt of the appropriate ligand (mole ratio 1:2). The yellow precipitates, which formed immediately, were filtered off, washed with 50% aqueous ethanol and then absolute ethanol, and dried by pumping for several hours. The solutions used were thoroughly degassed and the reactions carried out under a nitrogen atmosphere [3]. The complexes with HL = ligands 10 and 15 were similarly prepared by the reaction of an Mn^{II} salt with the respective sodium dithiocarbamate in water or 50% aqueous ethanol [4].

A single crystal X-ray determination was performed for the complex with ligand 2, $Mn(C_5H_{10}NS_2)_2$. The compound crystallizes in the triclinic system, space group $P\bar{1}$-C_i^1 (No. 2), with the lattice constants a = 11.039(5), b = 10.055(5), c = 7.529(2) Å, α = 70.9(1)°, β = 83.7(1)°, γ = 82.2(1)°; Z = 2. The structure was solved up to R = 0.062. Positional parameters and isotropic and anisotropic thermal parameters are presented in the paper. The calculated density is D = 1.23 g/cm³. The structure of $Mn(C_5H_{10}NS_2)_2$ consists of two independent centrosymmetric molecules, linked into infinite chains by the sharing of some of the S atoms between the two Mn atoms (see **Fig. 15**, p. 140). The manganese atoms are coordinated by four coplanar sulfur atoms of the chelate rings, mean Mn–S distance 2.55 ± 0.05 Å, and two additional sulfur atoms from neighboring molecules at a mean distance of 2.74 ± 0.04 Å, completing a distorted octahedron. Bond lengths (in Å) and angles (in °) are:

Mn(1)–S(1)	2.562(4)	S(1)–Mn(1)–S(2)	71.2(1)	S(3)–Mn(2)–S(4)	70.7(1)
Mn(1)–S(2)	2.530(3)	S(4′)–Mn(1)–S(2)	93.3(1)	S(1)–Mn(2)–S(3″)	94.7(1)
Mn(1)–S(4)	2.778(3)	S(4′)–Mn(1)–S(1)	87.5(1)	S(1)–Mn(2)–S(4″)	88.2(1)
Mn(2)–S(4)	2.605(4)	S(1)–Mn(1)–S(2′)	108.7(1)	S(4″)–Mn(2)–S(3)	109.3(1)

Mn(2)–S(3)	2.511(3)	S(4)–Mn(1)–S(1)	92.5(1)	S(1)–Mn(2)–S(4)	91.8(1)
Mn(2)–S(1)	2.704(3)	S(4)–Mn(1)–S(2)	86.7(1)	S(1)–Mn(2)–S(3)	85.3(1)
		Mn(1)–S(4)–Mn(2)	88.7(1)	Mn(1)–S(1)–Mn(2)	91.2(1)

The structure of $Mn(C_5H_{10}NS_2)_2$ is different from those of the analogous Ni^{II}, Cu^{II}, and Zn^{II} complexes [1]. This is in contrast to earlier publications [5, 6]. For the Mn^{II} compounds with the other ligands magnetic moment values (see below) also indicate some degree of polymerization [1, 3]. The complexes with ligands 1 and 4, which are insoluble in ethanol, probably have a three-dimensional polymeric structure [1]. Obviously, substituents have a marked effect on the structure [3].

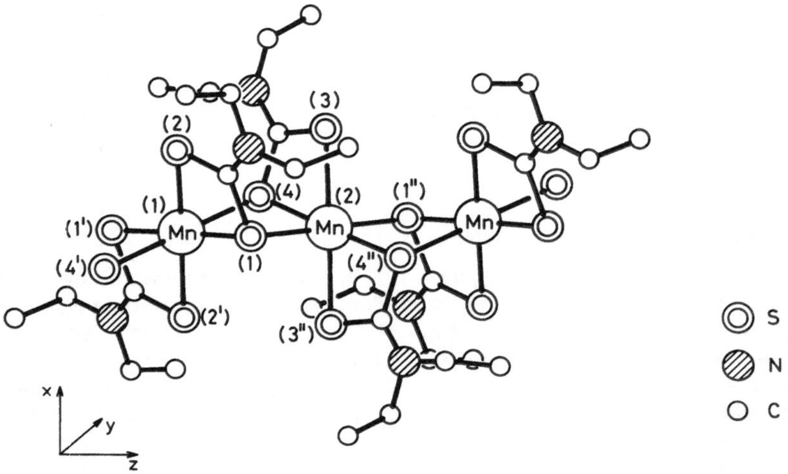

Fig. 15. Structure of the manganese(II) diethyldithiocarbamate, $Mn(C_5H_{10}NS_2)_2$ [1]. H atoms omitted for clarity.

The magnetic properties of the complexes have been extensively studied. The magnetic moments vary with temperature. The decrease in magnetic moments with decrease in temperature can be ascribed to antiferromagnetic intrachain interactions of high-spin ($S = {}^5/_2$) Mn^{II} centers. Magnetic moments, μ_{eff} in μ_B, of the complexes at various temperatures and parameters, resulting from the analysis of the variation of magnetic susceptibility with temperature, are summarized below (Θ = Weiss constant in K; J = exchange constant in cm^{-1}):

ligand	complex	T in K	μ_{eff}	$-\Theta$	$-J$	g	Ref.
1	$Mn(C_3H_6NS_2)_2$	295	5.58	—	3.90	1.99	[3]
		90	5.13	—	—	—	[3]
2	$Mn(C_5H_{10}NS_2)_2$	293	5.65	—	—	—	[1]
		84	5.15	—	—	—	[1]
		300	5.38	—	—	—	[2]
		2	0.39	—	—	—	[2]
		295	5.61	28	4.10	1.98	[3]
		90	5.09	—	—	—	[3]

For numbers, substituents, and formulas of ligands, see p. 139

ligand	complex	T in K	μ_{eff}	$-\Theta$	$-J$	g	Ref.
3	$Mn(C_7H_{14}NS_2)_2$	295	4.85	—	19.7	2.06	[3]
		90	2.88	—	—	—	[3]
4	$Mn(C_7H_{14}NS_2)_2$	294	5.30	—	—	—	[1]
6	$Mn(C_9H_{18}NS_2)_2$	295	5.47	35	3.90	1.93	[3]
		90	4.93	—	—	—	[3]

The magnetic susceptibility of the diethyldithiocarbamate, $Mn(C_5H_{10}NS_2)_2$, shows a broad maximum at 65 K. The low-temperature magnetic behavior can be correlated with an intrachain antiferromagnetic exchange, $J/K = -7.5$ K, with $g = 2.0$ and $D = 0$ (D = zero-field splitting parameter) [2]. The Néel point of the complex with ligand 3, $Mn(C_7H_{14}NS_2)_2$, is at 142 K [3]. The ESR spectra of the complexes with ligands 2, 10, and 15, $Mn(C_5H_{10}NS_2)_2$, $Mn(C_{13}H_{10}NS_2)_2$, and $Mn(C_9H_{10}NS_2)_2$, respectively, have been measured at 285 K and analyzed in terms of high-spin ($S = 5/2$) Mn^{II} complexes: $g = 2.002$ for all the complexes and the hyperfine coupling constant $A_0 = -88.2$, -87.7, and -88.0×10^{-4} cm^{-1} for the complexes with ligands 2, 10, and 15, respectively [4].

In the IR spectrum of $Mn(C_5H_{10}NS_2)_2$ (in KBr) bands at ~1430, 1240, and 880 cm^{-1} were assigned to $\nu(CN)$, $\nu_{as}(CS)$, and $\nu_s(CS)$ vibrations, respectively (corresponding bands of $Na(C_5H_{10}NS_2)$ are observed at 1450, 1260, and 890 cm^{-1}) [7].

For the diethyldithiocarbamate, $Mn(C_5H_{10}NS_2)_2$, bonding energies of p electrons have been determined by X-ray photoelectron spectroscopy: $E = 652.3$ eV for $2p_{1/2}$ electrons and $E = 641.0$ eV for $2p_{3/2}$ electrons [20]. Semiempirical CNDO/2 calculations have been used to determine bonding and electronic structures of the complexes with ligands 1 and 2, $Mn(C_3H_6NS_2)_2$ and $Mn(C_5H_{10}NS_2)_2$ [23, 24, 25], assuming a square-planar configuration of the donor atoms, an overall D_{2h} symmetry, and a low-spin state of the hypothetical isolated molecule [23, 24]. It is shown that the highest occupied MO is localized almost completely on the d orbitals of Mn, while the lowest unoccupied MO is delocalized over the ligand, primarily on the C–N bond [25]. The results indicate that the contribution of π bonding to the energy of the Mn–S bond amounts to 10 to 14% and that a significant transannular Mn–C interaction takes place which may account in part for the stability of the four-membered chelate rings [23, 24].

Half-wave potentials for the irreversible reduction of MnL_2 complexes at the dropping mercury electrode in DMF are (0.1 M tetraethylammonium perchlorate, TEAP, with an Hg electrode as reference electrode in DMF, 0.1 M TEAP): −1.72, −1.79, −1.83, −1.94 V at 25°C for the complexes with ligands 1, 2, 3, and 5, respectively. The products of the reduction are Mn^0 and the free ligand anions [10, 21].

The yellow substances are immediately oxidized by air oxygen to the dark violet Mn^{III} compounds, when moist or in solution [1 to 3, 8]. When dry they are stable in air for a few hours to a few weeks. The most readily oxidizable compound is the diethyldithiocarbamate, $Mn(C_5H_{10}NS_2)_2$, which darkens instantaneously in the presence of traces of oxygen [1]. The order of autoxidizability of various Mn^{II} dithiocarbamates and heterocyclic N-carbodithioates [26] is given in the "General Aspects" on pp. 134/5. The precipitate of $Mn(C_5H_{10}NS_2)_2$, obtained by addition of $Na(C_5H_{10}NS_2)$ to an aqueous Mn^{II} salt solution, dissolves on heating [8]. The complexes with ligands 2 and 14 are slightly soluble in methanol, ethanol, acetone, and dichloromethane. Those with ligands 1 and 4 are insoluble in these solvents [1]. All the complexes are readily soluble in pyridine, from which they are precipitated as bisadducts (p. 142). $Mn(C_5H_{10}NS_2)_2$ forms bisadducts also with other N-heterocyclic amine ligands (p. 142). The MnL_2 complexes react with gaseous NO in a solvent to give $[MnL_2(NO)]^+$ complexes. The same reaction in the presence of Cl^- ions yields $[MnL_2(NO)Cl]$ complexes (p. 144).

The complex with ligand 1, $Mn(C_3H_6NS_2)_2$, is an effective fungicide [27]. An Mn^{II} ion-selective electrode was prepared by dispersing the complex with ligand 11, $Mn(C_{15}H_{14}NS_2)_2$, in silicon rubber, poly(vinylchloride), or poly(vinylacetate). The properties of the electrode were studied [22].

$[Mn(C_5H_{10}NS_2)_2L'_2]$ with L' = pyridine (= C_5H_5N), 3-methylpyridine (= C_6H_7N), quinoline (= C_9H_7N), or morpholine (= C_4H_9NO). The complex with L' = pyridine was precipitated by addition of hexane to a pyridine solution of the diethyldithiocarbamate, $Mn(C_5H_{10}NS_2)_2$ [1]. The other complexes were obtained by the reaction of freshly prepared $Mn(C_5H_{10}NS_2)_2$ and the amine ligands (mole ratio 1:2) in absolute ethanol. The reaction mixture was refluxed for 30 min and then cooled to yield crystalline compounds. The complexes were isolated, washed with ethanol, followed by diethyl ether, and dried in vacuum. The melting points of the white crystalline solids are 186, 198, and 215°C for L' = 3-methylpyridine, quinoline, and morpholine, respectively. The complexes are nonelectrolytes in acetone solution [7]. The effective magnetic moments of all the compounds are between 5.92 and 5.95 μ_B at room temperature [1, 7]. In the IR spectra of the complexes in KBr, bands appear at ~1430, ~1240, and ~880 cm^{-1}, which were assigned to ν(CN), ν_s(CS), and ν_{as}(CS) vibrations, respectively. The ν(Mn–N) and ν(Mn–S) bands were located in the 380 to 350 cm^{-1} and 300 to 260 cm^{-1} regions. The electronic absorption spectra of the complexes in acetone show four maxima, in the regions 18.4 to 18.6, 22.6 to 22.8, 24.4 to 24.5, and at ~27.7 kK. The bands were assigned to the high-spin octahedral Mn^{II} transitions $^6A_{1g} \rightarrow {}^4T_{1g}(G)$, $\rightarrow {}^4T_{2g}(G)$, $\rightarrow {}^4E_g(G)$, and $\rightarrow {}^4T_{2g}(D)$, respectively [7].

$[MnL_2L']$ complexes with HL = ligand 1, 2, 5, or 9, and L' = bipyridine (bpy) or phenanthroline (phen) were prepared by the reaction of $MnCl_2$ and L' in water or aqueous ethanol, followed by the slow addition with stirring of an aqueous solution of NaL (mole ratio 1:3:1). The precipitates were isolated, washed with water and ethanol, and dried in vacuum over P_4O_{10}. The pink complexes were recrystallized from dichloromethane in a nitrogen atmosphere, washed with diethyl ether, and dried in vacuum [9 to 11]. The magnetic moments of both complexes with ligand 2, $[Mn(C_5H_{10}NS_2)_2bpy]$ and $[Mn(C_5H_{10}NS_2)_2phen]$, are μ_{eff} = 6.0 μ_B [9]. Similar values were reported in [11] for the phenanthroline adducts of the complexes with ligands 2 and 9 (5.9 and 6.1 μ_B), but different values were found for the bipyridine adducts of these complexes (5.6 and 5.5 μ_B, respectively) at room temperature. The X-band ESR spectra of the complexes with ligands 2 and 9 were measured on polycrystalline samples at room temperature and on solutions in frozen chloroform at 78 K, and the data obtained at 78 K were analyzed. The results are comparable to those calculated using the zero-field splitting parameters D = 0.19 cm^{-1}, λ (= E/D) = 0.04, and g_0 = 2.00 for magnetic field directions parallel to the principal D-tensor axes [11].

The IR spectra of the complexes with ligand 2, $[Mn(C_5H_{10}NS_2)_2bpy]$ and $[Mn(C_5H_{10}NS_2)_2phen]$, show bands at ~1500 and ~1480 cm^{-1}, which were both assigned to ν(CN) vibrations. The ν(Mn–S) band was localized at ~390 and the ν(Mn–N) band at ~300 cm^{-1} [28]. The reflectance spectra of these complexes show a shoulder at 14300 cm^{-1} which was assigned to the octahedral Mn^{II} transition $^6A_{1g} \rightarrow {}^4T_{1g}$ [9].

The complexes show four polarographic reduction waves at the dropping mercury electrode. The first two waves correspond to the reversible one-electron reductions $Mn^{II} \rightarrow Mn^I$ and $Mn^I \rightarrow Mn^0$. The third and fourth waves correspond to the two reductions of the aromatic amine ligand. Small effects on the metal reductions of the ligand substituents are noted. Bipyridine stabilizes the Mn^{II} state to a greater extent than phenanthroline in the complexes. The half-wave potentials (in V vs. Hg electrode at 25°C; supporting electrolyte 0.1 M $N(C_2H_5)_4ClO_4$) and

For numbers, substituents, and formulas of ligands, see p. 139

the charge-transfer rate constants, k (in 10^{-2} cm/s), for the first two reduction waves are collected below [10]:

ligand	complex	$Mn^{II} \rightarrow Mn^{I}$ $E_{1/2}$	k	$Mn^{I} \rightarrow Mn^{0}$ $E_{1/2}$	k
1	$[Mn(C_3H_6NS_2)_2bpy]$	−1.32	2.42±0.81	−1.83	0.85±0.21
	$[Mn(C_3H_6NS_2)_2phen]$	−1.25	2.12±0.75	−1.72	0.98±0.31
2	$[Mn(C_5H_{10}NS_2)_2bpy]$	−1.35	2.25±0.65	−1.87	0.92±0.31
	$[Mn(C_5H_{10}NS_2)_2phen]$	−1.27	2.34±0.78	−1.74	0.96±0.24
5	$[Mn(C_9H_{18}NS_2)_2bpy]$	−1.27	2.51±0.83	−1.82	0.95±0.32
	$[Mn(C_9H_{18}NS_2)_2phen]$	−1.27	2.44±0.81	−1.76	0.92±0.19

The solid complexes with ligand 2 appear to be air-stable, in marked contrast to the MnL_2 complexes themselves [9]. The complexes with ligands 2 and 9 are insoluble in water and ethanol. They are highly soluble in common organic solvents, such as chloroform and acetone [11].

[MnLL′] with HL′ = 8-quinolinol (= C_9H_7NO) for HL = ligand 1 and 2, and HL′ = Hacac and picolinic acid (= $C_6H_5NO_2$) for HL = ligand 2. $[Mn(C_5H_{10}NS_2)(acac)]$ was prepared by the reaction of a $CHCl_3$ solution of $[Mn^{III}(acac)_3]$ with an equimolar amount of $Na(C_5H_{10}NS_2)$. The original dark color of $[Mn(acac)_3]$ was discharged immediately and a grayish white silky compound separated out. The reaction mixture was refluxed for a few minutes to ensure complete reaction. $[Mn(C_5H_{10}NS_2)(C_6H_4NO_2)]$ was prepared by the reaction of a hot ethanolic solution of $[Mn(C_6H_4NO_2)_3]$ and an ethanolic solution of an equimolar amount of $Na(C_5H_{10}NS_2)$. The reaction mixture was refluxed for one hour and then cooled to yield shining chocolate-colored compounds. The products were isolated, washed with ethanol, followed by diethyl ether, and dried in vacuum. The same compounds were obtained by the reaction of $[Mn(C_5H_{10}NS_2)_3]$ with an equimolar amount of acetylacetone in the presence of aqueous ammonia, or by refluxing $[Mn(C_5H_{10}NS_2)_3]$ with picolinic acid in ethanol for 30 min. $[Mn(C_5H_{10}NS_2)(C_9H_6NO)]$ was prepared by the reaction of $[Mn(C_5H_{10}NS_2)_3]$ and an equimolar amount of 8-quinolinol in acetone. The reaction mixture was refluxed for about 30 min, whereupon microcrystalline compounds separated out. The green complex was isolated, washed with acetone, followed by diethyl ether, and dried in vacuum. $[Mn(C_5H_{10}NS_2)(acac)]$ melts at 290°C. The other two compounds do not melt up to 300°C. All the compounds are nonelectrolytes in nitrobenzene. The magnetic moments of the solid complexes are in the range between 5.80 and 5.96 μ_B at room temperature. Important bands (in cm^{-1}) in the IR and electronic spectra of the complexes (in Nujol) are summarized below (electronic transitions from $^6A_1$):

complex	ν(CN)	$\nu_{as}(CS)$	$\nu_s(CS)$	$\rightarrow {}^4T_1(G)$	$\rightarrow {}^4T_2(G)$	$\rightarrow {}^4E, {}^4A_1(G)$
$[Mn(C_5H_{10}NS_2)(acac)]$	1520	1000	700	20410	22730	23570
$[Mn(C_5H_{10}NS_2)(C_6H_4NO_2)]$	1520	1002	738	20000	22730	23570
$[Mn(C_5H_{10}NS_2)(C_9H_6NO)]$	1500	1020	710	20200	22470	23530

The positions of the ν(CN) and ν(CS) bands of the dithiocarbamate group indicate that coordination occurs through both sulfur atoms. Bands of the second ligands (presented) show that these ligands are uninegative and are coordinated in a bidentate manner as well. The electronic spectra suggest a tetrahedral geometry for all the complexes. Unlike $Mn(C_5H_{10}NS_2)_2$, the mixed ligand complexes are very stable to air oxidation. The complexes are not appreciably soluble in organic solvents [12]. $[Mn(C_3H_6NS_2)(C_9H_6NO)]$ is effective as a fungicide. A thermoanalytic study is reported in [29].

$MnL_2(NO)^+$ and $[MnL_2(NO)Cl]$. The formation of $MnL_2(NO)^+$ ions with HL = ligand 2 or 6 by passing NO gas through a toluene solution of $[Mn^{III}(C_5H_{10}NS_2)_3]$ or $[Mn^{III}(C_9H_{18}NS_2)_3]$, followed by argon was reported [13, 14]. $[MnL_2(NO)Cl]$ complexes with HL = ligand 2, 4, 5, 6, 10, or 15 were prepared by passing NO gas through toluene-water mixtures of $[Mn^{III}L_3]$ complexes (obtained in situ by air oxidation of the MnL_2 complexes) in the presence of Cl^- ions. The complexes were extracted into toluene and, after evaporation, dark brown solids were obtained. For the ESR studies of $[MnL_2(NO)Cl]$ complexes, toluene extracts from reaction mixtures were used directly [4, 15 to 17]. Analysis of the ESR spectra yielded the following parameters for $MnL_2(NO)^+$ complexes (T in Kelvin, hyperfine coupling constant A in G [14], in 10^{-4} cm^{-1} [13]):

ligand	complex	T	$g_{\parallel}$	$g_{\perp}$	g_0	$-A_{\parallel}$	$-A_{\perp}$	$-A_0$	Ref.
2	$Mn(C_5H_{10}NS_2)_2(NO)^+$	77	1.989	2.016	—	166.5	46.1	—	[14]
		243	—	—	2.005	—	—	84.0	[14]
6	$Mn(C_9H_{18}NS_2)_2(NO)^+$	125	1.992	2.023	—	152.1	43.1	—	[13]
		235	—	—	2.013	—	—	79.4	[13]

The parameters $g_{\parallel} = \sim 1.993$, $g_{\perp} = \sim 2.023$, $K = (81.4 \text{ to } 82.4) \times 10^{-4}$ cm^{-1}, and $-T_{zz} = (72.7 \text{ to } 73.4) \times 10^{-4}$ cm^{-1} ($K = -A_0 + (g_0 - 2.002)P$ with $P = 1.87 \times 10^{-4}$ cm^{-1} for ^{55}Mn; g_0 = isotropic g-factor, A_0 and T_{zz} = isotropic and anisotropic hyperfine coupling constants) have been evaluated for $[MnL_2(NO)Cl]$ complexes [4, 15 to 17]. The ESR parameters are consistent with low-spin ($S = 1/2$) complexes and linear coordinated NO groups [4, 13]. For the $[MnL_2(NO)Cl]$ complexes ^{14}N super-hyperfine coupling constants of $A_{\perp} = (-5.4 \text{ to } -5.5) \times 10^{-4}$ cm^{-1} were obtained. The ^{14}N super-hyperfine interaction probably arises from spin-polarization mechanisms involving π orbitals [4]. The solutions of the $MnL_2(NO)^+$ complexes are only poorly stable at room temperature. Therefore all experiments were performed at reduced temperatures [13, 14] and in an argon atmosphere [13]. For complexes with ligand 4, 5, or 6 changes in the ESR parameters were determined, when the Cl^- ion was replaced by a sulfur-containing ligand, such as a thiourea derivative or a thiol [16, 17].

$MnML_4$ complexes with HL = ligand 12 or 13 for M = Zn, Hg and HL = ligand 2, 12, or 13 for M = Cd were prepared by reaction of an aqueous acetone solution (20:80) of manganese(II) nitrate with an equimolar amount of Na_2ML_4. Violet precipitates formed and were isolated immediately. The products were washed with aqueous acetone (20:80) and with diethyl ether, followed by drying in vacuum [18, 19].

The magnetic moment of $MnCd(C_5H_{10}NS_2)_4$ is $\mu_{eff} = 4.40$ μ_B, from susceptibility measurements by the Cahn-Faraday method at room temperature [19]. The magnetic moment values of the other complexes (not given in detail) are around 4.0 μ_B. The magnetic data are consistent with a quartet ($S = 3/2$) ground state and eliminate the possibility of any dominant antiferromagnetic type of exchange interaction [18]. Important bands in the IR spectra of the compounds (in KBr or Nujol) and decomposition temperatures (t_{dec} in °C) are summarized below [18, 19]:

ligand	complex	ν(CN)	ν(CS)	ν(Mn–S)	ν(M–S)	t_{dec}
2	$MnCd(C_5H_{10}NS_2)_4$	—	—	—	—	102
12	$MnZn(C_8H_{14}NS_2)_4$	1470	1000	350	380	177
	$MnCd(C_8H_{14}NS_2)_4$	1480	995	340	420	170
	$MnHg(C_8H_{14}NS_2)_4$	1495	1000	365	400	173

For numbers, substituents, and formulas of ligands, see p. 139

ligand	complex	ν(CN)	ν(CS)	ν(Mn–S)	ν(M–S)	t_{dec}
13	$MnZn(C_{10}H_{18}NS_2)_4$	1485	990	370	400	188
	$MnCd(C_{10}H_{18}NS_2)_4$	1500	1000	375	385	175
	$MnHg(C_{10}H_{18}NS_2)_4$	1490	1005	360	410	182

The complexes are fairly stable at room temperature. The diethyldithiocarbamate, $MnCd(C_5H_{10}NS_2)_4$, is insoluble in water and common organic solvents but slightly soluble in nitrobenzene and chloroform [19]. The other complexes are also insoluble in water, but soluble in methanol, benzene, acetone, nitrobenzene, DMSO, and halogenated hydrocarbons [18]. The complexes are nonelectrolytes in nitrobenzene [18, 19]. The physical properties are consistent with a linear polymeric structure with bridging CS_2 groups, as shown below [18, 19].

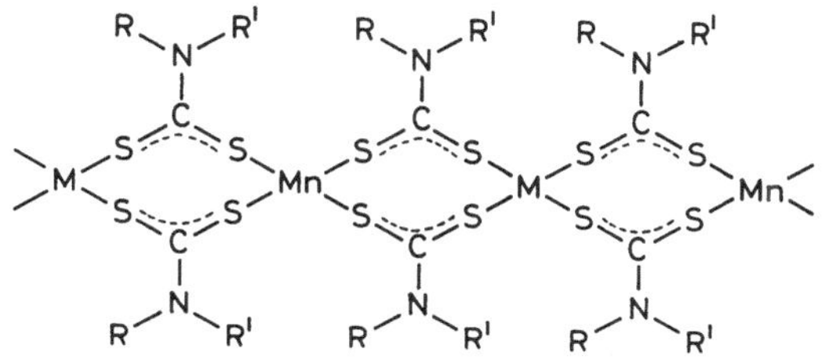

References:

[1] Ciampolini, M.; Mengozzi, C.; Orioli, P. (J. Chem. Soc. Dalton Trans. **1975** 2051/4).
[2] Eisman, G. A.; Reiff, W. M. (Inorg. Chem. **20** [1981] 3481/3).
[3] Hill, D. M.; Larkworthy, L. F.; O'Donoghue, M. W. (J. Chem. Soc. Dalton Trans. **1975** 1726/8).
[4] Jezierski, A. (J. Mol. Struct. **115** [1984] 11/4).
[5] Fackler, J. P.; Holah, D. G. (Inorg. Nucl. Chem. Letters **2** [1966] 251/5).
[6] Lahiry, S.; Anand, V. K. (J. Chem. Soc. D **1971** 1111/2).
[7] Saku, B. K.; Mohapatra, B. K. (J. Indian Chem. Soc. **56** [1979] 825/6).
[8] Malissa, H.; Miller, F. F. (Mikrochem. Ver. Mikrochim. Acta **40** [1952] 63/75).
[9] Holah, D. G.; Murphy, C. N. (Can. J. Chem. **49** [1971] 2726/31).
[10] Toropova, V. F.; Budnikov, G. K.; Ulakhovich, N. A. (Zh. Obshch. Khim. **45** [1975] 380/5; Russ. J. Gen. Chem. **45** [1975] 368/72).

[11] Pandeya, K. B.; Singh, R.; Mathur, P. K.; Singh, R. P. (Transition Metal Chem. [Weinheim] **11** [1986] 347/50).
[12] Pradhan, B.; Ramana Rao, D. V. (J. Indian Chem. Soc. **55** [1978] 226/8).
[13] Yordanov, N. D.; Iliev, V.; Shopov, D.; Jezierski, A.; Jeżowska-Trzebiatowska, B. (Inorg. Chim. Acta **60** [1982] 9/15).
[14] Garifyanov, N. S.; Luchkina, S. A. (Izv. Akad. Nauk SSSR Ser. Khim. **1969** 471/2; Bull. Acad. Sci. [USSR] **1969** 421/2).
[15] Jeżowska-Trzebiatowska, B.; Jezierski, A. (J. Mol. Struct. **46** [1978] 197/202).
[16] Jezierski, A.; Jeżowska-Trzebiatowska, B. (Bull. Acad. Polon. Sci. Ser. Sci. Chim. **27** [1979] 481/7).
[17] Jezierski, A.; Jeżowska-Trzebiatowska, B. (Nouv. J. Chim. **4** [1980] 599/602).
[18] Johri, K. N.; Kaushik, N. K.; Bajaj, R. K.; Sharma, A. K. (Acta Chim. Hung. **113** [1983] 325/9).
[19] Aggarwal, R. C.; Singh, B.; Singh, M. K. (J. Indian Chem. Soc. **59** [1982] 269/72).
[20] Vetchinkin, S. I.; Zimont, S. L.; Ioffe, M. S.; Borod'ko, Yu. G. (Khim. Fiz. **3** [1984] 635/42).

[21] Toropova, V. F.; Budnikov, R. G. K.; Ulakhovich, N. A. (Talanta **25** [1978] 263/7).
[22] Motonaka, J.; Nishioka, H.; Ikeda, S.; Tanaka, N. (Bull. Chem. Soc. Japan **59** [1986] 39/42).
[23] Pilipenko, A. T.; Savranskii, L. I.; Zubenko, A. I.; Kobylyashny, V. P. (J. Mol. Struct. **86** [1981] 155/62).
[24] Pilipenko, A. T.; Savranskii, L. I.; Zubenko, A. I. (Koord. Khim. **7** [1981] 1613/21; Soviet J. Coord. Chem. **7** [1981] 804/12).
[25] Miertus, S.; Frecer, V. (Collection Czech. Chem. Commun. **49** [1984] 2744/50).
[26] Gleu, K.; Schwab, R. (Angew. Chem. **62** [1950] 320/4).
[27] Bradley, C. E. (U.S. 2765327 [1952/56]; C.A. **1956** 5824).
[28] Malik, W. U.; Bembi, R.; Bhardwaj, V. K. (J. Indian Chem. Soc. **61** [1984] 379/80).
[29] Gardi, I.; Cserhat, T. (Novenyvedelem [Budapest] **21** [1985] 114/7 from C.A. **102** [1985] No. 216735).

35.10.2.2 Manganese(III) Compounds

35.10.2.2.1 Formation in Solution

The stepwise stability constants for the formation of the diethyldithiocarbamate, $Mn(C_5H_{10}NS_2)_3$, in dimethyl sulfoxide, I = 0.1 mol/L ($NaClO_4$), at 25°C, are: $\log K_1 = 4.9$, $\log K_2 = 4.3$, $\log K_3 = 3.3$, determined potentiometrically using an $Ag | C_5H_{10}NS_2^-$ electrode [1]. In ethyl acetate the values $\log K_3 = 4.32$ and 4.55 were obtained by two different spectrophotometric methods [2].

References:

[1] Labuda, J.; Skatuloková, M.; Németh, M.; Gergely, Š. (Chem. Zvesti **38** [1984] 597/605).
[2] Usatenko, Yu. I.; Fedash, N. P. (Tr. Komis. Analit. Khim. Akad. Nauk SSSR Inst. Geokhim. Analit. Khim. **14** [1963] 183/90; C.A. **59** [1963] 13397).

35.10.2.2.2 Isolated [MnL_3] Compounds

35.10.2.2.2.1 Preparation

Three main procedures are reported:

A) The complexes with ligands 2 through 6 were prepared by the reaction of a methanol solution of hydrated manganese(III) acetate and the respective dithiocarbamate. After stirring the solution, deep violet crystals formed [1, 2]. The products were isolated, washed with aqueous methanol (75 vol%), and dried in vacuum over P_4O_{10} [1]. The complex with ligand 1, [$Mn(C_3H_6NS_2)_3$], which initially could not be obtained by this method [1], was later prepared similarly but using cooled methanol as the solvent [2].

B) The complexes with ligands 2 through 9, 11, 14, and 15 were prepared by the reaction of aqueous solutions of manganese(II) sulfate and the respective sodium dithiocarbamate in air [3, 4]. In another preparation of the complexes with ligands 11 and 14, absolute ethanol was used as the solvent instead of water. The complexes crystallized as air-sensitive burgundy needles [6].

C) The complexes with ligands 1 through 6, 9 through 11, 14, 15 were also prepared by air or peroxide oxidation of aqueous solutions of $MnCl_2 \cdot 4H_2O$ and the dithiocarbamate, prepared in situ from the respective amine by reaction with CS_2 in aqueous alkali. The solids which precipitated were washed with cold water and dried in vacuum [4, 7, 8].

For numbers, substituents, and formulas of ligands, see p. 139

The dimethyldithiocarbamate, $[Mn(C_3H_6NS_2)_3]$, was also prepared by electrolysis of manganese in the presence of tetramethylthiuram disulfide in acetone under argon in a special apparatus. The solid precipitated, or was obtained by evaporation of the solvent in vacuum, followed by the addition of petroleum ether (b.p. 40 to 60°C). The compound was washed with petroleum ether and dried in vacuum [9].

The diethyldithiocarbamate, $[Mn(C_5H_{10}NS_2)_3]$, was also obtained by the reaction of $Na(C_5H_{10}NS_2)$ with the bromination product of Mn^{III} acetylacetonate (formulated as $Mn(HacacBr)Br \cdot Br_3$) [10, 11]. This complex crystallizes in two polymorphs, depending on the solvent (see below) [5, 14].

The complexes were purified by dissolving them in chloroform and reprecipitation with methanol, ethanol [3, 12], or petroleum ether (b.p. 40 to 70°C) [4], or cyclohexane [5]. Or, they were recrystallized from acetone [14], dichloromethane-petroleum ether (80 to 100°C), or dichloromethane-ethanol mixtures [8]. Recrystallization was done under a nitrogen atmosphere, which prevented air decomposition [6]. The complexes were dried in vacuum over silica gel [4, 8]. Determination of the molecular weights of the complexes with ligands 2 to 6 and 16 in benzene show that the complexes are monomeric [1, 13].

References:

[1] Prabhakaran, C. P.; Patel, C. C. (Indian J. Chem. **7** [1969] 1257/60).
[2] Novoselov, R. I.; Sokolovskaya, I. P. (Izv. Sibirsk. Otd. Akad. Nauk SSSR Ser. Khim. Nauk **1974** 60/3; C.A. **82** [1975] No. 8022).
[3] Golding, R. M.; Healy, P.; Newman, P.; Sinn, E.; Tennant, W. C.; White, A. H. (J. Chem. Phys. **52** [1970] 3105/9).
[4] Cambi, L.; Cagnosso, A. (Atti Accad. Nazl. Lincei Classe Sci. Fis. Mat. Nat. Rend. [6] **14** [1931] 71/4; C.A. **1936** 2172).
[5] Healy, P. C.; White, A. H. (J. Chem. Soc. Dalton Trans. **1972** 1883/7).
[6] Que, L., Jr.; Pignolet, L. H. (Inorg. Chem. **13** [1974] 351/6).
[7] Muthuswamy, S.; Venkappayya, D. (J. Inst. Chem. [India] **56** [1984] 129/32).
[8] Hendrickson, A. R.; Martin, R. L.; Rohde, N. M. (Inorg. Chem. **13** [1974] 1933/9).
[9] Casey, A. T.; Vecchio, A. M. (Transition Metal Chem. [Weinheim] **11** [1986] 366/8).
[10] Paike, N. H. (Rev. Roumaine Chim. **30** [1985] 517/24).

[11] Paike, N. H. (Bull. Pure Appl. Sci. **1** [1982/83] 115/22).
[12] Yasuda, H.; Suga, K.; Aoyagui, S. (J. Electroanal. Chem. Interfacial Electrochem. **86** [1978] 259/70).
[13] Kumar, S.; Kaushik, N. K.; Mittal, I. P. (Therm. Anal. Proc. 6th Intern. Conf., Bayreuth, FRG, 1980, Vol. 2, pp. 137/46; C.A. **94** [1981] No. 24225).
[14] Elliot, R. L.; West, B. O.; Snow, M. R.; Tiekink, E. R. T. (Acta Cryst. C **42** [1986] 763/4).

35.10.2.2.2.2 Molecular Structure

The molecular structures of the two polymorphs of the diethyldithiocarbamate, $[Mn(C_5H_{10}NS_2)_3]$, have been determined by X-ray diffraction. Polymorph I crystallizes from chloroform-cyclohexane in the monoclinic system, space group $P2_1/a\text{-}C_{2h}^5$ (No. 14), with a=18.04(2), b=8.34(2), c=15.88(3) Å, β=96.7(1)°; Z=4. Atomic positional parameters and isotropic temperature factors for the nonhydrogen atoms and anisotropic temperature factors for the manganese and sulfur atoms are presented in the paper. The structure was solved up to R=0.11 [1]. Lattice plane distances from X-ray powder patterns are summarized in [2]. The experimental density is D_{exp}=1.37(5) g/cm³, the calculated density is D_{calc}=1.40 g/cm³ [1]. The environment of the manganese atom in polymorph I consists of three approximately opposing and equivalent pairs of Mn–S bonds (see **Fig. 16** a). The distortion of the molecule is attributed to the electronic ground state of the high-spin d^4 Mn^{III} (^5E_g), which is susceptible to Jahn-Teller distortions. The average Mn–S bond length, 2.46 Å, is consistent with an Mn^{III}-thiolate linkage, while the average C–N bond length, 1.36 Å, is consistent with an appreciable multiple bond between the atoms [1].

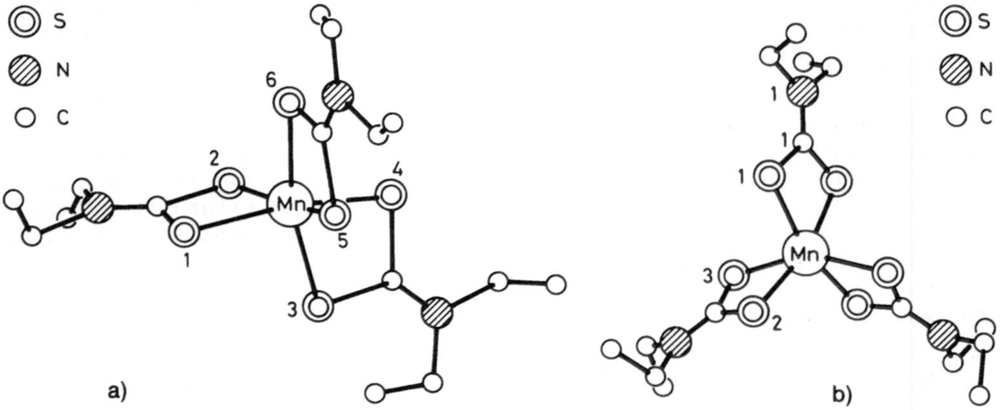

Fig. 16. Molecular structure of the diethyldithiocarbamate, $[Mn(C_5H_{10}NS_2)_3]$, a) of polymorph I [1], b) of polymorph II [3]. H atoms omitted for clarity.

Selected bond lengths (in Å) and angles (in °) in polymorph I are [1]:

Mn–S(1)	2.431(7)	Mn–S(3)	2.373(9)	Mn–S(5)	2.557(7)
Mn–S(2)	2.541(7)	Mn–S(4)	2.428(7)	Mn–S(6)	2.383(8)
S(1)–Mn–S(2)	72.4(2)	S(2)–Mn–S(3)	97.9(3)	S(3)–Mn–S(5)	100.1(3)
S(1)–Mn–S(3)	92.3(3)	S(2)–Mn–S(4)	98.4(3)	S(3)–Mn–S(6)	166.1(3)
S(1)–Mn–S(4)	162.6(3)	S(2)–Mn–S(5)	159.6(3)	S(4)–Mn–S(5)	95.5(3)
S(1)–Mn–S(5)	97.6(3)	S(2)–Mn–S(6)	92.0(2)	S(4)–Mn–S(6)	94.9(3)
S(1)–Mn–S(6)	100.0(3)	S(3)–Mn–S(4)	74.1(2)	S(5)–Mn–S(6)	71.9(2)

Polymorph II crystallizes from acetone in the monoclinic system as well, but space group I2/a-C^6_{2h} (No. 15) with a = 17.171(4), b = 10.188(2), c = 14.822(4) Å, β = 112.13(2)°; Z = 4. The experimental density is D_{exp} = 1.40(1) g/cm³, the calculated density is D_{calc} = 1.382 g/cm³ at 293 K. The structure was solved up to R = 0.023. Atomic positional parameters for the non-hydrogen atoms, interatomic distances, and bond angles are presented in the paper. The molecular structure of polymorph II (Fig. 16b) is in essential agreement with that reported for polymorph I, the major difference being that in polymorph II the molecule is constrained to C_2 symmetry. Mn(1), C(1), and N(1) all lie on a twofold axis. The disparity in the Mn–S bond distances formed by two of the ligand molecules is ascribed to a Jahn-Teller distortion. Selected bond lengths (in Å) and angles (in °) in polymorph II are [3]:

Mn–S(1)	2.382(1)	S(1)–Mn–S(1′)	74.2(1)	S(2)–Mn–S(2′)	102.3(1)
Mn–S(2)	2.389(1)	S(1)–Mn–S(2)	92.2(1)	S(2)–Mn–S(3)	71.9(1)
Mn–S(3)	2.582(1)	S(1)–Mn–S(3)	99.3(1)	S(3)–Mn–S(3′)	153.3(1)

For numbers, substituents, and formulas of ligands, see p. 139

References:

[1] Healy, P. C.; White, A. H. (J. Chem. Soc. Dalton Trans. **1972** 1883/7).
[2] Malissa, H.; Kolbe-Rohde, H. (Talanta **8** [1961] 841/5).
[3] Elliot, R. L.; West, B. O.; Snow, M. R.; Tiekink, E. R. T. (Acta Cryst. C **42** [1986] 763/4).

35.10.2.2.2.3 Magnetic Properties

The Mn^{III} complexes are paramagnetic and show a magnetic moment which is consistent with a high-spin d^4 ion. The table below lists the experimental magnetic moments for the solids at various temperatures. The ambient temperature moments generally are in the range 4.7 to 5.0 μ_B and are independent of temperature from 4.2 to 352.8 K [1 to 7]. A comparison of the experimental magnetic moments with calculated values suggests that the energy separation between the electronic ground state, ^5E_g, and the lowest-lying excited state, $^3T_{1g}$ [9], is of the order of 2000 cm^{-1} [1]. The absence of variation of the moment with temperature suggests that the separation between the electronic ground state, ^5E_g, and the $^3T_{2g}$ state is >4000 cm^{-1} [2].

ligand	complex	μ_{eff} in μ_B a)	T in K	Ref.
1	$[Mn(C_3H_6NS_2)_3]$	4.82±0.06	100 to 300	[1]
2	$[Mn(C_5H_{10}NS_2)_3]$	4.81±0.10	100 to 300	[1]
		4.8	4.2 to 300	[2]
		4.90	298	[3]
3	$[Mn(C_7H_{14}NS_2)_3]$	4.88	298	[3]
4	$[Mn(C_7H_{14}NS_2)_3]$	4.80	298	[3]
5	$[Mn(C_9H_{18}NS_2)_3]$	4.83	298	[3]
6	$[Mn(C_9H_{18}NS_2)_3]$	23.8 b)	293	[4]
9	$[Mn(C_{13}H_{22}NS_2)_3]$	4.83	room temp.	[6]
11	$[Mn(C_{15}H_{14}NS_2)_3]$	4.75±0.08	100 to 300	[1]
		4.93	296	[5]
		25.1 b)	293	[4]
14	$[Mn(C_8H_8NS_2)_3]$	4.99	296	[8]
15	$[Mn(C_9H_{10}NS_2)_3]$	4.82±0.06	100 to 300	[1]
16	$[Mn(C_{10}H_{12}NS_2)_3]$	4.80	—	[7]

a) Magnetic moments from susceptibility measurements at the given temperatures by the Gouy method [1, 6, 7] or by the Faraday method [2, 5]. – b) Weiss magnetons.

The ESR spectra of pure Mn^{III} dithiocarbamates at 77 K show a very broad isotropic peak at about g = 2 which may arise from the d^4 manganese. Powder or single crystal spectra of the complexes (those with ligands 1, 2, 11, 14, and 15 were measured) doped into cobalt analogs (1%) at 77 K show a pattern of three resonances with a 5 G line width at g = 2.122 to 2.143, 2.082 to 2.126, and 2.032 to 2.076. The spectra do not arise from the paramagnetism of manganese(III) but from electron delocalization to or near the ligand sulfur atoms [8].

The ^1H NMR spectra of the paramagnetic Mn^{III} complexes in solution show contact-shifted ligand resonances. The downfield shifts are caused by a spin polarization mechanism [9]. The table on p. 150 lists the proton hyperfine coupling constants (in 10^4 Hz), which were derived from the temperature dependence of the contact shifts [10] for $[MnL_3]$ complexes in deuterochloroform at 303 K (A_a for protons on C adjacent to N (=C_a); A_b for protons on C next to C_a (=C_b); A_c for protons on C next to C_b (=C_c); A_d for aromatic protons):

ligand	complex	A_a	A_b	A_c	A_d	Ref.
1	$[Mn(C_3H_6NS_2)_3]$	12.9*)	—	—	—	[11]
2	$[Mn(C_5H_{10}NS_2)_3]$	5.87	−0.39	—	—	[9]
3	$[Mn(C_7H_{14}NS_2)_3]$	5.84	−0.39	−0.13	—	[9]
5	$[Mn(C_9H_{18}NS_2)_3]$	5.89	0.16	−0.15	—	[9]
6	$[Mn(C_9H_{18}NS_2)_3]$	5.85	0.50	—	—	[9]
7	$[Mn(C_{11}H_{22}NS_2)_3]$	5.88	0.00	0.00	—	[9]
8	$[Mn(C_{11}H_{22}NS_2)_3]$	6.01	0.21	0.04	—	[9]
11	$[Mn(C_{15}H_{14}NS_2)_3]$	4.69	—	—	0.00	[9]
14	$[Mn(C_8H_8NS_2)_3]$	15.2*)	—	—	8.00	[11]
15	$[Mn(C_9H_{10}NS_2)_3]$	6.6*)	−3.0	—	7.0	[11]

*) At 90 MHz, all the rest at 60 MHz.

The shift of the CH_2 protons of the complex with ligand 11, $[Mn(C_{15}H_{14}NS_2)_3]$, is −9.44 ppm relative to the internal standard $CHDCl_2$ in CD_2Cl_2 solution at 31°C. The shift of the CH_3 protons of the complex with ligand 14, $[Mn(C_8H_8NS_2)_3]$, is −29.18 ppm. The spectra show isotropic shifts which vary in a linear manner with 1/T in the −80 to 31°C temperature range. The low temperature spectrum reveals the presence of *cis-trans* geometrical isomers for $[Mn(C_8H_8NS_2)_3]$. The temperature dependence of the spectrum of $[Mn(C_{15}H_{14}NS_2)_3]$ is consistent with a slow metal-centered optical inversion. Thus the geometrical and optical isomerizations are slow on the NMR time scale below −60°C. The activation parameters for the manganese-centered inversion are: $\Delta H^{\neq}=11.0\pm1$, $\Delta G^{\neq}=10.6\pm0.2$ kcal/mol (at −35°C), $\Delta S^{\neq}=1.5\pm5$ cal·mol⁻¹·K⁻¹ for $[Mn(C_{15}H_{14}NS_2)_3]$; $\Delta H^{\neq}=9.8\pm1.0$ kcal/mol (from $\Delta G^{\neq}$ and assuming $\Delta S^{\neq}=3.0$ $cal \cdot mol^{-1} \cdot K^{-1}$), $\Delta G^{\neq}=9.1\pm0.5$ kcal/mol (at −50°C) for $[Mn(C_8H_8NS_2)_3]$. A trigonal twist mechanism is assigned as the primary rearrangement pathway [5].

References:

[1] Golding, R. M.; Healy, P. C.; White, A. H. (Trans. Faraday Soc. **67** [1971] 1672/7).
[2] Figgis, B. N.; Toogood, G. E. (J. Chem. Soc. Dalton Trans. **1972** 2177/82).
[3] Prabhakaran, C. P.; Patel, C. C. (Indian J. Chem. **7** [1969] 1257/60).
[4] Cambi, L.; Szegö, L. (Ber. Deut. Chem. Ges. B **64** [1931] 2591/8).
[5] Que, L., Jr.; Pignolet, L. H. (Inorg. Chem. **13** [1974] 351/6).
[6] Muthuswamy, S.; Venkappayya, D. (J. Inst. Chem. [India] **56** [1984] 129/32).
[7] Kumar, S.; Kaushik, N. K.; Mittal, I. P. (Therm. Anal. Proc. 6th Intern. Conf., Bayreuth, FRG, 1980, Vol. 2, pp. 137/46; C.A. **94** [1981] No. 24225).
[8] Golding, R. M.; Sinn, E.; Tennant, W. C. (J. Chem. Phys. **56** [1972] 5296/9).
[9] Golding, R. M.; Healy, P.; Newman, P.; Sinn, E.; Tennant, W. C.; White, A. H. (J. Chem. Phys. **52** [1970] 3105/9).
[10] Golding, R. M. (Pure Appl. Chem. **32** [1972] 123/35).

[11] Golding, R. M.; Healy, P. C.; Colombera, P.; White, A. H. (Australian J. Chem. **27** [1974] 2089/97).

For numbers, substituents, and formulas of ligands, see p. 139

35.10.2.2.2.4 Infrared Spectra

The IR spectra of the complexes show a number of characteristic absorption bands which are assigned to vibrations originating primarily in the ligand and those that arise because of the manganese(III)-sulfur interaction. The most prominent bands from the spectra of a variety of complexes are listed below. The spectra were taken from paraffin or mineral oil mulls [1, 2], Nujol mulls [3 to 6], or from KBr disks or pellets [6, 7, 9 to 11], in a dry N_2 atmosphere [5]; band positions in cm^{-1}.

ligand	complex	ν(CN)	$ν_{as}$(CS)	$ν_s$(CS)	ν(Mn–S)	Ref.
1	$[Mn(C_3H_6NS_2)_3]$	1500	—	—	355	[1]
2	$[Mn(C_5H_{10}NS_2)_3]$	1502	1000	630	—	[3]
		1492	—	—	—	[2]
		1487	—	—	380, 370	[7]
		—	—	—	388, 376	[4]
		—	—	—	376, 315	[5]
		—	—	—	376, 364	[6]
		1480	—	—	395, 368	[1]
		1498	996, 915 a)	—	—	[11]
3	$[Mn(C_7H_{14}NS_2)_3]$	1495	990	620	—	[3]
4	$[Mn(C_7H_{14}NS_2)_3]$	1490	950	610	—	[3]
5	$[Mn(C_9H_{18}NS_2)_3]$	1493	970	610	—	[3]
6	$[Mn(C_9H_{18}NS_2)_3]$	1487	998	630	—	[3]
9	$[Mn(C_{13}H_{22}NS_2)_3]$	1440 b)	970 b)	—	—	[9]
16	$[Mn(C_{10}H_{12}NS_2)_3]$	—	—	—	380	[10]

a) This band was assigned to $ν_s(CS_2)$ in [11]. – b) Free ligand bands also at 1440 and 970 cm^{-1}.

The position of the ν(CN) ligand band suggests that the bond order of the CN bond is close to two and that the multiple bond character is higher in the complex than in the free ligand. The position of the ν(CS) bands is consistent with a bidentate chelate structure [3, 9 to 11]. The frequency of the ν(Mn–S) band suggests a moderate manganese(III)-sulfur interaction. The force constant for the Mn–S bond in $[Mn(C_5H_{10}NS_2)_3]$ is calculated to be k = 2.34 mdyn/Å for $ν_{as}$(Mn–S) at 376 cm^{-1} and $ν_s$(Mn–S) at 315 cm^{-1} [5]. Additional bands in the spectrum of $[Mn(C_5H_{10}NS_2)_3]$ are reported in [2, 11], and in the far-IR spectrum of $[Mn(C_5H_{10}NS_2)_3]$ in [4, 5, 8]. Assignments are given in [5, 11].

References:

[1] Brown, K. L.; Golding, R. M.; Healy, P. C.; Jessop, K. J.; Tennant, W. C. (Australian J. Chem. **27** [1974] 2075/81).

[2] Chatt, J.; Duncanson, L. A.; Venanzi, L. M. (Suomen Kemistilehti B **29** [1956] 75/84; C.A. **1957** 5559).

[3] Prabhakaran, C. P.; Patel, C. C. (Indian J. Chem. **7** [1969] 1257/60).

[4] Hall, G. R.; Henrickson, D. N. (Inorg. Chem. **15** [1976] 607/18).

[5] Kellner, R.; Nikolov, G. (J. Inorg. Nucl. Chem. **43** [1981] 1183/8).

[6] Aravamudan, G.; Brown, D. H.; Venkappayya, D. (J. Chem. Soc. A **1971** 2744/7).

[7] Kellner, R. (Anal. Chim. Acta **63** [1973] 277/84).

[8] Kellner, R. (Anal. Chim. Acta **68** [1974] 49/60).

[9] Muthuswamy, S.; Venkappayya, D. (J. Inst. Chem. [India] **56** [1984] 129/32).

[10] Kumar, S.; Kaushik, N. K.; Mittal, I. P. (Therm. Anal. Proc. 6th Intern. Conf., Bayreuth, FRG, 1980, Vol. 2, pp. 137/46; C.A. **94** [1981] No. 24225).

[11] Sokolskii, D. V.; Kurashvili, L. M.; Zavorokhina, I. A. (Izv. Akad. Nauk Kaz.SSR Ser. Khim. **21** [1971] 10/4; C.A. **76** [1972] No. 78768).

35.10.2.2.2.5 Electronic Absorption and Mass Spectra

In the electronic spectra of the complexes up to 9 maxima are observed. Assignments of the bands are not unequivocal. Wavenumbers of the first 5 maxima for a variety of complexes and solvents are summarized in Table 1 (ν in kK, log ε in parentheses).

Table 1
Absorption Maxima*) in the Spectra of Manganese(III) dialkyl- and diaryldithiocarbamates in Organic Solvents.

ligand	complex	solvent	1	2	3	4	5	Ref.
1	$[Mn(C_3H_6NS_2)_3]$	$CHCl_3$	6.0	16.0	—	—	—	[5]
2	$[Mn(C_5H_{10}NS_2)_3]$	C_6H_6	—	—	—	19.8(3.56)	—	[6]
		CCl_4	—	—	—	19.8(3.57)	28.2(3.98)	[7]
		CCl_4	5.5(2.01)	16.0(2.40)	—	19.7(3.10)	—	[3]
		CCl_4	—	15.8(2.98)	17.1(2.97)	19.4(3.40)	27.8(4.0)	[4]
		CCl_4	5.9	—	—	—	—	[6]
3	$[Mn(C_7H_{14}NS_2)_3]$	CCl_4	—	15.6(2.91)	16.4(2.95)	19.4(3.39)	27.8(3.98)	[4]
4	$[Mn(C_7H_{14}NS_2)_3]$	CCl_4	—	15.2(2.96)	16.7(3.10)	20.0(3.58)	27.8(3.96)	[4]
5	$[Mn(C_9H_{18}NS_2)_3]$	CCl_4	—	15.5(2.9)	16.7(2.9)	19.2(3.40)	27.9(3.92)	[4]
6	$[Mn(C_9H_{18}NS_2)_3]$	CCl_4	—	15.4(2.9)	16.1(2.9)	19.6(3.53)	27.8(3.88)	[4]
9	$[Mn(C_{13}H_{22}NS_2)_3]$	CH_3COCH_3	—	16.0	—	20.0	27.88	[8]
16	$[Mn(C_{10}H_{12}NS_2)_3]$	CH_3COCH_3	—	15.0	—	19.9	24.8	[2]

*) Assignments are reported as follows: for band 1, $^5B_{1g}\rightarrow{}^5A_{1g}$ [1], charge transfer [3]; for band 2, $^5B_{1g}\rightarrow{}^5B_{2g}$ [1], $^5B_{1g}\rightarrow{}^5A_{1g}$ [2], $^5E_g\rightarrow{}^5T_{2g}$ [3], $^5E\rightarrow{}^5E'$ [4]; for band 3, $^5E\rightarrow{}^6A$ [4]; for band 4, $^5B_{1g}\rightarrow{}^5E_g$, $a_{1g}\rightarrow\pi^*$ [1], $^5B_{1g}\rightarrow{}^5B_{2g}$ [2], $^5E_g\rightarrow{}^5T_{2g}$ [3], $\pi\rightarrow e_g$ [4]; for band 5, $\pi\rightarrow e_g$, $b_{2g}\rightarrow\pi$ [1], $^5B_{1g}\rightarrow{}^5E_g$ [2], $\pi\rightarrow t_{2g}$ [4].

The following unassigned bands for the diethyldithiocarbamate, $[Mn(C_5H_{10}NS_2)_3]$, deviate from the above pattern: 10.4, 18.5, 21.1, 24.1, 35.7, 40.3, 44.2 for the solid complex and 18.2(2.70), 21.2(2.95), 25.0(3.11), 35.0(4.0), 40.0(4.0), 45.4(4.11) for the complex in acetonitrile. Bands at similar positions were observed for the complex in ethanol and heptane [9].

The negative ion mass spectrum of the diethyldithiocarbamate has been determined, using the desorption chemical ionization method and the complementary in-beam direct insertion method. The negative charge added to the complex enters metal-based orbitals to give a reduced Mn complex. The desorption chemical ionization method gives a small fraction of the complexes, 0.4% $[Mn(C_5H_{10}NS_2)_3]^-$ and 0.6% $[Mn(C_5H_{10}NS_2)_2]^-$, with the decomposition of the complex yielding the majority of ions, 85% $(C_5H_{10}NS_2)^-$ and 14% $(C_5H_{10}NS_2)$–CH_4 [10].

For numbers, substituents, and formulas of ligands, see p. 139

References:

[1] Nikolov, G. S. (Inorg. Chim. Acta **4** [1970] 610/2).
[2] Kumar, S.; Kaushik, W. K.; Mittal, I. P. (Therm. Anal. Proc. 6th Intern. Conf., Bayreuth, FRG, 1980, Vol. 2, pp. 137/46; C.A. **94** [1981] No. 24225).
[3] Dingle, R. (Acta Chem. Scand. **20** [1966] 33/44).
[4] Prabhakaran, C. P.; Patel, C. C. (Indian J. Chem. **7** [1969] 1257/60).
[5] Ewald, A. H.; Martin, R. L.; Ross, I. G.; White, A. H. (Proc. Roy. Soc. [London] A **280** [1964] 235/57, 251).
[6] Yamamoto, D.; Nishimoto, M. (Meiji Daigaku Nogakubu Kenkyu Hokoku No. 66 [1984] 39/43 from C.A. **102** [1985] No. 124564).
[7] Bode, H. (Z. Anal. Chem. **144** [1955] 165/86, 178/9).
[8] Muthuswamy, S.; Venkappayya, D. (J. Inst. Chem. [India] **56** [1984] 129/32).
[9] Kurashvili, L. M.; Zavorokhina, N. A. (Zh. Prikl. Spektrosk. **21** [1974] 676/9; J. Appl. Spectrosc. [USSR] **21** [1974] 1353/5).
[10] Gregor, I. K.; Guilhaus, M. (Org. Mass Spectrom. **17** [1982] 575/9).

35.10.2.2.2.6 Electrochemistry

Nonaqueous solutions of the Mn^{III} complexes display a rich redox chemistry. For example, the one-electron reversible oxidation of $[MnL_3]$ yields the Mn^{IV} complexes, $[MnL_3]^+$, while the one-electron reversible reduction yields the Mn^{II} complexes, $[MnL_3]^-$ [1, 2]. The half-wave potentials for the metal oxidation state changes from normal pulse voltammetry (A), alternating current voltammetry (B), cyclic voltammetry (C), and polarography at the dropping mercury electrode (D) are collected in the following table.

Table 2
Half-Wave Potentials for Redox Reactions of $[MnL_3]$ Complexes (in V).

ligand	solvent[a]	method	reference electrode	t in °C	$Mn^{IV} \rightarrow Mn^{III}$	$Mn^{III} \rightarrow Mn^{II}$	Ref.
1	acetone	A	Ag|AgCl (0.1 M LiCl)	22	0.425	−0.035	[1]
	DMF	D	Hg	25	—	−0.35	[5]
	DMF	D	Hg (DMF, 0.1 M TEAP)	25	—	−0.55	[6]
2	acetone	A	Ag|AgCl (0.1 M LiCl)	22	0.409	−0.074	[1]
	acetone	C	Ag|AgCl (0.1 M LiCl)	?	—	−0.11	[3]
	DMF	D	Hg (DMF, 0.1 M TEAP)	25	—	−0.56	[6]
	CH_3CN	C, D	Ag (0.1 M $AgNO_3$)	25	0.160	−0.540	[4]
	CH_3OH-C_6H_6[b] 1:1 (v/v)	D	SCE	20	—	−0.54	[7]
	CH_3OH-C_6H_6[c] 1:1 (v/v)	D	SCE	20	—	−0.56	[7]

Table 2 (continued)

ligand	solvent[a)]	method	reference electrode	t in °C	$Mn^{IV} \rightarrow Mn^{III}$	$Mn^{III} \rightarrow Mn^{II}$	Ref.
4	acetone	A	Ag|AgCl (0.1 M LiCl)	22	0.288	−0.186	[1]
	acetone	C	Ag|AgCl (0.1 M LiCl)	?	—	−0.13	[3]
	DMF	D	Hg (DMF, 0.1 M TEAP)	25	—	−0.56	[6]
	CH_3CN	C, D	Ag (0.1 M $AgNO_3$)	25	0.305	−0.665	[4]
5	acetone	A	Ag|AgCl (0.1 M LiCl)	22	0.429	−0.056	[1]
	acetone	C	Ag|AgCl (0.1 M LiCl)	?	—	−0.13	[3]
	DMF	D	Hg	25	—	−0.38	[5]
	DMF	D	Hg (DMF, 0.1 M TEAP)	25	—	−0.58	[6]
6	acetone	A	Ag|AgCl (0.1 M LiCl)	22	0.366	−0.13	[1]
9	acetone	A	Ag|AgCl (0.1 M LiCl)	22	0.251	−0.230	[1]
10	acetone	B	Ag|AgCl (0.1 M LiCl)	22	0.495[d)]	0.020[d)]	[1]
11	acetone	A	Ag|AgCl (0.1 M LiCl)	22	0.530	0.074	[1]
14	acetone	A	Ag|AgCl (0.1 M LiCl)	22	0.461	−0.007	[1]
15	acetone	A	Ag|AgCl (0.1 M LiCl)	22	0.457	−0.022	[1]

[a)] Supporting electrolyte 0.1 M tetraethylammonium perchlorate (TEAP), if not otherwise stated. – [b)] Supporting electrolyte 0.5 M LiCl. – [c)] Supporting electrolyte 0.5 M $LiClO_4$. – [d)] Peak potentials.

The potentials for the change of oxidation state of the metal are dependent on the electronic and steric properties of the substituents on the nitrogen atom. An increase in chain length and branching affords easier oxidation and more difficult reduction, while the introduction of phenyl groups produces the reverse trend. Thus, the cyclohexyl derivative is easiest to oxidize and hardest to reduce, while the benzyl derivative is hardest to oxidize and easiest to reduce [1, 2]. The potentials are also dependent on the solvent (cf. results of studies on the pyrrolidinecarbodithioato Mn^{III} complex, $[Mn(C_5H_8NS_2)_3]$, in various solvents reported on p. 173). The simple relationship between the hyperfine interaction constants of the α protons and the half-wave reduction potentials stated for the complexes with ligands 2, 3, and 5 suggests that reduction of the complexes takes place at the nitrogen atom [3].

For numbers, substituents, and formulas of ligands, see p. 139

The standard electron transfer rate constants for the system Mn^{IV}/Mn^{III} are k = 0.4 cm/s for the diethyl- and diisopropyldithiocarbamates. For the system Mn^{III}/Mn^{II}, k = 0.6 and 0.3 cm/s, respectively, for the two complexes [4]. Lower values for the Mn^{III}/Mn^{II} system, $k \cdot 10^2 = 1.63 \pm 0.41$, 1.80 ± 0.55, and 1.54 ± 0.36 cm/s for the complexes with ligands 1, 2, and 5, respectively, are reported in [5]. The relatively slow electron transfer reaction is assumed to proceed via an outer-sphere mechanism without change of the metal high-spin state [4]. The electron exchange entropies for the reduction reactions $Mn^{IV} \rightarrow Mn^{III}$ and $Mn^{III} \rightarrow Mn^{II}$ for the diethyldithiocarbamate, $[Mn(C_5H_{10}NS_2)_3]$, are 11.7 and 2.9 $cal \cdot mol^{-1} \cdot K^{-1}$, respectively. The reactions are largely enthalpic since the variation in entropy can only account for a small part of the variation in redox potentials. The charge distribution in the complex and the solvent symmetry play a role for the value of the entropy change [8].

The complexes with ligands 2 to 6 are nonelectrolytes in benzene and acetonitrile [9].

References:

[1] Hendrickson, A. R.; Martin, R. L.; Rohde, N. M. (Inorg. Chem. **13** [1974] 1933/9).
[2] Chant, R.; Hendrickson, A. R.; Martin, R. L.; Rohde, N. M. (Australian J. Chem. **26** [1973] 2533/6).
[3] Golding, R. M.; Lehtonen, K. (Australian J. Chem. **27** [1974] 2083/7).
[4] Yasuda, H.; Suga, K.; Aoyagui, S. (J. Electroanal. Chem. Interfacial Electrochem. **86** [1978] 259/70).
[5] Toropova, V. F.; Budnikov, G. K.; Ulakhovich, N. A. (Zh. Obshch. Khim. **45** [1975] 380/5; J. Gen. Chem. [USSR] **45** [1975] 368/72).
[6] Toropova, V. F.; Budnikov, G. K.; Ulakhovich, N. A. (Talanta **25** [1978] 263/7).
[7] Cordova, R.; Oliva, A.; Schrebler, R. (Talanta **24** [1977] 259/61).
[8] Schmitz, J. E. J.; van der Linden, J. G. M. (Inorg. Chem. **23** [1984] 3298/303).
[9] Prabhakaran, C. P.; Patel, C. C. (Indian J. Chem. **7** [1969] 1257/60).

35.10.2.2.2.7 Chemical Reactions

Note: Reactions of complexes formulated as Mn^{II} compounds in [5 to 7, 11] are introduced in this section. These complexes are obviously Mn^{III} compounds, because of the presence of air during the experiments.

On storing, the diisopropyldithiocarbamate retained its violet color, whereas the complexes with ligands 2, 3, 5, and 6 turned somewhat gray in about a year, showing that the former is more stable than the others [1]. The complex with ligand 6, $[Mn(C_9H_{18}NS_2)_3]$, oxidizes slowly in air, forming dithiuram disulfide [2]. The complex with ligand 16, $[Mn(C_{10}H_{12}NS_2)_3]$, decomposes in a nitrogen atmosphere at 205 to 250°C to yield the Mn^{II} complex, $Mn(C_{10}H_{12}NS_2)_2$, which subsequently forms $Mn(NCS)_2$ at 250 to 340°C, and then further decomposition occurs at 460 to 810°C. Thermal analysis yields an activation energy for the Mn^{III}/Mn^{II} change of $E_A = 60.03$ kcal/mol [3]. TGA studies under static conditions in air (heating rate 6°C/min) show decomposition of the complexes with ligands 9 and 14, $[Mn(C_{13}H_{22}NS_2)_3]$ and $[Mn(C_8H_8NS_2)_3]$, in the temperature interval 120 to 300°C. The final decomposition to Mn_3O_4 proceeds through the intermediate formation of MnS [4].

The complex with ligand 6, $[Mn(C_9H_{18}NS_2)_3]$, is soluble in boiling absolute ethanol, while that with ligand 11, $[Mn(C_{15}H_{14}NS_2)_3]$, is insoluble in ethanol [2]. Both complexes [2], as well as that with ligand 16, $[Mn(C_{10}H_{12}NS_2)_3]$ [3], are soluble in other common organic solvents, such as acetone, benzene, and chloroform [2, 3]. The water solubility of the diethyldithiocarbamate, $[Mn(C_5H_{10}NS_2)_3]$, is 1.45×10^{-4} mol/L. This value has been calculated [5] from the solubility product of the compound, $L = 7.94 \times 10^{-12}$, obtained by emf measurements at $I = 0.1$ mol/L [6]. Approximate solubilities (in g/100 mL solvent) for the complexes with ligands 2 and 5 in various solvents are given below [7]:

ligand	complex	$CHCl_3$	CCl_4	C_6H_6	$CH_3CO_2C_4H_9$
2	$[Mn(C_5H_{10}NS_2)_3]$	12.6	7.0	6.7	14.0
5	$[Mn(C_9H_{18}NS_2)_3]$	22.2	29.6	7.4	18.5

The $[MnL_3]$ complexes with HL = ligands 1, 2, 4 to 6, 9 to 11, 14, and 15 are relatively unstable in solution and are even light-sensitive [12].

The self-reduction of the complexes with ligands 1, 2, and 5 (with or without free ligand) is first-order with respect to the complex. Rate constants, k in $10^{-4} s^{-1}$, and activation energies, E_A in kcal/mol, for the three complexes are at 25°C: 5.24 and 6.5±1.2 for ligand 1; 2.14 and 16.6±2.4 for ligand 2; 1.38 and 14.4±1.4 for ligand 5. The reaction probably proceeds by the rapid formation of MnL_2^+ as an intermediate complex, with its subsequent reduction to form MnL_2, and oxidation of the released ligand [8]. The diethyldithiocarbamate, $[Mn(C_5H_{10}NS_2)_3]$, in CCl_4 decomposes to Mn_3O_4, when exposed to ozone at ambient temperature. Four mol of O_3 per mol of ligand are deactivated by the oxidation of the sulfur atoms. The second-order rate constant for the reaction is $k = 10^4$ $L \cdot mol^{-1} \cdot cm^{-1}$ [9]. The complexes $[MnL_3]$ with HL = ligands 2, 11, and 14, $[Mn(C_5H_{10}NS_2)_3]$, $[Mn(C_{15}H_{14}NS_2)_3]$, and $[Mn(C_8H_8NS_2)_3]$, in CH_2Cl_2-acetone react with $Cu(BF_4)_2$ to yield CuL_2BF_4 and Mn^{II}, which is then spontaneously reoxidized to Mn^{III} [10]. The complexes with ligands 1 and 15 react with the iodine azide ion to give manganese azide and the free ligand anion. A stability series was established for the complexes of 20 metals with the two ligands [11].

References:

[1] Prabhakaran, C. P.; Patel, C. C. (Indian J. Chem. **7** [1969] 1257/60).
[2] Cambi, L.; Cagnosso, A. (Atti Accad. Nazl. Lincei Classe Sci. Fis. Mat. Nat. Rend. [6] **14** [1931] 71/4; C.A. **1936** 2172).
[3] Kumar, S.; Kaushik, N. K.; Mittal, I. P. (Therm. Anal. Proc. 6th Intern. Conf., Bayreuth, FRG, 1980, Vol. 2, pp. 137/46; C.A. **94** [1981] No. 24225).
[4] Muthuswamy, S.; Venkappayya, D. (J. Indian Chem. Soc. **64** [1987] 571/3).
[5] Usatenko, Yu. I.; Barkalov, V. S.; Tulyupa, F. M. (Zh. Analit. Khim. **25** [1970] 1458/61; J. Anal. Chem. [USSR] **25** [1970] 1257/9).
[6] Still, E. (Finska Kemistsamfundets Medd. **73** [1964] 90/106; C.A. **63** [1965] 63).
[7] Barkalov, V. S.; Tulyupa, F. M.; Usatenko, Yu. I. (Ukr. Khim. Zh. **35** [1969] 590/3; Soviet Progr. Chem. **35** No. 6 [1969] 25/7).
[8] Novoselov, R. I.; Sokolovskaya, I. P. (Izv. Sibirsk. Otd. Akad. Nauk SSSR Ser. Khim. Nauk **1974** No. 5, pp. 60/3; C.A. **82** [1975] No. 8022).
[9] Yordanov, N. D.; Karadzhov, Yu. (Transition Metal Chem. [Weinheim] **10** [1985] 15/8).
[10] Ondo, B. M.; Barbier, J. P.; Hugel, R. P. (Inorg. Chim. Acta **77** [1983] L211/L212).

[11] Kurzawa, Z.; Karska, B. (Chem. Inz. Chem. No. 15 [1980] 41/5; C.A. **95** [1981] No. 50432).
[12] Hendrickson, A. R.; Martin, R. L.; Rohde, N. M. (Inorg. Chem. **13** [1974] 1933/9).

35.10.2.2.2.8 Analytical Chemistry

In most cases of analytical application of manganese(III) dithiocarbamates, the water insoluble diethyldithiocarbamate, $Mn(C_5H_{10}NS_2)_2$, is extracted into an organic solvent with subsequent determination of the Mn content in the extract.

For numbers, substituents, and formulas of ligands, see p. 139

A few applications are based only on the formation of a deeply colored precipitate: A sensitive specific detection method for Mn in the presence of many other metals is based on the solubility of $Mn(C_5H_{10}NS_2)_2$ in hot solvent, while the other metal diethyldithiocarbamates are precipitated. On shaking the filtered, cooling solution with air, the characteristic violet-black-brown precipitate of the Mn^{III} complex appears [43]. The red-brown $[Mn(C_5H_{10}NS_2)_3]$ complex which precipitates from an aqueous NH_4OH and tartrate solution of pH 9, can be used for the visual titration of Mn^{II}. It does not form in the presence of EDTA, nor KCN, at pH 12 [1]. $[Mn(C_5H_{10}NS_2)_3]$ can also be used for the sensitive microdetermination of Mn^{II} with spot tests on porcelain, glass plates, or filter paper [2].

Extraction of $[Mn(C_5H_{10}NS_2)_3]$. Two different extraction methods were used:

A) Addition of an excess of a salt of diethyldithiocarbamic acid, most generally $Na(C_5H_{10}NS_2)$, to the mildly acidic or basic aqueous Mn^{II} salt solution yields the Mn^{II} complex, which can be immediately extracted. The initially colorless organic phase then turns brown-violet by air oxidation of the $Mn(C_5H_{10}NS_2)_2$ complex during extraction [3]. Or, air oxidation of the Mn^{II} complex is accomplished within 1 min by vigorous shaking of the aqueous phase, and the violet Mn^{III} complex formed is extracted into the organic phase [3 to 5, 6].

B) Extraction is conducted with a solution of diethyldithiocarbamic acid [7], or a salt of it, in the organic solvent [8, 9]. The asymmetric forms of extraction and back-extraction curves [8], the violet color, and the absorption maximum [9] of the complex in the extract, indicate oxidation of the initially formed Mn^{II} complex to the Mn^{III} complex during extraction in this case, also [8]. Nevertheless, the formation of the $Mn(C_5H_{10}NS_2)_2$ complex in the organic phase is assumed in a number of publications [7, 9 to 12].

The most frequently used solvents are summarized below together with the pH ranges of maximum extraction:

solvent	method	pH	extraction in %	Ref.
CCl_4	A	6 to 9	100[a]	[13, 14]
	A	6 to 9	50	[4]
	B	>6	100	[8]
CCl_4 + isoamyl alcohol	A	3 to 9	~88	[4]
$CHCl_3$	B	>7	100	[8]
methyl isobutyl ketone	B	>6	100	[8]
ethyl acetate	A	~8	~90	[4]
isobutyl acetate	A	6 to 8	~90	[4]
amyl acetate	A	6.5 to 8	~97	[4]
benzene	B	>5	100	[8]
	B	9	100[b]	[15]

[a] Tartaric acid and NH_4Cl were added to the solution of the Mn^{II} salt before addition of $Na(C_5H_{10}NS_2)$ to prevent formation of hydroxide and carbonate.
[b] In the presence of 20% citrate.

Extraction of Mn with diethyldithiocarbamate into methyl isobutyl ketone is highly pH-dependent. It was satisfactory only at pH 6 [42]. For the extraction of Mn^{II} with diethylammonium diethyldithiocarbamate in CCl_4, the pH of half-extraction, $pH_{1/2}$, is 4.3 [8].

The distribution constant, log P, calculated from the solubility of the diethyldithiocarbamate complex in the aqueous and organic phase was calculated for the systems

H_2O-$CHCl_3$, H_2O-CCl_4, H_2O-benzene, and H_2O-butyl acetate. It is 3.40, 3.15, 3.11, and 3.45, respectively [10]. Extraction constants were determined for the systems H_2O-CCl_4 [9, 10, 12], H_2O-$CHCl_3$ [7, 12], H_2O-benzene [12], and H_2O-butyl acetate [12], but all are based on the assumption of the $Mn(C_5H_{10}NS_2)_2$ complex in the organic phase [7, 9, 10, 12]. Precipitation and extraction of $[Mn(C_5H_{10}NS_2)_3]$ are uneffected by cyanide at pH 8 to 9 and by citrate, tartrate, and phosphate ions. By addition of pyrophosphate and reducing agents, such as SO_2 and hydroxylamine, precipitation and extraction are prevented [6]. Extraction of $[Mn(C_5H_{10}NS_2)_3]$ (and other metal diethyldithiocarbamates) from the aqueous phase by means of sulfonated kerosene was investigated. The organic phase was heterogeneous, because flotation took place in addition to extraction. The carbamates were decomposed by means of HNO_3-HCl, and the regenerated kerosene reused [16]. The stability of the $[Mn(C_5H_{10}NS_2)_3]$ complex in the organic phase is dependent on the solvent. Measurements of the optical density of extracts containing ammonium perchlorate show no alteration within 15 min for CCl_4, $CHCl_3$, and butyl acetate, whereas alterations were observed for octanol, isoamyl alcohol, 2-octanol, isobutyl acetate, and diethyl ether [17]. The stability of the color of the $CHCl_3$ extract depends strongly on the pH of the aqueous phase. Extracts obtained in the pH range 5.7 to 6.7 were stable for longer times [5]. The $[Mn(C_5H_{10}NS_2)_3]$ complex was unstable in CCl_4, but was stable for at least 15 min when 12% nitrobenzene was added to the extract, and for 1 h when 80% nitrobenzene was added [8]. Extraction of microamounts of Mn (and other metals), as their diethyldithiocarbamates by nonpolar solvents in the presence of a high excess of the reagent, proceeds slowly and incompletely at pH 6. The use of polar solvents (ethyl acetate) and of mixed solvents increases the extraction rate of the dithiocarbamate and promotes fuller extraction [18]. Salts, especially sodium acetate, dissolved in the aqueous phase also increase the completeness and the rate of extraction, and thus reduce the number of successive extractions required. The salts coagulate the precipitated complexes, promote their wetting by nonpolar solvents, and also prevent hydrolysis [19].

For the back extraction from CCl_4 into aqueous HCl, the pH of half extraction, $pH_{1/2}$, is equal to 3.6 [8]. Back extraction from $CHCl_3$ into aqueous solution is $>90\%$ for 1 N HNO_3, 4 N HCl, and 1 N NaOH [20].

The efficient extraction of the $[Mn(C_5H_{10}NS_2)_3]$ complex into organic solvents has been utilized to separate traces of Mn from various mixtures, and for the quantitative determination of Mn by spectroscopic techniques. Thus, microamounts of Mn (and other metals) were extracted into $CHCl_3$, CCl_4, or amyl acetate from aqueous solutions of pH 3 to 5 containing an Al salt [21, 22], or into $CHCl_3$ at pH 7.5 to 10 from aqueous Sr salt solutions [23], or into isoamyl alcohol at pH 1 to 4 from aqueous Ca and Mg salt solutions [24]. Microamounts of Mn in sulfidic ores and in steels were determined by $CHCl_3$ extraction at pH 5.7 to 6.7 and subsequent spectrophotometry [5], or in the presence of Cu in seawater by $CHCl_3$ extraction and subsequent atomic absorption spectrometry [25]. Mn was extracted by diethylammonium diethyldithiocarbamate into CCl_4 and was spectrophotometrically determined in the extract at 505 nm [13]. Sub-ppm amounts of Mn in standard reference materials were determined by activation analysis after selective CCl_4 extraction of Mn and In in the presence of citrate, NH_3, and KCN and back extraction of the Mn into the aqueous phase [26]. Flash volatilization of a solution of the complex in CCl_4 enhances the sensitivity of atomic absorption spectrophotometry [27]. Extraction of $[Mn(C_5H_{10}NS_2)_3]$ into benzene, and subsequent polarographic analysis of the benzene extract, was used for the determination of Mn in the presence of other metals [15]. Spectrophotometric analysis of the benzene extract was performed for the determination of Mn (in the presence of other metals) in beryllium [28]. Isobutyl acetate was used as the extraction solvent in the determination of Cu and Mn in the presence of each other. The Mn was spectrophotometrically determined at 345 nm [17]. Sometimes a combination of diethyldithiocarbamate with other complexing agents was used. Thus, traces of Mn (and other

For numbers, substituents, and formulas of ligands, see p. 139

metals) in water were extracted with a combination of diethyldithiocarbamate, 8-quinolinol, and dithizone in $CHCl_3$ [29], or traces of Mn and other metals in $NaBH_4$ were extracted with a combination of diethyldithiocarbamate and 8-quinolinol in isoamyl alcohol-$CHCl_3$ (1:4) [30] and subsequently determined by emission spectroscopy [29, 30]. The use of extracts of $[Mn(C_5H_{10}NS_2)_3]$ in $CHCl_3$, CCl_4, and $CH_3CO_2C_5H_{11}$ in flame photometry was investigated [31].

Extraction of Other $[MnL_3]$ Complexes. Because of the good solubility of the dibutyldithiocarbamate, $[Mn(C_9H_{18}NS_2)_3]$, in organic solvents, the extraction of this complex was proposed for the separation of Mn from large quantities of other metals [11]. Mn in waste waters and Al samples was determined by extraction as the dibenzyldithiocarbamate, $[Mn(C_{15}H_{14}NS_2)_3]$, into methyl isobutyl ketone at pH 5 to 9 and subsequent analysis of the extract by atomic absorption spectrophotometry. The effect on extraction of pH, concentration of $Na(C_{15}H_{14}NS_2)$, and shaking time were studied [32].

Metal Exchange. Aqueous solutions of various cations, when mixed with $[Mn(C_5H_{10}NS_2)_3]$ in CCl_4, give a metal exchange reaction to form aqueous Mn^{3+} ions and the corresponding metal complex of ligand 2 in the organic phase. At pH 11 the cations Hg^{2+}, Pd^{2+}, Ag^+, Cu^{2+}, Tl^{3+}, Ni^{2+}, Pb^{2+}, Co^{2+}, Cd^{2+}, Tl^+, and Zn^{2+} exchange, while at pH 8.5 the cations In^{3+}, Sb^{3+}, Fe^{3+}, and Te^{4+} exchange [33]. At pH 5.6 aqueous Zn will quantitatively displace Mn [34]. The Hg^{II}/Mn exchange can be followed by radiometric techniques [35]. Equilibrium constants were determined for the exchange of aqueous Cu^{2+}, Pb^{2+}, Zn^{2+}, Cd^{2+}, Ni^{2+}, Co^{2+}, and Bi^{3+} ions with the Mn complexes of ligand 2 or 5 in $CHCl_3$, CCl_4, benzene, and butyl acetate. Negative values (tabulated) show that manganese is displaced in all cases [36].

Chromatography. Thin-layer chromatography of $[Mn(C_5H_{10}NS_2)_3]$ on silica gel G with various eluents yielded the following R_f values [37] (C_6H_6 = benzene, c-C_6H_{12} = cyclohexane):

C_6H_6	toluene	xylene	$CHCl_3$–c-C_6H_{12} (1:1)	$CHCl_3$–c-C_6H_{12}–$(C_2H_5)_2NH$ (30:65:5)	C_6H_6–c-C_6H_{12}–$(C_2H_5)_2NH$ (60:35:5)
0.76	0.35	0.21	0.33	0.53	0.68

High-pressure liquid chromatography of $[Mn(C_5H_{10}NS_2)_3]$ was used for the simultaneous separation and determination of Mn [38, 39] with Lichrosorb SI 60 [39] or cyanopropylsilane on silica gel [38] as the stationary phase and chloroform-cyclohexane (10:90 v/v) [39] or dichloromethane-hexane (15:85 v/v) [38] as the mobile phase [38, 39]. Trace analysis of metals in various natural samples can be achieved using high performance liquid chromatography. With benzene as mobile phase, 5 μm diameter hypersil silica gel as stationary phase, 1.34 mL/min flow rate, and 288 psi pressure, the parameters were: 4.10 min retention time, 5.49 mL retention volume, and 10 ng detection limit [40].

X-ray fluorescence spectroscopy can also be used to determine Mn in natural waters, after preconcentration of $[Mn(C_5H_{10}NS_2)_3]$ by adsorption on $[Zn(C_5H_{10}NS_2)_2]$ in an ion-exchange column. The detection limit is 2 μg Mn/L [41].

References:

[1] Wickbold, R. (Z. Anal. Chem. **152** [1956] 259/62).
[2] Malissa, H.; Miller, F. F. (Mikrochem. Ver. Mikrochim. Acta **40** [1953] 63/75).
[3] Specker, H.; Hartkamp, H.; Kuchtner, M. (Z. Anal. Chem. **143** [1954] 425/31).
[4] Usatenko, Yu. I.; Fedash, N. P. (Tr. Komis. Analit. Khim. Akad. Nauk SSSR Inst. Geokhim. Analit. Khim. **14** [1963] 183/90; C.A. **59** [1963] 13397).
[5] Specker, H. (Arch. Eisenhüttenw. **26** [1955] 267/70).
[6] Bode, H. (Z. Anal. Chem. **144** [1955] 165/86).
[7] Shen, L. H.; Yeh, S. J.; Lo, J. M. (Anal. Chem. **52** [1980] 1882/5).

[8] Honjo, T.; Imura, H. (Bull. Chem. Soc. Japan **53** [1980] 1753/4).
[9] Stary, J.; Kratzer, K. (Anal. Chim. Acta **40** [1968] 93/100).
[10] Usatenko, Yu. I.; Barkalov, V. S.; Tulyupa, F. M. (Zh. Analit. Khim. **25** [1970] 1458/61; J. Anal. Chem. [USSR] **25** [1970] 1257/9).

[11] Barkalov, V. S.; Tulyupa, F. M.; Usatenko, Yu. I. (Ukr. Khim. Zh. **35** [1969] 590/3; Soviet Progr. Chem. **35** No. 6 [1969] 25/7).
[12] Tulyupa, F. M.; Usatenko, Yu. I.; Barkalov, V. S. (Izv. Vysshikh Uchebn. Zavedenii Khim. Khim. Tekhnol. **14** [1971] 1200/4; C.A. **75** [1971] No. 144356).
[13] Bode, H.; Neumann, F. (Z. Anal. Chem. **172** [1960] 1/21).
[14] Bode, H. (Z. Anal. Chem. **143** [1954] 182/95).
[15] Córdova, R.; Oliva, A.; Schrebler, R. (Talanta **24** [1977] 259/61).
[16] Rublev, V. V.; Martynov, A. V.; Kamadeeva, N. N.; Ushakov, K. P. (Zhidk. Ekstr. Tr. 3rd Vses. Nauchn. Tekhn. Soveshch., 1967 [1969], pp. 413/4; C.A. **73** [1970] No. 124122).
[17] Shah, S. M.; Paul, J. (Microchem. J. **17** [1972] 119/24).
[18] Babko, A. K.; Freger, C. V.; Ovrutskii, M. I.; Lisetskaya, G. S. (Zh. Analit. Khim. **22** [1967] 670/4; J. Anal. Chem. [USSR] **22** [1967] 580/3).
[19] Freger, S. V.; Ovrutskii, M. I.; Lisetskaya, G. S. (Zh. Analit. Khim. **29** [1974] 19/22; J. Anal. Chem. [USSR] **29** [1974] 12/5).
[20] Kuroha, T.; Shibuya, S. (Bunseki Kagaku **21** [1972] 1505/10; C.A. **78** [1973] No. 62957).

[21] Miller, F. F.; Gedda, K.; Malissa, H. (Mikrochem. Ver. Mikrochim. Acta **40** [1953] 373/82).
[22] Smolik, M. (Zesz. Nauk. Politech. Slask. Chem. No. 96 [1981] 91/9 from C.A. **96** [1982] No. 75338).
[23] Smolik, M. (Zesz. Nauk. Politech. Slask. Chem. No. 108 [1983] 77/87 from C.A. **102** [1985] No. 33592).
[24] Cheng, K. L.; Melsted, S. W.; Bray, R. H. (Soil Sci. **75** [1953] 37/40 from C.A. **1953** 11070).
[25] Weiss, H. V.; Kenis, P. R.; Korkisch, J.; Steffan, I. (Anal. Chim. Acta **104** [1979] 337/43).
[26] Ravnik, V.; Dermeli, M.; Kosta, L. (J. Radioanal. Chem. **20** [1974] 443/53).
[27] Hilderbrand, D. C.; Pickett, E. E. (Anal. Chem. **47** [1975] 424/7).
[28] Motojima, K.; Tamura, N. (Bunseki Kagaku **14** [1965] 1150/3 from C.A. **65** [1966] 19302).
[29] Pohl, F. A. (Z. Anal. Chem. **139** [1953] 241/9).
[30] Rezchikov, V. G.; Murzina, O. I.; Gruzdeva, T. M. (Zavodsk. Lab. **51** [1985] 34/5 from C.A. **102** [1985] No. 159713).

[31] Eshelman, H. C.; Armentor, J. (Develop. Appl. Spectrosc. **3** [1963] 190/5).
[32] Ichijo, O. (Bunseki Kagaku **32** [1983] 339/41; C.A. **99** [1983] No. 63420).
[33] Bode, H.; Tusche, K.-J. (Z. Anal. Chem. **157** [1957] 414/27).
[34] Eckert, G. (Z. Anal. Chem. **155** [1957] 23/35).
[35] Braun, T.; Ladányi, L. (Acta Chim. [Budapest] **55** [1968] 361/3).
[36] Barkalov, V. S.; Tulyupa, F. M.; Usatenko, Yu. I. (Khim. Tekhnol. [Kharkov] **1971** No. 24, pp. 33/8; C.A. **77** [1972] No. 668815).
[37] Soundararajan, G.; Subbaiyan, M. (Indian J. Chem. A **22** [1983] 399/401).
[38] Gaetani, E.; Laureri, C. F.; Mangia, A. (Ann. Chim. [Rome] **69** [1979] 181/7).
[39] Liška, O.; Lehotay, J.; Brandšteterová, E.; Guiochon, G.; Colin, H. (J. Chromatog. **172** [1979] 384/7).
[40] Edward-Inatimi, E. B. (J. Chromatog. **256** [1983] 253/66).

[41] Lengar, Z.; Hudnik, V.; Gomiscek, S. (Vestn. Sloven. Kem. Drustva **28** [1981] 379/88 from C.A. **96** [1982] No. 168382).
[42] Mansell, R. E. (Atomic Absorption Newsletter **4** [1965] 276/7).
[43] Gleu, K.; Schwab, R. (Angew. Chem. **62** [1950] 320/4).

35.10.2.2.3 Other Manganese(III) Compounds

$[Mn(C_5H_{10}NS_2)_2Cl]$ was prepared by the reaction of the diethyldithiocarbamato complex, $[Mn(C_5H_{10}NS_2)_3]$, in chloroform with a 0.1 M solution of HCl in dichloromethane. With vigorous shaking a dark-colored microcrystalline complex separated out. The product was isolated, washed with diethyl ether, and dried in vacuum. The complex is monomeric. Its melting point is 135°C. The magnetic moment is $\mu_{eff} = 4.92\ \mu_B$, confirming the three oxidation state. The IR spectrum shows strong bands at 1490, 1000, and 725 cm^{-1}, which were assigned to ν(CN), $\nu_{as}(CS)$, and $\nu_s(CS)$, respectively. Bands in the electronic spectrum at ~12800, ~13300 (doublet), ~16000, and ~22700 cm^{-1} were assigned to $^5B_{1g} \rightarrow {}^5A_{1g}$, $^5B_{1g} \rightarrow {}^5B_{2g}$, and $^5B_{1g} \rightarrow {}^5E_g$ transitions, respectively. The spectrum indicates pentacoordination with, presumably, a square-pyramidal configuration (C_{4v} symmetry). Strong Jahn-Teller distortions are assumed. Hydrochloric acid reduces the complex to give a white solid Mn^{II} complex [1].

$[Mn(C_5H_{10}NS_2)_2\{C_2S_2(CF_3)_2\}]$ was prepared by the reaction of bis(trifluoromethyl)-1,2-dithietene with $Mn^{II}(C_5H_{10}NS_2)_2$, which was prepared in situ from $MnCl_2 \cdot 4H_2O$ and sodium diethyldithiocarbamate in degassed ethanol under nitrogen. The reaction mixture, which immediately turned brown-violet, was stirred for 2 h at room temperature and then filtered. The black precipitate was dissolved in dichloromethane, and the solution filtered to remove NaCl. Addition of hexane to the filtrate gave a crystalline solid. The complex was purified by chromatography on a Florisil column (wrapped in foil to prevent photodecomposition), using benzene as an eluent. A red-brown eluate was collected and the solvent removed. The residue was recrystallized under nitrogen from dry dichloromethane-heptane to give a 9% yield of black needles. The melting point of the compound is 155 to 157°C. The magnetic moment of the solid is $\mu_{eff} = 4.00\ \mu_B$ at 22°C, from susceptibility measurements by the Faraday method. The IR spectrum of the complex in bromoform solution shows the ν(CN) band at 1511 cm^{-1}. The similarity of the IR spectrum to that of the iron analog suggests equivalent structures of the two compounds, i.e., hexacoordination of the metal ion by four S atoms from the uninegative dithiocarbamate anions and two S atoms from the ring-opened dithietene ligand. X-ray determination of the interatomic distances in the Fe compound suggests an average oxidation level of −1 per dithietene ligand. This means that on coordination approximately one electron has been transferred from Mn^{II} to each ring-opened dithietene ligand, forming a resonance-stabilized five-membered chelate ring. The electronic spectrum of the complex in oxygen-free chloroform shows the following maxima (ν_{max} in cm^{-1}, ε in $L \cdot mol^{-1} \cdot cm^{-1}$ in parentheses): 9750 (256), 15400 (sh, 1300), 20800 (sh, 5600), 25300 (10850), 33900 (sh, 19500), 37500 (36600). The complex is photosensitive, especially in solution, and is unstable in moist air [2].

$(C_5H_{10}NS_2)_2Mo(O)OMn(C_5H_{10}NS_2)_2$ was prepared by the reaction of $Mn(C_5H_{10}NS_2)_2$ with $Mo(O)_2(C_5H_{10}NS_2)_2$ in deoxygenated tetrahydrofuran. The reaction mixture was stirred at room temperature for 2 h. Then the complex was precipitated with hexane. It was washed with hexane and dried in vacuum. The magnetic properties of the compound (not reported) indicate the presence of weakly coupled heterodinuclear combinations. However, samples from different preparations showed varying elemental analytical and magnetic data. Bands in the IR spectrum of the complex in Nujol were assigned as follows: 1512, 1485 cm^{-1} to ν(CN); 915 cm^{-1} to ν(Mo=O); and 786, 732 cm^{-1} to Mn–O–Mo vibrations. The mass spectrum shows the following peaks: 426, $Mo(O)_2(C_5H_{10}NS_2)_2$; 410, $Mo(O)(C_5H_{10}NS_2)_2$; 351, $Mn(C_5H_{10}NS_2)_2$; 149, $C_5H_{11}NS_2$; 116, $(C_2H_5)NCS$ [3].

References:

[1] Kanungo, B. K.; Pradhan, B.; Ramana Rao, D. V. (Indian J. Chem. A **21** [1982] 625/7).
[2] Pignolet, L. H.; Lewis, R. A.; Holm, R. H. (J. Am. Chem. Soc. **93** [1971] 360/71).
[3] Elliot, R. L.; Nichols, P. J.; West, B. O. (Australian J. Chem. **39** [1986] 975/85, 985).

35.10.2.3 Manganese(IV) Compounds

Note: Besides the $[MnL_3]X$ and $[MnL_2X_2]$ compounds reported in [2 to 7] the formation of a compound of composition $Mn(C_5H_{10}NS_2)_4$(?) on keeping $[Mn(C_5H_{10}NS_2)_3]$ in its mother liquor in the presence of air and excess ligand was discussed in [1].

$[MnL_3]ClO_4$ (HL = ligand 1, 2, or 9). The complexes with ligand 1 or 2 were prepared by slow addition of an acetone solution of hydrated manganese(II) perchlorate to the corresponding $[MnL_3]$ complexes in benzene in the presence of oxygen. The complexes were crystallized from $CHCl_3$ or $CHCl_3$-$C_6H_5NO_2$ mixtures by slow addition of benzene [2]. The complex with ligand 9, $[Mn(C_{13}H_{22}NS_2)_3]ClO_4$, was prepared by the reaction of $[Mn(C_{13}H_{22}NS_2)_3]$ in benzene and hydrated manganese(II) perchlorate in ethanol (mole ratio 2:1). The ethanol solution was added dropwise with vigorous stirring to the benzene solution. The reaction mixture was further stirred for 1 h, filtered, and the solid dissolved in dichloromethane. Addition of benzene and evaporation of the solution gave dark red crystals in a yield of 54%. The compound was dried in vacuum. Its melting point is 218 to 255°C [3]. The magnetic moment of $[Mn(C_5H_{10}NS_2)_3]ClO_4$ in $CHCl_3$ solution at 293 K is $\mu_{eff} = 3.95\ \mu_B$. The magnetic moment of the solid increases from $\mu_{eff} = 3.71\ \mu_B$ at 103 K to 4.08 μ_B at 290 K. The ESR spectrum of the complex in frozen $CHCl_3$ solution at 113 K yielded the parameters: $g_0 = 1.990 \pm 0.005$, $A_0 = 95 \pm 1$ G, $D_0 = 137 \pm 5$ G. The IR spectra of the complexes in Nujol mulls show the characteristic bands: ν(CN) at 1560, $\nu(ClO_4)$ at 1080 and 620, ν(Mn–S) at 380 cm^{-1} for $[Mn(C_3H_6NS_2)_3]ClO_4$, and ν(CN) at 1530, $\nu(ClO_4)$ at 1085 and 620, ν(Mn–S) at 405 and 382 cm^{-1} for $[Mn(C_5H_{10}NS_2)_3]ClO_4$ [2].

The complexes are explosive and are decomposed in air [2]. The complex with ligand 9, $[Mn(C_{13}H_{22}NS_2)_3]ClO_4$, in acetonitrile is irreversibly photoreduced by exciting light in the range 670 to 425 nm to give Mn^{II} compounds and thiuram disulfide with $[Mn(C_{13}H_{22}NS_2)_3]$ as an intermediate [3]. The complex with ligand 2 is a 1:1 electrolyte in nitrobenzene [2].

$[MnL_3]BF_4$ (HL = ligand 2, 4, or 9) and **$[Mn(C_5H_{10}NS_2)_3]BF_4 \cdot CHCl_3$**. The complexes with ligands 2 and 4, $[Mn(C_5H_{10}NS_2)_3]BF_4$ and $[Mn(C_7H_{14}NS_2)_3]BF_4$, were prepared by the reaction of $[Mn(C_5H_{10}NS_2)_3]$ and $[Mn(C_7H_{14}NS_2)_3]$ with gaseous BF_3 in dichloromethane in the open air. After bubbling BF_3 through the complex solution for 3 min, the mixture was stirred for about 8 min, followed by an additional 0.5 min bubbling of BF_3 gas. The resulting solution was then evaporated to dryness under vacuum. The black oily residue solidified upon treatment with diethyl ether. The solid was washed well with benzene to remove unreacted material, and then dissolved in dichloromethane. The extract was filtered, and addition of diethyl ether to the filtrate caused crystallization. The recrystallization process was repeated four times. The product was isolated and dried under vacuum at room temperature for 7 d [4]. $[Mn(C_5H_{10}NS_2)_3]BF_4$ was also obtained by the reaction of $[Mn(C_5H_{10}NS_2)_3]$ with $[Cu^{III}(C_5H_{10}NS_2)_2]BF_4$ in dichloromethane. Concentration of the solution and addition of diethyl ether yielded the solid compound [7]. The complex with ligand 9, $[Mn(C_{13}H_{22}NS_2)_3]BF_4$, was prepared by the reaction of an excess of boron trifluoride etherate, $BF_3 \cdot (C_2H_5)_2O$, added dropwise, with $[Mn(C_{13}H_{22}NS_2)_3]$ in benzene, in the presence of a vigorous stream of air. After addition of $BF_3 \cdot (C_2H_5)_2O$ was completed, bubbling was continued for 3 h. The resulting suspension was dried, neutralized with Na_2CO_3, and filtered. The solid was extracted repeatedly with dichloromethane. Careful addition of benzene or diethyl ether to the combined extracts gave the complex as dark crystals in a 75% yield. $[Mn(C_7H_{14}NS_2)_3]BF_4$ was also prepared by this procedure in a 60% yield [5]. $[Mn(C_5H_{10}NS_2)_3]BF_4 \cdot CHCl_3$ was prepared by slow addition of $Mn(BF_4)_2 \cdot 6H_2O$ in acetone to $[Mn(C_5H_{10}NS_2)_3]$ in benzene. The complex was recrystallized from chloroform by slow addition of benzene [2].

The magnetic moments of $[Mn(C_5H_{10}NS_2)_3]BF_4$ and $[Mn(C_7H_{14}NS_2)_3]BF_4$ in dichloromethane are 3.74 and 3.59 μ_B, respectively, at 36°C, measured by the NMR method with tetramethyl-

For numbers, substituents, and formulas of ligands, see p. 139

silane as standard [4]. The magnetic moment of solid $[Mn(C_5H_{10}NS_2)_3]BF_4$ is 4.3 μ_B at room temperature [7]. That of solid $[Mn(C_5H_{10}NS_2)_3]BF_4 \cdot CHCl_3$, resulting from susceptibility measurements by the Gouy method increases from 3.92 μ_B at 98 K to 4.05 μ_B at 295 K [2]. The IR spectra of $[Mn(C_5H_{10}NS_2)_3]BF_4$, taken from paraffin or halocarbon mulls [2] or Nujol mulls [4], show characteristic bands (in cm^{-1}) which were assigned as follows: 1535 [2], 1525 [4] to ν(CN); 1060 [2], 1040 [4] to $\nu(ClO_4)$; 405, 382 to ν(Mn–S) [2]. Other unassigned IR bands for $[Mn(C_5H_{10}NS_2)_3]BF_4$ and for $[Mn(C_7H_{14}NS_2)_3]BF_4$ are reported in [4]. The electronic spectra of the two complexes in dichloromethane show seven bands (not assigned) in the range between 200 and 500 nm [4]. The complexes are 1:1 electrolytes in nitrobenzene [2] and nitromethane [4]. The complexes in acetonitrile are thermally unstable and photosensitive. A photokinetic study of 10^{-4} mol/L $[Mn(C_5H_{10}NS_2)_3]BF_4$ in acetonitrile with exciting light in the range between 425 and 670 nm showed that it was decomposed within ~100 s [3].

$[Mn(C_{13}H_{22}NS_2)_3]PF_6$ was prepared by the reaction of the Mn^{III} complex with ligand 9, $[Mn(C_{13}H_{22}NS_2)_3]$, with KPF_6 (in a slight excess) in the presence of a little H_2SO_4 in chloroform-acetone 1:1 (v/v). After passing a stream of dry air through the brown reaction mixture for 5 h, Na_2SO_4 was added, and the suspension was filtered. The filtrate was reduced to dryness in vacuum and the residue taken up in dichloromethane. Addition of benzene to the solution (filtered once more) and cautious evaporating gave black crystals in a 52% yield. The melting point of the solid is 200°C (dec.). The complex is irreversibly photoreduced by exciting light, as described for $[Mn(C_{13}H_{22}NS_2)_3]ClO_4$ on p. 162 [3].

$[Mn(C_5H_{10}NS_2)_2X_2]$ (X = Cl or Br). The complex with ligand 2 and X = Cl was prepared by the reaction of dry gaseous Cl_2 with a suspension or solution of $[Mn(C_5H_{10}NS_2)_3]$ in CCl_4. The reaction mixture was shaken constantly until the color changed. The solution was kept at 100°C to remove excess solvent. The lustrous black residue was washed with diethyl ether and dried in vacuum. The complex with X = Br was prepared by the reaction of bromine with $[Mn(C_5H_{10}NS_2)_3]$ in CS_2. The dilute bromine solution was added dropwise to the solution of $[Mn(C_5H_{10}NS_2)_3]$ with vigorous stirring until the color changed. The solvent was removed at 60°C, and the greenish black residue was washed with ether and finally dried in vacuum. The complexes are monomeric. They are nonelectrolytes in ethanol. Melting points are: 110°C for X = Cl and 62°C for X = Br. The IR spectra of the complexes in Nujol mull show the characteristic bands: ν(CN) at 1515, $\nu_{as}(CS)$ at 1000, and ν(Mn–S) at 398 and 378 cm^{-1}. For $[Mn(C_5H_{10}NS_2)_2Cl_2]$ a band at 285 cm^{-1} was assigned to the ν(Mn–Cl) vibration. The electronic absorption spectra of the complexes in ethanol show maxima at 16000, 21740, and 24690 cm^{-1} for X = Cl and 16000, 21740, and 24390 cm^{-1} for X = Br, which were assigned to the electronic transitions $^4A_{2g} \rightarrow {}^2E_g, {}^2T_{2g}$, $^4A_{2g} \rightarrow {}^4T_{2g}$, $^4A_{2g} \rightarrow {}^4T_{1g}$ of octahedrally surrounded Mn^{IV} [6].

References:

[1] Usatenko, Yu. I.; Fedash, N. P. (Tr. Komis. Analit. Khim. Akad. Nauk SSSR Inst. Geokhim. Analit. Khim. **14** [1963] 183/90).
[2] Brown, K. L.; Golding, R. M.; Healy, P. C.; Jessop, K. J.; Tennant, W. C. (Australian J. Chem. **27** [1974] 2075/81).
[3] Eckstein, P.; Hoyer, E. (Z. Anorg. Allgem. Chem. **487** [1982] 33/43).
[4] Saleh, R. Y.; Straub, D. K. (Inorg. Chem. **13** [1974] 3017/9).
[5] Hendrickson, A. R.; Martin, R. L.; Rohde, N. M. (Inorg. Chem. **13** [1974] 1933/9).

[6] Pradhan, B.; Ramana Rao, D. V. (J. Indian Chem. Soc. **58** [1981] 733/5).
[7] Ondo, B. M.; Barbier, J.-P.; Hugel, R. P. (J. Chem. Res. S **1985** 302/3).

35.10.3 With Other Disubstituted Dithiocarbamic Acids

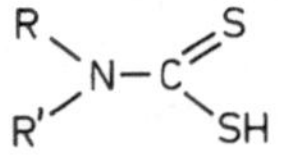

ligand	R	R'	formula
1	HOC_2H_4	HOC_2H_4	$C_5H_{11}NO_2S_2$
2	CH_3	$HOOCCH_2$	$C_4H_7NO_2S_2$
3	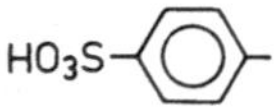	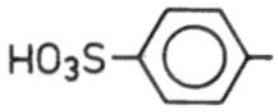	$C_{13}H_{11}NO_6S_4$
4	$C_6H_5CH{=}NCH_2CH_2$	$C_6H_5CH{=}NCH_2CH_2$	$C_{19}H_{21}N_3S_2$
5 to 7	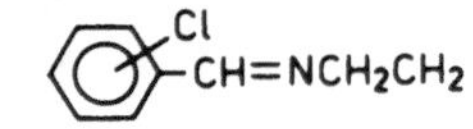 ligand 5 with 2-Cl; ligand 6 with 3-Cl; ligand 7 with 4-Cl	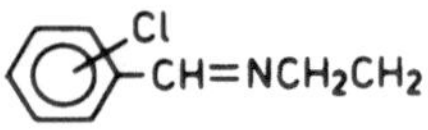	$C_{19}H_{19}Cl_2N_3S_2$
8	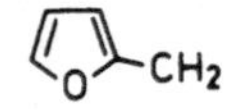	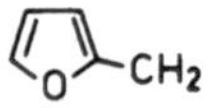	$C_{11}H_{11}NO_2S_2$

The polarographic behavior of the Mn^{II} complex with ligand 1, $\mathbf{Mn^{II}(C_5H_{10}NO_2S_2)_2}$, in acidic aqueous-organic medium was studied. With the aid of the catalytic limiting current, the stability of this complex in the layer adjacent to the electrode was assessed [1]. Extraction of Mn^{II}, as its complex with ligand 2, by O-containing solvents occurs by an ion-pair mechanism [2]. Mn^{2+} ions are quantitatively precipitated from aqueous solutions of $<$pH 9 by the sodium salt of ligand 3. The white precipitate is more stable to air oxygen than the diethyldithiocarbamato Mn^{II} complex (p. 141) [3].

The yellow $\mathbf{[Mn^{II}(C_5H_{10}NO_2S_2)_2L']}$ complexes with L' = bipyridine (bpy) or phenanthroline (phen) were prepared as described for $[MnL_2L']$ complexes on p. 142, but using the sodium or ammonium salt of ligand 1. The magnetic moment of both solid complexes is $\mu_{eff} = 5.5\ \mu_B$ at room temperature. The X-band ESR spectra of polycrystalline samples of the complexes and of solutions in frozen chloroform show features similar to those of the $[MnL_2L']$ complexes described on p. 142. The compounds are insoluble in water and ethanol [4].

$\mathbf{[Mn^{III}L(OH)X]}$ complexes with $X = H_2NCS_2^-$ and HL = ligands 4 to 7 were prepared by the reaction of a methanolic solution of $MnCl_2$ with an ammoniacal-methanolic solution of one of the ligands prepared in situ from the corresponding Schiff base and carbon disulfide. All operations were carried out under nitrogen. The yellow precipitates, which formed immediately, were washed with methanol and ether. The compounds easily oxidized in air to the brown Mn^{III} complexes. Melting points (in °C), magnetic moments (μ_{eff} in μ_B) from susceptibility measurements by the Faraday technique at room temperature, and characteristic IR bands of the Mn^{III} complexes (in KBr disks) are summarized on p. 165:

ligand	m.p.*)	μ_{eff}	ν_{as}(CN)	ν_s(CN)	ν_{as}(CS)	ν_s(CS)	ν(Mn–S)
4	75	4.64	1490	1430	980	965	360
5	103	4.70	1475	1435	985	960	370
6	95	4.80	1480	1430	980	—	375
7	65	4.61	1500	1440	970	—	375

*) With decomposition.

The ν(CN) band of the azomethine group, appearing in the range between 1660 and 1640 cm^{-1} for all the complexes, was the same as that of the respective free ligand. However, the positions of the listed ν(CN) and ν(CS) bands are those of bidentate S-coordinated dithiocarbamato groups. A band at ~3500 cm^{-1} was assigned to the ν(OH) vibration of coordinated hydroxy groups. An octahedral geometry of the complexes dimerized through bridging of two hydroxy groups is assumed. The electronic spectra of the complexes in dimethylformamide or dimethyl sulfoxide show 4 bands between 265 and 650 nm which were assigned to intraligand bands. The complexes are soluble in dimethylformamide and dimethyl sulfoxide, slightly soluble in chloroform but insoluble in other common organic solvents. They show antibacterial and antifungal activity, the complex with ligand 5 (2-Cl derivative) being more active than the others [5].

$Mn(C_{11}H_{10}NO_2S_2)_2$(?). A compound of this composition, however, assumed to be an Mn^{III} compound, was prepared by the reaction of an Mn salt with the potassium or ammonium salt of ligand 8 (solvent not given). The black precipitate was dissolved in methanol, reprecipitated with hexane, separated, and dried in vacuum at ~100°C. The compound melts at 131 to 132°C with decomposition. The half-wave potential in the anodic oxidation of the complex on a rotating Pt electrode in acetonitrile (0.1 mol/L $N(C_2H_5)_4ClO_4$) is E = 0.38 V vs. SCE at 25°C [6].

References:

[1] Budnikov, G. K.; Toropova, V. F.; Ulakhovich, N. A.; Medyantseva, E. P.; Frolova, V. P. (Zh. Analit. Khim. **32** [1977] 212/7; J. Anal. Chem. [USSR] **32** [1977] 172/6).
[2] Byr'ko, V. M.; Karashvili, L. G.; Shepel, L. I.; Tikhonova, T. I.; Levina, Z. I. (Deposited Doc. VINITI-6896-73 [1973] 1/9 from C.A. **86** [1977] No. 111694).
[3] Malissa, H.; Miller, F. F. (Mikrochem. Ver. Mikrochim. Acta **40** [1953] 63/75, 68).
[4] Pandeya, K. B.; Singh, R.; Mathur, P. K.; Singh, R. P. (Transition Metal Chem. [Weinheim] **11** [1986] 347/50).
[5] Manoussakis, G. E.; Bolos, C. A. (Inorg. Chim. Acta **108** [1985] 215/20).
[6] Gol'dfarb, Ya. L.; Ostapenko, E. G.; Vinogradova, V. G.; Zverev, A. N.; Polyakov, A. V.; Yanovskii, A. I.; Yufit, D. S.; Struchkov, Yu. T. (Khim. Geterotsikl. Soedin. No. 7 [1987] 902/9; Chem. Heterocycl. Compounds [USSR] **7** [1987] 740/7).

35.10.4 With N-Heterocyclic N-Carbodithioic Acids

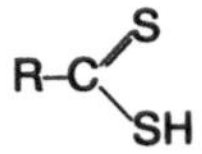

ligand	R	formula
1	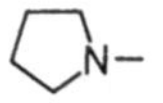	$C_5H_9NS_2$
2	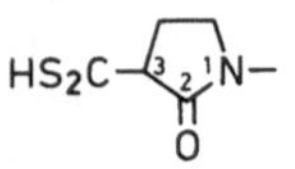	$C_6H_7NOS_4$
3	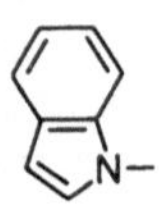	$C_9H_7NS_2$
4	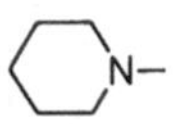	$C_6H_{11}NS_2$
5 to 7	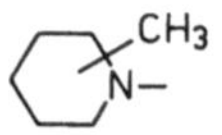 5 with 2-CH_3, 6 with 3-CH_3, 7 with 4-CH_3	$C_7H_{13}NS_2$
8	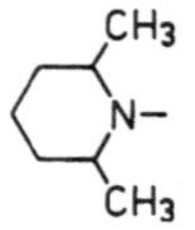	$C_8H_{15}NS_2$
9	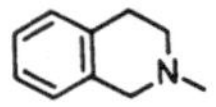	$C_{10}H_9NS_2$
10	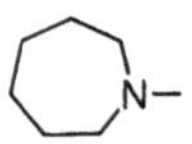	$C_7H_{13}NS_2$
11	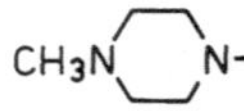	$C_6H_{12}N_2S_2$
12	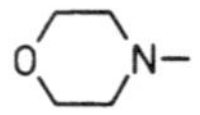	$C_5H_9NOS_2$
13	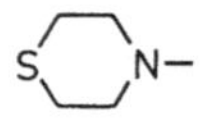	$C_5H_9NS_3$

35.10.4.1 Manganese(II) Compounds

The $Mn(C_5H_8NS_2)_3^-$ Ion in Solution. Coulometric and titrimetric studies of the manganese(II)-pyrrolidinecarbodithioate system in acetone indicate the existence of the colorless $Mn(C_5H_8NS_2)_3^-$ species. The ion has a reasonable stability in acetone. It reacts with additional Mn^{2+} ions to form the solid $Mn(C_5H_8NS_2)_2$ complex (see below) and is oxidized to give the $[Mn(C_5H_8NS_2)_3]$ complex (see p. 169) [1].

$Mn(C_5H_8NS_2)_2$ was prepared by the reaction of hydrated manganese(II) perchlorate (treated with a 3:1 mixture of acetone and 2,2-dimethoxypropane) with $Na(C_5H_8NS_2)\cdot 2H_2O$ in approximately 1:1 molar ratio. The reaction was carried out in a Schlenck tube under argon. When the precipitation was complete, the crystalline yellow solid was isolated, washed with oxygen-free solvent, and pumped dry. The compound is air-sensitive and is manipulated in a glove box [1].

Another preparation makes use of solutions of manganese(II) acetate in 50% aqueous ethanol and $Na(C_5H_8NS_2)$ in a 1:2 molar ratio. The ligand solution was slowly added to the metal solution under nitrogen. The yellow precipitate, which formed immediately, was isolated, washed with 50% aqueous ethanol and then absolute ethanol, and dried by pumping for several hours [2].

A third preparation of the complex involves the reaction of aqueous $MnCl_2$ and $Na(C_5H_8NS_2)$ under anaerobic conditions, as was described on p. 139 for dialkyl- and diaryldithiocarbamates of the MnL_2 type [3].

Measurements of the magnetic susceptibility yielded the magnetic moment $\mu_{eff}=5.19\ \mu_B$ at 90 K, 5.55 μ_B at 295 K [2], or 5.60 μ_B at 292 K [3]. In another work 4.91 μ_B was found, indicating an Mn^{III} compound [4]. The variation of the magnetic moment with temperature is ascribed to antiferromagnetic interaction, and the experimental data could be fit to an expression for the interaction in a dinuclear compound with $J=-2.60\ cm^{-1}$ and $g=1.92$. The Curie-Weiss law is obeyed with $\Theta=-16$ K [2]. The compound is insoluble in methanol, ethanol, acetone, or dichloromethane. It is easily dissolved in pyridine, from which mixed-ligand dithiocarbamato pyridine complexes are precipitated by addition of hexane [3]. The dimeric or polymeric compound reacts with additional $C_5H_8NS_2^-$ ions in solution to form the $Mn(C_5H_8NS_2)_3^-$ ion [1].

$Mn_2(C_6H_5NOS_4)_2$ was prepared by the reaction of a solution of $Li_2(C_6H_5NOS_4)$ in aqueous methanol and an equimolar amount of manganese(II) acetate in methanol. The dark green reaction mixture was stirred for one hour at room temperature. The precipitate was isolated, washed with methanol, and dried under vacuum. The magnetic moment, $\mu_{eff}=5.28\ \mu_B$ at room temperature, indicates metal-metal interaction. The IR spectrum shows the ν(CO) band at 1650 cm^{-1}. The band is not changed with respect to the free ligand. Assumed is a planar dinuclear complex structure involving two bridging O atoms and four S atoms from two ligand molecules in coordination [5].

[MnL_2bpy] and **[MnL_2phen]** complexes with HL = ligand 1, 4, 10, or 12 were prepared as described for the corresponding dialkyldithiocarbamato complexes on p. 142, but using the sodium or ammonium salts of the ligands [6, 13]. The complexes with HL = ligand 10, [$Mn(C_7H_{12}NS_2)_2$bpy] (brick red) and [$Mn(C_7H_{12}NS_2)_2$phen] (saffron), melt at 120 and 210°C, respectively [13]. The magnetic moments of the complexes with ligands 4, 10, and 12 (between 5.7 and 6.1 μ_B) indicate them to be high-spin octahedral [6, 13]. Low-temperature magnetic susceptibility measurements (down to 77 K) show the complexes with ligand 10 to obey the Curie-Weiss law [13]. The X-band ESR spectra of the complexes with ligand 1, 4, or 12 are similar to those of the [MnL_2bpy] and [MnL_2phen] complexes on p. 142 [6]. In the electronic spectra of the complexes with ligand 10 (in Nujol or chloroform) maxima are observed in the regions 16670, 21050 to 21740, 27780 to 28570, and 36100 to 37040 cm^{-1}, which were assigned to high-spin octahedral Mn^{II} and charge-transfer transitions [13]. Half-wave potentials (in V, vs. Hg electrode) and charge-transfer rate constants k (in 10^{-2} cm/s) at 25°C (supporting electrolyte 0.1 M $N(C_2H_5)_4ClO_4$), from polarographic studies at the dropping mercury electrode are as follows [7]:

ligand	complex	$Mn^{II}\rightarrow Mn^{I}$		$Mn^{I}\rightarrow Mn^{0}$	
		$E_{1/2}$	k	$E_{1/2}$	k
1	[$Mn(C_5H_8NS_2)_2$bpy]	1.34	1.54 ± 0.48	1.76	0.81 ± 0.19
	[$Mn(C_5H_8NS_2)_2$phen]	1.15	1.05 ± 0.30	1.62	0.70 ± 0.17
4	[$Mn(C_6H_{10}NS_2)_2$bpy]	1.36	1.62 ± 0.35	1.82	0.75 ± 0.22
	[$Mn(C_6H_{10}NS_2)_2$phen]	1.26	1.03 ± 0.25	1.75	0.82 ± 0.20
12	[$Mn(C_5H_8NOS_2)_2$bpy]	1.33	1.22 ± 0.34	1.82	0.72 ± 0.18
	[$Mn(C_5H_8NOS_2)_2$phen]	1.21	1.08 ± 0.30	1.70	0.65 ± 0.16

Additional two waves correspond to reductions of the aromatic amine ligand [7]. The complexes with ligands 4 and 12 are insoluble in water and ethanol. They are highly soluble in common organic solvents, such as chloroform and acetone [6]. [$Mn(C_7H_{12}NS_2)_2$phen] shows bactericidal and fungicidal activities [13].

[$Mn(C_6H_{10}NS_2)$acac] and **[$Mn(C_6H_{10}NS_2)C_6H_4NO_2$]** complexes with deprotonated ligand 4 and the acetylacetonate or picolinate ions and **[$MnL(C_9H_6NO)$]** compounds with deprotonated ligand 4 or 12 and the 8-quinolinolate ion were prepared in a manner analogous to the [$Mn^{II}(C_5H_{10}NS_2)L'$] complexes on p. 143, and have properties similar to those compounds [8].

$Mn(C_5H_8NS_2)_2(NO)^+$ and **[$MnL_2(NO)Cl$]** complexes were prepared, as described for the corresponding dialkyl- or diaryldithiocarbamato complexes on p. 144, using the appropriate ligands [9 to 11]. Analysis of ESR spectra, taken from toluene extracts of the reaction mixtures, yielded the following parameters for $Mn(C_5H_8NS_2)_2(NO)^+$ (T in Kelvin, hyperfine coupling constant A in 10^{-4} cm^{-1}):

ligand	complex	T	$g_{\parallel}$	$g_{\perp}$	g_0	$-A_{\parallel}$	$-A_{\perp}$	$-A_0$	Ref.
1	$Mn(C_5H_8NS_2)_2(NO)^+$	125	1.993	2.021	—	154.5	44.1	—	[9]
		235	—	—	2.011	—	—	80.9	[9]

The parameters $g_{\parallel} = \sim 1.993$, $g_{\perp} = \sim 2.022$, $K = (81.5 \text{ to } 82.6) \times 10^{-4}$ cm^{-1}, and $-T_{zz} = (72.7 \text{ to } 73.6) \times 10^{-4}$ cm^{-1} at 125 K (explanations of the parameters are given on p. 144) have been obtained for [$MnL_2(NO)Cl$] complexes with deprotonated ligands 1, 4, and 12 [10, 11]. The studies indicate low-spin ($S = 1/2$) complexes with structural and chemical properties similar to those reported for $MnL_2(NO)^+$ and [$MnL_2(NO)Cl$] complexes on p. 144 [9 to 11].

[$MnCd(C_6H_{10}NS_2)_4$] was prepared in a manner analogous to the [$MnML_4$] complexes described on p. 144. The magnetic moment of the complex is $\mu_{eff} = 3.85$ μ_B at room temperature. The complex decomposes at 165°C. The other physical and chemical properties are similar to those of the [$MnML_4$] complexes, indicating the polymeric structure shown on p. 145 [12].

References:

[1] Hendrickson, A. R.; Martin, R. L.; Rohde, N. M. (Inorg. Chem. **13** [1974] 1933/9).
[2] Hill, D. M.; Larkworthy, L. F.; O'Donoghue, M. W. (J. Chem. Soc. Dalton Trans. **1975** 1726/8).
[3] Ciampolini, M.; Mengozzi, C.; Orioli, P. (J. Chem. Soc. Dalton Trans. **1975** 2051/4).
[4] Siddiqi, K. S.; Shah, M. A. A.; Zaidi, S. A. A. (Bull. Soc. Chim. France **1983** 49/51).
[5] Vigato, P. A.; Caselato, U.; Acone, B.; Vidali, M.; Fenton, D. E. (Inorg. Nucl. Chem. Letters **16** [1980] 489/94).
[6] Pandeya, K. B.; Singh, R.; Mathur, P. K.; Singh, R. P. (Transition Metal Chem. [Weinheim] **11** [1986] 347/50).
[7] Toropova, V. F.; Budnikov, G. K.; Ulakhovich, N. A. (Zh. Obshch. Khim. **45** [1975] 380/5; J. Gen. Chem. [USSR] **45** [1975] 368/72).
[8] Pradhan, B.; Ramana Rao, D. V. (J. Indian Chem. Soc. **55** [1978] 226/8).
[9] Yordanov, N. D.; Iliev, V.; Shopov, D.; Jezierski, A.; Jeżowska-Trzebiatowska, B. (Inorg. Chim. Acta **60** [1982] 9/15).
[10] Jeżowska-Trzebiatowska, B.; Jezierski, A. (J. Mol. Struct. **46** [1978] 197/202).

[11] Jezierski, A.; Jeżowska-Trzebiatowska, B. (Nouv. J. Chim. **4** [1980] 599/602).
[12] Aggarwal, R. C.; Singh, B.; Singh, M. K. (J. Indian Chem. Soc. **59** [1982] 269/72).
[13] Singh, A. K.; Puri, B. K.; Rawlley, R. K. (Indian J. Chem. A **27** [1988] 430/3).

For numbers, substituents, and formulas of ligands, see p. 166

35.10.4.2 Manganese(III) Compounds

35.10.4.2.1 $[MnL_3]$ and $[MnL_3]\cdot S$ Complexes ($S = CH_2Cl_2$, $CHCl_3$, or benzene)

Preparation. The complexes with HL = ligand 1, 4, or 12, $[Mn(C_5H_8NS_2)_3]$, $[Mn(C_6H_{10}NS_2)_3]$, and $[Mn(C_5H_8NOS_2)_3]$, have been prepared by procedures analogous to those described for the corresponding dialkyl- and diaryldithiocarbamato complexes on p. 146: by method A) [1, 2], by method B) [3, 4], and by method C) [5]. Method C) was also used for the preparation of the complexes with ligand 5, 7, or 8 [5]. Method B), but starting with manganese(II) acetate, was used for the preparation of the complexes with ligands 4, 11, 13 [6], and 5, 6 or 7 [7]. Method B), but starting with $MnCl_2$, was also used for the complexes with ligand 4 or 12 [8, 46]. Instead of the sodium salts of the ligands [1, 2, 6, 7, 46], the potassium [3], ammonium [4], or morpholinium salts [8, 42] also were used. After precipitation $[Mn(C_5H_8NOS_2)_3]$ was left for 2 h at 0°C, filtered off, washed with cold water, and dried under vacuum for 2 h [42]. $[Mn(C_5H_8NOS_2)_3]$ was also prepared by exposing the morpholine complexes of manganese nitrate, chloride, or isothiocyanate (Mn oxidation state not given) to carbon disulfide in a closed vessel; the reaction was followed by chemical elemental analysis and X-ray powder diffraction [8]. The complexes with ligand 4, 11, or 13 were recrystallized from chloroform under anhydrous nitrogen [6]. For the rest, solvents mentioned on p. 147 were used for recrystallization [3 to 5]. Crystals suitable for an X-ray structure determination were grown from dichloromethane-ethanol [3].

Molecular Structure. Single crystal X-ray structure determinations have been carried out for the 1-morpholinecarbodithioato complexes $[Mn(C_5H_8NOS_2)_3]\cdot CH_2Cl_2$ [3] and $[Mn(C_5H_8NOS_2)_3]\cdot CHCl_3$ [9]. Crystal data for the benzene solvate, $[Mn(C_5H_8NOS_2)_3]\cdot C_6H_6$, have been obtained from precession photographs [10]. Lattice plane distances from X-ray powder photographs of the complex with ligand 1, $[Mn(C_5H_8NS_2)_3]$, are tabulated in [11]. Crystal data and structural information derived from the above studies are summarized below:

complex	$[Mn(C_5H_8NOS_2)_3]\cdot CH_2Cl_2$	$[Mn(C_5H_8NOS_2)_3]\cdot CHCl_3$	$[Mn(C_5H_8NOS_2)_3]\cdot C_6H_6$
crystal system	triclinic	triclinic	monoclinic
space group	$P\bar{1}$-C_i^1 (No. 2)	$P\bar{1}$-C_i^1 (No. 2)	$C2/c$-C_{2h}^6 (No. 15)
a in Å	13.067(3)	13.60(1)	19.2
b in Å	10.824(6)	11.009(5)	17.9
c in Å	11.457(5)	11.733(5)	12.4
α in °	116.23(5)	117.09(4)	—
β in °	104.06(3)	99.89(6)	53.5
γ in °	100.19(4)	107.72(5)	—
Z	2	2	—
D_{exp} in g/cm³	1.55	1.56	—
D_{calc} in g/cm³	1.61	1.60	—
Ref.	[3]	[9]	[10]

Atomic positional and isotropic thermal parameters are presented in the papers. The structures have been solved up to R = 0.04 [3, 9]. As an example, the distorted octahedral structure of $[Mn(C_5H_8NOS_2)_3]\cdot CHCl_3$ is shown in **Fig. 17**, p. 170, which also displays the numbering scheme for the various S atoms.

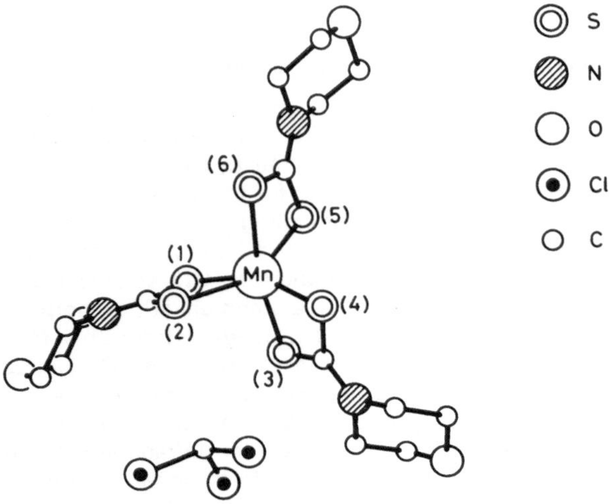

Fig. 17. Molecular structure of the 1-morpholinecarbodithioato complex, $[Mn(C_5H_8NOS_2)_3]\cdot CHCl_3$ [9].

Bond distances and angles within the coordination sphere of $[Mn(C_5H_8NOS_2)_3]\cdot CHCl_3$ are [9]:

Mn–S(1)	2.483(1)	Mn–S(3)	2.344(1)	Mn–S(5)	2.527(1)
Mn–S(2)	2.584(1)	Mn–S(4)	2.433(1)	Mn–S(6)	2.365(1)

S(1)–Mn–S(2)	70.72(4)	S(2)–Mn–S(3)	100.40(5)	S(3)–Mn–S(5)	98.89(5)
S(1)–Mn–S(3)	87.72(4)	S(2)–Mn–S(4)	99.57(5)	S(3)–Mn–S(6)	165.72(5)
S(1)–Mn–S(4)	157.32(5)	S(2)–Mn–S(5)	153.72(5)	S(4)–Mn–S(5)	102.88(5)
S(1)–Mn–S(5)	92.48(5)	S(2)–Mn–S(6)	91.36(4)	S(4)–Mn–S(6)	96.59(4)
S(1)–Mn–S(6)	103.92(5)	S(3)–Mn–S(4)	73.63(4)	S(5)–Mn–S(6)	72.73(4)

Quite similar bond distances and angles were found for $[Mn(C_5H_8NOS_2)_3]\cdot CH_2Cl_2$ [3]. Mean opposite Mn–S bond lengths are: 2.38(1), 2.47(1), and 2.53(1) Å for $[Mn(C_5H_8NOS_2)_3]\cdot CH_2Cl_2$ and 2.35(1), 2.46(1), and 2.56(1) Å for $[Mn(C_5H_8NOS_2)_3]\cdot CHCl_3$ [9]. The trigonal distortion of the MnS_6 octahedron for the latter compound, $\Phi = 35.5°$, is between that for octahedral symmetry ($\Phi = 60°$) and that for the trigonal prism ($\Phi = 0°$) [12]. The complex structures show, in addition to the trigonal distortion, a tetrahedral distortion, which can be attributed to the Jahn-Teller effect. Different opposite Mn–S bond lengths in the z direction may be due to interaction between the S atom of the elongated Mn–S bond with a solvent molecule. Different bond lengths in the xy plane, also, will ensure that no two of the d orbitals will have the same energies [3, 9].

A distorted octahedral structure with bidentate coordination through both S atoms of the ligand anions was also concluded from spectral data for the complexes with HL = ligands 1, 4, and 9 by [2], with HL = ligands 5, 6, and 7 by [7], with HL = ligands 4, 11, and 13 by [6], and with HL = ligands 4 and 12 by [1]. In addition to the dissymmetry caused by the chelating agent, strong Jahn-Teller distortion is expected to be operative, and hence the present complexes may be considered to have C_2 symmetry [2].

Magnetic Properties. For manganese(III) N-heterocyclic N-carbodithioates of the $[MnL_3]$ type the text on pp. 149/50 on magnetic moments, ESR, and ^1H NMR spectra of corresponding dialkyl- and diaryldithiocarbamates is also valid. Gouy magnetic susceptibility measurements on the solids at various temperatures are summarized on p. 171:

For numbers, substituents, and formulas of ligands, see p. 166

ligand	complex	μ_{eff} in μ_B	T in K	Ref.
1	$[Mn(C_5H_8NS_2)_3]$	4.87	room temperature	[2]
4	$[Mn(C_6H_{10}NS_2)_3]$	4.8	300	[46]
		4.86±0.06	100 to 300	[13]
		4.90	room temperature	[1]
		4.9	room temperature	[6]
		5.07	room temperature	[2]
		24.4*)	293	[14]
5	$[Mn(C_7H_{12}NS_2)_3]$	4.86	room temperature	[7]
6	$[Mn(C_7H_{12}NS_2)_3]$	4.82	room temperature	[7]
7	$[Mn(C_7H_{12}NS_2)_3]$	4.81	room temperature	[7]
9	$[Mn(C_{10}H_{10}NS_2)_3]$	4.97	room temperature	[2]
11	$[Mn(C_6H_{11}N_2S_2)_3]$	4.9	room temperature	[6]
12	$[Mn(C_5H_8NOS_2)_3]$	4.45	room temperature	[8]
		5.01	room temperature	[1]
		5.2	300	[46]
13	$[Mn(C_5H_8NS_3)_3]$	4.8	room temperature	[6]

*) Weiss magnetons.

Proton hyperfine coupling constants (in 10^4 Hz), for $[MnL_3]$ complexes in deuterochloroform at 303 K are (A_a for protons on C adjacent to N (= C_a); A_b for protons on C next to C_a (= C_b); A_c for protons on C next to C_b):

ligand	complex	A_a	A_b	A_c	Ref.
1	$[Mn(C_5H_8NS_2)_3]$	16.5*)	0.1	—	[15]
4	$[Mn(C_6H_{10}NS_2)_3]$	6.27	0.05	1.12	[15, 16]
12	$[Mn(C_5H_8NOS_2)_3]$	8.18	0.20	—	[15, 16]

*) At 90 MHz, all the rest at 60 MHz.

Infrared Spectra. Positions of the characteristic IR bands of N-heterocyclic N-carbodithioato $[MnL_3]$ complexes are quite similar to those of the corresponding dialkyl- and diaryldithiocarbamato complexes (see p. 151). The spectra have been taken from the complexes in KBr [46], in hexachlorobutadiene, KBr, or CsI [17], in Nujol or KBr [6, 7, 18], or in Nujol or hexachlorobutadiene [1]. Atmospheric water was removed by flushing the spectrometer housing with dry N_2 [6, 7]. The following bands (in cm^{-1}) have been recorded:

ligand	complex	ν(CN)	ν_{as}(CS)	ν_s(CS)	ν(Mn–S)	Ref.
1	$[Mn(C_5H_8NS_2)_3]$	1478a), 1479b)	—	—	320b)	[17]
		—	—	—	317c), 320d)	[18]
		1480	960	—	—	[2]
4	$[Mn(C_6H_{10}NS_2)_3]$	1470	960	—	—	[2]
		1492	984	—	—	[1]
		1495, 1475	970	—	360	[6]
		1515	1000	700	—	[46]

ligand	complex	ν(CN)	ν_{as}(CS)	ν_s(CS)	ν(Mn–S)	Ref.
5	$[Mn(C_7H_{12}NS_2)_3]$	1427	955	537	373, 353	[7]
6	$[Mn(C_7H_{12}NS_2)_3]$	1425	953	535	370, 357	[7]
7	$[Mn(C_7H_{12}NS_2)_3]$	1439	968	539	364, 354	[7]
11	$[Mn(C_6H_{11}N_2S_2)_3]$	1480	992	—	358	[6]
12	$[Mn(C_5H_8NOS_2)_3]$	1460	995	—	368, 353	[8]
		1498	992	—	—	[1]
		1500	990	695	—	[46]
13	$[Mn(C_5H_8NS_3)_3]$	1472	998	—	370	[6]

a) In hexachlorobutadiene. – b) In KBr or CsI. – c) In KBr. – d) In Nujol.

Additional far-IR bands are given for the complex with ligand 1 in [8, 18] and for the complexes with ligands 4, 11, and 13 in [6].

Electronic Spectra. The electronic absorption spectra of the N-heterocyclic carbodithioato $[MnL_3]$ complexes show up to 6 maxima. Wavenumbers of the first four maxima for the complexes in the solid state or in various solvents are summarized below; ν in cm^{-1}, log ε in parentheses (ε in $L \cdot mol^{-1} \cdot cm^{-1}$):

ligand	complex	state	1	2	3	4	Ref.
1	$[Mn(C_5H_8NS_2)_3]$	in CH_2Cl_2	—	—	20000 (3.50)	28169 (3.97)	[5]
4	$[Mn(C_6H_{10}NS_2)_3]$	solid	—	17050 sh	20530	—	[6]
		solid	—	17285	19352	26520	[1]
		in $CHCl_3$	—	15822 h	20491	—	[46]
		in CCl_4	5600 (2.12)	16100 (~2.40)	19600 (3.15)	—	[19]
		in benzene	—	—	19399	—	[35]
5	$[Mn(C_7H_{12}NS_2)_3]$	solid	—	16130 sh	19420	27600	[7]
6	$[Mn(C_7H_{12}NS_2)_3]$	solid	—	16530 sh	20200	28000	[7]
7	$[Mn(C_7H_{12}NS_2)_3]$	solid	—	16530 sh	19610	27600	[7]
11	$[Mn(C_6H_{11}N_2S_2)_3]$	solid	—	17000 sh	19800	—	[6]
12	$[Mn(C_5H_8NOS_2)_3]$	solid	—	16000	20000	—	[8]
		solid	—	17242	19230	26316	[1]
		in $CHCl_3$	—	15920 sh	20322	—	[46]
13	$[Mn(C_5H_8NS_3)_3]$	solid	—	16650 sh	20200	—	[6]

Band 1 was assigned to a charge transfer transition [19]. In most publications bands 2 and 3 have been assigned to a d-d transition based on O_h symmetry. Assignments are contradictory: $^5E_g \rightarrow {}^5T_{2g}$ [8, 46], d-d [2, 19] to band 2; $^5E_g \rightarrow {}^5T_{2g}$ [6, 7], d-d [2], charge transfer [8, 46] to band 3. The occurrence of more than one d-d band indicates lowering of symmetry from O_h. The complexes may be considered to have C_2 symmetry. The bands observed at ~16000 and ~20000 cm^{-1} in the complex spectra can be assigned to transitions involving the split components of ^5E_g and $^5T_{2g}$ in a low-symmetry ligand field [2]. Three bands in the 27000 to 40000 cm^{-1} range can be attributed to the NCS_2 chromophore of the coordinated ligand [7].

For numbers, substituents, and formulas of ligands, see p. 166

Electrochemistry. The one-electron reversible oxidation of N-heterocyclic carbodithioato $[MnL_3]$ complexes yields the Mn^{IV} complexes, $[MnL_3]^+$, and the one-electron reversible reduction yields the Mn^{II} complexes, $[MnL_3]^-$ (as stated for the corresponding dialkyl- and diaryldithiocarbamates on p. 153). Half-wave potentials for oxidation state changes of the metal in nonaqueous solution have been determined by (A) normal pulse voltammetry, (B) cyclic voltammetry, and (C) polarography at the dropping mercury electrode:

ligand	solvent[a)]	method	reference electrode	t in °C	$Mn^{IV} \rightarrow Mn^{III}$	$Mn^{III} \rightarrow Mn^{II}$	Ref.
1	acetone	A	Ag|AgCl (0.1 M LiCl)	22	0.487	0.017	[5]
	CH_2Cl_2[b)]	A	Ag|AgCl (0.1 M LiCl)	22	0.510	0.092	[5]
	DMF	C	Hg	25	—	−0.35	[20]
	CH_3CN[c)]	B, C	Ag|AgCl (0.1 M $AgNO_3$)	25	0.120	−0.495	[21]
	MIBK[d)]	C	SCE	room temp.	—	−0.56	[22]
4	acetone	A	Ag|AgCl (0.1 M LiCl)	22	0.395	0.062	[5]
	acetone	B	Ag|AgCl (0.1 M LiCl)	22	—	−0.10	[23]
	DMF	C	Hg	25	—	−0.38	[20]
	DMF	C	Hg (DMF, 0.1 M TEAP)	25	—	−0.58	[24]
5	acetone	A	Ag|AgCl (0.1 M LiCl)	22	0.367	−0.098	[5]
6	acetone	A	Ag|AgCl (0.1 M LiCl)	22	0.395	−0.056	[5]
8	acetone	A	Ag|AgCl (0.1 M LiCl)	22	0.350	−0.114	[5]
12	acetone	A	Ag|AgCl (0.1 M LiCl)	22	0.487	0.055	[5]
	DMF	C	Hg	25	—	−0.31	[20]

a) Supporting electrolyte 0.1 M tetraethylammonium perchlorate (TEAP), if not otherwise stated. – b) Supporting electrolyte 0.1 M tetrabutylammonium tetrafluoroborate. – c) Supporting electrolyte 0.15 M TEAP. – d) Supporting electrolyte tetrabutylammonium perchlorate.

Except for substituent effects (which are discussed for dialkyl- and diaryldithiocarbamates on p. 154) the potentials of the oxidation state changes of the metal are dependent on the solvent. A voltammetric study of the complex with ligand 1, $[Mn(C_5H_8NS_2)_3]$, suggests that the variation in half-wave potential with different solvents is not dependent on the solvation energy of the ions formed by oxidation or reduction but is dependent on the solvation energy of the neutral complexes [25]. The standard electron transfer rate constant for the reduction reaction $Mn^{IV} \rightarrow Mn^{III}$ is $k=0.3$ cm/s and for the reduction reaction $Mn^{III} \rightarrow Mn^{II}$, $k=0.4$ cm/s for the complex with the pyrrolidine derivative, $[Mn(C_5H_8NS_2)_3]$ [21]. An appreciably lower value, $(1.25 \pm 0.27) \times 10^{-2}$ cm/s for the reaction $Mn^{III} \rightarrow Mn^{II}$, was found for the complex with the piperidine derivative, $[Mn(C_6H_{10}NS_2)_3]$ [20].

Conductivity measurements on solutions of the complexes with ligands 1, 4, 9, and 12 in acetonitrile and nitrobenzene [1, 2] and of the complexes with ligands 4 to 7, 11, and 13 in DMF [6, 7] show that the complexes are nonelectrolytes in these solvents [1, 2, 6, 7].

Chemical Reactions. Contradictory data are reported for melting points of the complexes with ligands 4 and 12, $[Mn(C_6H_{10}NS_2)_3]$ and $[Mn(C_5H_8NOS_2)_3]$: 140 [46] and 230°C [1] for $[Mn(C_6H_{10}NS_2)_3]$; 183 [46] and 205°C [1] for $[Mn(C_5H_8NOS_2)_3]$. Decomposition of $[Mn(C_5H_8NOS_2)_3]$ at 100°C was stated by [42]. Thermogravimetric (TG) analyses of $[Mn(C_5H_8NS_2)_3]$ (ligand 1), $[Mn(C_6H_{10}NS_2)_3]$ (ligand 4), and $[Mn(C_{10}H_{10}NS_2)_3]$ (ligand 9) show that the complexes are stable up to 200°C. The decomposition occurs in one stage [2]. TG and DTG analyses of the complexes with HL = ligands 5, 6, and 7, $[Mn(C_7H_{10}NS_2)_3]$, indicate that the complexes undergo transformation to polysulfides of the type Mn_xS_y. Beyond 800°C Mn_2O_3 is obtained [7]. The complexes with ligands 1, 4, 9, 11, and 13 are soluble in chloroform, carbon tetrachloride, benzene, acetone, pyridine, DMF, and acetonitrile. They are less soluble or insoluble in water, methanol, ethanol, acetone, and carbon disulfide [2, 4, 6]. The $[MnL_3]$ complexes with HL = ligands 1, 4, 5, 7, 8, and 12 are relatively unstable in solution and even light-sensitive, as has been reported for the corresponding dialkyl- and diaryldithiocarbamates [5, 35]. The complex with HL = ligand 1, $[Mn(C_5H_8NS_2)_3]$, in dichloromethane-acetone reacts with $Cu(BF_4)_2$ to yield $Cu(C_5H_8NS_2)_2BF_4$ and Mn^{II}, which is then spontaneously reoxidized to Mn^{III} [26]. The complexes with HL = ligands 4 and 12, $[Mn(C_6H_{10}NS_2)_3]$ and $[Mn(C_5H_8NOS_2)_3]$, react with the iodine-azide ion to give manganese azide and the ligand anions. The results provided a stability series for the complexes of 20 metals with the two ligands [27].

Analytical Chemistry. The N-heterocyclic N-carbodithioato $[MnL_3]$ complexes are utilized for the microdetermination of Mn^{II} in a manner similar to that described on pp. 156/9 for the diethyldithiocarbamato complex, $[Mn(C_5H_{10}NS_2)_3]$. Mainly, method A) described on p. 157, was used for the extraction of the complexes. The reagent for precipitating Mn^{II} (the ligand HL and the cation of the ligand salt used), the conditions, the method, and the matrix in which Mn^{II} was determined are summarized below:

HL No.	cation	optimum pH	solvent	method[a)]	matrix	Ref.
1	NH_4	2.2 to 2.8	$CH_3C(O)$-*i*-C_4H_9	A, C	aqueous solution	[28]
	NH_4	various [29], 3 [47]	$CH_3C(O)$-*i*-C_4H_9	A [47], B [29]	aqueous solution	[29, 47]
	NH_4	3.5	$CH_3C(O)$-*i*-C_4H_9	B	natural waters	[30]
	NH_4	2 to 6	$CH_3C(O)$-*i*-C_4H_9	A	potable waters	[31]
	—	various studied	$CHCl_3$	—	soil extracts	[32]
	NH_4	—	$CHCl_3$, C_6H_6	A	aqueous KCl solution	[33]
	Na	6.0 to 6.6	$CHCl_3$	B at 578 nm	aqueous solution	[34]
4	Na	6.0 to 6.6	$CHCl_3$	B at 578 nm	aqueous solution	[34]
	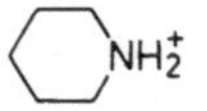	6.5 to 9.2	benzene	B at 450 nm	organosilicon resins	[35]
10	b)	5 to 6	alcohols[c)] ketones[d)]	A	aqueous solution	[36]
	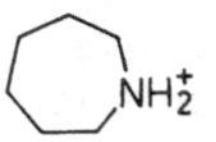	various	$CHCl_3$[e)]	A	aqueous solution	[37]

a) A = atomic absorption spectrometry, B = spectrometry, C = flame emission spectrometry. – b) Not reported. – c) Butyl, isobutyl, amyl, or isoamyl alcohol. – d) Amyl methyl ketone, butyl ethyl ketone, or diisopropyl ketone. – e) Additional extraction with 8-quinolinol and $CHCl_3$-isoamyl alcohol 4:1.

The stability of the chloroform extracts strongly depends on the pH of the aqueous phase. Extracts of the complexes with ligands 1 and 4, $[Mn(C_5H_8NS_2)_3]$ and $[Mn(C_6H_{10}NS_2)_3]$, are only stable for longer periods of time, if the pH of the aqueous phase is between 6.0 and 6.6 [34]. The stability of the extract of $[Mn(C_5H_8NS_2)_3]$ into methyl isobutyl ketone was increased by addition of 20% (v/v) acetone [29] or of 2.5% ethanolic solution of ammonium pyrrolidinecarbodithioate (ethanol solution:extract = 1:4) [30] or by rinsing the extract with high-purity water. Both stability and sensitivity were improved by back extraction of the chelates from the methyl isobutyl ketone phase into a smaller volume of an acidified aqueous solution (pH ≦ 2) [31]. The benzene extract of $[Mn(C_6H_{10}NS_2)_3]$ was stabilized by addition of 4% pyridine [35].

For numbers, substituents, and formulas of ligands, see p. 166

The sensitivity of the atomic absorption-spectrophotometric determination was increased by a factor of 2 with respect to aqueous solution by the use of alcohols and amyl methyl ketone as solvents for the complex with ligand 10, $[Mn(C_7H_{12}NS_2)_3]$, and by a factor of 3 to 4 by the use of butyl ethyl ketone or diisopropyl ketone [36]. Extraction constants for the formation of the MnL_2 complex (?) with HL = ligands 1 [38] and 12 [39] were determined for the H_2O-$CHCl_3$ system [38, 39]. Equilibration time in the latter case was 0.5 h [39].

In addition to extraction-spectrophotometric methods, N-heterocyclic N-carbodithioato $[MnL_3]$ complexes have been applied to the determination of microamounts of manganese: by precipitation of the $[Mn(C_5H_8NS_2)_3]$ complex from natural waters by the ammonium salt of ligand 1 at pH 5 to 9 and adsorption of the complex on active carbon, prior to neutron activation analysis [30]; by amperometric titration of Mn at a rotating Pt electrode, using an aqueous solution of the sodium salt of ligand 4 [41]; by gravimetric and titrimetric determination of manganese by use of the morpholinium salt of ligand 12 [42]; and by spot tests on a glass plate, using the sodium salt of ligand 3 [43].

The Mn^{3+} ion is displaced from the 1-piperidinecarbodithioate, $[Mn(C_6H_{10}NS_2)_3]$, in organic medium by other cations in aqueous solution, as described for the diethyldithiocarbamato complex on p. 159. The exchange series is the same as established for that complex [35].

Thin-layer chromatography (TLC) of $[Mn(C_6H_{10}NS_2)_3]$ on silica gel G with various solvents yielded the following R_f values (C_6H_{14} = cyclohexane) [44]:

benzene	toluene	xylene	$CHCl_3$–C_6H_{14} 1:1 (v/v)	$CHCl_3$–C_6H_{14}–$NH(C_2H_5)_2$ 30:65:5 (v/v)	benzene–C_6H_{14}–$NH(C_2H_5)_2$ 60:35:5 (v/v)
0.26	0.16	0.17	0.20	0.47	0.50

TLC of the 1-pyrrolidinecarbodithioate, $[Mn(C_5H_8NS_2)_3]$, on silica gel H with the elution solvent C_6H_{14}–$CHCl_3$–$NH(C_2H_5)_2$ (60:30:5) gave the R_f value 0.72. Extraction of the spot from the plate with $CHCl_3$ yielded 75.8% of the complex, which could be identified by IR spectroscopy [45].

References:

[1] Das, A. K.; Ramana Rao, D. V. (Indian J. Chem. **13** [1975] 620/1).
[2] Sarasukutty, S.; Prabhakaran, C. P. (Current Sci. [India] **46** [1977] 799/800).
[3] Butcher, R. J.; Sinn, E. (J. Chem. Soc. Dalton Trans. **1975** 2517/22).
[4] Cambi, L.; Cagnosso, A. (Atti Reale Accad. Lincei [6] **14** [1931] 71/4; C.A. **1936** 2172).
[5] Hendrickson, A. R.; Martin, R. L.; Rohde, N. M. (Inorg. Chem. **13** [1974] 1933/9).
[6] Preti, C.; Tosi, G.; Zannini, P. (J. Mol. Struct. **65** [1980] 283/92).
[7] Fabretti, A.C.; Franchini, G. C.; Preti, C.; Tosi, G.; Zannini, P. (Transition Metal Chem. [Weinheim] **10** [1985] 284/7).
[8] Brown, D. H.; Aravamudan, G.; Venkappayya, D. (J. Chem. Soc. A **1971** 2744/7).
[9] Butcher, R. J.; Sinn, E. (J. Am. Chem. Soc. **98** [1976] 5159/68).
[10] Butcher, R. J.; Sinn, E. (J. Am. Chem. Soc. **98** [1976] 2440/9).

[11] Malissa, H.; Kolbe-Rohde, H. (Talanta **8** [1961] 841/5).
[12] Butcher, R. J.; Ferraro, J. R.; Sinn, E. (J. Chem. Soc. Chem. Commun. **1976** 910/2).
[13] Golding, R. M.; Healy, P. C.; White, A. H. (Trans. Faraday Soc. **67** [1971] 1672/7).
[14] Cambi, L.; Szegö, L. (Ber. Deut. Chem. Ges. **64** [1931] 2591/8).
[15] Golding, R. M.; Healy, P. C.; Colombera, P.; White, A. H. (Australian J. Chem. **27** [1974] 2089/97).

[16] Golding, R. M.; Healy, P.; Newman, P.; Sinn, E.; Tennant, W. C.; White, A. H. (J. Chem. Phys. **52** [1970] 3105/9).
[17] Kellner, R. (Anal. Chim. Acta **63** [1973] 277/84).
[18] Kellner, R. (Anal. Chim. Acta **68** [1974] 49/60).
[19] Dingle, R. (Acta Chem. Scand. **20** [1966] 33/44).
[20] Toropova, V. F.; Budnikov, G. K.; Ulakhovich, N. A. (Zh. Obshch. Khim. **45** [1975] 380/5; J. Gen. Chem. [USSR] **45** [1975] 368/72).

[21] Yasuda, H.; Suga, K.; Aoyagui, S. (J. Electroanal. Chem. Interfacial Electrochem. **86** [1978] 259/70).
[22] Ichimura, A.; Morimoto, Y.; Kitamura, H.; Kitagawa, T. (Bunseki Kagaku **33** [1984] E503/E510).
[23] Golding, R. M.; Lehtonen, K. (Australian J. Chem. **27** [1974] 2083/7).
[24] Toropova, V. F.; Budnikov, G. K.; Ulakhovich, N. A. (Talanta **25** [1978] 263/7).
[25] Ichimura, A.; Kitagawa, T. (Bull. Chem. Soc. Japan **53** [1980] 2528/30).
[26] Ondo, B. M.; Barbier, J. P.; Hugel, R. P. (Inorg. Chim. Acta **77** [1983] L211/L212).
[27] Kurzawa, Z.; Karska, B. (Sb. Vys. Sk. Chem. Technol. Praze K Chem. Inz. **15** [1980] 41/5; C.A. **95** [1981] No. 50432).
[28] Ramirez-Munoz, J. (Flame Notes **1** [1966] 8/9; C.A. **66** [1967] No. 108853).
[29] Jenne, E. A.; Ball, J. W. (At. Absorption Newsletter **11** [1972] 60/1; C.A. **77** [1972] No. 134769).
[30] Roberts, R. F. (Anal. Chem. **49** [1977] 1862/3).

[31] Subramanian, K. S.; Meranger, J. C. (Analyst **105** [1980] 620/4).
[32] Lakanen, E. (At. Absorption Newsletter **5** [1966] 17/8).
[33] Mazonska, D. (Zesz. Nauk. Politech. Slask. Chem. No. 88 [1979] 148/9; C.A. **92** [1980] No. 170120).
[34] Specker, H.; Hartkamp, H.; Kuchtner, M. (Z. Anal. Chem. **143** [1954] 425/31).
[35] Dostal, P.; Chermak, J.; Kartous, J. (Collection Czech. Chem. Commun. **33** [1968] 1539/48).
[36] Kotova, N. B.; Tereshchenko, A. P. (Mater. 5th Rab. Soveshch. Vop. Krugovarota Veshchestv Zamknutoi Sist. Osn. Zhiznedeyatel. Nizshkih Organizmov, Kiev 1967 [1968], pp. 67/9; C.A. **72** [1970] No. 50572).
[37] Novikova, N. N.; Tereshchenko, A. P.; Byr'ko, V. M. (Mater. 5th Rab. Soveshch. Vop. Krugovorata Veshchestv Zamknutoi Sist. Osn. Zhiznedeyatel. Nizshikh Organizmov, Kiev 1967 [1968], pp. 57/9; C.A. **72** [1970] No. 63495).
[38] Shen, L. H.; Yeh, S. J.; Lo, J. M. (Anal. Chem. **52** [1980] 1882/5).
[39] Sastri, V. S.; Aspila, K. I.; Chakrabarti, C. L. (Can. J. Chem. **47** [1969] 2320/3).
[40] Wijkstra, J.; Van der Sloot, H. A. (J. Radioanal. Chem. **46** [1978] 379/88).

[41] Usatenko, Yu. I.; Uvarova, K. A. (Zavodsk. Lab. **26** [1960] 1098/101; C.A. **56** [1962] 9385).
[42] Sakla, A. B.; Helmy, A. A.; Beyer, W.; Harhash, F. E. (Talanta **26** [1979] 519/22).
[43] Malissa, H.; Miller, F. F. (Mikrochem. Ver. Mikrochim. Acta **40** [1953] 63/75).
[44] Soundararajan, G.; Subbaiyan, M. (Indian J. Chem. A **22** [1983] 399/401).
[45] Malissa, H.; Kellner, R.; Prokopowski, P. (Anal. Chim. Acta **63** [1973] 225/9).
[46] Garg, B. S.; Singh, A. L.; Dixit, R. (Transition Metal Chem. [Weinheim] **13** [1988] 351/5).
[47] Mansell, R. E. (At. Absorption Newsletter **4** [1965] 276/7).

35.10.4.2.2 $[MnL_2X]$ and $[MnLX_2]$ Complexes

$[Mn(C_7H_{12}NS_2)_2OH]$. The violet-brown complex with ligand 10 was obtained by the reaction of $MnCl_2$ with a 4- to 5-fold excess of $Na(C_7H_{12}NS_2)$ in aqueous solution. The melting point is 98°C. Variable temperature magnetic measurements on polycrystalline samples yielded magnetic moments from 4.35 to 4.9 μ_B. The low-temperature magnetic behavior indicates antiferromagnetic metal-metal interactions. A band at 962 cm^{-1} in the IR spectrum was assigned to the bending mode of a bridging OH group. The electronic spectrum displays a shoulder at ~16500 cm^{-1}, assignable to the $^5E_g \rightarrow {}^5T_{2g}$ transition, and three charge transfer bands from 19500 to 37000 cm^{-1}. A dinuclear, octahedral structure with two bridging OH groups was proposed [1].

$[Mn(C_6H_{10}NS_2)_2Cl]$ and **$[Mn(C_5H_8NOS_2)_2Cl]$.** The complexes with ligands 4 and 12 were prepared in a manner analogous to that described for the corresponding diethyldithiocarbamato complex on p. 161 and have properties and structures similar to that compound. Melting points are 125 and 170°C. The magnetic moments are 4.80 and 4.95 μ_B, respectively. The ν(CN), $\nu_{as}(CS)$, and $\nu_s(CS)$ bands in the IR spectrum are located at 1500, 1005, and 730 cm^{-1} for $[Mn(C_6H_{10}NS_2)_2Cl]$ and at 1520, 1025, and 720 cm^{-1} for $[Mn(C_5H_8NOS_2)_2Cl]$ [2].

$[MnL_2acac]$ and **$[MnL(acac)_2]$.** The complexes with HL = ligand 4 or 12 were prepared by the reaction of an aqueous solution of $MnCl_2 \cdot 4H_2O$, NaL, and an ethanol solution of 2,4-pentanedione in the required stoichiometric ratio. The compounds separated on shaking. They were washed with water and ethanol and dried in vacuum over P_4O_{10}. Melting points (in °C), magnetic moments (μ_{eff} in μ_B), and pertinent bands (in cm^{-1}) in the IR spectra of the dark violet complexes (in KBr disks) are summarized below:

complex	m.p.	μ_{eff}	ν(CN)	$\nu_{as}(CS)$	$\nu_s(CS)$	ν(CO)	ν(C=C)
$[Mn(C_6H_{10}NS_2)_2acac]$	204	5.3	1500	1010	715	1620	1550
$[Mn(C_6H_{10}NS_2)(acac)_2]$	212	5.0	1500	1005	720	1625	1550
$[Mn(C_5H_8NOS_2)_2acac]$	182	5.3	1505	1005	710	1600	1535
$[Mn(C_5H_8NOS_2)(acac)_2]$	230	4.6	1505	1000	—	1630	1540

The electronic spectra display a shoulder at ~16000 cm^{-1} and an intense maximum between 20000 and 25000 cm^{-1}, assignable to the $^5E_g \rightarrow {}^5T_{2g}$ and charge transfer transitions, respectively. The spectral data and additional ESR results indicate a *cis*-octahedral structure. Thermogravimetric analyses show decomposition of the complexes in the temperature range 473 to 1193°C. From this data the decomposition is first order. The activation energies are: E_A = 17.84 kJ/mol for $[Mn(C_6H_{10}NS_2)_2acac]$ and 21.96 kJ/mol for $[Mn(C_6H_{10}NS_2)(acac)_2]$. The complexes are soluble in chloroform, and the solutions decolorize on standing [3].

$[MnL_2(C_2H_4NO_2)]$ and **$[MnL(C_2H_4NO_2)_2]$.** The complexes with HL = ligand 4 or 12 and glycine (= $C_2H_5NO_2$) were prepared by the reaction of $MnCl_2$, NaL, and glycine in the required stoichiometric ratio. The dark brown (for ligand 4) or violet (for ligand 12) complexes separated on stirring. They were washed with water and ethanol and dried in vacuum over P_4O_{10}. Melting points (in °C), magnetic moments (in μ_{eff}), and important IR bands of the complexes (in KBr disks) are summarized below:

complex	m.p.	μ_{eff}	ν(CN)	$\nu_{as}(CS)$	$\nu_s(CS)$	$\nu(NH_2)$	$\nu_{as}(COO)$	$\nu_s(COO)$
$[Mn(C_6H_{10}NS_2)_2(C_2H_4NO_2)]$	194	4.8	1495	990	710	3345	1665	1375
$[Mn(C_6H_{10}NS_2)(C_2H_4NO_2)_2]$	190	4.5	1490	982	700	3345	1668	1380
$[Mn(C_5H_8NOS_2)_2(C_2H_4NO_2)]$	202	4.7	1490	995	715	—	1665	1375
$[Mn(C_5H_8NOS_2)(C_2H_4NO_2)_2]$	208	4.8	1495	990	715	3340	1670	1370

Maxima in the electronic spectra, the proposed structure, and the temperature range of decomposition are similar to those of [MnL_2acac] and [MnL(acac)$_2$] complexes described on p. 177. The activation energies are E_A = 29.74, 39.65, 26.14, and 21.35 kJ/mol for the complexes in the above order. The complexes are soluble in chloroform, and the solutions decolorize upon standing [3].

References:

[1] Singh, A. K.; Puri, B. K.; Rawlley, P. K. (Indian J. Chem. A **27** [1988] 430/3).
[2] Kanungo, B. K.; Pradhan, B.; Ramana Rao, D. V. (Indian J. Chem. A **21** [1982] 625/7).
[3] Garg, B. S.; Singh, A. L.; Dixit, R. (Transition Metal Chem. [Weinheim] **13** [1988] 351/5).

35.10.4.3 Manganese(IV) Compounds

[Mn($C_5H_8NS_2$)$_3$]ClO_4 and **[Mn($C_6H_{10}NS_2$)$_3$]ClO_4·n$CHCl_3$** (n = 0 or 1). The complex with ligand 1 was prepared by reaction of an ethanol solution of hexaaquamanganese(II) perchlorate, added slowly dropwise to a vigorously stirred benzene solution of excess [Mn($C_5H_8NS_2$)$_3$]. The stirring of the reaction mixture was continued for 1 h in air, and a finely divided solid formed. The dark crystals were isolated and recrystallized from either CH_3NO_2 or CH_2Cl_2 by slow addition of benzene or diethyl ether. The yield was approximately 60% [1]. The complex with ligand 4 was prepared by slowly adding hexaaquamanganese(II) perchlorate in acetone to [Mn($C_6H_{10}NS_2$)$_3$] in benzene. The complex was recrystallized from chloroform or chloroform-nitromethane mixtures by slow addition of benzene [2].

The molecular structure of [Mn($C_6H_{10}NS_2$)$_3$]ClO_4·$CHCl_3$ has been determined using single crystal X-ray diffraction techniques. The crystal data are: monoclinic system, space group $P2_1/c$-C^5_{2h} (No. 14) with the lattice constants a = 6.401(5), b = 20.596(2), c = 25.500(2) Å, β = 109.66(1)°; Z = 4. The structure was solved up to R = 0.071. Positional parameters, bond distances, and bond angles are presented in the paper. The calculated density is D_{calc} = 1.58 g/cm^3, the experimental density is D_{exp} = 1.60 g/cm^3. The manganese atom is coordinated by six sulfur atoms from the three chelated ligands in a manner similar to Fig. 17, p. 170, for the Mn^{III} complex, [Mn($C_5H_8NOS_2$)$_3$]·$CHCl_3$, the coordination sphere having approximate D_3 symmetry. The angle of twist is ~40°. Average Mn–S distances for the three chelated ligands are: 2.339(3), 2.336(3), and 2.326(3) Å. The Mn–S distances are significantly shorter than those of [Mn($C_6H_{10}NS_2$)$_3$]. The $CHCl_3$ molecule shows only nonbonding contacts with the cation [3].

[Mn($C_6H_{10}NS_2$)$_3$]ClO_4·$CHCl_3$ is a 1:1 electrolyte in nitrobenzene. The magnetic moment of the complex in $CHCl_3$ solution at 293 K is μ_{eff} = 4.05 μ_B. The moment of the solid increases from 3.97 μ_B at 95 K to 4.06 μ_B at 326 K. The ESR spectrum of [Mn($C_6H_{10}NS_2$)$_3$]ClO_4 in frozen $CHCl_3$ at 113 K yields the parameters g_0 = 1.990 ± 0.005, A_0 = 95 ± 1 G, D_0 = 133 ± 5 G. The magnetic properties are consistent with a high-spin Mn^{IV} complex. The IR spectrum of [Mn($C_6H_{10}NS_2$)$_3$]ClO_4 in Nujol shows the characteristic bands: ν(CN) at 1530 and ν(Mn–S) at 395 cm^{-1}. Bands at 1080, 747, and 620 cm^{-1} were assigned to vibrations of the ClO_4^- ion [2]. For [Mn($C_5H_8NS_2$)$_3$]ClO_4 (in Nujol) an anion band was observed at 1090 cm^{-1}. The electronic absorption spectrum of this complex in CH_2Cl_2 solution shows the prominent maxima: 485 nm (log ε = 4.07), 428 nm, 407 nm (4.12). Of these, the highest and lowest energy bands were assigned to charge-transfer transitions [1].

[Mn($C_5H_8NS_2$)$_3$]BF_4 and **[Mn($C_6H_{10}NS_2$)$_3$]BF_4·n$CHCl_3$** (n = 0 or 0.5). The pyrrolidinecarbodithioato complex was prepared in a manner analogous to that described for the dicyclohexyldithiocarbamato complex [Mn($C_{13}H_{22}NS_2$)$_3$]BF_4 on p. 162. The yield of [Mn($C_5H_8NS_2$)$_3$]BF_4 was ~45% [1]. The piperidinecarbodithioato complex was prepared by slowly adding Mn(BF_4)$_2$ ·6H_2O in acetone to [Mn($C_6H_{10}NS_2$)$_3$] in benzene. The complex was recrystallized from

For numbers, substituents, and formulas of ligands, see p. 166

chloroform by slow addition of benzene. $[Mn(C_6H_{10}NS_2)_3]BF_4 \cdot 0.5\ CHCl_3$ is a 1:1 electrolyte in nitrobenzene [2].

The room temperature magnetic moment of $[Mn(C_5H_8NS_2)_3]BF_4$ is 3.8 μ_B [1]. The moment of solid $[Mn(C_6H_{10}NS_2)_3]BF_4 \cdot 0.5\ CHCl_3$ increases from 4.05 μ_B at 100 K to 4.19 μ_B at 298 K. Bands in the IR spectrum of this compound in Nujol were assigned as follows: 1535 cm^{-1} to ν(CN); 400 and 375 cm^{-1} to ν(Mn–S) [2]. Bands at 1060 [2] and 1050, 522, and 515 cm^{-1} [1] were assigned to vibrations of the BF_4^- ion [1, 2]. The electronic spectrum of $[Mn(C_5H_8NS_2)_3]BF_4$ shows prominent maxima at 485 (log ε = 4.07), 425(sh), and 407 (4.12) nm [1].

$[Mn(C_5H_8NOS_2)_3]I_5 \cdot 0.5\,CH_2Cl_2$ was obtained by slow evaporation of a dichloromethane solution (100 mL) containing $[Mn(C_5H_8NOS_2)_3]$ (100 mg) and iodine (150 mg). The shiny dark brown crystals were suitable for an X-ray structure determination. The complex crystallizes in the monoclinic system, space group $P2_1/c-C_{2h}^5$ (No. 14) with a = 20.451(9), b = 18.910(5), c = 18.714(5) Å, β = 109.68°; Z = 8. The calculated density is D_{calc} = 2.376 g/cm^3. The structure was solved up to R = 0.0570. Atomic positional parameters for the nonhydrogen atoms and selected bond distances and angles are presented in the paper. The structure of $[Mn(C_5H_8NOS_2)_3]I_5 \cdot 0.5\ CH_2Cl_2$ is made up of anionic layers lying approximately parallel to (010), interspersed with $[Mn(C_5H_8NOS_2)_3]^+$ cations differently inclined (**Fig. 18**). The slightly puckered layers are composed of chains of V-shaped I_5^- units running parallel to [001]. The layers contain a substructure of 12-membered, square, and nonplanar rings, each side of which consists of four iodine atoms. There are two similar but independent $[Mn(C_5H_8NOS_2)_3]^+$ cations: one of which, containing Mn(1), lies parallel to the anionic layers; the other, containing Mn(2), occupies the channels formed by the square rings of the layers. In both cations the manganese has a trigonally-distorted octahedral environment. The coordination around the manganese and the Mn–S bond distances, ranging from 2.305(9) to 2.351(7) Å, agree with those found for $[Mn(C_6H_{10}NS_2)_3]ClO_4 \cdot CHCl_3$ (see p. 178). The dichloromethane molecule shows only nonbonding contacts with the cations [4].

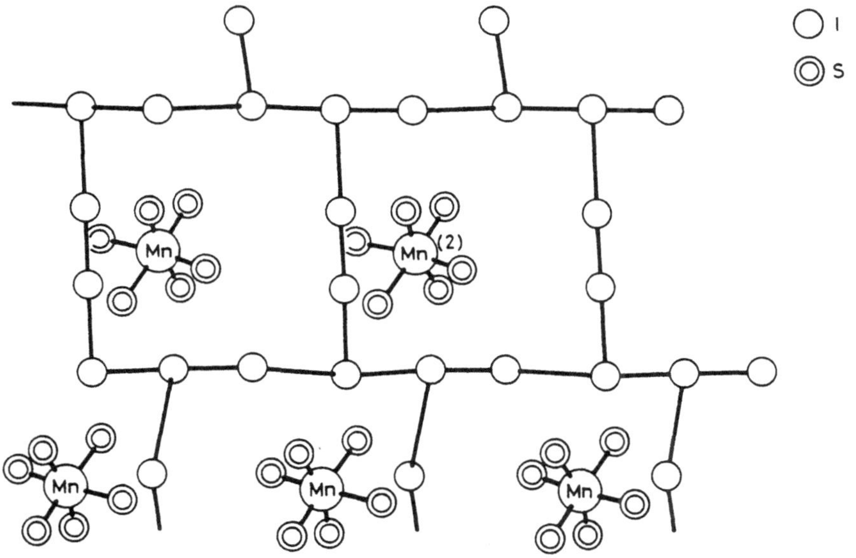

Fig. 18. Projection of the anionic layer in $[Mn(C_5H_8NOS_2)_3]I_5 \cdot 0.5CH_2Cl_2$ viewed along b showing the cation $[Mn(C_5H_8NOS_2)_3]^+$ containing Mn(2). The morpholine rings and adjacent C atoms are omitted for clarity [4].

[Mn($C_5H_8NOS_2$)$_2$X$_2$] (X = Cl or Br). The dark gray complexes with ligand 12 were prepared in a manner analogous to that described for the corresponding diethyldithiocarbamato complexes on p. 163 and have properties similar to those compounds. Melting points are: 95°C for the complex with X = Cl, 90°C for that with X = Br. The IR spectra show the characteristic bands: ν(CN) at 1530 and ν_{as}(CS) at 992 cm^{-1} for both complexes (in Nujol mulls). Bands in the electronic spectra are at almost the same positions as those of diethyldithiocarbamato complexes [5].

References:

[1] Hendrickson, A. R.; Martin, R. L.; Rohde, N. M. (Inorg. Chem. **13** [1974] 1933/9).
[2] Brown, K. L.; Golding, R. M.; Healy, P. C.; Jessop, K. J.; Tennant, W. C. (Australian J. Chem. **27** [1974] 2075/81).
[3] Brown, K. L. (Cryst. Struct. Commun. **3** [1974] 493/7).
[4] Deplano, P.; Trogu, E. F.; Bigoli, F.; Pellinghelli, M. A. (J. Chem. Soc. Dalton Trans. **1987** 2407/10).
[5] Pradhan, B.; Ramana Rao, D. V. (J. Indian Chem. Soc. **58** [1981] 733/5).

35.10.5 With Alkylene- or Azaalkylenebis(dithiocarbamic Acids) and 1,4-Piperazinebis-(carbodithioic Acid)

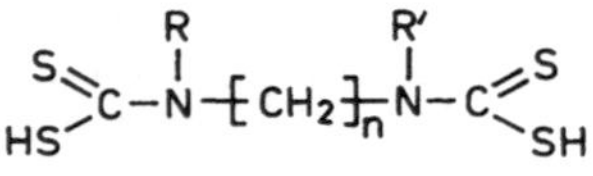

ligand	n	R	R'	formula
1	2	H	H	$C_4H_8N_2S_4$
2	2	H	CH_2CH_2OH	$C_6H_{12}N_2OS_4$
3	3	H	H	$C_5H_{10}N_2S_4$
4	6	H	H	$C_8H_{16}N_2S_4$

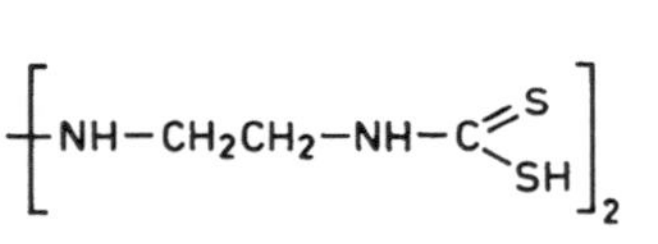

ligand 5 (= $C_6H_{14}N_4S_4$)

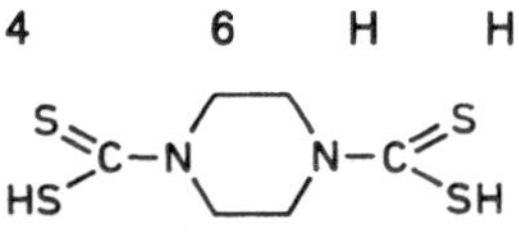

ligand 6 (= $C_6H_{10}N_2S_4$)

Mn^{II}L and Mn^{II}L·nH$_2$O. The complexes with ligands 1, 2, 4, and 6 were prepared by the reaction of an Mn^{II} salt with the sodium salt of the ligand in aqueous solution [1 to 7]. The precipitates were washed with deionized water and with absolute ethanol (for the complexes with ligands 1 and 4) [1], or with water, methanol, ethanol, or acetone and diethyl ether (for the complex with ligand 6 [5 to 7]). They were dried in vacuum [1, 5, 7] over P_4O_{10} [5, 7] or at 70°C [6]. All the operations were carried out in a glove bag under an atmosphere of prepurified dry nitrogen, and all solvents were deaerated with prepurified dry nitrogen [1, 6]. For Mn($C_4H_6N_2S_4$) (ligand 1) the reaction proceeded to completion within 15 min; however, for Mn($C_8H_{14}N_2S_4$) (ligand 4) [1] and Mn($C_6H_8N_2S_4$) (ligand 6) [6] 45 min and 1 h, respectively, were needed. In the preparation of the complex with ligand 1, the dihydrate, Mn($C_4H_6N_2S_4$)·2H_2O, arose first and was then converted into the anhydrous complex on drying over P_4O_{10} [2, 8]. While the anhydrous compound, Mn($C_6H_8N_2S_4$), was obtained by two authors in the case of the complex with ligand 6 [5, 6], the monohydrate was reported in spite of drying over P_4O_{10} by [7].

X-ray powder diffraction studies of Mn($C_4H_6N_2S_4$) (ligand 1) show that the complex is crystalline, but has a low degree of symmetry [2, 9]. A table of interplanar distances for Mn($C_4H_6N_2S_4$) and its dihydrate is presented in [2]. S atoms at opposite ends of each ligand coordinate to different Mn atoms to form an infinite chain: This structure is supported by diffraction and IR data for $Mn(C_4H_6N_2S_4)$ [9], and by its magnetic properties for $Mn(C_6H_8N_2S_4)$ (ligand 6) [5, 6].

Below are summarized the colors, the magnetic moments from the Faraday method at 296 K [1] or the Gouy method at 305 K [5], and the important bands in the IR spectra of the complexes in KBr [1], in Nujol [5], and in Nujol, Fluorolube, or KBr [6].

ligand	complex	color	μ_{eff}	ν(CN)	ν(CS)	ν(Mn-S)	Ref.
1	$Mn(C_4H_6N_2S_4)$	yellow	5.8 [1]	—	1029, 995, 959 [9]	369 [1]	[1, 9]
4	$Mn(C_8H_{14}N_2S_4)$	light pink	5.75	—	—	359	[1]
6	$Mn(C_6H_8N_2S_4)$	—	5.42	1418	995	360	[5]
		yellow[a)]	—	1476[b)]	994[c)]	—	[6]

a) Turns to brown on contact with air. – b) Na salt at 1451 cm^{-1}. – c) Na salt at 998 cm^{-1}.

Magnetic susceptibility data obtained from 77 to 350 K fit the Curie-Weiss law with the Curie constant C = 4.19 and the Weiss constant $\Theta = -45$ K for $Mn(C_4H_6N_2S_4)$, and C = 4.09 and $\Theta = -36$ K for $Mn(C_8H_{14}N_2S_4)$. Negative Θ values indicate antiferromagnetic metal-metal interactions for both compounds [1].

IR bands at 3220 and 1608 cm^{-1} indicate the presence of water in $Mn(C_4H_6N_2S_4)\cdot 2H_2O$. The removal of the water, which is considered to be water of crystallization, splits the band at ~958 cm^{-1} into two, at 951 and 970 cm^{-1}. In addition, a new band and a shift are observed in the 1430 to 1340 cm^{-1} range. These changes are attributed to a phase transition of $Mn(C_4H_6N_2S_4)$ on dehydration [8]. The electronic spectrum of the complex with ligand 6, $Mn(C_6H_8N_2S_4)$, shows a charge transfer band at 450 nm and a d-d band at 570 nm (17544 cm^{-1}) [7].

The complexes with ligands 1, 4, and 6 decompose at 131, 134 [1], and 140°C [10], respectively. They are practically insoluble in water and common organic solvents [1, 6, 8]. The solubility product of $Mn(C_4H_6N_2S_4)$ in aqueous solution, I = 0.1 mol/L ($NaClO_4$) at 25°C, was determined by potentiometric titration: $\log K_S = -6.56 \pm 0.24$ (concentration in mol/L) [11]. The hydrolysis of this complex in aqueous solution depends on the pH. Rate constants of the first-order reaction decrease from 8.3 h^{-1} at pH 3.8 to 1.16×10^{-3} h^{-1} at pH 11.2. Decomposition products are: monomeric and polymeric ethylenethiuram monosulfide, 2-imidazolidinethione (ethylenethiourea), and sulfur [12]. The complex with ligand 6 is resistent to concentrated H_2SO_4 at room temperature and to 50% NaOH at slightly elevated temperature [10].

Samples of $Mn(C_4H_6N_2S_4)$ and $Mn(C_4H_6N_2S_4)\cdot 2H_2O$ containing from 5 to 80 mol% of Zn were prepared by the reaction of a mixed Mn^{II} and Zn^{II} salt solution with an aqueous solution containing $(C_4H_6N_2S_4)^{2-}$ ions at 40 to 45°C. X-ray powder patterns of $Mn(C_4H_6N_2S_4)$, $Zn(C_4H_6N_2S_4)$, and a mixture of both compounds containing 20% Mn and 2.5% Zn (common names mancozep or dithane M-45) show that mancozep is not a chemical individual [2]. Hydrolysis of mancozep is slower than that of the pure Mn^{II} complex. First-order rate constants decrease from 5.9 h^{-1} at pH 3.8 to 8.2×10^{-4} h^{-1} at pH 11.2 [12].

$Mn(C_4H_6N_2S_4)$ (common names maneb, dithane M-22, or manzate) is an important fungicide with the main application being the control of fungal diseases of potatoes and tomatoes [16]. Fungicidal effects were also observed for the ammonia [14] and dodecylamine adducts [17] of maneb and for mancozep, the fungicidal effectiveness of which is higher than that of maneb and zineb ($Zn(C_4H_6N_2S_4)$) alone [2, 13]. Maneb and products containing maneb are relatively unstable. A study of the factors effecting the rate of degradation of commercial products containing maneb or mancozep is presented in [18]. Solid formulations [19, 20] or aqueous solutions of maneb [21, 22] were stabilized by the addition of, e.g., formaldehyde [21], sodium hydroxymethanesulfonate [20], guanidine derivatives [19], or dimethylolurea [22]. The inflammability of powdered maneb is suppressed by the addition of nitrosophenols [23] or by coating maneb particles with $Cu(C_4H_6N_2S_4)$ during preparation [15]. An ethylenebis(dithiocarbamato) complex containing Mn and Cu has fungicidal and acaricidal activity [25]. It was also found to be an effective component in antifouling coating compositions [24]. $Mn(C_4H_6N_2S_4)$ has rubber vulcanizing properties in addition to its fungicidal activity [15]. Bactericidal effects are observed for the dodecylamine adduct of maneb, $Mn(C_4H_6N_2S_4)\cdot C_{12}H_{25}NH_2$ (m.p. 170°C with decomposition) [17], and $Mn(C_6H_{10}N_2OS_4)$ (ligand 2) [4].

Toxicological studies of maneb have been carried out with newborn rats [29] and newts [26, 27]. The complex changes the enzymatic activity of alkaline phosphatase in blood serum and in the liver of rats [30]. Exposure to maneb as a health hazard to forestry workers was studied [31]. General references concerning toxicological aspects of maneb are [32, 36, 37, 38]. General references concerning the agricultural application of maneb or mancozep are [28, 33 to 35].

$Mn^{II}ZnL_2$ and $Mn^{II}ZnLL'$. The compound with ligand 1, $Mn^{II}Zn(C_4H_6N_2S_4)_2$, and the mixed-ligand compound with HL = ligand 1 and HL' = ligand 3, $Mn^{II}Zn(C_4H_6N_2S_4)(C_5H_8N_2S_4)$, were prepared by the addition of the mixed Na, NH_4 salt of ligand 1 or of ligands 1 and 3 to a solution of $MnSO_4$ and $ZnSO_4$ in hot water. $Mn^{II}ZnL_2$ complexes containing the ligands with n (number of methylene groups) = 3, 4, 5, or 6 were similarly prepared [3].

$Mn_2^{III}(C_6H_8N_2S_4)_3\cdot 4H_2O$ was prepared by slow dropwise addition of a methanolic solution of ligand 6 to a solution of manganese(III) acetate dihydrate dissolved in the minimum amount of methanol and vigorous stirring. The precipitate was washed with methanol and dried over P_4O_{10}. The magnetic moment is 5.01 μ_B at room temperature, indicating a high-spin complex. IR bands in the 3400 to 3200 and 1625 to 1615 cm^{-1} ranges were assigned to ν(OH) and δ(OH) vibrations of water. A broad band extending from 600 to 650 nm in the electronic spectrum was assigned to the $^5E_g \rightarrow {}^5T_{2g}$ transition, the only allowed one for an octahedral Mn^{III} complex. The complex starts decomposing on heating between 80 and 100°C. DTG peaks between 90 and 110°C and the corresponding DTA peak (around 100°C) indicate the elimination of lattice-held water molecules [7].

References:

[1] Kwoka, W.; Moyer, R.O.; Lindsay, R. (J. Inorg. Nucl. Chem. **37** [1975] 1889/92).
[2] Gergely, S.; Garaj, J. (Chem. Zvesti **42** [1988] 229/34).
[3] Nash, L.H. (U.S. 3259643 [1963/66] from C.A. **65** [1966] 20022).
[4] Tokyo Organic Chemical Industries, Ltd. (Japan. 81-139451 [1980/81] from C.A. **96** [1982] No. 103693).
[5] Bhoon, Y.K.; Singh, R.P. (Acta Ciencia Indica Ser. Chem. **7** [1981] 111/5).
[6] Tombeux, J.; Van Poucke, L.C.; Eeckhout, Z. (Spectrochim. Acta A **28** [1972] 1943/7).
[7] Aravindakshan, K.K. (Indian J. Chem. A **25** [1986] 592/4).
[8] Kozlov, G. A.; Volchek, S.I. (Zh. Neorgan. Khim. **14** [1969] 756/62; Russ. J. Inorg. Chem. **14** [1969] 396/9).
[9] Genchev, M.; Davarski, K. (God. Vissh. Khim. Tekhnol. Inst. Burgas. **13** [1978] 39/44; C.A. **93** [1980] No. 195818).
[10] Szymik, Z. (Pr. Nauk. Univ. Slask. Katowicach No. 25 [1971] 1/70 from C.A. **78** [1973] No. 160137).

For numbers, substituents, and formulas of ligands, see p. 180

[11] Gergely, S.; Garaj, J. (Chem. Zvesti **42** [1988] 223/7).
[12] Klisenko, M.A.; Vekshtein, M.Sh. (Zh. Obshch. Khim. **41** [1971] 1125/30; Russ. J. Gen. Chem. **41** [1971] 1125/30).
[13] Golyshin, N.M.; Abelentsev, V.I.; Dymshakova, G.M.; Matveenko, T.M.; Yarykina, I.V. (Khim. Sel'sk. Khoz. **5** [1967] 427/30; C.A. **67** [1967] No. 90017).
[14] Windel, H.; Badische Anilin- & Soda-Fabrik A.-G. (Ger. 1202266 [1961/65] from C.A. **64** [1966] No. 3102).
[15] Lehureau, J.; Progil S. A. (South African 6804841 [1967/68] from C.A. **71** [1969] No. 37894).
[16] Ullmanns Encykl. Tech. Chem. 4th Ed. **12** [1976] 5.
[17] Crewe, F.; Ortner, G.; Farbenfabriken Bayer A.-G. (Belg. 656267 [1963/65] from C.A. **65** [1966] 7068).
[18] Sovljanski, R.; Zivanovic, B. (Arh. Hig. Rada Toksikol. **34** [1983] 233/8 from C.A. **101** [1984] No. 78745).
[19] Kikuno, T.; Yoshida, S.; Ouchi-Shinko Chemical Industrial Co., Ltd. (Japan. 71-09687 [1968/71] from C.A. **75** [1971] No. 5304).
[20] Yoshigaki, S.; Yoshida, S.; Takahashi, J.; Ouchi-Shinko Chemical Industrial Co., Ltd. (Japan. 73-43102 [1970/73] from C.A. **80** [1974] No. 82145).

[21] Adams, J. B., Jr.; E. I. du Pont de Nemours and Co. (Braz. 81-06574 [1981/83] from C.A. **99** [1983] No. 117879).
[22] Plancon, R.; Ozanne, G.; Soc. Usines Chimiques Rhone-Poulenc (Fr. 1482885 [1966/67] from C.A. **68** [1968] No. 21559).
[23] Yamakagawa, Y.; Shimizo, K.; Tanabe, Y.; Onishi, S.; Onouchi, H.; Kawaguchi Chem. Industry Co., Ltd. (Japan. 73-42862 [1969/73] from C.A. **81** [1974] No. 13154).
[24] Ikari, Koki; Tokyo Organic Chemical Industries, Ltd. (Japan. 86-176672 [1985/86] from C.A. **105** [1986] No. 193026).
[25] Roehm G.m.b.H. (Japan. Tokkyo Koho 81-35642 [1968/81] from C.A. **95** [1981] No. 199061).
[26] Zaffaroni, N. P.; Arias, E.; Capodanno, G.; Zavanella, T. (Bull. Environ. Contam. Toxicol. **20** [1978] 261/7 from C.A. **89** [1978] No. 210024).
[27] Zavanella, T.; Arias, E.; Zaffaroni, M. P. (Tumori **65** [1979] 163/7 from C.A. **91** [1979] No. 103374).
[28] Lyman, W. R. (Pestic. Terminal Residues Invited Papers Intern. Symp., Tel Aviv 1971, pp. 243/56 from C.A. **78** [1973] No. 38757).
[29] Schmandke, H.; Bleyl, D.W.R.; Seidler, H. (Nahrung **29** [1985] 421/4 from C.A. **103** [1985] No. 18079).
[30] Urbanek-Karlowska, B.; Fonberg-Broczek, M. (Przeglad. Lek. [2] **41** [1984] 227/9 from C.A. **101** [1984] No. 105442).

[31] Kangas, J.; Koskinen, A. (Vortr. Konf. Sicherheitstech. Landwirtsch., Budapest 1978, pp. 159/61 from C.A. **90** [1979] No. 191823).
[32] IARC Working Group (IARC Monogr. Eval. Carcinog. Risk. Chem. Man **12** [1976] 137/49).
[33] Wegler, R. (Chemie der Pflanzenschutz- und Schädlingsbekämpfungsmittel, Vol. 2, Springer, Berlin – Heidelberg – New York 1970, p. 66).
[34] Martin, H.; Worthing, C. R. (Pesticide Manual, 4th Ed., Brit. Crop Protection Council, Droitwich – Worchester 1974, p. 324).
[35] Perkow, W. (Wirksubstanzen der Pflanzenschutz- und Schädlingsbekämpfungsmittel, Parey, Berlin 1971).
[36] Fischbein, L. (in: Siegel, M. R.; Sister, H. D., Antifungal Compounds, Vol. 2, Dekker, New York – Basel 1977, pp. 537/98, 539/44).

[37] Engst, R.; Schnaak, W. (Residue Rev. **52** [1974] 45/67).
[38] Klimmer, O. R. (Pflanzenschutz- und Schädlingsbekämpfungsmittel. Abriß einer Toxikologie und Therapie von Vergiftungen, 2nd Ed., Hattingen, FRG, 1971).

35.11 Complexes with Hydrazinecarbothioic Acids and Derivatives

Remark. Complexes with condensation products of aldehydes or ketones and thiocarbazates or dithiocarbazates have already been described in "Manganese" D 6, 1988, pp. 361/4, together with other Schiff base complexes.

35.11.1 With O-Ethyl Thiocarbazates

$C_2H_5O-C(=S)-NH-NHR$

ligand	1	2	3
R	H	$C(O)C_6H_5$	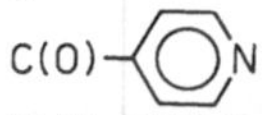
formula	$C_3H_8N_2OS$	$C_{10}H_{12}N_2O_2S$	$C_9H_{11}N_3O_2S$

$Mn^{II}(C_3H_7N_2OS)_2$ was prepared by the reaction of ethanol solutions of $MnCl_2$ and the hydrochloride of ligand 1 (mole ratio 1:2). After adjusting the pH of the reaction mixture to 5 to 6 by the addition of NH_3, a precipitate forms. The dirty yellow product was digested for a few minutes on a water bath, then washed with water and ethanol, and dried in vacuum. The solid compound melts above 300°C. The magnetic moment of the solid is $\mu_{eff}=5.90\ \mu_B$. IR spectral results show that the uninegative bidentate ligand binds to the Mn^{II} via the enethiolato group and the terminal nitrogen atom [1].

$Mn^{II}(C_{10}H_{11}N_2O_2S)_2$ and **$Mn^{II}(C_9H_{10}N_3O_2S)_2$** were prepared by adding manganese(II) acetate (5 mmol) dissolved in 50 mL water to the solution of ligand 2 or 3, respectively, in 50 mL ethanol, followed by addition of sodium acetate (~1 g). The yellow precipitates were handled as was the above complex. The melting points are 240°C and 274 to 276°C, respectively. The magnetic moments of both the solids are $\mu_{eff}=5.5\ \mu_B$. The IR spectra of the complexes in Nujol show only one ν(NH) band at 3230 or 3400 cm^{-1}, respectively. Negative shifts of the ligand amide I band (ν(CO) + ν(NCO)) from 1640 to 1580 or from 1680 to 1620 cm^{-1} are observed. The thioamide band IV (mainly due to ν(C=S)) at 910 cm^{-1} (ligand 2) or 920 cm^{-1} (ligand 3) was found to be absent in the spectrum of the complexes. A new band, due to ν(C–S) at 680 or 730 cm^{-1}, respectively, indicates thioenolization and subsequent bonding to manganese. Positive shifts of the thioamide I (δ(NH) + ν(CN)), thioamide II (ν(CN) + δ(NH)), and the ν(N–N) vibration modes are observed, but no change for the pyridine ring deformation mode. The uninegative tridentate ligands are therefore assumed to be coordinated to Mn^{II} through the carbonyl oxygen, the terminal nitrogen, and the sulfur atoms. Bands observed at 390 or 380 cm^{-1} were tentatively assigned to ν(Mn–N), at 360 cm^{-1} to ν(Mn–O), and at 290 or 310 cm^{-1} to ν(Mn–S) vibration modes. The complexes are insoluble in water and common organic solvents, very slightly soluble in DMF or DMSO [2].

References:

[1] Singh, N. K.; Srivastava, S. C.; Aggarwal, R. C. (Indian J. Chem. A **22** [1983] 704/7).
[2] Singh, N. K.; Srivastava, S. C.; Aggarwal, R. C. (Syn. React. Inorg. Metal-Org. Chem. **13** [1983] 989/1002).

35.11.2 With Dithiocarbazic Acid or Benzyl Dithiocarbazate

$$HS-\underset{\underset{S}{\|}}{C}-NH-NH_2 \qquad\qquad C_6H_5CH_2S-\underset{\underset{S}{\|}}{C}-NH-NH_2$$

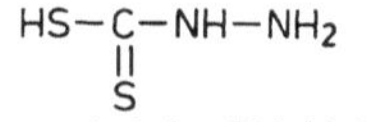

ligand 1 ($=CH_4N_2S_2$) ligand 2 ($=C_8H_{10}N_2S_2$)

Complexes in Solution. The stability constants of the complexes with ligand 2 were determined pH potentiometrically in 40 to 80% (v/v) ethanol-water and dioxane-water mixtures at ionic strength $I = 0.1M$ (KNO_3). For mole fractions of ethanol or dioxane between 0.15 and 0.5, $\log \beta$ increases linearly from 5.6 to 6.4 in ethanol-water, and from 6.1 to 7.2 in dioxane-water [1].

$Mn^{II}(CH_3N_2S_2)_2$ was prepared by adding, dropwise with stirring, the hydrazinium salt of ligand 1 (4 mmol) in 15 mL water to the solution of hydrated manganese(II) acetate (2 mmol in 25 mL water containing 1 mL of 5 M acetic acid). The mixture was stirred for 10 min. The precipitated complex was washed with water, alcohol, and dried in vacuum over KOH [2]. The complex was also obtained by treating the aqueous solution of manganese(II) sulfate with the stoichiometric amount of sodium salt of ligand 1 dissolved in methanol. The briskly stirred mixture precipitated a light brown complex, which was dried at 90°C [3]. The IR and Raman spectra of the complex in CsI pellets or Nujol mulls are recorded in the range 4000 to 33 cm^{-1}. Band assignments are made on the basis of the metal-isotope substitution (^{64}Zn-^{68}Zn) and N-deuteration data. The Mn^{II} complex (as well as those of Zn^{II}, Cd^{II}, Hg^{II}) is of a pseudo-tetrahedral structure. The data suggest that the uninegative bidentate ligand is coordinated via the amine nitrogen and the enethiolate sulfur atoms [2]. $Mn(CH_3N_2S_2)_2$ is soluble in water. The complex is active as an insecticide, a fungicide, and a bactericide [3].

References:

[1] Bose, R. N.; Akbar Ali, M. (J. Inorg. Nucl. Chem. **42** [1980] 148/9).
[2] Fąk-Michalska, D.; Kędzia, B. B. (Bull. Acad. Polon. Sci. Ser. Sci. Chim. **26** [1978] 63/71).
[3] Nash, L. H. (U.S. 3661983 [1972]; C.A. **77** [1972] No. 74834).

35.11.3 With Other Hydrazinecarbothioates

$$C_2H_5O-\underset{\underset{S}{\|}}{C}-NH-NHR$$

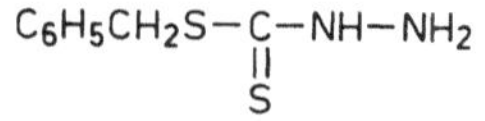

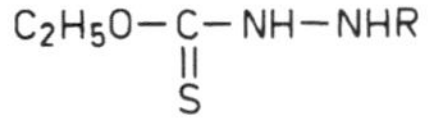

ligand 1 with R $=C(O)OC_2H_5$ ($=C_6H_{12}N_2O_3S$)
ligand 2 with R $=C(S)OC_2H_5$ ($=C_6H_{12}N_2O_2S_2$)

$Mn^{II}(C_6H_{10}N_2O_3S)$ and $Mn^{II}(C_6H_{10}N_2O_2S_2)$ were prepared by the reaction of hydrated manganese(II) acetate dissolved in water and an ethanol solution of an equimolar amount of ligand 1 or 2, respectively. Addition of sodium acetate and a few drops of dilute ammonia precipitated light yellow solids. The magnetic moments of the solid compounds are $\mu_{eff} = 5.88$ and $5.75\ \mu_B$, respectively. The IR spectral results indicate that the dinegative tetradentate ligand 1 coordinates to Mn^{II} in a polymeric solid via the two hydrazide nitrogens, enethiolate sulfur, and the enolate oxygen, whereas ligand 2 is bonded through the two hydrazide nitrogens and both enethiolate sulfur atoms.

Reference:

Singh, N. K.; Srivastava, S. C.; Aggarwal, R. C. (Indian J. Chem. A **23** [1984] 664/7).

35.12 Complexes with Thiourea or Related Compounds

General Reference:

Murzubraimov, B.; Coordination Compounds of d- and f-Metals with Amide Ligands, Koord. Khim. **11** [1985] 1587/605; Soviet J. Coord. Chem. **11** [1985] 903/20.

General Aspects. The preparation of complexes with thiourea or its derivatives in higher oxidation states of manganese is unlikely since the ligands reduce Mn^{VII} successively through Mn^{III} to Mn^{II} [1]. The reduction of Mn^{III} by thiourea in acidic aqueous solution occurs via a rapid inner-sphere electron transfer mechanism with the rate-determining step being the substitution of the sulfur-bonded ligand for a water molecule at the metal center. The second-order rate constant for the reaction of the Mn^{III} center with the protonated thiourea or its N,N′-dimethyl or -diethyl derivatives at 25°C are given in [2].

Complexes of the type MnL_4X_2 or MnL_2X_2 are mainly formed by reaction of a manganese(II) salt with the appropriate amount of thiourea or its alkyl and phenyl derivatives in alcoholic or aqueous alcoholic solution. An MnL_6X_2 complex ($X = ClO_4$) is known only with methylthiourea. Mixed ligand complexes of composition $[MnL_4L'_2]X_2$ and $[MnL_4L'']X_2$ are obtained by reaction of the thiourea complex, $[Mn(CH_4N_2S)_4(H_2O)_2]Cl_2$, with L′ = aniline, pyridine, or quinoline, and with L″ = ethylenediamine, 2,2′-bipyridine, or 1,10-phenanthroline, respectively. $[Mn(CH_4N_2S)_4py_2X_2]$ complexes with X = Cl or Br, where Mn^{II} is assumed to be octacoordinated, are described on p. 193. $MnL \cdot H_2O$ complexes are obtained with the acyl derivatives of thiourea, see p. 199. Complexes with deprotonated ligands are also formed with amidino derivatives of thiourea or 1,2-hydrazinebis(carbothioamide), see p. 200.

The magnetic moments of the thiourea complexes are generally in the 5.9 to 6.0 μ_B range expected for a high-spin Mn^{II} complex. Proton magnetic measurements on $[Mn(CH_4N_2S)_4Cl_2]$ show a transition from the paramagnetic state to an antiferromagnetic state below 0.56 K. The IR spectra of the complexes in comparison to that of free thiourea reveal that coordination of the ligand is always realized by the sulfur atom. The IR spectrum of thiourea in the solid state is fairly complicated. For the observed changes in the ν(CN) and ν(CS) vibration modes on complexation, see p. 188. The IR data of the $MnL \cdot H_2O$ complexes with acyl derivatives of thiourea suggest that the solids contain a polymeric structure with an octahedral coordination of Mn^{II} to the dinegative ligands via the oxygen, nitrogen, and sulfur atoms. Bidentate coordination through the nitrogen and sulfur atoms is assumed for the amidino derivatives of thiourea or the dithiobiuret molecule, see p. 199.

A complex with the condensation product of thiourea and acetylacetone has been described in "Manganese" D 6, 1988, p. 219, together with similar Schiff base complexes.

References:

[1] Bessarabova, I. M.; Kemeleva, N. G.; Bekturova, G. B. (Khim. Khim. Tekhnol. [Alma-Ata] No. 19 [1976] 164/70 from C.A. **91** [1979] No. 48895).
[2] Davis, G. (Inorg. Chem. **11** [1972] 2488/94).

35.12.1 With Thiourea $H_2N{-}C({=}S){-}NH_2$ $(=CH_4N_2S)$

35.12.1.1 Formation in Solution

Formation constants of $Mn(CH_4N_2S)_n^{2+}$ complexes with n = 1 to 3 result from solubility measurements: log $K_1 = 1.48$, log $\beta_2 = 3.58$, log $\beta_3 = 5.20$ at 25°C.

Reference:

Stancheva, P. (Nauchn. Tr. Vissh. Ped. Inst. Plovdiv Mat. Fiz. Khim. Biol. **8** [1970] 103/11; C.A. **74** [1971] No. 60364), see also Perrin, D. D. (Stability Constants B IUPAC Chem. Data Ser. No. 22 [1979] 19).

35.12.1.2 $Mn^{II}(CH_4N_2S)_4X_2$ and $Mn^{II}(CH_4N_2S)_4X_2 \cdot nH_2O$ Compounds

The existence of $Mn(CH_4N_2S)_4X_2$ complexes (with X = Cl or ClO_4) as solid phases was shown by solubilities of the ternary $MnCl_2$–CH_4N_2S–H_2O system at 20°C [1], 25°C [2 to 4], 30°C [5], 35°C [6], 50 and 75°C [2 to 4], and of the $Mn(ClO_4)_2$–CH_4N_2S–H_2O system at 25°C [7], respectively. The ternary system, $MnCl_2$–CH_4N_2S–H_2O, does not show the existence of the $Mn(CH_4N_2S)_{10}Cl_2$ complex described in [8]. Neither was there evidence for a sulfato complex in the $MnSO_4$–CH_4N_2S–H_2O system [9, 10], although the $Mn(CH_4N_2S)_4SO_4$ complex was prepared and characterized by [11]. The $Mn(CH_3COO)_2$–CH_4N_2S–H_2O system was found to be eutonic. No complex formation with thiourea was found on the isotherms at 30°C [12], but monohydrates of acetato complexes, $[Mn(CH_4N_2S)_n(CH_3COO)_2] \cdot H_2O$ with n = 4 or 2, have been prepared recently by [33]; see pp. 189 and 191.

$[Mn(CH_4N_2S)_4X_2]$ complexes with X = Cl [13, 14, 16], Br [15], and NO_3 or ½SO_4 [11] were prepared by reaction of MnX_2 with thiourea (mole ratio 1:4) in ethanol solution. According to [14], a saturated ethanol solution of $MnCl_2$ was refluxed with the ligand in ethanol for 2 h. The reacton mixture was then concentrated to yield crystals. The chloro complex was washed with ethanol, followed by diethyl ether, and dried in vacuum [13], or recrystallized from ethanol [16]. The acetato complex, $[Mn(CH_4N_2S)_4(CH_3COO)_2]$, was obtained by heating the monohydrate, $[Mn(CH_4N_2S)_4(CH_3COO)_2] \cdot H_2O$ (see p. 189), at ~80 to 100°C in air [33].

The enthalpies of formation of crystalline and gaseous $[Mn(CH_4N_2S)_4Cl_2]$ have been measured by reaction calorimetry: ΔH_f°(cryst) = −207.0 ± 0.2 kcal/mol, ΔH_f°(g) = −175.2 kcal/mol. The enthalpy of sublimation is 31.8 kcal/mol [18]. The complex melts at 468 K [18, 29] and decomposes at 482 K [18]. A mean coordinate bond dissociation energy, $\bar{D}$(Mn–S) = 30.0 kcal/mol, results from thermochemical measurements [18]. The white bromo complex, $[Mn(CH_4N_2S)_4Br_2]$, melts at 179°C [15].

$[Mn(CH_4N_2S)_4Cl_2]$ crystallizes from the $MnCl_2$–CH_4N_2S–H_2O system at 25°C as yellow glassy-transparent bitruncated pyramids [2] or at 35°C in the form of white crystals [6]. According to [19], the complex forms very pale green bipyramidal crystals. X-ray data indicate the tetragonal space group $P4_2/n$-C_{4h}^4 (No. 86) with the crystal parameters: a = 13.76, c = 9.07 Å; Z = 4. The value a = 13.76 Å corresponds to the diagonal of the unit cell assumed in earlier investigations [20], where a = 9.68 ± 0.02, c = 8.97 ± 0.01 Å; Z = 2 was found. The complex is isotypic with the tetragonal $[Ni(CH_4N_2S)_4Cl_2]$ which shows a six-coordinate metal with *trans*-axial positions of the Cl^- ions and coordination of the ligand via the sulfur atom [20]. The K-absorption spectrum of the complex reveals two chlorine band edges with a line at 4.3860 Å, due to the transition of one electron from the K-shell into the M-shell, and a second line at 4.3826 Å, because of the complete dissociation of one electron from the K-shell [17].

The density of $[Mn(CH_4N_2S)_4Cl_2]$, determined at 19°C in bromoform or benzene, is D = 1.686 g/cm³ and corresponds to the calculated value D_{calc} = 1.7 g/cm³ [20]; D = 1.731 was determined in o-xylene at 35°C [6].

The magnetic moments of solid $[Mn(CH_4N_2S)_4Cl_2]$ and $[Mn(CH_4N_2S)_4Br_2]$, μ_{eff} = 5.9 [14] and 6.03 μ_B [15], indicate high-spin Mn^{II} (d^5) compounds. Interpolated crystal and powder magnetic data for $[Mn(CH_4N_2S)_4Cl_2]$ in the temperature range 300 to 80 K are given in [21]: μ_{eff} = 5.99 μ_B results from the average magnetic susceptibility, $\bar{\chi}_m$ = 14.940 × 10^{-6} cm³/mol at 300 K; μ_{eff} =

6.16 μ_B from $\bar{\chi}_m = 59.230 \times 10^{-6}$ cm³/mol at 80 K. The measurements show that the magnetic anisotropy $\Delta\chi$ ($=\chi_{\parallel} - \chi_{\perp}$) increases from 34×10^{-6} cm³/mol at 300 K to 418×10^{-6} cm³/mol at 80 K [21].

The magnetic susceptibility follows a Curie-Weiss law in the paramagnetic region with $\chi_{001} < \chi_{110}$. Below 0.56 K χ_{110} decreases very rapidly, while χ_{001} falls off more slowly. Magnetic susceptibility measurements made along the [100], [110], and [001] crystal axes indicate that the sublattice magnetization vectors do not lie near the [001] axis, but near the (001) plane [19, 22].

Proton magnetic resonance measurements on $[Mn(CH_4N_2S)_4Cl_2]$ in the range 4.2 to 0.4 K show a paramagnetic state above 0.56 K, while below this temperature there is an antiferromagnetic state [19]. Specific heat measurements from 3.75 to 0.47 K show a λ type magnetic transition with a Néel temperature of 0.58 ± 0.02 K and $\Delta S = 2.76$ cal·mol⁻¹·K⁻¹ [22].

The IR spectral data of $[Mn(CH_4N_2S)_4Cl_2]$ and $[Mn(CH_4N_2S)_4Br_2]$ are consistent with coordination of the neutral ligand to Mn^{II} via the sulfur atom. A decrease in the ν(CS) vibration mode and an increase in ν(CN), without appreciable changes in the frequencies of the ν(NH) vibrations, were observed on complexation. Main absorption bands of thiourea and $[Mn(CH_4N_2S)_4Cl_2]$ in KBr are shown below with assignments:

assignment	ν(NH)	Ref.	$\delta(NH_2)$	Ref.	ν(NCS)	Ref.
thiourea	3380, 3274, 3174	[23]	1620	[23]	1478	[23]
			1610	[16]	1471	[16]
$[Mn(CH_4N_2S)_4Cl_2]$	3462, 3390, 3280, 3188	[23]	1628, 1606	[23]	1485, 1444	[23]
	3390, 3285, 3185	[24]	1626, 1587	[16]	1480	[24]
			1620	[24]	1467	[16]

assignment	ν(CS) + ν(NCS)	Ref.	ν(NCS)	Ref.	ν(CS)	Ref.
thiourea	1418	[23]	1094	[23]	734	[23]
	1412	[16]	1087	[16]	733	[16]
$[Mn(CH_4N_2S)_4Cl_2]$	1433, 1391	[16]	1098	[16]	725	[16]
	1420	[24]	1090	[24]	714	[23]
	1394	[23]			735, 635	[24]

Similar bands were observed in [25, 26]. The IR spectrum of thiourea is fairly complicated. On the basis of experimental and theoretical work involving the analysis of the normal modes of vibration, Kharitonov et al. [24] proposed a substantial interpretation of the vibrational spectra of free and coordinated thiourea. Only the vibrations of the NH bonds may be considered characteristic. Other vibrations of thiourea are complex stretching-deformation vibrations; e.g., the band at 1420 cm⁻¹ was assigned to a combination of CS, CN, NCN, NCS, and HNC vibration modes on the basis of experimental and theoretical work involving the analysis of the normal modes of vibration. It was found in particular, that free thiourea and thiourea bound in complexes have no IR vibrations which would be completely localized in the C=S bond. This bond is significantly deformed with frequencies of about 1417, 734, and 634 cm⁻¹ (for free thiourea). Therefore, the formation of a metal-sulfur bond in the complexes can be established on the basis of the changes in the values of these frequencies [24, 27]. Bands observed in the far-IR spectrum of $[Mn(CH_4N_2S)_4Cl_2]$ in Nujol were assigned as follows: ν(Mn–S) at 227 cm^{-1}, δ(Mn–S) at 162 cm^{-1}, δ(S–Mn–S) at 112 cm^{-1}, ν(Mn–Cl) at 187 and 134 cm^{-1} [28]; see also [29].

The electronic reflectance spectrum of polycrystalline $[Mn(CH_4N_2S)_4Cl_2]$ shows maxima at 15600, 20000, and 25000 cm^{-1}, assigned to the transitions $^6A_{1g} \rightarrow {}^4T_{1g}, \rightarrow {}^4T_{2g}$, and $\rightarrow {}^4A_{1g}$ in the distorted octahedral crystal field. The spectra of aqueous solutions of manganese(II) chloride and $[Mn(CH_4N_2S)_4Cl_2]$ have identical absorption bands, indicating complete hydration of the thiourea complex [30].

In the crystal powder of $[Mn(CH_4N_2S)_4Cl_2]$ a macroscopic second harmonic generation (SHG) effect was observed, due to micro environmental distortion present in the octahedron. This property, mainly observed with $Cd(CH_4N_2S)_2ClI$, is investigated as a model for frequency-doubling in crystals of coordination compounds [31, 32].

$[Mn(CH_4N_2S)_4Cl_2]$ and $[Mn(CH_4N_2S)_4Br_2]$ are nonelectrolytes in nitrobenzene [14] or acetone [15], respectively. The conductivity of the chloro complex in aqueous solution corresponds to a 1:2 electrolyte, probably due to formation of $[Mn(H_2O)_6]Cl_2$ [13, 14]. The formation of thiocyanate ions by reaction of $[Mn(CH_4N_2S)_4Cl_2]$ or other thiourea complexes with nitrous acid in buffered acidic medium (pH 4) was investigated in [16].

The heating curve of $Mn(CH_4N_2S)_4(ClO_4)_2$ has four clearly expressed effects. The endothermic effect at 135°C corresponds to fusion of the compound. At 175°C the transparent melt is converted into a suspension. The endothermic effect at 188°C is due to the partial loss of thiourea. Explosive decomposition was observed at 257°C [7]. For the thermolysis of $[Mn(CH_4N_2S)_4(CH_3COO)_2]$, see [33].

$[Mn(CH_4N_2S)_4(H_2O)_2]Cl_2$ was prepared by the reaction of hydrated manganese(II) chloride with a 3% aqueous methanol solution of the ligand, in a 1:4 mole ratio. The reaction mixture was refluxed for 3 h and cooled to yield crystals. The product was isolated, washed with small portions of ethanol, and dried under vacuum. The complex melts at 131°C and shows a magnetic moment, $\mu_{eff} = 5.98\ \mu_B$, at 300 ± 1 K. The IR spectrum confirms the presence of coordinated water molecules by a band at 3460 cm^{-1}. The position of IR bands due to coordinated thiourea is similar to those observed for the mixed ligand complexes, see p. 192. Bands observed in the electronic reflectance spectrum of the solid compound are collected in the table on p. 193 [34].

$[Mn(CH_4N_2S)_4(CH_3COO)_2]\cdot H_2O$. A hot aqueous solution containing $Mn(CH_3COO)_2\cdot 4H_2O$ and thiourea (mole ratio 1:4) was acidified with acetic acid to pH 4, then heated for 20 min, and evaporated slowly in a vacuum desiccator over anhydrous $CaCl_2$. The pale pink crystals precipitated after ~7 d, were separated, quickly washed with cold water, acetone, and ether, then dried and recrystallized from water. Interplanar distances and relative intensities from X-ray diffraction patterns and IR data are given in the paper. Coordination of thiourea by means of the sulfur atom is indicated by small decreases of the vibrational frequencies at ~1090 to 1070, 740 to 720, and ~640 to 620 cm^{-1} in comparison with uncoordinated thiourea. The vibrational frequencies with significant contributions from changes in the CN bands are slightly increased on complexation. The complex is dehydrated on heating in air at ~80 to 100°C. A scheme for the thermolysis up to 1000°C is given. The compound is soluble in water, poorly soluble in ethanol, and insoluble in common organic solvents. The acetato groups are assumed to be coordinated directly to Mn in the inner sphere. As shown by conductivity measurements and the interaction with $AgNO_3$, these groups are split off on dissolution [33].

References:

[1] Zumaliev, S.; Rysmendeev, K. R. (Sb. Statei Aspir. Kirg. Gos. Univ. Fiz. Mat. Estestv. Nauk No. 3 [1970] 126/34; C.A. **74** [1971] No. 131069).

[2] Druzhinin, I. G.; Parilova, O. I. (Zh. Prikl. Khim. **41** [1968] 1612/5; J. Appl. Chem. [USSR] **41** [1968] 1531/4).
[3] Parilova, O. I. (Uch. Zap. Mosk. Obl. Ped. Inst. **193** No. 2 [1968] 26/32; C.A. **71** [1969] No. 95459).
[4] Parilova, O. I.; Vidyakina,, L. V.; Druzhinin, I. G. (Uch. Zap. Mosk. Obl. Ped. Inst. **193** No. 2 [1968] 33/44; C.A. **71** [1969] No. 95626).
[5] Rysmendeev, K.; Druzhinin, I. G. (Tr. Kirg. Gos. Univ. Ser. Khim. Nauk **1968** 3/7; C.A. **72** [1970] No. 83468).
[6] Siddhanta, S. K.; Swaminathan, K. (Proc. Symp. Chem. Coord. Compounds, Agra, India, 1959 [1960], Vol. 2, pp. 91/6; C.A. **1961** 6235).
[7] Karnaukhov, A. S.; Runov, N. N.; Goryunov, Yu. A. (Zh. Neorgan. Khim. **18** [1973] 2266/8; Russ. J. Inorg. Chem. **18** [1973] 1199/200).
[8] Walter, G.; Storfer, E. (Monatsh. Chem. **65** [1934/35] 53/8).
[9] Kydynov, M. K.; Dzhumabaeva, S. (Vzaimodiestvie Tiomocheviny Mocheviny Mineral'n. Solyami **1965** 34/7 from C.A. **65** [1966] 4718).
[10] Drushinin, I. G.; Duishenalieva, N.; Ismanova, B. (Izv. Akad. Nauk Kirg.SSR **1974** No. 6, pp. 38/40; C.A. **82** [1975] No. 90774).

[11] Stancheva, P. (Nauchn. Tr. Vissh. Ped. Inst. Plovdiv Mat. Fiz. Khim. Biol. **8** [1970] 103/11; C.A. **74** [1971] No. 60362).
[12] Ergeshbaev, D.; Murzubraimov, B.; Sulaimankulov, K.; Rysmendeev, K. (Zh. Neorgan. Khim. **18** [1973] 1406/9; Russ. J. Inorg. Chem. **18** [1973] 744/7).
[13] Rosenheim, A.; Meyer, V. J. (Z. Anorg. Allgem. Chem. **49** [1906] 17/25).
[14] Dash, K. C.; Ramana Rao, D. V. (Indian J. Chem. **3** [1965] 514/6).
[15] Dash, K. C.; Ramana Rao, D. V. (Current Sci. [India] **35** [1966] 203/4).
[16] Shibutani, Y.; Tsuboi, M.; Matsumoto, C.; Shinra, K. (Nippon Kagaku Kaishi **1981** 1861/6; C.A. **96** [1982] No. 62111).
[17] Stelling, O. (Z. Physik. Chem. B **24** [1934] 282/92).
[18] Ashcroft, S. J. (J. Chem. Soc. A **1970** 1020/4).
[19] Au, R.; Cowen, J. A.; Spence, R. D.; van Till, H. (Low Temp. Phys. Proc. 9th Intern. Conf., Columbus, Ohio, 1964 [1965] B, pp. 877/9; C.A. **65** [1966] 11526).
[20] Nardelli, M.; Cavalca, L.; Braibanti, A. (Gazz. Chim. Ital. **86** [1956] 867/77, 875).

[21] Gerloch, M.; Lewis, J.; Smail, W. R. (J. Chem. Soc. Dalton Trans. **1972** 1559/65).
[22] Forstat, H.; Love, N. D.; McElearny, J. (J. Phys. Soc. Japan **21** [1966] 808).
[23] Dremyatskaya, L. D.; Lyubimova, N. B.; Beskov, S. D. (Zh. Fiz. Khim. **43** [1969] 850/3; Russ. J. Phys. Chem. **43** [1969] 470/2).
[24] Kharitonov, Yu. Ya.; Brega, V. D.; Ablov, A. V.; Proskina, N. N. (Zh. Neorgan. Khim. **19** [1974] 2166/77; Russ. J. Inorg. Chem. **19** [1974] 1187/8).
[25] Swaminathan, K.; Irving, H. M. N. H. (J. Inorg. Nucl. Chem. **26** [1964] 1291/4).
[26] Dash, K. C.; Ali, M.; Patel, R. N.; Ramana Rao, D. V. (J. Indian Chem. Soc. **44** [1967] 246/8).
[27] Murzubraimov, B. (Koord. Khim. **11** [1985] 1587/605; Soviet J. Coord. Chem. **11** [1985] 903/20).
[28] Adams, D. M.; Cornell, J. B. (J. Chem. Soc. A **1967** 884/9).
[29] Flint, C. D.; Goodgame, M. (J. Chem. Soc. A **1966** 744/7).
[30] Dremyatskaya, L. D.; Lyubimova, N. B.; Zelentsov, V. V. (Zh. Neorgan. Khim. **15** [1970] 2115/7; Russ. J. Inorg. Chem. **15** [1970] 1090/1).

[31] Xing, Guangcai; Jiang, Minhua; Shao, Zongshu; Xu, Dong (Zhongguo Jiguang **14** [1987] 302/8; C.A. **107** [1987] No. 144386).

[32] Xu, Dong; Jiang, Minhua; Tao, Xutang; Shao, Zongshu (Rengong Jingti **16** No. 1 [1987] 1/8; C.A. **107** [1987] No. 225623).
[33] Kharitonov, Yu. Ya; Ambroladze, L. N.; Tkavadze, L. M. (Koord. Khim. **14** [1988] 367/72; Soviet J. Coord. Chem. **14** [1988] 196/201).
[34] Sankla, D. S.; Misra, Sudhindra N. (J. Indian Chem. Soc. **57** [1980] 300/3).

35.12.1.3 $Mn^{II}(CH_4N_2S)_2X_2$ and $Mn^{II}(CH_4N_2S)_2X_2 \cdot H_2O$ Compounds

$Mn^{II}(CH_4N_2S)_2X_2$ (X = NO_3, NCS, NCSe). The pale green $Mn(CH_4N_2S)_2(NO_3)_2$ was precipitated from an aqueous alcoholic solution of Mn^{II} nitrate and thiourea in the mole ratio 1:2 [1]. $Mn(CH_4N_2S)_2(NCS)_2$ was prepared by addition of thiourea to an aqueous solution of $Mn(NCS)_2$. Only about one third of the calculated quantity of thiourea was used. The reaction mixture was heated to dissolve the solid, then filtered, and evaporated until crystallization started. The complex was dried at 50°C in vacuum [2]. From an alcoholic solution containing $Mn(NCS)_2$ and the ligand, excess thiourea precipitated first, before the isothiocyanato complex, $Mn(CH_4N_2S)_2(NCS)_2$, crystallized in the form of yellowish white needles [3]. The isoselenocyanato complex, $Mn(CH_4N_2S)_2(NCSe)_2$, was prepared by the reaction of the ligand in hot (60°C) dry ethanol with a dry ethanol solution of $Mn(NCSe)_2$ in a 2:1 mole ratio. The white precipitate, which is immediately produced upon mixing the solutions, was filtered using a Schlenk apparatus, and dried in vacuum. The complex is sensitive to moisture [4].

The long pale rose platelets of $Mn(CH_4N_2S)_2(NCS)_2$ belong to the triclinic system, space group $P\bar{1}$-C_i^1 (No. 2) with the lattice parameters: a = 3.95, b = 7.72, and c = 10.04 Å, α = 93.9°, β = 99.1°, and γ = 106.3°; Z = 1. The compound is isostructural with the corresponding complexes of Cd^{II}, Ni^{II}, and Co^{II} [5]. The crystal structure is probably similar to that of $Ni(CH_4N_2S)_2(NCS)_2$, for which a refinement has shown that each Ni atom lies on a symmetry center and is octahedrally surrounded by four sulfur atoms from thiourea molecules and two nitrogen atoms from the NCS^- groups. The octahedra are linked in chains with each thiourea sulfur atom bridging two adjacent Ni atoms [6].

The magnetic susceptibility of $Mn(CH_4N_2S)_2(NCS)_2$ follows the Curie-Weiss law over the temperature range 370 to 170 K with $\Theta = -1$ K. Below 170 K the susceptibility is lower than would be predicted, suggesting that there may be a transition at this temperature to a phase showing negative magnetic interactions. Magnetic moments of 5.79, 5.78, and 5.34 μ_B have been calculated at 362, 174, and 78 K, respectively [2].

Bands observed at 1140 and 1095 cm^{-1} in the IR spectrum of $Mn(CH_4N_2S)_2(NO_3)_2$ in Nujol were assigned to ν(NCN) + ν(CS) vibrations [1]. The spectrum of $Mn(CH_4N_2S)_2(NCS)_2$ shows bands in the range normally ascribed to N-bonded SCN groups. Vibrational modes of thiourea, which involve appreciable CS stretching character, are shifted slightly to lower energy in comparison to the uncoordinated ligand. A band at 286 cm^{-1} was assigned to ν(Mn–NCS), bands at 189, 181, 166, and 140 cm^{-1} to skeletal vibration modes. The IR spectrum of $Mn(CH_4N_2S)_2(NCSe)_2$ shows ν(NH) bands at 3350, 3250, and 3160 cm^{-1}. A band at 1610 cm^{-1} was assigned to ν(CS), (?) a band at 264 cm^{-1} to the ν(Mn–NCSe) vibration [4]. The electronic reflectance spectrum of the isothiocyanato complex with band maxima at 18600, 22500, and 26800 cm^{-1} is similar to those of other octahedral Mn^{II} complexes [2].

$Mn(CH_4N_2S)_2(CH_3COO)_2 \cdot H_2O$ was prepared by mixing hot aqueous solutions of $Mn(CH_3COO)_2 \cdot 4H_2O$ and thiourea (mole ratio 1:2) in a minimum quantity of water. The preparation procedure and the properties are similar to those of $[Mn(CH_4N_2S)_4(CH_3COO)_2]\cdot H_2O$, see p. 189. Interplanar distances and intensities from X-ray diffraction patterns and IR data are given in the paper. The complex is dehydrated on heating in air at ~80 to 90°C. A scheme is given for the thermolysis up to 1000°C. The complex is soluble in water and methanol, poorly soluble in ethanol, and insoluble in common organic solvents. The conductivity of a freshly prepared 10^{-3}M methanol solution increases on standing from 98 to 198 $cm^2\cdot\Omega^{-1}\cdot mol^{-1}$ [7].

References:

[1] Stancheva, P. (Nauchn. Tr. Vissh. Ped. Inst. Plovdiv Mat. Fiz. Khim. Biol. **8** No. 1 [1970] 103/11; C.A. **74** [1971] No. 60364).
[2] Flint, C. D.; Goodgame, M. (Inorg. Chem. **8** [1969] 1833/6).
[3] Rosenheim, A.; Meyer, V. J. (Z. Anorg. Allgem. Chem. **49** [1906] 13/27, 25).
[4] Benson, C. G.; Little, M. G.; McAuliffe, C. A. (Inorg. Chim. Acta **87** [1984] 169/75).
[5] Nardelli, M.; Cavalca, L.; Braibanti, A. (Gazz. Chim. Ital. **87** [1957] 917/22).
[6] Nardelli, M.; Fava Gasparri, G.; Giraldi Batistini, G.; Domiano, P. (Acta Cryst. **20** [1966] 349/53).
[7] Kharitonov, Yu. Ya.; Ambroladze, L. N.; Tkavadze, L. M. (Koord. Khim. **14** [1988] 367/72; Soviet J. Coord. Chem. **14** [1988] 196/201).

35.13.1.4 $Mn^{II}(CH_4N_2S)C_2O_4\cdot H_2O$

The complex was prepared by refluxing a suspension of $Mn(C_2O_4)\cdot 3H_2O$ with excess thiourea in ethanol for 4 to 6 d. The compound obtained was washed with ethanol, ether, and dried in vacuum. The magnetic moment is $\mu_{eff} = 5.94\ \mu_B$ at 30°C. Coordination of thiourea is indicated by the IR band of ν(CS), shifted from 740 to 705 cm^{-1} on complexation, and a new band at 245 cm^{-1}, assigned to the ν(Mn–S) vibration. The oxalato group is assumed to be bidentate and bridging, due to the positive shift of ν_{as}(OCO) and negative shift of ν_s(OCO) on complexation and a band observed at 790 cm^{-1}. A polymeric octahedral structure is proposed. The electronic reflectance spectrum of the complex shows bands at 16110, 22500, 26800 cm^{-1}. The compound is almost insoluble in organic solvents.

Reference:

Pandey, B. D.; Rupainwar, D. C. (Current Sci. [India] **49** [1980] 336/9).

35.12.1.5 Mixed Ligand Compounds

$[Mn(CH_4N_2S)_4L'_2]Cl_2$ and $[Mn(CH_4N_2S)_4L'']Cl_2$. Complexes with L' = aniline (= C_6H_7N), pyridine (py), or quinoline (= C_9H_7N) and L'' = ethylenediamine (en), bipyridine (bpy), or 1,10-phenanthroline (phen) were prepared by the reaction of a dry ethanolic solution of $[Mn(CH_4N_2S)_4(H_2O)_2]Cl_2$ with the appropriate ligand (mole ratio 1:2 for unidentate ligands and 1:1 for bidentate ligands). Upon cooling the reaction mixture the complexes separated out. The isolated compounds were washed with diethyl ether and dried in vacuum over anhydrous $CaCl_2$. Melting points of the complexes and their magnetic moments at 300±1 K are listed on p. 193 [1].

complex	m.p. in °C	μ_{eff} in μ_B
$[Mn(CH_4N_2S)_4(C_6H_7N)_2]Cl_2$	110	6.00
$[Mn(CH_4N_2S)_4py_2]Cl_2$	144	5.95
$[Mn(CH_4N_2S)_4(C_9H_7N)_2]Cl_2$	83	6.21
$[Mn(CH_4N_2S)_4en]Cl_2$	102	6.16
$[Mn(CH_4N_2S)_4bpy]Cl_2$	*)	6.14
$[Mn(CH_4N_2S)_4phen]Cl_2$	*)	6.12

*) Decomposition without melting at 350°C.

The IR data support sulfur coordination of the thiourea ligands: the high frequency NH absorption bands shift to higher wavenumbers, the strong band at 730 cm^{-1} to lower wavenumbers on complexation. The greater double bond character of the CN bond is indicated by an upward shift of the ligand band at 1473 to ~1480 cm^{-1}. Characteristic absorption bands of the second ligand L′ or L″ are also modified, demonstrating their coordination to the metal atom. Absorption bands of the solid complexes with L′ or L″ observed in the reflectance spectrum with calculated ligand field parameters Dq, B, C, F_2, and F_4 (in cm^{-1}), the nephelauxetic ratio (β) and nephelauxetic parameter for the coordinated ligand (h_x) are tabulated below together with corresponding values for the aqua complex:

L′ or L″	transition $^6A_{1g}\rightarrow$ $^4T_{1g}(G)$	$^1A_{1g}$, $^4E_g(G)$	$^4E_g(D)$	$^4T_{1g}(P)$	Dq	B	C	F_2	F_4	β	h_x
H_2O	20618	24390	29411	32258	763	717.0	3444	1194	95	0.91	1.25
C_6H_7N	20000	24096	28328	33333	850	604.5	3610	1120	103	0.76	3.92
py	20000	24875	24875(?)	27855(?)	850	425.7	4123	1034	117	0.54	6.54
C_9H_7N	19801	24096	28818	32786	873	674.5	3470	1170	99	0.85	2.02
en	19531	24154	27932	33003	901	539.5	3751	1075	107	0.68	4.47
bpy	19607	23809	28169	32768	895	622.5	3516	1124	100	0.79	2.96
phen	19801	24271	27777	33557	873	500.8	3852	1050	110	0.63	5.18

Octahedral coordination at Mn^{II} is assumed also in the light colored aqueous solutions of the complexes showing an extinction of weak intensity [1].

$[Mn(CH_4N_2S)_4py_2X_2]$ complexes with X = Cl or Br were reported to form by reaction of a methanol solution of $[Mnpy_2Cl_2]$ or $[Mnpy_2Br_2]$, respectively, with excess thiourea. The reaction mixture was refluxed for 6 h then cooled to give white platelets.The complexes were washed with methanol, petroleum ether, and finally dried in vacuum. The chloro complex melts at 177°C, the bromo complex at 167°C. The magnetic moments are μ_{eff} = 6.6 μ_B (X = Cl) and 6.1 μ_B (X = Br). The IR spectra are similar to those of $[Mn(CH_4N_2S)_4Cl_2]$ or $[Mn(CH_4N_2S)_4Br_2]$ complexes with respect to the thiourea vibrations. The spectra indicate that the pyridine molecules are also coordinated to manganese. Due to the nonelectrolyte character of the complexes in acetone Mn^{II} was assumed to be octa-coordinated. The complexes are soluble also in alcohol or nitrobenzene [2].

$K_2[Mn(CH_4N_2S)_2(C_7H_8(COO)_2)_2]$ was prepared in 98.8% yield from manganese(II) chloride, thiourea, and the potassium salt of 5-norbornene-2,3-dicarboxylic acid, $C_7H_8(COOH)_2$, in aqueous solution (mole ratio 1:2:2). The white complex precipitating from the cooled mixture was washed with cold water, alcohol, ether, and dried over P_4O_{10} in vacuum. The complex decomposes on heating. The IR spectrum reveals coordination of thiourea by the sulfur atom due to the upward shift of ν(NH)+ν(NCN) and the downward shift of δ(NH) and ν(CS) vibrations. The distance of the band positions ν_{as}(COO) at 1553 and ν_s(COO) at 1421 cm^{-1} indicate the monomeric nature of the octahedral complex with bidentate carboxylato groups. The low antiviral activity of the free ligand is enhanced by the formation of the Mn^{II} complex [3].

References:

[1] Sankhla, D. S.; Mishra, S. N. (J. Indian Chem. Soc. **57** [1980] 300/3).

[2] Dash, K. C.; Rao, D. V. R. (Z. Anorg. Allgem. Chem. **350** [1967] 207/10).

[3] Demeter, E. S.; Buzash, G. Yu.; Butsko, S. S.; Zubairov, M. M.; Lagutkin, N. L.; Stepanova, G. Yu.; Danilenko, G. I. (Fiziol. Akt. Veshchestva No. 13 [1981] 20/4; C.A. **96** [1982] No. 134765).

35.12.2 With N-Methyl- or N,N'-Dimethylthiourea

RNH–C(=S)–NHCH$_3$

ligand 1 with R = H ($=C_2H_6N_2S$)
ligand 2 with R = CH_3 ($=C_3H_8N_2S$)

$[Mn^{II}(C_2H_6N_2S)_6](ClO_4)_2$ and **$[Mn^{II}(C_3H_8N_2S)_4](ClO_4)_2$** complexes were prepared by the reaction of hydrated manganese(II) perchlorate with the appropriate amount of ligand 1 or 2, respectively, in 2-propanol. The mixture was heated for 30 to 60 min, then cooled in ice with vigorous stirring to help initiate crystallization. Ether or benzene was added to effect precipitation. The white $[Mn(C_2H_6N_2S)_6](ClO_4)_2$, recrystallized three times from 2-propanol, was obtained in 60% yield. It melts at 159°C. The yellowish white $[Mn(C_3H_8N_2S)_4](ClO_4)_2$, recrystallized from 2-propanol (yield 53%), melts at 128°C. Magnetic measurements at room temperature and lower temperatures reveal that the Curie-Weiss law is obeyed. Magnetic moments, $\mu_{eff}=5.76$ and 5.73 μ_B, respectively, were calculated at room temperature; $\Theta=-4$ K was determined for the complex with ligand 1. IR data suggest that the thiourea ligands are sulfur-bonded. An octahedral structure is assumed for the complex $[Mn(C_2H_6N_2S)_6](ClO_4)_2$, a tetrahedral for $[Mn(C_3H_8N_2S)_4](ClO_4)_2$. Both complexes behave as 1:2 electrolytes in nitromethane [1, 2].

$[Mn^{II}(C_2H_6N_2S)_2Cl_2]$ and **$[Mn^{II}(C_3H_8N_2S)_2X_2]$** (X = Cl or Br). The chloro complex with ligand 1 was prepared by the reaction of manganese(II) chloride and methylthiourea (mole ratio 1:2) in hot acetone. The reaction mixture was stirred while hot for 1 h, then chilled over ice with constant stirring while diethyl ether was added until a permanent turbidity appeared. The solution was stored in a refrigerator for 1 d, then concentrated. The complex separated after chilling with ice. The yellowish white powder was washed with cold acetone and ethanol and dried in a vacuum. The compound, obtained in 47% yield, melts at 176°C. The chloro complex with ligand 2 was prepared by the reaction of $MnCl_2$ with an anhydrous methanol solution of dimethylthiourea in the mole ratio 1:2. The white precipitate was recrystallized from 2-propanol. The product, obtained in 50% yield, melts at 205°C [1]. $[Mn(C_3H_8N_2S)_2Cl_2]$ was also prepared by the reaction of hydrated manganese(II) chloride in ethanol. The reaction mixture was refluxed for 30 min and filtered hot. Chilling the filtrate gave crystals, which were dried in a vacuum [2]. The bromo complex was obtained from $MnBr_2$ and dimethylthiourea in 2-propanol and recrystallized from 2-propanol in 25% yield. It melts at 157°C [1, 2].

The X-ray powder diffraction patterns suggest that the complexes $[Mn(C_3H_8N_2S)_2Cl_2]$ and $[Mn(C_3H_8N_2S)_2Br_2]$ are isostructural with their cobalt analogs, for which tetrahedral geometry is assumed, because of the magnetic and electronic spectral data. For d spacings and intensities, see the paper. The magnetic moments of $[Mn(C_2H_6N_2S)_2Cl_2]$ and $[Mn(C_3H_8N_2S)_2Cl_2]$ at room temperature, $\mu_{eff}=5.73$ and 5.74 μ_B, respectively, were calculated from the relation

$\mu = 2.84\ (T - \Theta)^{1/2}$ with $\Theta = -1$ K. For the bromo complex, $\mu_{eff} = 5.96\ \mu_B$ was calculated, but Θ was not determined [1, 2]. IR data are given in [2]. $[Mn(C_3H_8N_2S)_2Cl_2]$ is nonconducting in nitromethane [1, 2].

The thermal decomposition of $[Mn(C_3H_8N_2S)_2Cl_2]$ was studied by thermogravimetric analysis. In a vacuum, the complex decomposes in the range 110 to 252°C to a mixture of $MnCl_2$, MnS and an organic residue of N,N′-dimethylthiourea, methyl isothiocyanate, HCl, and polymerized material. In air, the complex loses one ligand at 191 to 262°C and another in the range 292 to 387°C. During the first step, the remaining coordinated ligand is probably oxidized to N,N′-dimethylurea. Formation of the complex $Mn(C_3H_8N_2O)Cl_2$ as an intermediate is assumed [3].

References:

[1] Askalani, P.; Bailey, R. A. (Can. J. Chem. **47** [1969] 2275/82).

[2] Askalani, P. (Diss. Rensselaer Polytech. Inst. 1969, pp. 1/170 from Diss. Abstr. Intern. B **30** [1969] 1563; C.A. **72** [1970] No. 128217).

[3] Bailey, R. A.; Tangredi, W. J. (J. Inorg. Nucl. Chem. **38** [1976] 2221/5).

35.12.3 With N-Phenylthiourea or Its N′-(4-Bromophenyl) Derivative

$C_6H_5NH-C(=S)-NHR$ ligand 1 with R = H $(= C_7H_8N_2S)$

ligand 2 with R = 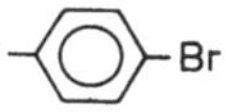 $(= C_{13}H_{11}BrN_2S)$

$[Mn^{II}(C_{13}H_{11}BrN_2S)_4X_2]$ complexes with X = Cl or Br were prepared by the reaction of hot ethanolic solutions of $MnCl_2$ or $MnBr_2$ with ligand 2 in the mole ratio 1:4 [1, 2]. The reaction mixture was refluxed for 3 to 4 h. The excess solvent was distilled off and the resulting solids were washed with water and 10% aqueous ethanol and dried in a vacuum. The chloro complex melts at 111°C [1], the bromo complex at 115°C [2]. Magnetic moments, $\mu_{eff} = 5.67\ \mu_B$ [1] and $5.88\ \mu_B$ [2], respectively, result from susceptibility measurements at room temperature (Faraday method). The IR spectrum of the bromo complex in Nujol shows a shift of the ν(NCN) ligand band from 1540 to 1560 cm^{-1} upon complexation and negative shifts (in cm^{-1}) of the thioamide ligand bands observed in the complex at 1230 ($\Delta\nu = -20$), 1015 ($\Delta\nu = -5$), and 720 cm^{-1} ($\Delta\nu = -55$) [2].

The complexes are soluble in nitrobenzene. The low molar conductivities of the solutions demonstrate that the complexes are nonelectrolytes [1, 2]. The electronic spectrum of $[Mn(C_{13}H_{11}BrN_2S)_4Br_2]$ in methanol does not show any d-d transition bands in the visible region, rather intraligand bands are observed at 31250 and 35714 cm^{-1} [2].

$[Mn^{II}(C_{13}H_{11}BrN_2S)_2X_2]$ complexes with X = Cl, Br, or NCS were obtained from the MnX_2 salt and ligand 2 in ethanol with a 1:2 mole ratio in a manner similar to the procedure used above. The chloro and isothiocyanato complexes melt at 110°C, the bromo complex melts at 121°C [1, 2]. The magnetic moments, μ_{eff}, at room temperature are 5.80, 5.85, and 5.88 μ_B for the complexes with X = Cl [1], Br, and NCS, respectively [2]. The IR spectra resemble those of the complexes above. In the spectrum of $[Mn(C_{13}H_{11}BrN_2S)_2(NCS)_2]$ a ν(CN) vibration mode at 2065 cm^{-1} indicates bonding of the NCS^- ion to manganese. The compounds, for which T_d symmetry is suggested, are nonelectrolytes in nitrobenzene [1, 2].

$[Mn^{II}(C_{13}H_{11}BrN_2S)Cl_2]$. This complex, possibly dimeric, was obtained by the reaction of $MnCl_2$ with ligand 2 in ethanol (mole ratio 1:1). For details of the preparation, see above. The compound melts at 120°C. The magnetic moment at room temperature is $5.71\ \mu_B$. A T_d symmetry is assumed for the complex; it shows no conductivity in nitromethane [2].

$[Mn^{II}(C_7H_8N_2S)_2X_2]\cdot H_2O$ (X = Cl, ½SO_4, CH_3COO) and **$[Mn^{II}(C_7H_8N_2S)_2X_2]\cdot nC_2H_5OH$** (n = 1 for X = NO_3, n = 2 for X = NCS). The complexes were obtained by the reaction of ligand 1 with $MnCl_2\cdot 4H_2O$, $Mn(CH_3COO)_2\cdot 4H_2O$, $Mn(NO_3)_2\cdot 6H_2O$, or $Mn(NCS)_2$ in ethanol (mole ratio 1:2). For preparation of the sulfato complex the ethanol solution of N-phenylthiourea was added to an aqueous solution of $MnSO_4\cdot 5H_2O$. The reaction mixtures were acidified to pH 3 with the corresponding HX acid, then refluxed for about 4 h. The crystalline complexes obtained from the solution were washed rapidly with cold water, acidified with the HX acid (pH 3), with alcohol (in the cases of X = Cl and NCS), acetone (except the complex with X = NCS), and ether, then dried over anhydrous $CaCl_2$ [3, 4]. X-ray diffraction patterns of the ligand and the complexes are given in the paper. The IR spectra of the complexes in liquid paraffin differ little from those of the free ligand. The ν(CS) bands are displaced slightly to lower wavenumbers on complexation. The infrared spectrum of the sulfato complex indicates that the sulfato group is in the inner sphere of the complex and is apparently bidentate. It may also have a bridging function. The spectra of the nitrato, acetato, and isothiocyanato compounds reveal that the anions are in the inner sphere of the complexes. The NCS^- ions are coordinated through the nitrogen atom [3, 4].

The complexes with X = Cl, CH_3COO, or NCS are stable in air; those with X = NO_3 or ½SO_4 are hygroscopic. The analysis of the thermograms and thermogravimetric curves show that the acetato and nitrato complexes have the lowest thermal stability. Dehydration of the acetato complex, $[Mn(C_7H_8N_2S)_2(CH_3COO)_2]\cdot H_2O$, occurs at ~60 to 100°C, loss of N-phenylthiourea occurs above 100°C. $[Mn(C_7H_8N_2S)_2(NO_3)_2]\cdot C_2H_5OH$ releases the ethanol molecule between 40 and 100°C. Decomposition to MnO_2 is observed between 100 and 310°C, to Mn_3O_4 is observed above 310°C. Dehydration of $[Mn(C_7H_8N_2S)_2Cl_2]\cdot H_2O$ occurs from ~60 to 70°C, loss of 0.5 mol of N-phenylthiourea takes place between 100 and 170°C, and complete loss of the ligand in the range 170 to 360°C. $[Mn(C_7H_8N_2S)_2SO_4]\cdot H_2O$ is dehydrated with simultaneous loss of 0.5 mol of ligand between ~50 to 100°C. Complete loss of the ligand molecules is observed between ~200 and 380°C. $[Mn(C_7H_8N_2S)_2(NCS)_2]\cdot 2C_2H_5OH$ decomposes in the range 90 to 240°C with loss of ethanol and N-phenylthiourea molecules [3].

The molar electrical conductivity of the aqueous and methanol solutions was measured at 20°C and at a dilution of 1000 L/mol. The conductivity of the sulfato complex in aqueous solution corresponds to that of a two-ion electrolyte; the conductivity of the other compounds agrees with that of a three-ion electrolyte. The electrical conductivity of the complexes in freshly prepared methanol solutions coincides with that of nonelectrolytes, but gradually increases to values characteristic of a three-ion electrolyte [3].

References:

[1] Chaurasia, M. R.; Saxena, S. K.; Khattri, S. D. (J. Indian Chem. Soc. **58** [1981] 1099/100).
[2] Chaurasia, M. R.; Saxena, S. K.; Khattri, S. D. (J. Nepal Chem. Soc. **1** [1981] 89/96).
[3] Kharitonov, Yu. Ya.; Ambroladze, L. N. (Zh. Neorgan. Khim. **32** [1987] 657/64; Russ. J. Inorg. Chem. **32** [1987] 368/73).
[4] Kharitonov, Yu. Ya.; Ambroladze, L. N. (Koord. Khim. **8** [1982] 1287).

35.12.4 With 2-, 3-, or 4-(N'-Phenylthioureido)benzoic Acids

$C_6H_5NH-C(=S)-NHC_6H_4COOH$ $(=C_{14}H_{12}N_2O_2S = HL)$

The stability constants of **$Mn^{II}L^+$** complexes with HL = 2-, 3-, or 4-(N'-phenylthioureido)benzoic acid, log K_1 = 1.70, 2.05, and 2.09, respectively, were determined pH-potentiometrically in 50 vol% aqueous dioxane solution at 30°C, I = 0.1 M ($NaClO_4$). The formation of a chelate complex was ruled out. The ligands are probably oxygen-bonded.

Reference:

Jähnig, W.; Kanne, E. (Z. Chem. [Leipzig] **25** [1985] 381/2).

35.12.5 With N-(2-Benzothiazolyl)-N'-phenylthiourea

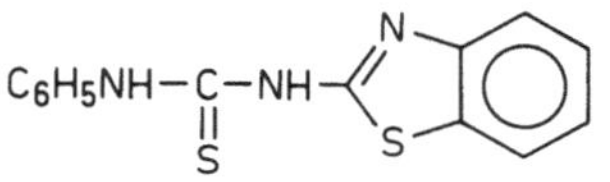

$(=C_{14}H_{11}N_3S_2 = HL)$

$[Mn^{II}(C_{14}H_{10}N_3S_2)_2(H_2O)_2]$ and **$[Mn^{II}(C_{14}H_{10}N_3S_2)_2(C_8H_8N_2)_2]$** ($C_8H_8N_2$ = 2-methyl-1H-benzimidazole). The aqua complex was prepared by refluxing a manganese(II) salt with the required amount of the ligand in ethanol. The solvent was distilled off after 2 h. The resulting complex was washed with water and 10% ethanol and dried in vacuum. It melts at 250°C. On refluxing the ethanol solution of the complex with the stoichiometric amount of 2-methyl-1H-benzimidazole for 3 h, the mixed ligand complex, $[Mn(C_{14}H_{10}N_3S)_2(C_8H_8N_2)_2]$, was formed. After evaporation of the excess ethanol, the compound was washed and dried as described above. It melts at 170°C. The magnetic moments of the complexes are in the range 6.0 to 6.10 μ_B. The IR spectra of the complexes in Nujol show shifts of the thioamide bands in comparison to the free ligand, which indicates Mn–S linkage by the thioamide sulfur atoms. The change in the ν(C=N) vibration mode of the thiazolyl part shows a coordinate bond from the nitrogen atom of the thiazolyl ring to manganese. The unchanged ν(C–S–C) vibration mode reveals no coordination through the sulfur atom of the heterocyclic ring. Bands at 795 to 750 cm^{-1} indicate the coordination of water molecules. An additional ν(NH) band in the spectrum of $[Mn(C_{14}H_{10}N_3S_2)_2(C_8H_8N_2)_2]$ shows coordination of the 2-methylbenzimidazole molecules through the tertiary nitrogen atom. The electronic spectrum of the aqua complex does not show any d-d transition in the visible region. $[Mn(C_{14}H_{10}N_3S_2)_2(C_8H_8N_2)_2]$ shows bands at 18518, 23204, 28160, and 32186 cm^{-1} which may be attributed to the transitions $^6A_{1g} \rightarrow {}^4T_{1g}(G)$, $\rightarrow {}^4E_g$, $^4A_{1g}(G)$, $\rightarrow {}^4E_g(D)'$, and $\rightarrow {}^4T_{1g}(P)$, respectively. The β_{35} value of 0.86 confirms the octahedral structure of the complex. The complexes are nonelectrolytes in nitrobenzene.

Reference:

Chaurasia, M. R.; Saxena, S. K. (Indian J. Chem. A **20** [1981] 741/3).

35.12.6 With N-Thiocarbamoylbenzamides or -ethylcarbamate

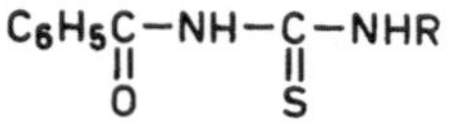

ligand 1 to 6

ligand	R	formula
1	C_6H_5	$C_{14}H_{12}N_2OS$
2	$C_6H_4OCH_3$-2	$C_{15}H_{14}N_2O_2S$
3	$C_6H_4NO_2$-2	$C_{14}H_{11}N_3O_3S$
4	C_6H_4Cl-2	$C_{14}H_{11}ClN_2OS$
5	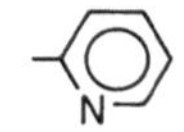	$C_{13}H_{11}N_3OS$
6	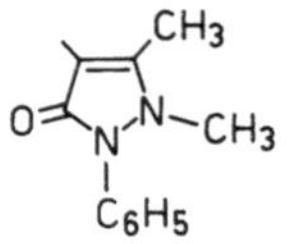	$C_{19}H_{18}N_4O_2S$

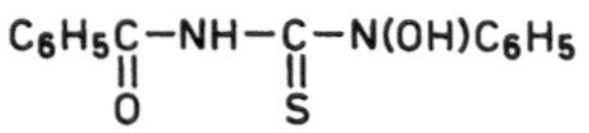

ligand 7 (= $C_{14}H_{12}N_2O_2S$)

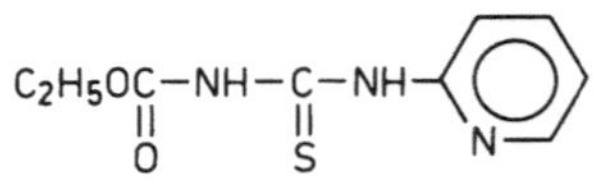

ligand 8 (= $C_9H_{11}N_3O_2S$)

General. The ligands 1 to 4 are believed to exist in equilibrium with thione and thiol tautomeric forms. Their electronic spectra show an intense band at ~250 nm and a weak band at 210 nm. These features are characteristic of a conjugated system and provide evidence in favor of the tautomeric forms below:

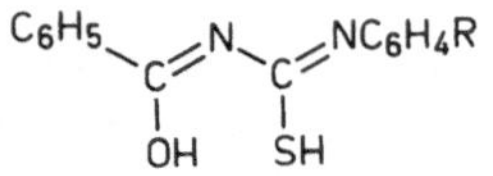

cis form

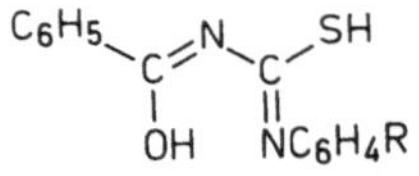

trans form

Considering that the *trans* structure would be more stable, and in the light of IR results, a polymeric structure was proposed for the $MnL \cdot H_2O$ complexes, where the twice deprotonated ligands 1 to 4 are assumed to be coordinated to manganese through nitrogen, sulfur, and oxygen atoms [1, 2]. The ligands 5 to 8 (= HL) are reported to form MnL^+ and MnL_2 complexes in solution. The bidentate ligands 5 and 8 are assumed to be coordinated via the carbonyl oxygen, the pyridine nitrogen, and the sulfur atoms [3].

Complexes in Solution. The stability constants of MnL^+ and MnL_2 complexes with the ligands HL = 5 to 8 were determined at 25°C in aqueous organic solutions by pH-potentiometric titrations. The measurements with ligand 7 were carried out under nitrogen atmosphere.

ligand	solvent	I in mol/L	log K_1	log K_2	Ref.
5	75 vol% ethanol	0.1 ($NaClO_4$)	4.19	3.71	[3]
6	70 vol% ethanol	0.1 (NaCl)	5.69	4.74	[4]
7	70 vol% dioxane	0.1 ($NaClO_4$)	4.77	3.68	[5]
8	75 vol% ethanol	0.1 ($NaClO_4$)	4.84	4.52	[3]

$MnL \cdot H_2O$ complexes with H_2L = ligands 1 to 4 were prepared by the reaction of equimolar amounts of Na_2L and a manganese(II) salt in aqueous solution. The precipitates that immediately formed were digested on a water bath for 30 to 60 min and washed several times with water and finally ethanol, and then dried in vacuum. The red $Mn(C_{14}H_{10}N_2OS) \cdot H_2O$ melts at 250°C and shows a magnetic moment at 32°C of $\mu_{eff} = 5.88 \mu_B$ [1]. The magnetic moment at 32°C of the peach-colored $Mn(C_{14}H_9N_3O_3S) \cdot H_2O$ is 5.86 μ_B, that of the brown $Mn(C_{15}H_{12}N_2O_2S) \cdot H_2O$ and $Mn(C_{14}H_9ClN_2OS) \cdot H_2O$ are 5.94 and 5.85 μ_B, respectively [2]. The ligands are bonded through their oxygen, nitrogen, and sulfur atoms. The spectra of the methoxy and chloro derivatives (ligands 2 and 4) show that the substituents $R = OCH_3$ and Cl are not involved in the coordination, but the NO_2 group (ligand 3) is coordinated to Mn through the oxygen atom. The presence of the water molecule is reflected by bands in the 3600 to 3200 cm^{-1} region, probably indicating hydrogen bonding. The strong band at ~750 cm^{-1} can be assigned to rocking modes of water [2]. The complexes are insoluble in water and most common organic solvents [1, 2].

References:

[1] Satpathy, K. C.; Mishra, H. P.; Mahana, T. D. (J. Indian Chem. Soc. **56** [1979] 248/50).

[2] Satpathy, K. C.; Mishra, H. P.; Panda, A. K.; Satpathi, A. K.; Tripathy, A. (J. Indian Chem. Soc. **56** [1979] 761/3).

[3] Shoukry, M. M.; Mahgoub, A. E. S.; Elnagdi, M. H. (J. Inorg. Nucl. Chem. **42** [1980] 1171/6).

[4] Shoukry, M. M.; Ghoneim, A. K.; Shoukry, E. M.; Elnagdi, M. H. (Syn. React. Inorg. Metal-Org. Chem. **12** [1982] 815/25).

[5] Sharma, B. K.; Mathur, S. P.; Dubey, S. P. (Chim. Acta Turc. **13** [1985] 287/90).

35.12.7 With 1-Amidinothiourea, 1-Amidino-2-ethylisothiourea, or Dithiobiuret

$$H_2N-\underset{\underset{S}{\|}}{C}-NH-\underset{\underset{NH}{\|}}{C}-NH_2 \qquad HN=\underset{\underset{SC_2H_5}{|}}{C}-NH-\underset{\underset{NH}{\|}}{C}-NH_2 \qquad H_2N-\underset{\underset{S}{\|}}{C}-NH-\underset{\underset{S}{\|}}{C}=NH_2$$

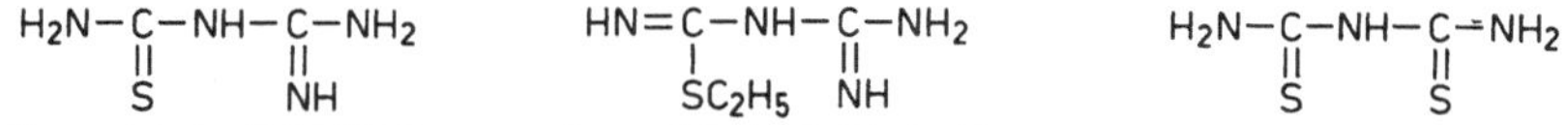

ligand 1 (= $C_2H_6N_4S$)   ligand 2 (= $C_4H_{10}N_4S$)   ligand 3 (= $C_2H_5N_3S_2$)

$Mn^{II}(C_2H_5N_4S)_2$ and **$Mn^{II}(C_4H_9N_4S)_2$** complexes were prepared by the reaction of an ethanolic solution of $MnCl_2$ and ligand 1 or 2 in the mole ratio 1:2. The reaction mixture was refluxed for 1 h. The crystalline brownish yellow compounds were washed with ethanol and dried in air. Magnetic moments of $\mu_{eff} = 5.85$ and 5.83 μ_B, respectively, result from susceptibility measurements at room temperature [1]. The IR spectrum of $Mn(C_2H_5N_4S)_2$ was investigated. Ligand 1 in its enethiol form coordinates through the sulfur and the $>C=N-$ nitrogen atom [2]. Both complexes are insoluble in chloroform, ether, benzene, nitrobenzene, sparingly soluble in ethanol, but soluble in methanol. In freshly prepared methanol solution the complexes are nonelectrolytes. The methanol solutions change their color from yellow to chocolate-red on standing [1].

$[Mn^{II}(C_2H_5N_4S)(C_2H_6N_4S)Cl]_2$ was obtained from a methanolic solution (100 mL) of $MnCl_2 \cdot 2H_2O$ (0.01 mol) and ligand 1 (0.04 mol). On standing fine yellow-green crystals precipitated from the filtered green solution within 2 h. The compound was washed with methanol and dried in vacuum. The subnormal moment $\mu_{eff} = 4.72 \mu_B$ may indicate the presence of bridging chloride ligands in a dimeric structure. The formation of MnS, when the complex is treated with hot concentrated aqueous alkali, may demonstrate that the ligand chelates to Mn^{II} via sulfur and nitrogen atoms. The compound is soluble and behaves as a nonelectrolyte in DMF and DMSO. The molecular weight of the complex, determined osmometrically in DMF solution (~650) corresponds to the calculated value (651) [3].

$[Mn^{II}(C_2H_5N_3S_2)_2SO_4]\cdot H_2O$ and $[Mn^{II}(C_2H_5N_3S_2)_2X_2]\cdot C_2H_5OH$ (X = NCS, CH_3COO) complexes were prepared by addition of stoichiometric amounts of ligand 3 in hot ethanolic solution to a hot aqueous solution of $MnSO_4\cdot 5H_2O$ or to hot ethanolic solutions of $Mn(NCS)_2$ or $Mn(CH_3COO)_2\cdot 4H_2O$, respectively. The mixtures were acidified to pH 3. At higher pH dark precipitates are formed, due to oxidation of the compounds. After slow evaporation of the solvent in a desiccator and standing for several days, the complexes separated from the reaction mixture. They were washed rapidly with cold water, acetone, ether, and dried in vacuum over $CaCl_2$. The pink sulfato complex decomposes on heating at 260°C. The yellow isothiocyanato and acetato complexes are hygroscopic; they decompose at 160 and 110°C, respectively. The d spacings and their intensities obtained from X-ray diffraction patterns of the complexes are given in the paper [4].

The IR spectra of the complexes show no shifts of ν(NH) vibrations upon complexation. Both sulfur atoms are assumed to be coordinated. The distorted octahedral geometry at Mn^{II} is accomplished by two monodentate isothiocyanato or acetato groups, whereas the wavenumbers of the sulfato group indicate its bridging nature. The thermolysis of $[Mn(C_2H_5N_3S_2)_2SO_4]\cdot H_2O$ reveals the release of the solvate molecule between 60 and 110, of one ligand between 220 and 270 and the second ligand between 270 and 410°C. The corresponding steps of $[Mn(C_2H_5N_3S_2)_2(NCS)_2]\cdot C_2H_5OH$ are at 70 to 105, 160 to 220, and 440 to 550°C. At higher temperatures $MnSO_4$ and Mn_3O_4 were obtained. $[Mn(C_2H_5N_3S_2)_2(CH_3COO)_2]\cdot C_2H_5OH$ is transformed into $[Mn(C_2H_5N_3S_2)_{1.5}(CH_3COO)_2]$ at ~50 to 130°C and into MnO at ~240 to 360°C. The electrical conductivities of the sulfato, isothiocyanato, and acetato complexes in freshly prepared aqueous solutions are $\Lambda = 92$, 240, and 250 $cm^2\cdot\Omega^{-1}\cdot mol^{-1}$, respectively. Dissolved in methanol the acetato complex reveals $\Lambda = 92\ cm^2\cdot\Omega^{-1}\cdot mol^{-1}$. The anion groups are probably displaced gradually by solvent molecules. The aqueous solution of the sulfato complex shows only slow (~1.5 h) precipitation of $BaSO_4$, after addition of Ba^{2+} ions. Addition of silver nitrate to the aqueous solutions of the isothiocyanato or acetato complex causes precipitation in 10 to 20 min. This behavior also indicates the inner sphere coordination of the anion groups [4].

References:

[1] Syamal, A. (J. Indian Chem. Soc. **44** [1967] 1084/5).
[2] Syamal, A. (Z. Naturforsch. **24b** [1969] 1192/3).
[3] Roy, N. K.; Saha, R. C. (Indian J. Chem. A **23** [1984] 484/7).
[4] Kharitonov, Yu. Ya.; Ambroladze, L. N. (Zh. Neorgan. Khim. **32** [1987] 1381/7; Russ. J. Inorg. Chem. **32** [1987] 831/5).

35.12.8 With 1,2-Hydrazinebis(carbothioamide) or Its N,N'-Diphenyl Derivative

RNH−C(=S)−NH−NH−C(=S)−NHR

ligand 1 R = H; $(= C_2H_6N_4S_2 = H_2L)$
ligand 2 R = C_6H_5; $(= C_{14}H_{14}N_4S_2 = H_2L)$

$Mn^{II}(C_2H_4N_4S_2)$ was prepared by the reaction of an ethanol solution of ligand 1 or its sodium salt in water with an aqueous solution of an equimolar amount of Mn^{II} halide. The pH of the reaction mixture is raised to 6 to 7 with the addition of dilute NaOH solution (0.1 M). The light brown precipitate was digested on a water bath for 30 min, then isolated, washed with water and ethanol and dried in vacuum [1]. A similar procedure was given in [2, 3]: A hot (70 to 80°C) solution of ligand 1 (1.34 g) in 50 mL of water was added to the aqueous solution of $MnCl_2\cdot 4H_2O$ (1 g) in 30 mL of water. The mole ratio MnX_2 salt : H_2L was 1:2. A saturated potassium hydroxide solution (~2 mL) raised the pH to 8. After heating the mixture on a water bath for 20 to 30 min, the light brown precipitate was separated from the cooled mixture. According to [4] the aqueous solution of $MnCl_2$ and ligand 1 (mole ratio 1:1) was brought up to pH 13 by sodium hydroxide. The formation of manganese(II) hydroxide was prevented even in a four- to fivefold excess of the ligand [5]. The complex melts at 240°C [1].

The magnetic moment of the solid complex at room temperature, $\mu_{eff} = 5.4\ \mu_B$, indicates a polymeric structure [1]. The IR spectra of the complex in KBr [1] or Nujol [3] show no shifts of the ligand bands between 3400 and 3150 cm^{-1}, i.e., the nitrogen atoms of the NH_2 groups are not involved in coordination. Significant differences in the spectrum of the complex in comparison to that of the free ligand were observed below 1350 cm^{-1} [3]. Bands and the assignments (free ligand in parentheses) are as follows: $\delta(NH_2)$ at 1600 (1635, 1605) [3], ν(CN) at 1520 (1545), ν(CS) at 1315 (1375), ν(CS) + ν(CN) at 730 (790) [1], ν(CS) at 745 (~830) [3], 730 cm^{-1} [5]. The ligand is assumed to be coordinated through the enethiolate sulfur atoms and the N(2) and N(3) nitrogen atoms of the tautomeric ligand, $H_2\overset{1}{N}-\underset{S^-}{C}=\overset{2}{N}-\overset{3}{N}=\underset{S^-}{C}-\overset{4}{N}H_2$, which forms five-membered rings with two manganese atoms in a polymeric structure [1 to 5]. The enethiolate bonding is confirmed by a band in the 200 to 210 nm range, observed in the electronic spectrum of the freshly prepared complex in saturated dioxane solution and in the spectrum of the potassium salt of ligand 1 [4]. A weak ligand field shoulder at 20000 and a charge transfer or intraligand transition at 24000 cm^{-1} were observed by [1]. Thermogravimetric analysis of $Mn(C_2H_4N_4S_2)$ shows an endothermic effect between 200 and 225°C, accompanied by a mass loss of ~12% [3]. The complex is sparingly soluble in water and ordinary organic solvents [1 to 5]; freshly prepared samples are slightly soluble in ethanol, acetone, and dioxane [1].

$Mn^{II}(C_{14}H_{12}N_4S_2)$ was prepared in alkaline aqueous solution (pH 13) by reaction of a manganese(II) salt and a four- to fivefold excess of ligand 2. The structural properties are similar to those assumed for $Mn(C_2H_4N_4S_2)$, see above. The complex is nearly insoluble in water and common organic solvents [5].

References:

[1] Satpathy, K. C.; Mahana, T. D. (J. Indian Chem. Soc. **56** [1979] 1173/6).

[2] Batyr, D. G.; Baloyan, B. M.; Popa, E. V.; Kharitonov, Yu. Ya. (Koord. Khim. **6** [1980] 968/9; C.A. **93** [1980] No. 106194).

[3] Batyr, D. G.; Baloyan, B. M.; Popa, E. V.; Kharitonov, Yu. Ya. (Koord. Khim. **7** [1981] 737/42; Soviet J. Coord. Chem. **7** [1981] 359/64).

[4] Vassilev, G. N.; Davarski, K. A.; Genchev, M. S. (Dokl. Bolg. Akad. Nauk **36** [1983] 461/4).

[5] Davarski, K.; Vangarova, V.; Todorov, T. (Sb. Dokl. 1st Nats. Konf. Mladite Nauchni Rab. Spets. Neft Khim., Burgas, Bulg., 1976, pp. 360/4; C.A. **94** [1981] No. 64672).

35.13 Complexes with Thiosemicarbazides or Related Compounds

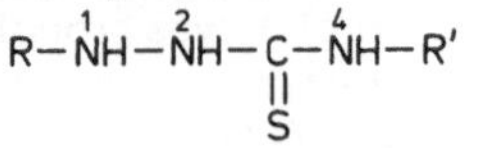

$$R-\overset{1}{N}H-\overset{2}{N}H-\underset{\underset{S}{\|}}{C}-\overset{4}{N}H-R' \qquad \text{ligand 1 to 12}$$

ligand	R	R'	formula	ligand	R	R'	formula
1	H	H	CH_5N_3S	10		C_6H_5	$C_{14}H_{13}N_3O_2S$
2	H	C_6H_5	$C_7H_9N_3S$				
3	C_6H_5	H	$C_7H_9N_3S$				
4	C_6H_5	C_6H_5	$C_{13}H_{13}N_3S$	11		C_6H_5	$C_{14}H_{12}ClN_3OS$
5	$CH_3C(O)$	C_6H_5	$C_9H_{11}N_3OS$				
6	$(CH_3)_3\overset{\oplus}{N}CH_2C(O)$	C_6H_5	$C_{12}H_{19}N_4OS^+$	12		C_6H_5	$C_{13}H_{18}N_4OS$
7	$NCCH_2C(O)$	$C(O)C_6H_5$	$C_{11}H_{10}N_4O_2S$				
8	$C_4H_9C(O)$	C_6H_5	$C_{12}H_{17}N_3OS$				
9	$C_6H_5C(O)$	C_6H_5	$C_{14}H_{13}N_3OS$				

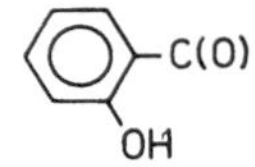

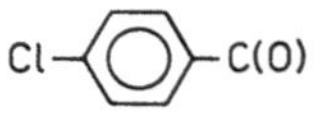

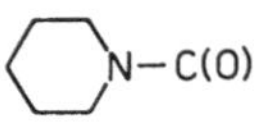

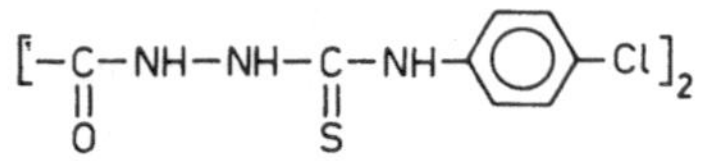

ligand 13 (= $C_{16}H_{14}Cl_2N_6O_2S_2$)

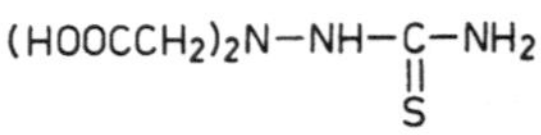

ligand 14 (= $C_5H_9N_3O_4S$)

35.13.1 Complexes in Solution

Stability constants of complexes with ligand 14, $Mn^{II}(C_5H_7N_3O_4S)$ and $Mn^{II}(C_5H_8N_3O_4S)^+$, were determined potentiometrically (glass electrode) in aqueous solution at 30°C and $I = 0.1M$ (KNO_3): $\log K_1 = 2.0$ and $\log K_{MnHL}^{Mn} = 1.5$ [1]. Thermodynamic data for the formation of the $Mn(C_5H_7N_3O_4S)$ complex are $\Delta G = -2.8$, $\Delta H = 7.2$ kcal/mol, $\Delta S = 33$ cal·mol^{-1}·K^{-1} [2]. The stability constants of Mn^{II} complexes with ligand 7, $Mn(C_{11}H_9N_4O_2S)^+$ and $Mn(C_{11}H_9N_4O_2S)_2$, were determined by pH potentiometric titrations in 70% ethanol-water mixture at 25°C and $I = 0.1M$ (KCl): $\log K_1 = 7.08$, $\log K_2 = 3.68$ [3].

References:

[1] Goddard, D. R.; Nwankwo, S. I. (J. Chem. Soc. A **1967** 1371/5).
[2] Goddard, D. R.; Nwankwo, S. I.; Staveley, L. A. K. (J. Chem. Soc. A **1967** 1376/8).
[3] Shoukry, M. M.; Darwish, N. A.; Morsi, M. A. (Gazz. Chim. Ital. **112** [1982] 301/5).

35.13.2 Manganese(II) Compounds

$Mn(C_{12}H_{16}N_3OS)_2$ and $Mn(C_{14}H_{11}ClN_3OS)_2$. The complexes were obtained by refluxing solutions of an Mn^{II} salt with the stoichiometric amounts of ligand 8 [1] or ligand 11 [2] in absolute ethanol. The precipitates were washed successively with ethanol and diethyl ether and finally dried in a vacuum desiccator over anhydrous $CaCl_2$. The pale yellow $Mn(C_{12}H_{16}N_3OS)_2$ melts at 186°C [1], the white $Mn(C_{14}H_{11}ClN_3OS)_2$ at 240°C [2]. The IR spectrum of $Mn(C_{12}H_{16}N_3OS)_2$ in Nujol shows no $\nu(N(1)-H)$ and unshifted $\nu(C=S)$ vibrations. The $\nu(N(2)-H)$ band at 3180 cm^{-1} is negatively shifted ($\Delta\nu = -90$ cm^{-1}) in comparison to the free ligand. New bands of $\nu(C=N)$ at 1600 and $\nu(C-O)$ at 1085 cm^{-1} were observed on complexation. This indicates that the bidentate ligand is coordinated via the N(2) nitrogen and the enolate oxygen atoms. The ν(Mn–O) and ν(Mn–N) vibration modes were found at 500 and 360 cm^{-1}, respectively [1]. The spectrum of $Mn(C_{14}H_{11}ClN_3OS)_2$ in Nujol indicates that the bidentate deprotonated ligand is coordinated through the N(1) nitrogen atom and the sulfur atom of the enethiolate group [2]. Both complexes are soluble in dimethylformamide, dimethyl sulfoxide, but insoluble in other organic solvents. The complexes are nonelectrolytes in dimethylformamide [1, 2].

$Mn(C_{14}H_{12}N_3OS)_2$ and **$Mn(C_{14}H_{12}N_3OS)_2 \cdot 2H_2O$**. The dihydrate was prepared by the reaction of ligand 9 and excess manganese(II) salt in aqueous ethanol. The pH of the reaction mixture was raised by the addition of solid sodium acetate and a yellow precipitate formed. The product was digested on a water bath for 1 h, washed with hot water and ethanol, and dried in a desiccator. The magnetic moments of the violet anhydrous complex and the yellow dihydrate are $\mu_{eff} = 5.73$ and 6.33 μ_B, respectively, at room temperature. The IR spectra of both complexes (in Nujol) show negatively shifted ν(NH) and positively shifted ν(N–N) bands in comparison with the spectrum of the free ligand, but no ν(C=O) mode. The new bands at 416 and 420 cm^{-1} are assigned to ν(Mn–O), at 320 and 335 cm^{-1} to ν(Mn–N) vibrations, respectively. The spectral results suggest that the uninegative ligand probably coordinates to Mn^{II} via the enolate oxygen and the nitrogen atoms. The electronic spectrum of $Mn(C_{14}H_{12}N_3OS)_2$ in Nujol shows charge transfer bands at 20000 and 33300 cm^{-1}, that of $Mn(C_{14}H_{12}N_3OS)_2 \cdot 2H_2O$ at 25000 and 34500 cm^{-1}. Additional bands observed at 31250 and 40800 cm^{-1} or 27780 and 40000 cm^{-1}, respectively, were assigned to intraligand transitions as they correspond to the 33000 and 40000 cm^{-1} bands of the ligand in THF solution. Both complexes are insoluble in chloroform, carbon tetrachloride, petroleum ether, very slightly soluble in nitrobenzene. In nitrobenzene the complexes are nonelectrolytes [3].

$Mn(C_{14}H_{11}N_3O_2S)$ was prepared by the reaction of ethanol or aqueous solutions of excess manganese(II) salt and ligand 10. The reaction mixture was digested on a water bath for 4 h to yield an ash-colored precipitate, which was washed with ethanol and dried in vacuum. It does not melt up to 350°C. The magnetic moment is $\mu_{eff} = 6.00$ μ_B. The IR spectrum of the complex in Nujol shows negative shifts of the ν(NH) and ν(CS) ligand bands. The phenolic ν(OH) and the ν(C=O) vibrations disappear on complexation. It was suggested that the dinegative ligand coordinates to Mn^{II} via the oxygen, nitrogen, and sulfur atoms. The complex is insoluble in chloroform and carbon tetrachloride, sparingly soluble in nitrobenzene, soluble in pyridine or ethylenediamine. In nitrobenzene it behaves as a nonelectrolyte [4].

$Mn(C_{16}H_{12}Cl_2N_6O_2S_2) \cdot 2H_2O$. An aqueous solution of $MnCl_2 \cdot 4H_2O$ (1 mmol) was mixed with ligand 13 (1 mmol) dissolved in dimethylformamide with continuous stirring and digesting on a steam bath for 1 h. The white complex formed was washed with water and dried in vacuum. It decomposes at 283°C. The magnetic moment of the solid is $\mu_{eff} = 4.90$ μ_B. Due to the disappearance of the ν(NH) ligand band in the IR spectrum of the complex (in KBr) the dinegative ligand is assumed to be coordinated to Mn^{II} by the N(1) nitrogen and the enethiolate sulfur atoms in a square-planar arrangement around Mn^{II}. The complex is insoluble in water and in organic solvents [5].

$Mn(C_{13}H_{16}N_4OS) \cdot 5H_2O$ was obtained by refluxing stoichiometric amounts of $MnCl_2 \cdot 4H_2O$ and ligand 12 in aqueous ethanol for 30 min. Solid sodium acetate was added as a buffer. The yellowish brown precipitate was washed with absolute ethanol and ether and dried in vacuum over $CaCl_2$. The complex does not melt up to 290°C. In the IR spectrum of the complex in Nujol the ν(N(1)–H) and ν(N(2)–H) modes of the ligand disappear. New bands appear and were assigned as follows: ν(C=N(1)) 1635, ν(C=N(2)) 1620, ν(C–O) 955, ν(C–S) 630, ν(Mn–O) 470, ν(Mn–N) 375, ν(Mn–S) 400 cm^{-1}. The data indicate a dimeric structure with bridging dinegative ligands coordinated by the N(1) nitrogen and the enethiolate sulfur atoms to one Mn^{II}, and by the enolate oxygen atom to the other Mn^{II}. The octahedral Mn^{II} environments are completed by coordinated water molecules. The complex is soluble only in dimethylformamide or dimethyl sulfoxide. In dimethylformamide it behaves as a nonelectrolyte [6]

$[Mn(CH_5N_3S)_3](ClO_4)_2$ was prepared by the reaction of ethanol solutions of hydrated manganese(II) perchlorate and ligand 1 (mole ratio 1:3) in the presence of a few drops of acetic acid. The reaction mixture was refluxed for 2 h and cooled. The white precipitate was washed with ethanol, diethyl ether, and dried in vacuum. The complex melts at 140°C. The magnetic moment of the solid compound at room temperature is $\mu_{eff} = 5.90\ \mu_B$. In comparison to the free ligand the IR spectrum of $[Mn(CH_5N_3S)_3](ClO_4)_2$ in Nujol shows the main bands to be negatively shifted: the $\delta(NH_2)$ vibration of the hydrazine residue at 1635 cm^{-1} ($\Delta\nu = -8\ cm^{-1}$), the amide II band at 1590 cm^{-1} ($\Delta\nu = -27\ cm^{-1}$), and the $\nu(C{=}S)$ mode at 705 cm^{-1} ($\Delta\nu = -100\ cm^{-1}$); the band at 1560 (due to $\nu(CN) + \delta(NH_2)$) is shifted to higher wavenumbers ($\Delta\nu = 28\ cm^{-1}$). A band of the ionic perchlorate group was observed at 1090 cm^{-1}. The bidentate ligand is assumed to be coordinated to Mn^{II} via the N(1) nitrogen and the sulfur atoms. The electrical conductivity of $[Mn(CH_5N_3S)_3](ClO_4)_2$ in 10^{-3} M nitromethane solution is $\Lambda = 224\ cm^2 \cdot \Omega^{-1} \cdot mol^{-1}$, indicating a 1:2 electrolyte [7].

$[MnL_2Cl_2]$ and **$[MnL_2Br_2]$** complexes with ligands 1 to 4 were prepared by the reaction of the appropriate manganese(II) halide and ligand 1, 2, 3, or 4 in ethanol or acetone containing a little triethyl orthoformate as a dehydrating agent. The reaction mixture was concentrated on a steam bath and cooled to yield the solid. In some cases addition of benzene or acetone was required to give precipitation. The products were washed with ethanol and dried [8]. The $[Mn(CH_5N_3S)_2X_2]$ complexes with X = Cl or Br were also prepared by mixing stoichiometric amounts of the manganese(II) halide in ethanol with ligand 1 in hot absolute ethanol. The mixture, containing a few drops of acetic acid, was refluxed for 2 h and then cooled. The precipitates were washed with absolute ethanol, followed by diethyl ether, and dried to yield the light pink chloro complex and the pale yellow bromo complex [7]. $[Mn(CH_5N_3S)_2Cl_2]$ was observed as a solid phase together with $MnCl_2 \cdot 4H_2O$ at 20 or 40°C in the ternary system $MnCl_2 \cdot 4H_2O$–CH_5N_3S–H_2O [9]. The decomposition points of the complexes and their magnetic moments at room temperature are as follows:

complex	MnL_2Cl_2						MnL_2Br_2				
ligand	1	1	1	2	3	4	1	1	2	3	4
t_{dec} in °C	178 [a]	>200	185	>200	>200	>200	180 [a]	>200	>200	>200	>200
μ_{eff} in μ_B	6.20	5.8	—	5.85	5.90	5.88	6.02	5.80	5.83	5.87	5.84
Ref.	[7]	[8]	[10]	[8]	[8]	[8]	[7]	[8]	[8]	[8]	[8]

[a] Melting point.

In the IR spectra of the complexes in Nujol [7, 8, 11] or KBr [12] the band positions of the $\nu(NH_2)$, $\nu(NH)$, and $\nu(CS)$ vibrations are shifted negatively upon complexation, that of $\nu(CN)$ is shifted positively, indicating the coordination of ligands 1 to 4 to Mn^{II} through the N(1) nitrogen and the sulfur atoms. The octahedral arrangement at Mn^{II} is accomplished by the halide atoms. The band positions are given in cm^{-1}:

complex	$\nu(NH_2)$, $\nu(NH)$	$\delta(NH)$ +$\nu(CN)$	$\nu(CN)$ +$\nu(CS)$	$\nu(CS)$	$\nu(Mn{-}N)$	$\nu(Mn{-}S)$	$\nu(Mn{-}X)$	Ref.
$[Mn(CH_5N_3S)_2Cl_2]$		1633		705				[7]
	3225 3140	1610	1275	725	400	330	280	[8]
	3320 to 3200	1610		703			353	[11]
	3166 3115							[11]

For numbers and formulas of ligands, see p. 202

complex	ν(NH_2), ν(NH)		δ(NH) +ν(CN)	ν(CN) +ν(CS)	ν(CS)	ν(Mn–N)	ν(Mn–S)	ν(Mn–X)	Ref.
$[Mn(C_7H_9N_3S)_2Cl_2]$ a)	3240	3145	1615	1275	730	405	340	275	[8]
$[Mn(C_7H_9N_3S)_2Cl_2]$ b)	3240	3150	1600	1270	730	400	330	275	[8]
$[Mn(C_{13}H_{13}N_3S)_2Cl_2]$		3115	1600	1270	725	410	335	270	[8]
$[Mn(CH_5N_3S)_2Br_2]$			1640		715				[7]
	3240	3150	1610	1270	720	400	335	250	[8]
$[Mn(C_7H_9N_3S)_2Br_2]$ a)	3245	3140	1610	1275	735	405	335	245	[8]
$[Mn(C_7H_9N_3S)_2Br_2]$ b)	3255	3140	1610	1270	735	410	330	240	[8]
$[Mn(C_{13}H_{13}N_3S)_2Br_2]$		3130	1610	1275	730	405	340	235	[8]

a) Ligand 2. – b) Ligand 3.

IR spectral data of $[Mn(CH_5N_3S)_2Cl_2]$ are also given in [11, 12] together with other transition metal complexes with semicarbazide or thiosemicarbazide. Maxima ($\tilde{\nu}_{max}$ in cm^{-1}) observed in the electronic reflectance spectra of the complexes, due to transitions from the ground state, $^6A_{1g}(S)$, and the resulting ligand field parameters (Dq, B, C in cm^{-1}) are:

complex	$^6A_{1g}\rightarrow{}^4T_{1g}(G)$	$\rightarrow{}^4T_{2g}(G)$	$\rightarrow{}^4A_{1g}$, $^4E_g(G)$	$\rightarrow{}^4T_{2g}(D)$	$\rightarrow{}^4E_g(D)$	Dq	B	C	β
$[Mn(CH_5N_3S)_2Cl_2]$	18200	22000	25220	27200	29750	834	758	3528	0.79
$[Mn(C_7H_9N_3S)_2Cl_2]$ a)	17900	21500	25000	26500	29200	821	746	3358	0.78
$[Mn(C_7H_9N_3S)_2Cl_2]$ b)	17900	21240	25100	26700	29400	821	746	3378	0.78
$[Mn(C_{13}H_{13}N_3S)_2Cl_2]$	17670	22000	25200	27100	29220	810	736	3380	0.77
$[Mn(CH_5N_3S)_2Br_2]$	18250	22100	25200	27200	29900	836	760	3520	0.79
$[Mn(C_7H_9N_3S)_2Br_2]$ a)	18000	21700	25200	27000	29450	825	750	3390	0.78
$[Mn(C_7H_9N_3S)_2Br_2]$ b)	18000	21500	25200	26700	29500	825	750	3390	0.78
$[Mn(C_{13}H_{13}N_3S)_2Br_2]$	17700	22100	25200	27100	29200	811	737	3417	0.77

a) Ligand 2. – b) Ligand 3.

The data agree with octahedral geometry at Mn^{II} coordinated by the bidentate ligand molecules 1 to 4 and the halide groups [8]. $[Mn(CH_5N_3S)_2Cl_2]$ and $[Mn(CH_5N_3S)_2Br_2]$ behave as nonelectrolytes in 10^{-3} M nitromethane solution [7]. The TGA curve of $[Mn(CH_5N_3S)_2Cl_2]$ shows thermal effects corresponding to fusion with decomposition at 185°C, release of N_2H_4 at 200°C, of H_2S at 240°C, of CS_2 at 350°C. Decomposition of remaining nitrogen carbide, C_3N_4, mixed with $MnCl_2$ occurs at 540°C [10].

$[Mn(C_9H_{11}N_3OS)Cl_2]$ was obtained by refluxing an equimolar mixture of ligand 5 and hydrated manganese(II) chloride in absolute ethanol for 0.5 to 3 h. The precipitate was washed with absolute ethanol and ether and dried in vacuum over $CaCl_2$. The yellowish complex does not melt up to 300°C. The IR spectrum in Nujol or a CsI disk shows the following bands (shifts in comparison to the free ligand (in cm^{-1}) are given in parentheses): ν(N(4)–H) 3360 (–20), ν(N(2)–H) 3260 (–10), ν(N(1)–H) 3160 (+10), ν(CO) 1650 (–45), ν(CS) 1225 (+5), 760 (–20), ν(NN) 1010 (+10), ν(Mn–O) 430, ν(Mn–N) 360, ν(Mn–Cl) 260 cm^{-1}. The data indicate that the carbonyl oxygen, the N(1) nitrogen atoms of the ligand, and the chloro ions are coordinated to Mn^{II}. The complex is insoluble in most common organic solvents. In dimethylformamide it behaves as a nonelectrolyte [13].

$[Mn(C_{12}H_{19}N_4OS)Cl_2]Cl$. Equimolar amounts of the hydrochloride of ligand 6 and Mn^{II} chloride in absolute ethanol were refluxed for about 1 h and then concentrated to yield a yellow precipitate, which was washed with hot ethanol and dried in vacuum. The compound melts at 258°C. In the IR spectrum of the complex the ν(NH) bands observed at 3470, 3230, and 3140 cm^{-1} have been shifted negatively by complexation; the ν(N–N) vibration at 1000 cm^{-1} is shifted positively. New bands at 1340 and 410 cm^{-1} are attributed to OH deformation modes, a band around 350 cm^{-1} to the ν(Mn–N) vibration. They indicate that the bidentate ligand is coordinated to Mn^{II} through the nondeprotonated enolized oxygen atom and the N(1) nitrogen atom. The Mn^{II} environment is completed by two chlorine atoms. The complex is partially soluble in most common organic solvents but readily soluble in dimethylformamide and dimethyl sulfoxide. The electrical conductivity of the complex, $\Lambda = 81\ cm^2 \cdot \Omega^{-1} \cdot mol^{-1}$ in dimethylformamide at 25°C, is in the range of a 1:1 electrolyte [14].

References:

[1] Bekheit, M. M.; Ibrahim, K. M. (Syn. React. Inorg. Metal-Org. Chem. **16** [1986] 1135/47).
[2] El-Asmy, A. A.; Ibrahim, K. M.; Bekheit, M. M.; Mostafa, M. M. (Syn. React. Inorg. Metal-Org. Chem. **14** [1984] 785/97).
[3] Aggarwal, R. C.; Yadav, R. B. S. (Transition Metal Chem. [Weinheim] **1** [1976] 139/42).
[4] Aggarwal, R. C.; Yadav, R. B. S. (Indian J. Chem. A **15** [1977] 462/3).
[5] Amna, M.; Rai, R. A.; Lakshmi, M. (J. Indian Chem. Soc. **63** [1986] 517/9).
[6] Khalifa, M. E.; El-Asmy, A. A.; Ibrahim, K. M.; Mostafa, M. M. (Syn. React. Inorg. Metal-Org. Chem. **16** [1986] 1305/17).
[7] Pradhan, B.; Rao, D. V. R. (J. Indian Chem. Soc. **54** [1977] 136/8).
[8] Srivastava, A. K. (Proc. Natl. Acad. Sci. India A **54** [1984] 45/51).
[9] Shtrempler, G. I.; Murzubraimov, B.; Rysmendeev, K. (Tr. Kirg. Gos. Univ. Ser. Khim. Nauk **1975** No. 1, pp. 31/7; C.A. **85** [1976] No. 8631).
[10] Shtrempler, G. I.; Murzubraimov, B.; Rysmendeev, K. (Russ. J. Inorg. Chem. **27** [1982] 442/4).

[11] Murzubraimov, B.; Toktomamatov, A. (Koord. Khim. **11** [1985] 596/602; C.A. **103** [1985] No. 61708).
[12] Murzubraimov, B.; Shtrempler, G. I. (Russ. J. Inorg. Chem. **27** [1982] 829/30).
[13] El-Asmy, A. A.; Bekheit, M. M.; Ibrahim, K. M.; Mostafa, M. M. (Acta Chim. Hung. **121** [1986] 391/402).
[14] El-Asmy, A. A.; Mostafa, M. M. (J. Coord. Chem. **12** [1983] 291/6).

35.13.3 Manganese(III) Compounds

$[Mn(CH_4N_3S)_2L]$. The mixed ligand complexes with the anions $C_5H_7O_2^-$ (acetylacetonate), $C_6H_4NO_2^-$ (picolinate), or $C_5H_{10}NS_2^-$ (diethyldithiocarbamate) were prepared by reaction of ethanol solutions of manganese(II) chloride with ligand 1 and the NaL salt of the second ligand in a 1:2:1 mole ratio. Ethanolic NaOH solution was added to the reaction mixture until it was faintly alkaline. When air was bubbled through the solution, the color darkened, and gray-brown precipitates formed. The complexes were washed with ethanol and diethyl ether, then dried in vacuum. The solids do not melt up to 300°C. The magnetic moments of the complexes, μ_{eff} = 5.05, 5.04, and 4.93 μ_B, respectively, indicate the $Mn^{III}(d^4)$ spin configuration. According to IR spectral data, the Mn^{III} has an octahedral environment with two deprotonated ligand molecules, coordinated via the N(1) nitrogen and the enethiolate sulfur atoms and the appropriate uninegative anion, L^-, which coordinates bidentately through its OO, ON, or SS donor atoms, respectively. The electronic absorption spectra of the complexes show maxima at 13500, 22900, and 25000 cm^{-1}, which can be assigned to the tetragonally-distorted octahedral environment of Mn^{III} with the transitions $^5B_{1g} \rightarrow {}^5A_{1g}, \rightarrow {}^5B_{2g}$, and $\rightarrow {}^5E_g$, respectively. The complexes are nonelectrolytes in nitrobenzene and monomeric in camphor.

Reference:

Pradhan, B.; Rao, D. V. R. (Indian J. Chem. A **20** [1981] 751/2).

35.14 Complexes with Thiocarbonohydrazide or Derivatives

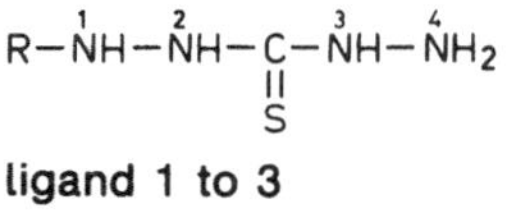

ligand 1 to 3

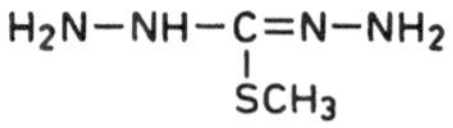

ligand 4 (= $C_2H_8N_4S$)

ligand	1	2	3
R	H	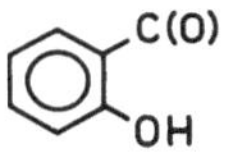	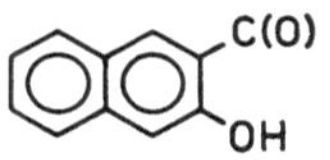
formula	CH_6N_4S	$C_8H_{10}N_4O_2S$	$C_{12}H_{12}N_4O_2S$

Complexes in Solution. The formation of a soluble complex with ligand 1 is mentioned in [1], but a stability constant for the complex is not reported. The stability constant, log K_1 = 2.022(7), for the complex with ligand 4, $Mn(C_2H_8N_4S)^{2+}$, was determined potentiometrically (glass electrode) at 25°C and I = 0.5 M (KCl) in a nitrogen atmosphere. The ligand is assumed to be coordinated through the N(1) nitrogen and the sulfur atoms, forming a five-membered chelate ring. The absence of mixed or polynuclear complexes was demonstrated [2].

$Mn^{II}(C_{12}H_{10}N_4O_2S)$. An ethanolic solution of manganese(II) chloride and ligand 3 (1:1 mole ratio) was adjusted to pH 6 to 7 by adding sodium acetate solution and then heated on a water bath for 30 min. The greenish yellow precipitate was washed with ethanol and dried. A magnetic moment of μ_{eff} = 5.91 μ_B results from susceptibility measurements at ambient temperature. The IR spectrum of the complex (in KBr) exhibits no phenolic ν(OH) band. The ν(CN) ligand band at 1580 cm^{-1} is shifted to 1590 cm^{-1} and the bands due to ν(CS) vibration modes are much lowered in intensity upon complexation. The shift of the amide IV band (due to the major contribution of NCO) from 680 to 670 cm^{-1} and its decrease in intensity confirms the participation of the amido nitrogen in coordination. The manganese atom is assumed to be coordinated through the phenolate oxygen, the N(1) nitrogen, and the enethiolate sulfur atoms in a distorted tetrahedral structure; the fourth position is occupied by a bridging oxygen or sulfur atom. DTA and TGA curves reveal the absence of coordinated or lattice water. The compound decomposes at 320°C. It is insoluble in common organic solvents and is a nonelectrolyte in dimethylformamide [7].

$[Mn^{II}(C_8H_8N_4O_2S)(H_2O)_3]\cdot H_2O$ was prepared by the reaction of equimolar amounts of aqueous solutions of manganese(II) chloride and ligand 2. The reaction mixture was concentrated and cooled to yield crystals, which were recrystallized from hot methanol and dried in a vacuum desiccator. The magnetic moment of the buff solid is $\mu_{eff}=6.00$ μ_B at 25°C. In the IR spectrum of the complex (in KBr) the broad band at 3480 to 3435 cm^{-1} probably results from the vibrational modes of the attached water molecule, whereas a weak band at 3675 to 3660 cm^{-1} confirms coordinated water molecules. The N(3) and N(4) nitrogen atoms seem to be not involved in bond formation, but the downward shift (~30 cm^{-1}) of the combined NCS bending, NH_2 rocking, and NH_2 deformation mode (at ~1200 cm^{-1}) indicates bonding of the N(2) nitrogen atom. No phenolic ν(OH) band was observed in the complex spectrum. The ν(CS) ligand bands observed at 1310, 1270, 1065, 670, and 605 cm^{-1} were shifted to lower frequencies indicating coordination of manganese to sulfur. Bands between 520 to 465 and 298 to 270 cm^{-1} are assigned to the ν(Mn–N) and ν(Mn–S) vibrations. The formation of the 1:1 complex was also confirmed by conductometric titrations [3].

$[Mn^{II}(C_8H_9N_4O_2S)_2]\cdot 4H_2O$ and **$[Mn^{II}(C_8H_9N_4O_2S)(OH)(H_2O)_2]\cdot 3H_2O$.** The tetrahydrate precipitated from an ethanolic solution containing $MnCl_2\cdot 4H_2O$ and ligand 2 (mole ratio 1:3) upon raising the pH to 4.0 (by dropwise addition of ethanolic NaOH). The mixture was digested on a water bath, the solid washed with ethanol, and dried in a desiccator. The hydroxo complex was obtained by a similar procedure (mole ratio 1:2.5) at pH 10. The mixture was evaporated on a water bath and kept in a refrigerator for a couple of days to yield crystals. The product was washed with small amounts of ethanol, then with diethyl ether and air-dried in a desiccator. The magnetic moment of the complexes at room temperature are $\mu_{eff}=5.92$ and 5.86 μ_B, respectively. The IR spectral data of the complexes suggest that in both compounds the uninegative tridentate ligand is coordinated to the octahedral Mn^{II} via the phenolate oxygen, the N(2) nitrogen, and the sulfur atoms. Bands observed in the electronic spectra of the complexes in a noncoordinating solvent ($\tilde{\nu}_{max}$ in cm^{-1}), assigned to transitions from the $^6A_{1g}$ ground state and ligand field parameters (Dq and B in cm^{-1}), are as follows:

complex	$^6A_{1g}\rightarrow{}^4T_{1g}(G)$	$\rightarrow{}^4T_{2g}(G)$	$\rightarrow{}^4E_g(G)$	$\rightarrow{}^4T_{2g}(D)$	$\rightarrow{}^4E_g(D)$
$[Mn(C_8H_9N_4O_2S)_2]\cdot 4H_2O$	18200	23100	24758	27620	30150
$[Mn(C_8H_9N_4O_2S)(OH)(H_2O)_2]\cdot 3H_2O$	18050	22950	24250	27600	29980

complex	$\rightarrow{}^4T_{1g}(P)$	Dq	B	β
$[Mn(C_8H_9N_4O_2S)_2]\cdot 4H_2O$	32508	833.8	758	0.82
$[Mn(C_8H_9N_4O_2S)(OH)(H_2O)_2]\cdot 3H_2O$	32800	828.3	753	0.82

The octahedral transitions agree with the magnetic moment of a high-spin $Mn^{II}(d^5)$ state. The electrical conductivities of the complexes in 10^{-3}M aqueous solution, $\Lambda=38$ and 23 $cm^2\cdot\Omega^{-1}\cdot mol^{-1}$, are those of nonelectrolytes [4].

$[Mn^{II}(CH_6N_4S)_3](NO_3)_2$ and **$[Mn^{II}(CH_6N_4S)_3](ClO_4)_2$.** The nitrate precipitated on adding 20 mL of ethanol to a solution of $Mn(NO_3)_2\cdot 6H_2O$ (2.88 g) and ligand 1 (1.06 g) in 5 mL of dimethyl sulfoxide containing a few drops of 1N HNO_3 at 70°C. The perchlorate was prepared by dissolving $Mn(ClO_4)_2\cdot 6H_2O$ (3.6 g) and ligand 1 (1.06 g) in 30 mL of warmed ethanol containing a few drops of 2N $HClO_4$. The filtered solution was cooled to yield the complex. For both white compounds a magnetic moment of $\mu_{eff}=5.8$ μ_B results from susceptibility measurements (Faraday method) at room temperature. Vibration modes (in cm^{-1}) observed in the IR spectra of the complexes (in KBr) are:

compound	ν(NH) + ν(NH_2)	δ(NH_2)	δ(NH) + ν(CN)	ν(CN) + ν(CS)	δ(NH_2)	ν(CS) + δ(CNN) + ν(CN)
$[Mn(CH_6N_4S)_3](NO_3)_2$	3350 3270 3175	1645	1550	1310	1175	950
	3330 3225	1625 sh	1520		1120	
$[Mn(CH_6N_4S)_3](ClO_4)_2$	3340 3265 3170	1650	1555	1315	1175	965
	3300 3220 3120	1620 sh	1510			
CH_6N_4S	3300 3270 3205	1642	1530	1287	1142	930
		1622	1495			

The band assigments are deduced from a normal coordinate analysis, based on the crystal structure analysis of $Cd(CH_6N_4S)_2Cl_2$ reported in [5]. The positions of the ν(NH_2) and ν(NH) vibrations indicate coordination of one amino nitrogen atom to Mn^{II} and probably some ligand hydrogen-bonding to both the anions. The bidentate coordination of ligand 1 through nitrogen and sulfur atoms in the octahedral environment of Mn^{II} is additionally confirmed by sulfur K_β X-ray spectral data in the range of 152 to 166 eV. The complexes are soluble in water and slightly soluble in ethanol [6].

References:

[1] Buu-Hoi, N. P.; Loc, T. B.; Xuong, N. D. (Bull. Soc. Chim. France **1955** 694/7).
[2] Braibanti, A.; Mori, G.; Dallavalle, F.; Leporati, E. (Inorg. Chim. Acta **6** [1972] 106/12).
[3] Bhardwaj, S.; Ansari, M. N.; Kanshik, R. D.; Jain, M. C. (J. Indian Chem. Soc. **61** [1984] 484/6).
[4] Swami, M. P.; Jain, P. C.; Srivastava, A. K. (Roczniki Chem. **47** [1973] 2013/9).
[5] Bigoli, F.; Braibanti, A.; Manotti Manfredi, A. M.; Tiripicchio, A.; Tiripicchio Camelli, M. (Inorg. Chim. Acta **5** [1971] 392/6).
[6] Savel'eva, Z. A.; Larionov, S. V.; Nikolaev, A. V.; Nasonova, L. I.; Dolenko, G. N. (Izv. Sibirsk. Otd. Akad. Nauk SSSR Ser. Khim. Nauk **1977** No. 4, pp. 73/8; C.A. **87** [1977] No. 192993).
[7] Rao, P. V. K.; Satyanarayana, A.; Rao, R. S. (Proc. Natl. Acad. Sci. India A **57** III [1987] 292/8).

35.15 Complexes with Carbonotrithioic or -perthioic Acids and Derivatives

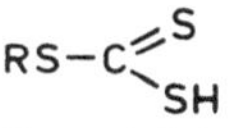

ligand 1 to 4

ligand	1	2	3	4
R	H	C_2H_5	C_4H_9	C_6H_5
formula	CH_2S_3	$C_3H_6S_3$	$C_5H_{10}S_3$	$C_7H_6S_3$

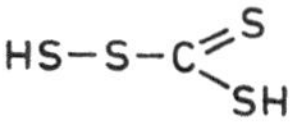

ligand 5 (= CH_2S_4)

Complexes in Solution. The formation of **$[Mn^{II}(CS_3)_2]^{2-}$** ions was studied colorimetrically at 536 nm in 10^{-4} to 10^{-5} M aqueous solution containing $BaCS_3$ and an Mn^{II} salt. The complex obeys the Lambert-Beer law but is oxidized rapidly. Therefore, the stability constant was not determined as in the case of the analogous Ni^{II}, Co^{II}, or Fe^{II} complexes [1]. A spectrophotometric determination of ultratrace amounts of Mn^{II} by its accelerating effect on the coloration of K_2CS_3 is reported in [2].

$Mn^{II}(C_3H_5S_3)_2$ and **$Mn^{II}(C_7H_5S_3)_2$** complexes were obtained by adding the potassium salt of ligand 2 or 4 to the aqueous solution of manganese(II) chloride at pH 7 under nitrogen. The yellow-brown precipitates were extracted into toluene directly from the reaction mixture. After drying, the complexes are insoluble in nonpolar solvents. The ESR spectra of the toluene solutions show at 285 K a six-line Mn^{II} pattern with g = 2.002 and a hyperfine coupling constant $A_0(^{55}Mn) = -87.6 \times 10^{-4}\ cm^{-1}$ for both complexes. In the extracted species manganese is probably six-coordinated by four coplanar sulfur atoms and two water molecules in apical positions. After several minutes polymerization and a decrease of the hyperfine splitting was observed. The complexes are readily oxidized in air to dark violet Mn^{III} compounds [3].

$[Mn^{II}L_2(NO)Cl]$ complexes with HL = ligands 2 to 4 were prepared by passing NO gas through toluene-water mixtures of MnL_3 compounds (obtained in situ by air oxidation of the MnL_2 complexes) in the presence of Cl^- ions. The complexes were extracted into toluene and obtained as dark brown solids after evaporation of the solvent under nitrogen [3]. The ESR spectra of the toluene solutions shows at 120 K a six-line pattern. The following parameters have been evaluated (K, T, and A in $10^{-4}\ cm^{-1}$, $K = -A_0 + (g_0 - 2.002)P$ with $P \approx 187 \times 10^{-4}\ cm^{-1}$).

complex	T in K	g_0	$g_{\parallel}$	$g_{\perp}$	$K(^{55}Mn)$	$T_{zz}(^{55}Mn)$	$A_{\parallel}(^{55}Mn)$	$A_{\perp}(^{14}N)$	c^2	Ref.
$[Mn(C_3H_5S_3)_2(NO)Cl]$	120	—	1.994	2.023	81.4	−72.4	—	−5.3	—	[3]
$[Mn(C_5H_9S_3)_2(NO)Cl]$	300 to 125	2.013	1.994	—	81.5	—	151.9	—	0.65	[4]
	125	—	1.994	2.023	81.4	−72.5	—	—	0.66	[5]
$[Mn(C_7H_5S_3)_2(NO)Cl]$	120	—	1.994	2.023	81.4	−72.5	—	−5.3	—	[3]
	125	—	1.994	2.023	81.4	−72.5	—	—	0.66	[6]

Interpretation of the results are identical with those given for Mn^{II} complexes with xanthic acids, see p. 132.

$[P(C_6H_5)_4]_2[Mn^{II}(CS_4)_2] \cdot dmf$ was prepared under anaerobic conditions: $[P(C_6H_5)_4][(S_5)Mn^{II}(S_6)]$ (see p. 92), 1 g dissolved in 30 mL of dimethylformamide, was combined with 20 mL of CS_2 and the mixture stirred for 5 min. To the red solution was added 10 mL of absolute ethanol and ether until the first signs of nucleation became apparent. Upon standing overnight the solution deposited orange-red crystal clusters. As shown for the corresponding unsolvated Ni^{II} species the anion is planar, and in the two bidentate CS_4^{2-} chelates the C=S groups are located *trans* to each other. The IR spectrum of the anion shows a ν(C=S) vibration mode at 905 cm^{-1}. Solutions of $[P(C_6H_5)_4]_2[Mn(CS_4)_2] \cdot dmf$ are extremely oxygen-sensitive, turning blue-green instantaneously upon exposure to air [7].

$[P(C_6H_5)_4]_2[Mn^{II}(CS_4)(S_6)]$. The complex was prepared by the same procedure as above, but with 40 mL CS_2. The solution of the orange-red complex is very air-sensitive. The electronic spectrum is somewhat different from the spectrum obtained for the $[Mn(CS_4)_2]^{2-}$ complex. Data are not given [7].

References:

[1] Bankovskii, Yu. A.; Jevinsch, A. (Latv. PSR Zinat. Akad. Vestis **1957** No. 3, pp. 123/31; C.A. **1958** 5193).
[2] Hiraoka, S.; Shinonaga, T.; Aso, T.; Yamamoto, D. (Bunseki Kagaku **33** [1984] 496/7; C.A. **102** [1985] No. 16674).
[3] Jezierski, A. (J. Mol. Struct. **115** [1984] 11/4).
[4] Jeżowska-Trzebiatowska, B.; Jezierska, J.; Jezierski, A. (Magn. Resonance Relat. Phenom. Proc. 20th Congr. AMPERE, Tallinn 1978 [1979], p. 301; C.A. **93** [1980] No. 123135).
[5] Jeżowska-Trzebiatowska, B.; Jezierski, A. (J. Mol. Struct. **46** [1978] 197/202).
[6] Jezierski, A.; Jeżowska-Trzebiatowska, B. (Nouv. J. Chim. **4** [1980] 599/602).
[7] Coucouvanis, D.; Patil, P. R.; Kanatzidis, M. G.; Detering, B., Baenziger, N. C. (Inorg. Chem. **24** [1985] 24/31).

35.16 Complexes with Carbothioic Acids

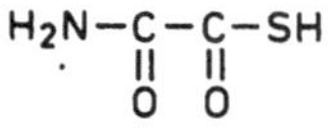
ligand 1
(= $C_2H_3NO_2S$)

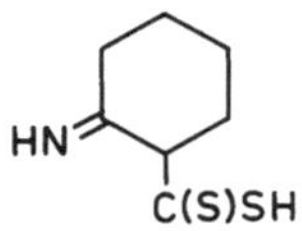
ligand 2
(= $C_7H_{11}NS_2$)

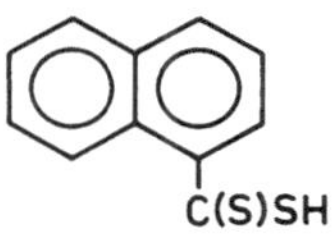
ligand 3
(= $C_{11}H_8S_2$)

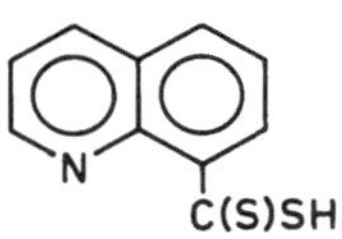
ligand 4
(= $C_{10}H_7NS_2$)

$Mn^{II}(C_7H_{10}NS_2)_2$ was prepared by dropwise addition of ligand 2 dissolved in ethanol to an ethanol solution of Mn^{II} halide (mole ratio 1:2). The reddish brown precipitate, obtained after stirring the solution for ~5 h and concentration, was washed with ethanol and dried in vacuum. The complex has a magnetic moment of 5.9 μ_B. IR and electronic spectral data, as well as ^1H NMR studies, are concordant with a bidentate coordination of the $-CS_2$ group to manganese [1].

$Mn^{II}(C_{11}H_7S_2)_2$ precipitated upon mixing aqueous solutions of a manganese(II) salt and a sodium salt of ligand 3 (0.01 M) at pH 0 to 3. The yellow-pink complex is readily soluble in pentanol, ethylacetate, chloroform, 1,2-dichloroethane, benzene, is soluble in ether, carbon tetrachloride, is slightly soluble in heptane, and is insoluble in decane [2].

$Mn^{II}(C_{10}H_6NS_2)_2$ was prepared by the reaction of stoichiometric amounts of manganese(II) nitrate and the tetraethylammonium salt of ligand 4 in aqueous solution. The complex was dried at 105°C for 2 h. The compound was tested as an ion-selective electrode in a water-chloroform mixture. $Mn(C_{10}H_6NS_2)_2$ is soluble in water, nitrobenzene, chloroform, and 1,2-dichloroethane [3].

$Mn^{II}(C_2H_2NO_2S)_2 \cdot 2H_2O$ was obtained on dropwise addition of an Mn^{II} salt (0.05 mol) dissolved in 70 mL H_2O to the sodium salt of ligand 1 (0.05 mol) in 25 mL H_2O. The crystals were washed with water. The IR spectrum of the complex in KBr exhibits characteristic bands (not assigned) at the following wavenumbers (shifts in comparison to the Ba salt of the ligand are given in parentheses): 3430(30), 3327(12), 1677(26), 1630(40), 1450(35), 1382(6), 1256(24), 942(14) cm^{-1}. The amino nitrogen atom of the ligand coordinates to Mn^{II}. DTA data show the release of coordinated water molecules at 133°C; decomposition starts at 192°C [4].

References:

[1] Sharma, Ram Rattan; Singh, Bijendra; Kapoor, Ramesh N. (Transition Metal Chem. [Weinheim] **12** [1987] 431/2).
[2] Janson, E.; Gertner, M.; Bankovski, J. (Uch. Zap. Latv. Gos. Univ. **57** [1964] 53/7; C.A. **63** [1965] 12296).
[3] Lebedeva, O. A.; Janson, E. (Latv. PSR Zinat. Akad. Vestis Kim. Ser. **1986** No. 4, pp. 423/7; C.A. **105** [1986] No. 180327).
[4] Kibbel, H. V.; Lampe, B.; Schütt, S.; Tesch, F. (Z. Anorg. Allgem. Chem. **480** [1981] 186/92).

35.17 Complexes with Thioamides or Thiohydrazides and Related Compounds

35.17.1 With N, N-Dimethylmethanethioamide

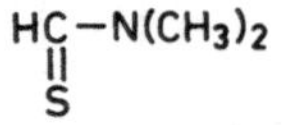

($=C_3H_7NS$)

$Mn^{II}(C_3H_7NS)_4(ClO_4)_2$ was prepared according to a procedure given in [2] by stirring hydrated manganese(II) perchlorate with triethyl orthoformate for 2 h under dry nitrogen, followed by the addition of dry ligand. Stirring of the suspension was continued for 1 h. The precipitate was separated under nitrogen, washed with dry diethyl ether containing ca. 10% triethyl orthoformate, and dried in vacuum [1, 2]. The visible spectrum of the complex in neat dimethylmethanethioamide reveals formation of the hexacoordinated **$[Mn(C_3H_7NS)_6]^{2+}$** ion. The ligand exchange reaction, $[Mn(C_3H_7NS)_6]^{2+} + {}^*C_3H_7NS \rightleftharpoons [Mn(C_3H_7NS)_5{}^*(C_3H_7NS)]^{2+} + C_3H_7NS$, was studied kinetically by high-pressure ^1H NMR spectroscopy in the 240 to 377 K temperature range and 0.1 to 200 MPa pressure range. The exchange rate constant at 298 K, $k_{ex} = 3.9(\pm 0.8) \times 10^7\ s^{-1}$, and the activation energy, $\Delta G^* = 29.7 \pm 0.5$ kJ/mol, were determined. The positive value of the activation volume at 302 K, $\Delta V^* = 11.5 \pm 1.5\ cm^3/mol$, indicates an expanded transition state; i.e., a dissociative reaction mechanism. This value contrasts with the negative ΔV^* values (-5 to $-7\ cm^3/mol$) obtained previously for exchange reactions of $[MnL_6]^{2+}$ complexes with $L = H_2O$, CH_3OH, or CH_3CN in the corresponding solvent. For these reactions an associative reaction mechanism was discussed. Thus in the case of N, N′-dimethylmethanethioamide a mechanistic change from an associative mechanism (I_a) to a dissociative mechanism (I_d) was suggested because of the increased molar volume [1].

References:

[1] Fielding, L.; Moore, P. (J. Chem. Soc. Chem. Commun. **1988** 49/52).
[2] McAteer, C. H.; Moore, P. (J. Chem. Soc. Dalton Trans. **1983** 353/7).

35.17.2 With Pyridinecarbothioamides or Derivatives

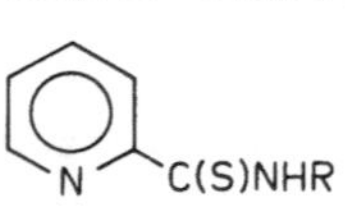

ligand 1 with R = H ($=C_6H_6N_2S$)
ligand 2 with $R = C_6H_5$ ($=C_{12}H_{10}N_2S$)
ligand 3 with $R = C_6H_4CH_3$-2 ($=C_{13}H_{12}N_2S$)

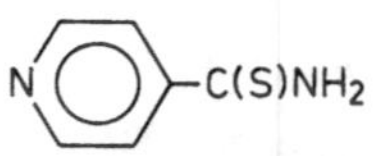

ligand 4 ($=C_6H_6N_2S$)

$Mn^{II}(C_{13}H_{11}N_2S)_2$ was prepared by the reaction of ligand 3 dissolved in ethanol with manganese(II) chloride in aqueous or ethanolic solution. The precipitate was freed from excess ligand by washings with ethanol and benzene [1, 2]. The light yellow complex melts at 130°C. In the IR spectrum the ν(NH) band position is hidden; the disappearance of the free ligand ν(CS) band (at 1360 cm^{-1}) indicates bonding to manganese in the enethiol form. The ESR spectrum of the solid complex reveals g = 2.013 and A = 480 G at room temperature [2]; g = 2.003 and A = 90 G were obtained from frozen solution in dimethylformamide [1]. The complex is soluble in ethanol, acetone, dimethylformamide, and insoluble in benzene, chloroform, carbon tetrachloride. The electronic spectra of ethanol solutions of both $Mn(C_{13}H_{11}N_2S)_2$ and the free ligand show two maxima at ~34000 and ~42000 cm^{-1} [2].

$[Mn(C_6H_6N_2S)_2Cl_2]$ precipitates immediately from hot ethanol solution containing $MnCl_2$ and ligand 4. The dark yellow complex was washed with ethanol and ether and dried in vacuum. A magnetic moment of 5.90 μ_B was found at ambient temperature. The IR spectrum of the complex in Nujol or KBr shows the characteristic bands of ν(CS) at 795 cm^{-1} and of pyridine ring vibrations at 675 (in plane) and 470 cm^{-1} (out of plane). Bands at 380, 325, and 300 cm^{-1} were assigned to ν(Mn–N), ν(Mn–S), and ν(Mn–Cl), respectively. The negative shift, $\Delta\nu(CS) = -15\ cm^{-1}$, and positive shifts of the ligand pyridine ring vibrations (55 and 66 cm^{-1}) observed upon complexation, indicate the bidentate nature of the ligand with bonding sites being the thiocarbonyl sulfur and the pyridine nitrogen. A polymeric structure is assumed. The compound does not melt below 360°C. It is insoluble in water or common organic solvents, slightly soluble in dimethylformamide or dimethyl sulfoxide and behaves as a nonelectrolyte in dimethylformamide [7].

$[Mn^{II}(C_6H_6N_2S)_2I_2] \cdot H_2O$. The monohydrate was formed by reaction of an ethanol solution of excess MnI_2 with ligand 1. The reaction mixture was refluxed for 10 min and then set aside to cool for several days, after which orange crystals separated. The hydrate was washed with a little cold ethanol then dried over P_4O_{10} in vacuum. The magnetic moment of $[Mn^{II}(C_6H_6N_2S_2)I_2]$ in the solid state is $\mu_{eff} = 5.95\ \mu_B$ at 20°C [3]. The IR spectrum displays bands characteristic of ν(NH) at 3150, δ(NH) at 1555, and ν(CS) at 1260 and 800 cm^{-1}. The ν(NH), δ(NH), and one ν(CS) band are shifted to lower wavenumbers in comparison to the corresponding free ligand bands at 3330, 1570, and 830 cm^{-1}. The spectral results suggested that the ligand is bonded to Mn^{II} via the pyridine and amine nitrogen atoms [4]. The complex in ethanol shows an absorption maximum at 377 nm. It behaves as a nonelectrolyte in dimethylformamide [3].

$Mn^{III}(C_{12}H_9N_2S)_3$ was prepared according to the method described in [5] for other complexes with ligand 2 (i.e., reaction of the metal salt in aqueous alkaline solution). The complex sensitizes the photo-oxidative degradation of polyisobutylene. The polymer degradation is initiated by the photochemical decomposition of the complex at 254 nm to form a free ligand radical and $Mn^{II}(C_{12}H_9N_2S)_2$ [6].

References:

[1] Solozhenkin, P. M.; Semenov, E. V.; Baratova, Z. R.; Rukhadse, E. G. (Tezisy Dokl. 12th Vses. Chugaevskoe Soveshch. Khim. Kompleksn. Soedin., Novosibirsk 1975, Vol. 2, pp. 330/1; C.A. **85** [1976] No. 186073).

[2] Solozhenkin, P. M.; Baratova, Z. R.; Semenov, E. V.; Voitkovskii, Yu. B.; Sidorov, S. V.; Rukhadse, E. G. (Izv. Akad. Nauk Tadzh. SSR Otd. Fiz. Mat. Geol. Khim. Nauk **1977** No. 4, pp. 35/41; C.A. **89** [1978] No. 33735).

[3] Sutton, G. J. (Australian J. Chem. **19** [1966] 2059/68).

[4] Sutton, G. J. (Australian J. Chem. **22** [1969] 2475/8).

[5] Bähr, G.; Scholz, E. (Z. Anorg. Allgem. Chem. **299** [1959] 281/91).

[6] Chandra, R.; Handa, S. P. (Polym. Photochem. **3** [1983] 391/406).

[7] Singh, B. (Syn. React. Inorg. Metal-Org. Chem. **17** [1987] 457/68).

35.17.3 With 4,4-Dimethyl-2,6-dioxo-N-phenyl-1-cyclohexanecarbothioamide

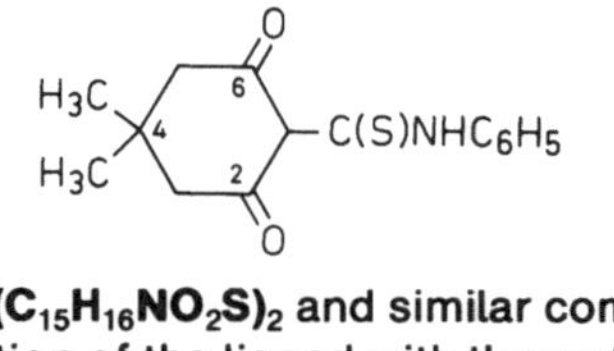

$(= C_{15}H_{17}NO_2S)$

$Mn^{II}(C_{15}H_{16}NO_2S)_2$ and similar complexes of Zn^{II}, Cd^{II}, Ni^{II}, Co^{II}, and Cu^{II} have been prepared by reaction of the ligand with the metal acetate (10% excess) in ethanol or methanol [1]. The IR data indicate that the metal replaces the hydrogen atom of the enolic form of the ligand and forms a chelate ring involving the sulfur atom [1, 2]. The strength of the metal-ligand bond increases as follows: Mn<Cd<Cu<Zn<Co<Ni. The composition and stability was studied by TG and DTA. The manganese complex melts at 220°C, according to the TG analysis. Steps of transition were observed at 270 and 355°C with the corresponding activation energies and orders of reaction being E_a=39 kcal/mol (n=1.35) and E_a=33 kcal/mol (n=0.8), respectively [3].

References:

[1] Fişel, S.; Mocanu, R. (Analele Stiint. Univ. Al. I. Cuza Iasi 1c [2] **15** [1969] 131/6).
[2] Mocanu, R.; Fişel, S.; Gabe, I. (Lucr. 3rd Conf. Natl. Chim. Anal., Bucharest, Rom., 1971, Vol. 3, pp. 187/90; C.A. **77** [1972] No. 11589).
[3] Mocanu, R.; Biró, A.; Fişel, S. (Analele Stiint Univ. Al. I. Cuza Iasi 1c [2] **17** [1971] 13/8; C.A. **75** [1971] No. 147331).

35.17.4 With N-Phenyl-1H-benzimidazole-2-carbothioamide

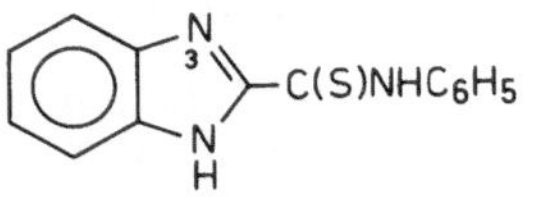

(= $C_{14}H_{11}N_3S$)

$Mn^{II}(C_{14}H_{10}N_3S)_2$ was prepared by refluxing a mixture of manganese(II) acetate with the ligand (mole ratio 1:2) in methanol. The white microcrystalline precipitate was washed several times with methanol and ether; yield is 25%. The IR spectra of the $M(C_{14}H_{10}N_3S)_2$ complexes (M=Zn, Mn, Ni, Co, Cu) suggest that the enethiol form of the bidentate ligand is coordinated to manganese through the sulfur atom and the N(3) nitrogen atom of the imidazole ring. $Mn(C_{14}H_{10}N_3S)_2$ melts at 311°C with decomposition. It is insoluble in water and most organic solvents, insoluble in mineral acids and alkaline medium, but soluble (with decomposition) in concentrated sulfuric acid.

Reference:

Bovykin, B. A.; Romanovskaya, L. G.; Ranskii, A. P.; Zanina, I. A.; Artyukhova, E. P. (Vopr. Khim. Khim. Tekhnol. No. 73 [1983] 22/4; C.A. **101** [1984] No. 203187).

35.17.5 With Dithiooxamide or Its Tetraethyl Derivative

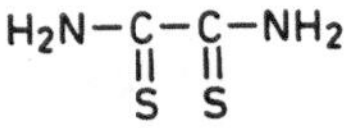

ligand 1 (= $C_2H_4N_2S_2$)

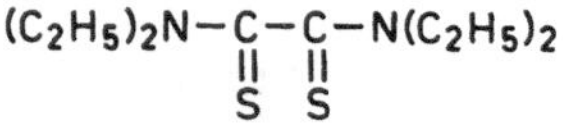

ligand 2 (= $C_{10}H_{20}N_2S_2$)

Complexes in Solution. The stepwise stability constants for the formation of complexes $Mn(C_2H_3N_2S_2)^+$ and $Mn(C_2H_3N_2S_2)_2$ with the monodeprotonated ligand 1 have been determined by potentiometric pH titration in 3:1 dioxane-water mixture under oxygen-free conditions. At 20°C and I=0.1 M($NaClO_4$) the stability constants are log K_1=6.29±0.07 and log K_2=7.02±0.07. The uncommon order, log K_2>log K_1, may be due to the π acceptor ability of the sulfur donor atoms of the ligand [1].

[Mn^{II}(C$_{10}$H$_{20}$N$_2$S$_2$)$_3$](ClO$_4$)$_2$ was prepared by dehydrating Mn(ClO$_4$)$_2$·6H$_2$O (5 mmol) with 10 mL of 2,2-dimethoxypropane. To this mixture ligand 2 was added (30 mmol dissolved in 20 mL of boiling acetone). The orange precipitate was washed with diethyl ether and dried over Drierite [2]. The complex was also obtained by mixing ligand 2 in hot (70 to 80°C) glacial acetic acid and manganese(II) perchlorate in 2,2-dimethoxypropane (mole ratio 4.5:1). It separated on cooling the mixture [3].

The magnetic moment is $\mu_{eff} = 5.97\ \mu_B$ at 25°C (Faraday method) [2], 5.78 μ_B from susceptibility measurements at room temperature (Gouy method). The Curie-Weiss law is obeyed over the temperature range 300 to 100 K with the Weiss constant, $\Theta = 3$ K [3]. The IR spectrum of the complex in Nujol mull shows bands due to ionic ClO$_4$ at 1100 and 625 cm^{-1} [2]. The spectrum in a KBr disk shows the characteristic bands: ν(CN) at 1544, ν(CS) at 858, and ligand bands at 485 and 407 cm^{-1}. The Nujol mull spectrum shows far-IR bands: ν(Mn–S) at 254, δ(Mn–S) at 175, and ligand bands at 374, 348, 298, and 80 cm^{-1}. The electronic absorption spectrum of a dimethylformamide solution of the complex shows a maximum at 18520 cm^{-1}, whereas the solid shows a maximum at 22220 cm^{-1}. The bands are assigned to the high-spin Mn^{II} octahedral transitions $^6A_{1g} \rightarrow {}^4T_{1g}$(G) and $\rightarrow {}^4T_{2g}$(G), respectively. Analysis of the spectral data yields the parameters Dq = 880 cm^{-1}, B = 770 cm^{-1} [3]. The spectral results suggest that the neutral bidentate ligand is bonded to the octahedral Mn^{II} via the sulfur atoms [2, 3]. The complex is a 1:2 electrolyte in 10^{-3} M nitromethane solution; the electrical conductivity is $\Lambda = 146$ cm$^2 \cdot \Omega^{-1} \cdot$mol^{-1} at 25°C [2].

[Mn^{II}(C$_2$H$_4$N$_2$S$_2$)$_2$X$_2$]. Complexes with X = Cl or NCS were prepared by the reaction of ethanol solutions of ligand 1 and a manganese(II) salt in the molar ratio 2:1. The reaction mixture was refluxed for 15 to 60 min and then concentrated by evaporation in air to give crystalline compounds. The product was washed with ethanol and then diethyl ether, and dried in a vacuum desiccator. The magnetic moments of the solids are $\mu_{eff} = 5.90$ and 5.85 μ_B for X = Cl and NCS, respectively. The IR spectra of the complexes in Nujol mulls show the following characteristic bands: ν(NH) at 3300, ν(CS) at 1200 cm^{-1}; for [Mn(C$_2$H$_4$N$_2$S$_2$)$_2$Cl$_2$]; ν(NH) at 3150, ν(CS) at 1190, and ν(NCS) at 2090 cm^{-1} for [Mn(C$_2$H$_4$N$_2$S$_2$)$_2$(NCS)$_2$] with N-bonded NCS$^-$ ions. The electronic absorption spectra of the complexes show high-spin Mn^{II} bands at 320, 380, 410, 450, 460 nm and ligand bands at 215 and 312 nm. Spectral results suggest that the octahedral Mn^{II} is bonded to the ligand via the nitrogen and sulfur atoms of the neutral bidentate ligand. The complexes are nonelectrolytes in acetone [4].

References:

[1] Burger, K.; Szántó-Horváth, G.; Papp-Molnár, E. (Acta Chim. [Budapest] 71 [1972] 127/36).
[2] Hart, D. M.; Rolfs, P. S.; Kessinger, J. M. (J. Inorg. Nucl. Chem. 32 [1970] 469/75).
[3] Peyronel, G.; Pellacani, G. C.; Pignedoli, A.; Benetti, G. (Inorg. Chim. Acta 5 [1971] 263/9).
[4] Mahapatra, B. B.; Panda, A.; Mishra, N. C.; Pujari, S. K. (J. Indian Chem. Soc. 58 [1981] 75/6).

35.17.6 With Thiohydroxamic Acids

$$R'-\underset{\underset{S}{\|}}{C}-N(OH)R \quad (= HL)$$

ligand	1	2	3	4	5	6
R	H	H	CH$_3$	CH$_3$	CH$_3$	C$_6$H$_5$
R'	C$_6$H$_5$	CH$_3$O—◯—	C$_6$H$_5$	CH$_3$O—◯—	C$_6$H$_5$CH$_2$	C$_6$H$_5$
formula ...	C$_7$H$_7$NOS	C$_8$H$_9$NO$_2$S	C$_8$H$_9$NOS	C$_9$H$_{11}$NO$_2$S	C$_9$H$_{11}$NOS	C$_{13}$H$_{11}$NOS

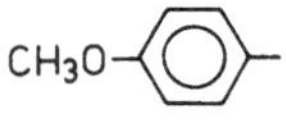

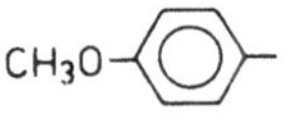

Complexes in Solution. The stability constants of $Mn(C_{13}H_{10}NOS)^+$ and $Mn(C_{13}H_{10}NOS)_2$ complexes with ligand 6, determined potentiometrically (glass electrode) in 1:1 dioxane-water mixture (v/v) at 25°C are log $K_1=5.20$, log $K_2=4.5$ [1]; and log $K_1=6.12$, log $K_2=6.6$ obtained in 3:1 dioxane-water mixture (v/v) at 30°C [2]. A yellow-green precipitate formed on addition of ligand 6 to aqueous Mn^{II} salt solution, and was extracted into $CHCl_3$ [1]. Extraction from buffered aqueous solution at ionic strength 0.1 M ($NaNO_3$) yields $Mn(C_{13}H_{10}NOS)_2$ in the $CHCl_3$ phase. The extraction parameters are $pH_{1/2}=6.30$ and log $K_{ex}=-7.8$ [3]. Manganese(II) can be determined colorimetrically with ligand 2 by measuring the absorption of the chloroform phase at 450 nm ($\varepsilon=1.06\times10^4$ L·mol^{-1}·cm^{-1}). During the extraction at pH 5.6 to 6.5 the complex $Mn^{III}(C_8H_8NO_2S)_3$ is formed, which shows an absorption maximum at 390 nm ($\varepsilon=1.253\times10^4$ L·mol^{-1}·cm^{-1}) and the extraction parameter, log $K_{ex}=-2.37$ [4]. The IR spectral data indicate that ligand 2 is coordinated by the deprotonated hydroxyl oxygen and the sulfur atoms [5].

$Mn^{II}(C_8H_8NO_2S)_2$ was prepared by the reaction of ligand 2 and manganese(II) sulfate in the presence of excess potassium acetate. The voluminous canary yellow precipitate was washed repeatedly with water and then ethanol. The magnetic moment of the solid is $\mu_{eff}=5.89$ μ_B. The complex is slightly soluble in alcohol, insoluble in nonpolar solvents [6].

$Mn^{II}(C_7H_6NOS)_2\cdot0.5H_2O$ was prepared by the reaction of an Mn^{II} salt with the sodium salt of ligand 1 in buffered aqueous solution (pH 6). The green precipitate was washed with water. It was purified by dissolution in a large amount of ethanol at room temperature and concentration in vacuum, followed by precipitation with petroleum ether. The complex decomposes at ~200°C. The compound is very slightly soluble in dilute ammonia, sparingly soluble in methanol, chloroform, ethyl acetate, and soluble in acetone or pyridine [7].

$[Mn^{III}L_3]$ Compounds (HL = ligands 2 to 5). Complexes with ligand 2, $[Mn(C_8H_8NO_2S)_3]$ [6], or with ligand 3, $[Mn(C_8H_8NOS)_3]$ [8], were prepared by the reaction of the corresponding thiohydroxamic acid in ethanol with a solution of $Mn(CH_3COO)_3\cdot2H_2O$ in acetic acid. The green precipitates were isolated, dissolved in chloroform and fractionally crystallized by the addition of petroleum ether. The products were dried in vacuum over P_4O_{10} [6, 8]. According to [9], $[Mn(C_8H_8NOS)_3]$ and the complexes with ligand 4 or 5, $[Mn(C_9H_{10}NO_2S)_3]$ and $[Mn(C_9H_{10}NOS)_3]$, respectively, were prepared by mixing stoichiometric amounts of $Mn(acac)_3$ with the corresponding ligand in ethanol. The reaction mixture was stirred at room temperature for 30 min and then evaporated to dryness. The residue was dissolved in benzene and then petroleum ether (60 to 80°C) was added slowly with stirring. The green precipitates that formed were isolated. The process of dissolution and precipitation was repeated three times to obtain the pure compounds [9].

$[Mn(C_8H_8NOS)_3]$, recrystallized from chloroform-ethanol solution, was subjected to single-crystal X-ray structure analysis. The refinement, using 3452 reflections with $F^2>3\sigma(F^2)$, gave $R=0.035$ and $R_w=0.038$. The crystals are monoclinic, space group $P2_1/n-C^5_{2h}$ (No. 14) with a = 11.604(3), b = 12.044(3), c = 18.990(6) Å, β = 94.87(2)°; Z = 4. The complex has a six-coordinate *cis* structure with the uninegative bidentate ligand 3 coordinating via the oxygen and sulfur atoms. The coordination sphere shows a considerable tetragonal distortion (10.2%). The axial bond lengths are Mn–O = 2.132, Mn–S = 2.584 Å, the equatorial ones Mn–O = 1.958(2), 1.941(2), Mn–S = 2.324, 2.354 Å. A comparison with similar complexes of Fe^{III} or Cr^{III} shows that the Mn^{III} Jahn-Teller distortion in the axial direction is similar to that for the $Fe^{III}(d^5)$ complex, and in the equatorial direction is similar to that for the $Cr^{III}(d^3)$ complex. The mean ligand bond distances in the chelate ring of $[Mn(C_8H_8NOS)_3]$ are N–O = 1.349(8), C–S = 1.707(6), N–C = 1.306(4) Å. The calculated and measured densities are D = 1.39 and 1.38(1) g/cm³, respectively [8]. The magnetic moment of the solid dark green $[Mn(C_8H_8NO_2S)_3]$ is $\mu_{eff}=4.74$ μ_B. The complex is soluble in nonpolar solvents [6].

For numbers, substituents, and formulas of ligands, see p. 215

$[Mn^{IV}L_3]PF_6$ complexes with HL = ligands 3 to 5 were prepared by passing dry chlorine gas through the acetonitrile solution of the corresponding MnL_3 complex until the green color changed to red-brown. The complexes precipitated upon addition of an excess of NH_4PF_6 and evaporation at room temperature. The dark compounds were washed thoroughly with water, dried in vacuum over P_4O_{10}, and recrystallized from acetonitrile. The magnetic moments of the solid at 298 K and the band maxima with extinction coefficients ε (in parentheses) of the electronic spectra in acetonitrile solution are:

complex	μ_{eff} in μ_B	λ_{max} in nm (ε in $L\cdot mol^{-1}\cdot cm^{-1}$)			
$[Mn(C_8H_8NOS)_3]PF_6$	4.08	575 (2370)	410 sh (4800)	285 sh (16940)	255 (20770)
$[Mn(C_9H_{10}NO_2S)_3]PF_6$	4.02	580 sh (3520)	410 (8980)	348 (45370)	—
$[Mn(C_9H_{10}NOS)_3]PF_6$	4.09	570 (1250)	460 sh (1100)	312 sh (6850)	260 (16140)

The ESR X-band spectra of the complexes in powder form, as well as in dilute frozen acetonitrile solution, show a strong signal at g = 2 and a weak signal at low field, g = 4. The strong signal shows manganese nuclear hyperfine structure with A ≈ 100 G. The spectra are consistent with a small zero-field splitting. The g values are listed in the following table together with the potentials E (in V) obtained from cyclic voltametry measurements:

complex	g at 298 K powder		g at 77 K powder		g at 77 K solution		E $Mn^{IV}\rightarrow Mn^{III}$	E $Mn^{III}\rightarrow Mn^{II}$
$[Mn(C_8H_8NOS)_3]PF_3$	2.013	4.441	1.991	4.386	1.983	4.330	0.31	−0.21
$[Mn(C_9H_{10}NO_2S)_3]PF_6$	2.007	4.361	2.009	4.400	1.986	4.226	0.21	−0.29
$[Mn(C_9H_{10}NOS)_3]PF_6$	2.009	4.226	2.016	4.236	2.009	4.324	0.21	−0.32

In acetonitrile solution at 298 K with $N(C_2H_5)_4ClO_4$ as supporting electrolyte the complexes show two one-electron reductions versus SCE. The first step for $Mn^{IV}\rightarrow M^{III}$ is ideally reversible, the second for $M^{III}\rightarrow M^{II}$ is somewhat less reversible. E is calculated as the average of anodic and cathodic peak potentials. In acetonitrile the complexes behave as 1:1 electrolytes [9].

References:

[1] Brydon, G. A.; Ryan, O. E. (Anal. Chim. Acta **35** [1966] 190/4).
[2] Dietzel, R.; Thomas, P. (Z. Anorg. Allgem. Chem. **381** [1971] 214/8).
[3] Uhlemann, E.; Maack, B.; Raab, M. (Anal. Chim. Acta **116** [1980] 403/6).
[4] Skorko-Trybuła, Z.; Debska, B. (Chem. Anal. [Warsaw] **13** [1968] 557/73).
[5] Skorko-Trybuła, Z. (Prace Nauk. Politech. Warsz. Chem. No. 8 [1973] 7/99; C.A. **83** [1975] No. 71070).
[6] Cambi, L.; Bacchetti, T.; Paglia, E. (Rend. Ist. Lombardo Sci. Lettere **90** [1956] 577/93; C.A. **1958** 12643).
[7] Nagata, K.; Mizukami, S. (Chem. Pharm. Bull. [Tokyo] **15** [1967] 61/9).
[8] Freyberg, D. P.; Abu-Dari, K.; Raymond, K. N. (Inorg. Chem. **18** [1979] 3037/43).
[9] Pal, S.; Ghosh, P.; Chakravorty, A. (Inorg. Chem. **24** [1985] 3704/6).

35.17.7 With 2-Furancarbothiohydrazides or Related Compounds

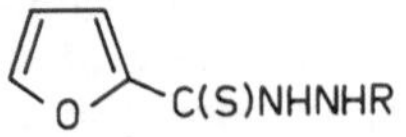

ligand	R	formula
1	H	$C_5H_6N_2OS$
2	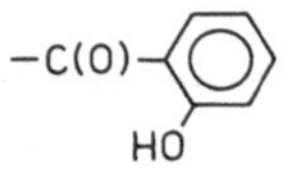	$C_{12}H_{10}N_2O_3S$
3	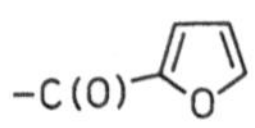	$C_{10}H_8N_2O_3S$
4	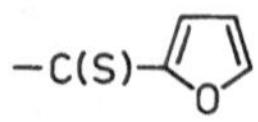	$C_{10}H_8N_2O_2S_2$

ligand	R'	formula
5	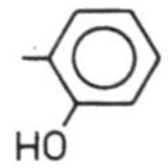	$C_{12}H_{10}N_2O_2S$
6	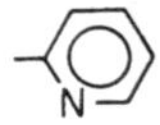	$C_{11}H_9N_3OS$

$Mn^{II}(C_{12}H_9N_2O_3S)_2$ was obtained by digesting for 15 min the mixture of a methanolic solution of hydrated manganese(II) chloride (1 mmol) and the stoichiometric amount of ligand 2 dissolved in 20 mmol methanolic potassium hydroxide. The light green complex was washed with hot ethanol and dried in vacuum. It does not melt below 300°C. The magnetic moment, resulting from susceptibility measurements (Faraday method) at room temperature, is $\mu_{eff} = 5.95\ \mu_B$. The absence of the ν(OH) ligand band and the presence of a ν(CO) band in the IR spectrum of the complex (in Nujol) indicate involvement of the phenolic oxygen and non-participation of the carbonyl oxygen in bonding with the manganese atom. Negative shifts (30 and 40 cm^{-1}) in ν(NH) + δ(CN) and ν(CN) + δ(NH) vibration modes may arise from coordination of one of the hydrazinic nitrogen and the sulfur atoms. Bands at 380, 320, and 275 cm^{-1} are assigned to ν(Mn–S), ν(Mn–O), and ν(Mn–N) vibrations. Thus, the ligand is assumed to have a uninegative tridentate function in the complex. The X-band ESR spectra of the complex in solid state and in chloroform solution at liquid nitrogen temperature are plotted. The solution spectrum reveals well-resolved signals at $g \approx 2$. A six-line pattern is suggestive of symmetrical coordination with near-octahedral geometry. The complex is soluble in water and common organic solvents, slightly soluble in methanol, ethanol, soluble in dimethylformamide and dimethyl sulfoxide.The electrical conductivity in dimethylformamide demonstrates the non-ionic nature of the complex [1].

$Mn^{II}(C_{12}H_9N_2O_2S)_2$ was prepared by adding the ethanol solution (20 mL) of ligand 5 (20 mmol) to the aqueous solution (20 mL) of manganese(II) acetate (10 mmol) in the presence of 0.5 g sodium acetate. The mixture was digested on a water bath for ~30 min. The brown precipitate was washed with water, then with ethanol, and dried in vacuum. The complex melts at 210°C and the magnetic moment is $\mu_{eff} = 6.02\ \mu_B$ (Faraday method, temperature not given). The IR spectrum of the complex in Nujol shows the noninvolvement of the phenolic oxygen atom in bonding. The absence of the ν(NH) band and a new band at 720 cm^{-1}, due to the ν(CS) vibration mode, indicate coordination of the ligand through the enethiol sulfur atom. The shifts of ν(C=N) and ν(N–N) vibrations upon complexation suggest additional bonding through the azomethine nitrogen atom. The complex is insoluble in water, soluble in dimethylformamide, dimethyl sulfoxide. In dimethylformamide it behaves as a nonelectrolyte. Antifungal studies show that the complex inhibits the cultures of Rhizopus oryzae more than the free ligand [2].

$Mn^{II}(C_{11}H_8N_3OS)_2$. The complex with ligand 6 was obtained by refluxing the ethanolic suspension of $[Mn(C_5H_5N_2OS)_2(H_2O)_2]$, see p. 219, with an excess of 2-pyridinecarbaldehyde for about 3 h. The solid was washed with ethanol and dried in vacuum. The violet complex does not melt below 300°C. The magnetic moment is $\mu_{eff} = 6.02\ \mu_B$ (Faraday method, temperature not given). The IR data suggest bonding of the ligand through the azomethine nitrogen, the pyridine nitrogen, and the enethiol sulfur atoms. Octahedral geometry at Mn^{II} is proposed. The complex is insoluble in water but slightly soluble in dimethylformamide and dimethyl sulfoxide. It behaves as a nonelectrolyte in dimethylformamide [3].

$[Mn^{II}(C_5H_5N_2OS)_2(H_2O)_2]$ was prepared by the reaction of ligand 1, dissolved in the minimum amount of ethanolic potassium hydroxide, and an aqueous solution of hydrated manganese(II) chloride (mole ratio 2:1). The brown precipitate was washed with water, ethanol, and ether, and dried in vacuum. The complex recrystallized from ethanol does not melt up to 300°C. The magnetic moment of the solid, $\mu_{eff} = 5.65\ \mu_B$ (Faraday method, temperature not given), suggests octahedral geometry for Mn^{II}. The IR spectral results show that the mesomeric form of the uninegative bidentate ligand is bonded to Mn^{II} via the enethiol sulfur and the terminal nitrogen atoms. The complex is insoluble in water, soluble in dimethylformamide and dimethyl sulfoxide. In dimethylformamide it behaves as a nonelectrolyte [4]. By refluxing the ethanolic suspension of the complex with an excess of 2-pyridinecarbaldehyde a Schiff base complex of composition $Mn(C_{11}H_8N_3OS)_2$ was formed, see p. 218 [3].

$[Mn^{II}(C_{10}H_6N_2O_3S)(H_2O)_2]$ and **$[Mn^{II}(C_{10}H_6N_2O_2S_2)(H_2O)_2]$** were prepared by the reaction of hydrated manganese(II) acetate and a methanol solution of an equimolar amount of ligand 3 or 4, respectively, in the presence of sodium acetate. The precipitates were washed several times with water, then with ethanol, and dried in vacuum. The solids do not melt up to 300°C. The magnetic moment of the yellow complex, $[Mn(C_{10}H_6N_2O_3S)(H_2O)_2]$, is $\mu_{eff} = 5.71\ \mu_B$, that of the orange complex, $[Mn(C_{10}H_6N_2O_2S_2)(H_2O)_2]$, is $\mu_{eff} = 5.80\ \mu_B$. IR spectral results suggest that the dinegative tetradentate ligands coordinate to the manganese via the two hydrazinic nitrogens and two sulfur atoms in the case of ligand 4, or one sulfur and one oxygen atom in the case of ligand 3. The magnitude of the positive shift of the ν(N–N) vibration suggests that the hydrazinic nitrogen atoms are involved in bridge formation. The band positions of the OH stretching and rocking modes are assigned to water molecules coordinated in the octahedral manganese(II) environment [5].

$Mn^{II}(C_{12}H_8N_2O_2S)\cdot H_2O$ was prepared by adding a solution of ligand 5 (10 mmol) dissolved in ethanol (30 mL containing 20 mmol of potassium hydroxide) to the hot ethanol solution of hydrated manganese(II) chloride (10 mmol). The green precipitate was washed with water, ethanol, and dried in vacuum. It does not melt below 300°C. The magnetic moment, $\mu_{eff} = 4.42\ \mu_B$ (Faraday method, temperature not given), may be lowered due to partial spin pairing, and suggests square-planar geometry of the complex. Its IR spectrum shows no phenolic ν(OH) vibration. The data suggest that the dinegative tridentate ligand is coordinated by the phenolic oxygen, the azomethine nitrogen, and the enethiol sulfur in a plane also containing a coordinated water molecule. Nonligand bands in the 405 to 260 cm^{-1} region are tentatively assigned to ν(Mn–X) vibration modes with X = O, N, S. The complex is insoluble in water, soluble in dimethylformamide or dimethyl sulfoxide. It behaves as a nonelectrolyte in dimethylformamide solution [2].

$Mn^{II}(C_5H_6N_2OS)_2Cl_2$. The yellow complex precipitated immediately after mixing the stoichiometric amounts of manganese(II) chloride dissolved in ethanol and ligand 1 dissolved in ethanol-ether. The complex melts at 198°C. The magnetic moment is $\mu_{eff} = 6.20\ \mu_B$, obtained from susceptibility measurements (Faraday method, temperature not given). The IR spectral results suggest that the bidentate ligand is bonded in the octahedral manganese environment via the sulfur and the terminal nitrogen atoms. The complex is insoluble in water, soluble in methanol, dimethylformamide, or dimethyl sulfoxide. In methanol solution it behaves as a 1:2 electrolyte [2].

References:

[1] Agrawal, S.; Singh, N. K. (Spectrochim. Acta A **42** [1986] 507/13).

[2] Singh, N. K.; Agrawal, S.; Aggarwal, R. C. (Syn. React. Inorg. Metal-Org. Chem. **15** [1985] 75/92).

[3] Singh, N. K.; Agrawal, S.; Aggarwal, R. C. (Polyhedron **3** [1984] 1271/6).).

[4] Singh, N. K.; Agrawal, S.; Aggarwal, R. C. (Indian J. Chem. A **21** [1982] 973/6).

[5] Singh, N. K.; Agrawal, S.; Aggarwal, R. C. (Indian J. Chem. A **23** [1984] 137/9).

35.18 Complexes with Thiazyl Trifluoride or Imidosulfuryl Fluorides

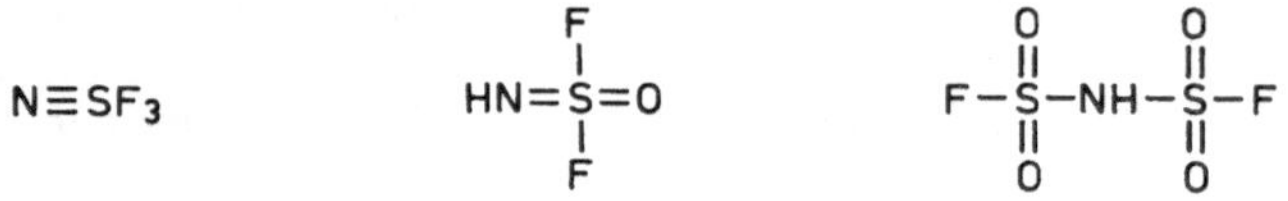

$$N \equiv SF_3$$

ligand 1 (= L)

$$HN=\underset{\underset{F}{|}}{\overset{\overset{F}{|}}{S}}=O$$

ligand 2 (= HL)

$$F-\underset{\underset{O}{||}}{\overset{\overset{O}{||}}{S}}-NH-\underset{\underset{O}{||}}{\overset{\overset{O}{||}}{S}}-F$$

ligand 3 (= HL)

[Mn^{II}(NSF$_3$)$_4$(AsF$_6$)$_2$]. The complex was prepared in quantitative yield by reaction of [Mn(SO$_2$)$_2$(AsF$_6$)$_2$], see p. 112, with thiazyl trifluoride: 30 mmol of ligand 1 and ca. 5 cm^3 of solvent SO$_2$ were condensed at $-196°C$ onto [Mn(SO$_2$)$_2$(AsF$_6$)$_2$] in an evacuated vessel. After being allowed to warm up slowly, the mixture was stirred for 5 h at room temperature, then the excess SO$_2$ and ligand were pumped off. The colorless compound may be recrystallized from liquid SO$_2$ or an SO$_2$ClF–SO$_2$ mixture [1, 2]. A single-crystal X-ray diffractometer study reveals a monoclinic lattice, space group P2$_1$/n-C$_{2h}^5$, No. 14 (standard setting P2$_1$/c) with the lattice parameters: a = 7.496(4), b = 10.378(5), c = 13.979(6) Å, β = 94.33(3)°, V = 1084.4 Å; Z = 2. Calculated density, D = 2.589 g/cm^3. The crystal structure was refined to R = 0.048 for 1226 reflections. Atomic coordinates are given in the paper. As shown in **Fig. 19**, the manganese ion is coordinated octahedrally by four NSF$_3$ and two AsF$_6^-$ ions in *trans* configuration. Selected interatomic distances and angles are given below [1]:

distance	in Å	angle	in °	angle	in °
Mn–F(1)	2.193(4)	F(1)–Mn–N(1)	90.6(2)	Mn–F(1)–As	150.6(2)
Mn–N(1)	2.193(6)	F(1)–Mn–N(2)	91.3(2)	Mn–N(1)–S(1)	160.1(4)
Mn–N(2)	2.181(5)	N(1)–Mn–N(2)	89.7(2)	Mn–N(2)–S(2)	160.2(3)

The unusual covalent interaction of the AsF$_6^-$ ion is indicated by the As–F bond lengths. The bond distances and angles of the AsF$_6$ group correspond closely to those of uncoordinated AsF$_6^-$ ions (average As–F distance beeing 1.673 Å), except for the bridging As–F(1) bond of 1.740(4) Å. The Mn–F(1) distance of 2.193(4) Å is comparable to that in MnF$_2$ (2.14 Å) [1]. The N–S and S–F bonds are considerably shorter than in the free thiazyl trifluoride, which is probably due to withdrawal by the metal of π-electron density from the ligand [1, 3].

Characteristic bands (in cm^{-1}) in the IR spectrum were assigned as follows:

vibration mode	ν(NS)	ν(SF$_3$)	ν(AsF$_6^-$)
NSF$_3$	1515	811, 775	—
[Mn(NSF$_3$)$_4$(AsF$_6$)$_2$]	1578	886, 835	721, 702, 674, 590

vibration mode	δ(NSF)	δ(FSF)	δ(AsF$_6^-$)
NSF$_3$	521, 342	492	—
[Mn(NSF$_3$)$_4$(AsF$_6$)$_2$]	548 to 540, 352	456 to 448	397

These data are consistent with the crystallographic results; i.e., ν(NS) and ν(SF) are higher in the complex than in the free ligand, in agreement with the shorter bond lengths in the complex. Splitting of the δ(FSF) deformation mode in the spectrum of the complex indicates a deviation from local C_{3v} symmetry of the NSF_3 ligands [1]. The force constant, f_{NS} = 1398 N/m, and the NS bond length, 139.7 pm, were calculated [4].

The complex is extremely sensitive to oxygen and moisture [1].

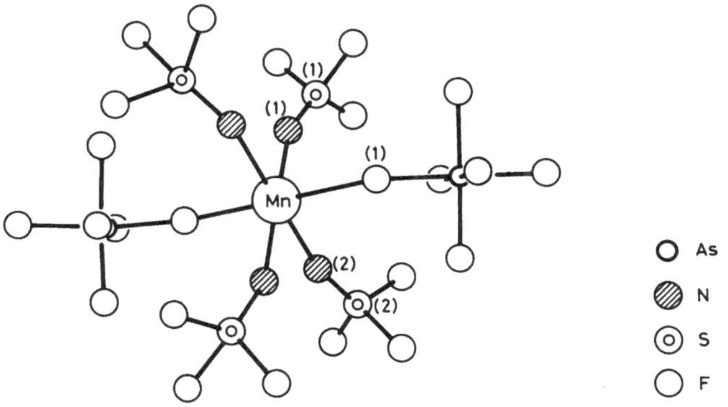

Fig. 19. Structure of the $[Mn(NSF_3)_4(AsF_6)_2]$ complex [1].

$[P(C_6H_5)_4]_2[Mn^{II}(NSOF_2)_4]$ was prepared by reacting a suspension of 0.005 mol $[P(C_6H_5)_4]_2$-$[MnBr_4]$ and 0.02 mol $[AgNSOF_2]$ in dry acetone. The mixture was stirred for 1 h, filtered, and the colorless complex precipitated with petroleum ether was recrystallized from a small amount of methanol. The IR spectrum of the complex shows several absorption bands, but their assignments are not given. The ^1H NMR spectrum in deuteroacetonic solution exhibits a broad signal with a maximum at 5.9 ppm. A tetrahedral structure with coordination through the nitrogen atoms is proposed for the metal anion [5].

$[N(CH_3)_4]_2[Mn^{II}Cl_2\{N(SO_2F)_2\}_2]$ was prepared by reacting equimolar amounts (10 mmol) of anhydrous $MnCl_2$ and $[N(CH_3)_4][N(SO_2F)_2]$ (obtained by solvolytic reaction of ligand 3 and $[N(CH_3)_4]Cl$) in ca. 40 mL CCl_4. The mixture was stirred magnetically at room temperature for 24 h. The product formed was filtered in a dry nitrogen atmosphere, washed with dry CCl_4, and finally dried in vacuum. The dirty white, high-melting compound reveals a magnetic moment of $\mu_{eff} = 5.93\ \mu_B$ at room temperature. The IR spectrum of the complex shows all the characteristic bands assigned to deprotonated ligand 3 and ν(Mn–Cl) at 290 and 275 cm^{-1}. The diffuse reflectance spectrum exhibits maxima at 450 and 340 nm; these are assigned to $^6A_{1g} \rightarrow {}^4E_g(G)$ and $6A_{1g} \rightarrow {}^4E_g(D)$ transitions in a tetrahedral environment around Mn^{II}. The formation of this mixed-complex anion and its stabilization by the large $[N(CH_3)_4]^+$ cation indicate the pseudohalogen character of ligand 3. The moisture-sensitive compound is insoluble in nonpolar solvents [6].

References:

[1] Buss, B.; Clegg, W.; Hartmann, G.; Jones, P. G.; Mews, R.; Noltemeyer, M.; Sheldrick, G. M. (J. Chem. Soc. Dalton Trans. **1981** 61/3).
[2] Mews, R. (J. Chem. Soc. Chem. Commun. **1979** 278/9).
[3] Glemser, O.; Mews, R. (Angew. Chem. **92** [1980] 904/21).
[4] Schnepel, F. M.; Mews, R.; Glemser, O. (J. Mol. Struct. **60** [1980] 89/92).
[5] Eisenbarth, R.; Sundermeyer, W. (Z. Naturforsch. **36b** [1981] 1343/4).
[6] Dhingra, P. L.; Verma, R. D. (Indian J. Chem. A **26** [1987] 139/41).

35.19 Complexes with S-Heterocycles or Their Derivatives

For the sake of congruity several complexes with ligands containing S-heterocyclic groups have been described in former volumes of the "Manganese" D series, or other chapters of the present volume, see the remarks below and those on pp. 225, 227, 228, 230, 233, 237, and 241. A polymeric complex with tetracyanotetrathiafulvalene is described in Chapter 33, p. 18, complexes with heterocyclic sulfoxides in Chapter 35.3, pp. 103/6.

35.19.1 With Derivatives of Thiophene

Remark. Complexes with 2-thienylalkanediones have already been described in "Mangan" D 1, 1979, pp. 115 and 122/4. For more recent data on complexes with 4,4,4-trifluoro-1-(2-thienyl)-1,3-butanedione, $C_4H_3SC(O)CH_2C(O)CF_3$ (= $C_8H_5F_3O_2S$), see [1 to 4]. Complexes with Schiff bases or hydrazones derived from thiophenecarbaldehyde or thienyl ketones are described in "Manganese" D 6, 1988, pp. 76/7, 92/4, 196, 263, 285/7, and 294/6.

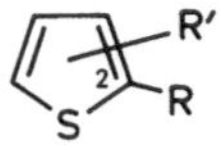

= HL

ligand	R	R'	formula
1	CH=NOH	H	C_5H_5NOS
2	CH=NOH	5-CH_3	C_6H_7NOS
3	C(O)OH	H	$C_5H_4O_2S$
4	$C(O)OC_2H_5$	3-OH	$C_7H_8O_3S$
5	$C(O)CH_3$	3-OH	$C_6H_6O_2S$
6	$C(O)NHNH_2$	H	$C_5H_6N_2OS$
7	$C(O)N(OH)C_6H_5$	H	$C_{11}H_9NO_2S$
8	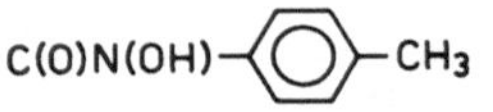	H	$C_{12}H_{11}NO_2S$

Complexes in Solution. Formation constants of the complex $[Mn^{II}(C_5H_3O_2S)]^+$ with 2-thiophenecarboxylic acid (ligand 3) in aqueous solution were determined pH-potentiometrically under nitrogen at 30, 40, and 50°C and I = 0.2 M ($NaClO_4$): log K_1 = 1.98, 2.17, 2.22, respectively. The calculated thermodynamic parameters are $-\Delta G°$ = 2.74, 3.11, and 3.78 kcal/mol at the corresponding temperatures, $\Delta H°$ = 5.56 kcal/mol and $\Delta S°$ = 27.7 cal·mol^{-1}·K^{-1} at 40°C [5].

Formation constants of $Mn^{II}L^+$ and $Mn^{II}L_2$ complexes with HL = ligands 1 to 5, 7, and 8 in H_2O-dioxane mixtures were determined potentiometrically (glass electrode) or spectrophotometrically under nitrogen at ionic strength I = 0.1 M ($NaClO_4$):

ligand	t in °C	method	vol% dioxane	log K_1	log K_2	Ref.
1	30	pot.	50	8.07	5.69	[6]
2	30	pot.	50	7.01	4.48	[6]
3	25	pot.	50	1.65	—	[7]
	30	pot.	50	2.28	—	[8]
4	25	spect.	10	3.79	—	[9]
5	25	pot.	10	3.0±0.2	—	[10]
	25	spect.	10	2.9±0.1	—	[10]
7	25	pot.	50	5.13	4.16	[11]
	25	pot.	50	5.26*)	4.46*)	[12]
	35	pot.	50	5.17*)	4.36*)	[12]

ligand	t in °C	method	vol% dioxane	log K_1	log K_2	Ref.
8	25	pot.	50	5.33*)	4.49*)	[12]
	35	pot.	50	5.21*)	4.37*)	[12]

*) At $I \rightarrow 0$.

For the stepwise formation of the complexes with ligand 7 in 50% dioxane the enthalpy values, $\Delta H_1^\circ = -3.79$ and $\Delta H_2^\circ = -4.21$ kcal/mol, have been calculated. Corresponding values for the complexes with ligand 8 are $\Delta H_1^\circ = -4.63$ and $\Delta H_2^\circ = -5.05$ kcal/mol [12].

Formation constants of mixed ligand complexes, $[Mn^{II}LL']^+$ with HL = ligand 7 and L' = bipyridine or phenanthroline, were determined potentiometrically in a 50 vol% dioxane-water mixture at 35°C: log $K^{MnL'}_{MnLL'} = 5.31$ and log $\beta^{Mn}_{MnLL'} = 11.02$ for $[Mn(C_{11}H_8NO_2S)bpy]^+$, log $K^{MnL'}_{MnLL'} = 5.43$ and log $\beta^{Mn}_{MnLL'} = 12.37$ for $[Mn(C_{11}H_8NO_2S)phen]^+$ [13].

$[Mn^{II}(C_5H_5NOS)_2X_2]$ complexes with X = NO_3, Cl, Br, I, ½SO_4, NCS, NCSe, CH_3COO were prepared by the reaction of ligand 1 and the appropriate hydrated manganese(II) salt (mole ratio 2:1) in a 3:2 (v/v) mixture of ethanol and 2,2-dimethoxypropane. Prior to mixing, the reactant solutions were heated separately for 20 to 30 min at 60 to 70°C with stirring. The combined mixture was heated under reflux for 15 to 30 min. The crystalline solids that formed on cooling the mixture were washed with ethanol and diethyl ether and dried in an evacuated desiccator over P_4O_{10}. For preparation of the sulfato or acetato complexes the manganese(II) salt was dissolved in a minimum amount of water. These precipitates were washed initially with water, then treated as described above [14]. The chloro and bromo complexes were also prepared by refluxing the appropriate hydrated manganese(II) halide with an ethanol solution of ligand 1 in a 1:4 mole ratio at 65 to 70°C with stirring for 2 h. The solutions were filtered and allowed to stand overnight at room temperature to yield pale yellow solids. The products were recrystallized from methanol [15]. The Mn^{II} complexes are isomorphous with the corresponding Ni^{II}, Co^{II}, and Fe^{II} complexes due to X-ray powder diffraction data [14].

The magnetic moments (μ_{eff} in μ_B) from susceptibility measurements (Faraday method) at room temperature and at 78 K, the absorption maxima in the diffuse reflectance spectra and the calculated ligand field parameters (in cm^{-1}, except β) are summarized below [14]. The magnetic moments in parentheses were determined by the Gouy method [15].

X	NO_3	Cl	Br	I	½SO_4	NCS	NCSe	CH_3COO
μ_{eff} at room temp.	5.92	5.92(6.05)	5.90(5.96)	5.91	5.86	5.93	5.92	5.88
μ_{eff} at 78 K	5.91	5.91	5.89	5.90	5.87	5.92	5.90	5.87
$^6A_{1g} \rightarrow {}^4T_{1g}(G)$	16280	16620	16260	16150	17070	16620	16600	18300
$\rightarrow {}^4T_{2g}(G)$	20920	21070	20800	20600	21400	21300	21280	22630
$\rightarrow {}^4A_{1g}, {}^4E_g(G)$	24740	23300 24700	23090 24400	24430	24900	25200	25200	25920
$\rightarrow {}^4T_{2g}(D)$	27800	27820	27600	27600	27880	28460	28470	28900
$\rightarrow {}^4E_g(D)$	29000	28970	28740	28780	29040	29670	29690	30070
Dq	1070	1086	1080	1070	1080	1095	1099	1078
B	608	610	620	622	591	638	640	593
C	3732	3720	3640	3641	3798	3764	3760	3997
β	0.634	0.635	0.645	0.648	0.615	0.664	0.667	0.617

The IR spectra of the $[Mn(C_5H_5NOS)_2X_2]$ complexes in KBr or CsI disks or Nujol mulls show ν(OH) bands in the 3260 to 3198 cm^{-1} region. The ν(C=N) and ν(NO) bands of the ligand at 1610 and 945 cm^{-1} are positively shifted to the regions 1655 to 1630 and 990 to 960 cm^{-1}, respectively. These positive shifts indicate that the oxime protone is not ionized and the C=NOH group is involved in coordination. The shift of the ring C=C out of plane deformation from 468 to 490 cm^{-1} is consistent with coordination of the sulfur atom. N-bonded coordination of the NCS or NCSe anions is suggested by the observation of ν(CN) at 2080 cm^{-1}. The sulfato complex shows bands at 900($ν_1$), 448($ν_2$), 1140, 1095, 1050($ν_3$), and 665, 600($ν_4$) cm^{-1} suggesting a bidentate coordinated sulfato group. Monodentate coordination of the acetato group is suggested by the observation of $ν_{as}$(COO) at 1645 and $ν_s$(COO) at 1390 cm^{-1}. The far-IR spectra show prominent bands of ν(Mn–S) in the regions 320 to 300 and 299 to 280 cm^{-1}, of ν(Mn–X) at 270 and 260 cm^{-1} for X = NO_3, Cl, NCS, NCSe [14]. The IR spectra of $[Mn(C_5H_5NOS)_2Cl_2]$ and $[Mn(C_5H_5NOS)_2Br_2]$ in Nujol show bands of ν(Mn–S) at 213 and 210 cm^{-1}, respectively, bands of ν(Mn–Cl) at 232, and ν(Mn–Br) below 200 cm^{-1} [15].

The physical properties of the complexes suggest that they have a *trans* tetragonal structure with coordinated anions and neutral bidentate ligands being bonded to the Mn^{II} via the sulfur and the nitrogen atoms in a distorted octahedron [14]. However, a polymeric structure is assumed by [15] for $[Mn(C_5H_5NOS)_2Cl_2]$ and $[Mn(C_5H_5NOS)_2Br_2]$ with Mn^{II} in a hexacoordinated environment with nitrogen-bonded monodentate ligands and bridging halide groups [15]. The complexes are insoluble in water and nonpolar solvents, partially soluble in polar solvents. The electrical conductivities of the complexes in ethanol or nitrobenzene solution are low, indicating the presence of nonelectrolytes [14].

$[Mn^{II}(C_5H_6N_2OS)_2Cl_2]$ was prepared by mixing ethanolic solutions of $MnCl_2$ (1 mmol in 10 mL) and ligand 6 (2 mmol in 15 mL). The white precipitate was washed repeatedly with ethanol, then with ether, and dried in vacuum. The X-ray powder diffraction lines have been indexed to tetragonal symmetry. The computed lattice parameters are a = 9.94, b = 12.03 Å; Z = 4. The densities are D_{calc} = 2.28 and D_{obs} = 2.20 g/cm^3. The complex melts at 220°C. The magnetic moment at room temperature is $μ_{eff}$ = 6.20 $μ_B$. Important IR bands of the complex in Nujol and the observed shifts (Δν) in comparison to the free ligand are: ν(CO) at 1630(−25), ν(CN) + δ(NH) at 1520(−10), δ(NH_2) at 1590(−40), ν(Mn–O) at 415, ν(Mn–N) at 340, ν(Mn–Cl) at 230 cm^{-1}. Bonding through the carbonyl oxygen atom and the primary nitrogen atom is suggested.

The complex is insoluble in water and common organic solvents, soluble in DMF or DMSO. In DMF solution it behaves as a nonelectrolyte. The data agree with an octahedral geometry at Mn^{II} [16].

References:

[1] Patel, K. S. (J. Inorg. Nucl. Chem. **42** [1980] 1235/40).
[2] Patel, K. S.; Adimado, A. A. (J. Inorg. Nucl. Chem. **42** [1980] 1241/6).
[3] Ueda, K.; Gokoh, M.; Yamamoto, Y. (Mem. Fac. Technol. Kanazawa Univ. **13** [1980] 81/91; C.A. **93** [1980] No. 102125).
[4] Saeed, M. M.; Cheemna, M. N.; Qureshi, I. H. (Radiochim. Acta **38** [1985] 107/9).
[5] Sandhu, S. S.; Kumaria, J. N.; Sandhu, R. S. (Indian J. Chem. A **14** [1976] 817/8).
[6] Chaudhury, A. K.; Da, B. (J. Indian Chem. Soc. **54** [1977] 565/7).
[7] Erlenmeyer, H.; Griesser, R.; Prijs, B.; Sigel, H. (Helv. Chim. Acta **51** [1968] 339/48).
[8] De, B.; Chaudhuri, A. K. (J. Indian Chem. Soc. **53** [1976] 783/6).
[9] Courtin, A.; Sigel, H. (Helv. Chim. Acta **48** [1965] 617/25).
[10] Petri, S.; Sigel, H.; Erlenmeyer, H. (Helv. Chim. Acta **49** [1966] 1612/6).

[11] Bag, S. P.; Lahiri, S. (Indian J. Chem. A **13** [1975] 1214/6).
[12] Abbasi, S. A. (Polish J. Chem. **51** [1977] 821/4).

[13] Abbasi, S. A. (Polish J. Chem. **58** [1984] 61/4).
[14] Mohan, M. (Gazz. Chim. Ital. **109** [1979] 655/61).
[15] Kuma, H.; Motobe, K.; Yamada, S. (Bull. Chem. Soc. Japan **51** [1978] 1101/3).
[16] Singh, B.; Singh, R. N.; Aggarwal, R. C. (Syn. React. Inorg. Metal-Org. Chem. **15** [1985] 853/72).

35.19.2 With Tetrahydrothiophene or Tetrahydro-2-thiophenecarboxylic Acid

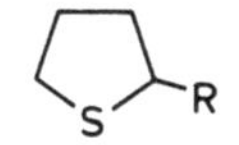

ligand 1 with R = H (= C_4H_8S)
ligand 2 with R = COOH (= $C_5H_8O_2S$)

Remark. Manganese(II) complexes with tetrahydrothiophene-1-oxide or -1,1-dioxide are described on pp. 103 and 112, respectively.

Complexes in Solution. The stability constants of Mn^{II} complexes in mixed solvent systems were determined potentiometrically (glass electrode) or spectrophotometrically by the competition method, using the corresponding Cu^{II} tetrahydrothiophene system:

complex	t in °C	method	medium	ionic strength	log K_1	Ref.
$Mn(C_4H_8S)^{2+}$	25	pot.	75% dioxane	0.1 M $NaClO_4$	~0.4	[1]
	25	spectr.	50% ethanol	1.0 M $NaClO_4$	−0.31	[2, 3]
	25	spectr.	96% DMF	1.0 M $NaClO_4$	0.0 ± 0.2	[4]
$Mn(C_5H_7O_2S)^{+}$	25	pot.	50% dioxane	0.1 M $NaClO_4$	1.80	[5]
	30	pot.	50% dioxane	0.1 M $NaClO_4$	2.36	[6]

[$Mn^{II}(C_4H_8S)_2X_2$] complexes with X = Cl, Br, I, NCS were prepared by the reaction of MnX_2 and an excess (10% over the stoichiometric amount) of ligand 1 in toluene. The mixture was refluxed overnight in an atmosphere of dry argon. The complex separated on cooling, the mixture was washed with toluene, and dried in vacuum. Color, magnetic moments at room temperature, IR bands of ν(Mn–S) or ν(Mn–X–Mn) vibration modes (in cm^{-1}), and absorption maxima observed in the diffuse reflectance spectra are shown below:

complex	color	μ_{eff} in μ_B	ν(Mn–S)	ν(Mn–X–Mn)	λ_{max} in nm
$Mn(C_4H_8S)_2Cl_2$	off-white	5.35	330	240	525, 442, 420, 390
$Mn(C_4H_8S)_2Br_2$	pink	5.55	320	190	528, 450, 410, 400sh
$Mn(C_4H_8S)_2I_2$	buff	5.80	320	130	535, 470, 430, 390
$Mn(C_4H_8S)_2(NCS)_2$	cream	5.41	315	—	510, 460, 450, 380

The magnetic moments of the complexes with X = Cl, Br, I are consistent with antiferromagnetic interaction. The complexes do not exhibit ESR spectra, further evidence for strong Mn–X–Mn interactions. The IR spectra (in Nujol) show no bands of terminal Mn–X stretching frequencies, but do show ν(Mn–X–Mn) vibrations. The complexes are therefore assumed to be pseudooctahedral polymers in which Mn^{II} is bonded by halogen bridges and by the sulfur atom of the tetrahydrothiophene molecules. The structure of the isothiocyanato complex is different. Only a single ν(CN) band was observed at 2100 cm^{-1}, the region for a terminal Mn–NCS linkage. ν(CS) bands occur at 790 cm^{-1} and δ(NCS) at 480 cm^{-1}. This complex exhibits an ESR spectrum with a single broad line in the g = 2 region at room temperature. A monomeric pseudotetrahedral structure is suggested [6].

References:

[1] Sigel, H.; McCormick, D. B.; Griesser, R.; Prijs, B.; Wright, L. D. (Biochemistry **8** [1969] 2687/95).
[2] Rheinberger, V. M.; Sigel, H. (Naturwissenschaften **62** [1975] 182).
[3] Sigel, H.; Rheinberger, V. M.; Fischer, B. E. (Inorg. Chem. **18** [1979] 3334/9).
[4] Sigel, H.; Scheller, K. H. (J. Inorg. Biochem. **16** [1982] 297/310).
[5] Sigel, H.; Griesser, R.; Prijs, B.; McCormick, D. B.; Joiner, M. G. (Arch. Biochem. Biophys. **130** [1969] 514/20).
[6] De, B.; Chaudhury, A. K. (J. Indian Chem. Soc. **53** [1976] 783/6).
[7] Barratt, D. S.; McAuliffe, C. A.; Stacey, C. C. (Inorg. Chim. Acta **102** [1985] L35/L36).

35.19.3 With Derivatives of 1,2-Dithiolane

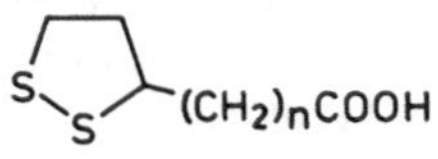

ligand	n	trivial name	formula
1	0	tetranorlipoic acid	$C_4H_6O_2S_2$
2	2	bisnorlipoic acid	$C_6H_{10}O_2S_2$
3	4	d-α-lipoic acid	$C_8H_{14}O_2S_2$
4	4	l-α-lipoic acid	$C_8H_{14}O_2S_2$
5	4	d,l-α-lipoic acid	$C_8H_{14}O_2S_2$

Complexes in Solution. Stability constants of $Mn^{II}L^+$ complexes with HL = ligands 1 to 5 in 50% aqueous dioxane were determined potentiometrically (glass electrode) at 25°C and I = 0.1 M ($NaClO_4$): log K_1 = 1.87, 2.11, 2.07, 1.93, and 2.06, respectively.

The stability of the complexes is determined by the basicity of the corresponding carboxylate group [1]. For the complex with tetrahydro-2-thiophenecarboxylic acid, $Mn(C_5H_7O_2S)^+$ (see p. 225), and that with tetranorlipoic acid, $Mn(C_4H_5O_2S_2)^+$, an equilibrium is expected between a chelated isomer (in which the carboxylato moiety and a sulfur atom are coordinated to Mn^{II}) and a simple carboxylato coordinated isomer. The calculated (dimensionless) equilibrium constant reveals ≦20% of the O,S-coordinated species for each complex. The ligand properties are of interest, because of their potential as binding sites for Mn^{2+} or Ca^{2+}, Zn^{2+}, Cu^{2+} ions in biological systems [2].

References:

[1] Sigel, H.; Prijs, B.; McCormick, D. B.; Shih, J. C. H. (Arch. Biochem. Biophys. **187** [1978] 208/14).
[2] Sigel, H.; Scheller, K. H.; Rheinberger, V. M.; Fischer, B. E. (J. Chem. Soc. Dalton Trans. **1980** 1022/8).

35.19.4 With Biotin and Related Compounds

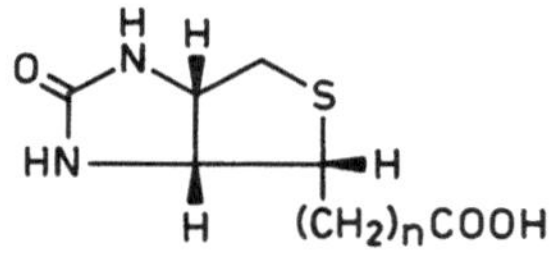

Remark. Manganese(II) complexes with biocytin, the biotinyllysine dipeptide, and its derivatives are described in "Manganese" D 4, 1985, p. 345.

Complexes in Solution. The ligands listed below form $Mn^{II}L^+$ complexes in 50% aqueous dioxane. Their stability constants have been determined potentiometrically (glass electrode) at 25°C and I = 0.1 M ($NaClO_4$):

ligand	d-biotin	biotin sulfoxide	biotin sulfone
n	4	4	4
formula	$C_{10}H_{16}N_2O_3S$	$C_{10}H_{16}N_2O_4S$	$C_{10}H_{16}N_2O_5S$
log K_1	2.07	1.97*), 198**)	2.06

*) With d-biotin d-sulfoxide. – **) With d-biotin l-sulfoxide.

ligand	homobiotin	bisnorbiotin	tetranorbiotin
n	5	2	0
formula	$C_{11}H_{18}N_2O_3S$	$C_8H_{12}N_2O_3S$	$C_6H_8N_2O_3S$
log K_1	2.00	1.80	1.66

The NMR spectrum suggests that Mn^{II} is bonded to the sulfur of d-biotin from below the plane of the tetrahydrothiophene ring in the same way as oxygen is bonded in d-biotin sulfoxide. For tetranorbiotin the NMR spectrum indicates mainly that a simple carboxylic acid complex is formed [1]. The formation of a 1:1 complex with d-biotin in 96% aqueous dimethylformamide at 25°C, and I = 0.1 ($NaClO_4$) was also demonstrated spectrophotometrically by use of the exchange equilibrium with the corresponding Cu^{II} complex. ^1H NMR measurements indicate that the Mn^{II} is coordinated to the sulfur atom *trans* to the urea ring [2].

References:

[1] Sigel, H.; McCormick, D. B.; Griesser, R.; Prijs, B.; Wright, L. D. (Biochemistry **8** [1969] 2887/95).
[2] Sigel, H.; Scheller, K. H. (J. Inorg. Biochem. **16** [1982] 297/310).

35.19.5 With Derivatives of Thiazole or Isothiazole

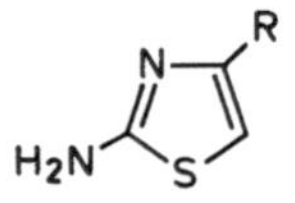

ligands 1 to 6

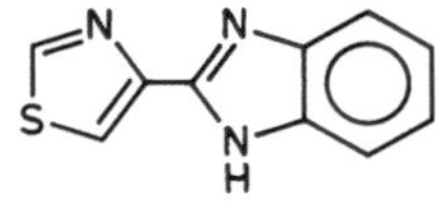

ligand 7 (= $C_{10}H_7N_3S$)

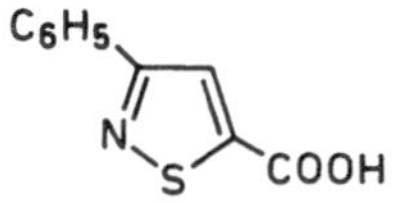

ligand 8 (= $C_{10}H_7NO_2S$)

ligand	1	2	3	4	5	6
R	H	CH_3	C_6H_5	C_6H_4F-4	C_6H_4Cl-4	C_6H_4Br-4
formula	$C_3H_4N_2S$	$C_4H_6N_2S$	$C_9H_8N_2S$	$C_9H_7FN_2S$	$C_9H_7ClN_2S$	$C_9H_7BrN_2S$

Remark. Manganese(II) complexes with thiazolylazo compounds were treated in "Manganese" D 5, 1987, pp. 312/7. For additional data, see [1]. An Mn^{II} Schiff base complex containing a thiazolyl group was described in "Manganese" D 6, 1988, pp. 15/6. For complexes with thiazolyl sulfonamides, see pp. 117/21 of this volume.

Complexes in Solution. Formation constants of complexes with ligand 1, $Mn^{II}(C_3H_4N_2S)^{2+}$ and $Mn^{II}(C_3H_4N_2S)_2^{2+}$, in aqueous solution were determined pH-potentiometrically at $I = 0.1$ ($NaNO_3$) and various temperatures [2]:

t in °C	25	30	35	40
log K_1	1.57	1.41	1.26	1.08
log β_2	5.34	5.30	5.26	5.23

The observed order for the stability constants of $M(C_3H_4N_2S)^{2+}$ complexes agrees with the Irving-Williams sequence, $Mn^{II} < Co^{II} < Ni^{II} < Cu^{II} < Zn$, whereas the order for the stability of $M(C_3H_4N_2S)_2^{2+}$ complexes deviates: $Mn^{II} > Cu^{II} > Co^{II} > Ni^{II} > Zn$. Ligand 1 is assumed to coordinate to Mn^{II} through the amino nitrogen. The thermodynamic parameters for the 1:1 complex at 25°C are $\Delta G = -2.2$ kcal/mol, $\Delta H = -13.9$ kcal/mol, $\Delta S = -39.4$ cal·mol^{-1}·K^{-1}. Values of $\Delta G_{\beta_2} = -7.3$ kcal/mol, $\Delta H_{\beta_2} = -3.1$ kcal/mol, and $\Delta S_{\beta_2} = 14.1$ cal·mol^{-1}·K^{-1} were calculated for the 1:2 complex [2].

For the complex with the deprotonated ligand 8, $Mn^{II}(C_{10}H_6NO_2S)^+$, a stability constant of log $K_1 = 1.51$ was determined pH-potentiometrically in 50 vol% aqueous dioxane at 25°C. The ligand coordinates through the carboxylate group only [3].

$[Mn^{II}(C_{10}H_7N_3S)_3](ClO_4)_2$. The complex was prepared by the reaction of a boiling ethanol solution of ligand 7 and triethyl orthoformate with hydrated manganese(II) perchlorate dissolved in hot ethanol (mole ratio $Mn(ClO_4)_2$: ligand = 1:4). The solution was evaporated in vacuum until crystals appeared; they were washed with dry ether and dried at 40°C in vacuum (1 mm Hg). According to X-ray powder diagrams, $Mn(C_{10}H_7N_3S)_3(ClO_4)_2$ is isomorphous with the corresponding Zn, Cd, and Co complexes. The white Mn^{II} complex does not melt below 250°C. The IR spectrum of the complex in Nujol shows no change in the free ligand bands except an imidazole ring vibration at 900 cm^{-1}, which was positively shifted. Bands at 1090, 613, and 460 cm^{-1} are attributed to noncoordinated perchlorate anions. Bands at 277 and 232 cm^{-1} were assigned to ν(Mn–N), at 144 cm^{-1} to δ(Mn–N) vibrations. The spectral results suggest an octahedral environment at Mn^{II} with bidentate ligands coordinated through the conjugated nitrogen atoms of the two rings. The complex is slightly soluble in water [5].

$[Mn^{II}(C_4H_6N_2S)_2Cl_2]$ and **$[Mn^{II}(C_9H_8N_2S)_2Cl_2]$** complexes were prepared by refluxing $MnCl_2$ with ligand 2 or 3 (mole ratio 1:2) in aqueous ethanolic medium for one hour. The complexes obtained on concentration of the solution were washed with water, ethanol, ether, and dried in vacuum. The magnetic moments at room temperature are 5.53 and 5.54 μ_B, respectively. The IR spectra of the complexes in KBr reveal negative shifts of the ν(NH) vibrations, unchanged ν(CN) bands (at 1630 cm^{-1}), and the disappearance of the ligand's ν(CS) band at 650 cm^{-1}. The ligands are coordinated to Mn^{II} through the amino nitrogen and the ring sulfur atoms. The complexes are nonelectrolytes in nitromethane solution at 20°C [9].

[$Mn^{II}L_2X_2$] complexes with L = ligands 4 to 6 and X = NO_3, Cl, ½ SO_4, CH_3COO were prepared by refluxing stoichiometric mixtures of Mn^{II} salt and ligand in ethanol for 1 h. The compounds separated after concentrating the reaction mixture to half of its volume and standing for some time. [$Mn(C_9H_7ClN_2S)_2X_2$] complexes are thermally stable up to 120°C [6]. The room temperature magnetic moments of [$Mn(C_9H_7FN_2S)_2X_2$] were found to be about 5.5 μ_B [7], of [$Mn(C_9H_7ClN_2S)_2X_2$] between 5.85 and 6.00 μ_B [6], of [$Mn(C_9H_7BrN_2S)_2X_2$] between 5.52 and 5.95 μ_B [8]. The IR spectra of all the complexes in Nujol show negatively shifted ν(NH) vibrations, but no change in the ν(CN) band at 1630 cm^{-1}. The ν_{as}(CS) vibration of the free ligand at 650 has shifted to lower frequency, the ν_s(CS) at 685 cm^{-1} has disappeared. The ligands are coordinated to Mn^{II} through the amino nitrogen and the ring sulfur atoms. Octahedral geometry of the complexes is supported by weak bands (in cm^{-1}) recorded in MgO at room temperature and assigned to the following transitions:

complex	$^6A_{1g} \rightarrow {}^4T_{1g}(G)$	$\rightarrow {}^4T_{2g}(G)$	$\rightarrow {}^4A_{1g}(G)$	$\rightarrow {}^4T_{2g}(D)$	Ref.
[$Mn(C_9H_7FN_2S)_2X_2$]	17500 to 18800	21200 to 22000	25000 to 26200	27900	[7]
[$Mn(C_9H_7ClN_2S)_2X_2$]	15151 to 16665	16949 to 19808	24912 to 25728	28500	[6]
[$Mn(C_9H_7BrN_2S)_2X_2$]	~19800	~23000	~28500	~34000	[8]

[$Mn(C_9H_7ClN_2S)_2X_2$] complexes behave as nonelectrolytes [6]. Antifungal activity was observed for the complexes lower than that of the free ligands and not appreciably influenced by the nature of the anions [6 to 8].

$Mn^{II}(C_{10}H_7N_3S)_2Cl_2 \cdot 1.5H_2O$ was prepared by addition of $MnCl_2 \cdot 4H_2O$ dissolved in hot ethanol or water to a solution of ligand 7 in boiling ethanol under nitrogen (mole ratio 1:2). The precipitate was recrystallized from dimethylformamide-chloroform and dried at 100°C in vacuum over P_4O_{10}. The magnetic moment of the white complex at 298°C is 5.91 μ_B. Bands of the IR spectrum (4000 to 1000 cm^{-1}) are listed without assignments in the paper. The electronic diffuse reflectance spectrum shows maxima at 7299, 6135, 5128, 4629, 4386, and 3933 cm^{-1}. Octahedral coordination at Mn^{II} is assumed. The electrical conductivity of 10^{-3} M complex in methanol is $\Lambda = 102.2\ cm^2 \cdot \Omega^{-1} \cdot mol^{-1}$, a value lower than expected for a 1:2 electrolyte [4].

[$Mn^{II}(C_{10}H_6N_3S)_2C_{10}H_7N_3S$]$\cdot 2H_2O$ was prepared by the reaction of an aqueous solution of hydrated manganese(II) chloride with ligand 7 dissolved in hydrochloric acid (mole ratio 1:3). The mixture was stirred under nitrogen and neutralized with sodium hydroxide solution to yield a light gray solid. The compound was extracted with hot ethanol and dried at 100°C in vacuum over P_4O_{10}. The magnetic moment of the complex at 295°C is $\mu_{eff} = 5.98\ \mu_B$. The main IR bands of the complex are given in the paper [4] without assignments. Absorption maxima of the diffuse reflectance spectrum appear at 6135, 4878, 4608, 4347, and 3936 cm^{-1}. No evidence was obtained as to what part of the ligand molecule is coordinated. Chelation analogous to that of 2,2'-bipyridine was suggested. The probably polymeric complex is hygroscopic and insoluble in water and common solvents [4].

[$Mn^{II}(C_{10}H_7N_3S)Cl_2$] was prepared by mixing ethanol solutions of manganese(II) chloride and ligand 7 in a 1:1 mole ratio. The mixture was boiled for 1 to 3 h, and the resulting precipitate was washed with ethanol and dried in air. X-ray diffraction data (interplanar distances and intensities) are listed in the paper. The complex does not melt below 300°C. The IR spectrum of the complex in Nujol or KBr, compared with that of the free ligand, show the ν(CN) + δ(NH) vibration positively displaced and the δ(NH) band shifted from 1575 to 1610 cm^{-1}. The ligand band at 1355 cm^{-1} disappeared upon complexation. A marked decrease in the strength of the NH deformation band at 1093 and the shift of the ν(NH) band from 3450 to 3195 cm^{-1} is attributed to the rupture of the strong hydrogen bonds. The electronic absorption spectrum of the complex in KBr disks is similar to that of the free ligand which is apparently planar. The X-ray photoelectron spectrum of $[Mn(C_{10}H_7N_3S)Cl_2]$ reveals the following binding energy values, E (in eV), and the half line width (in eV given in parentheses): N1s, 400.2(2.7); S$2p_{3/2}$, 164.9(2.6); Cl$2p_{3/2}$, 198.7(3.0); and for the core electrons of manganese, 642.5(3.7). The data agree with bidentate ligand coordination via the aromatic nitrogen atoms and inner sphere bonding of the two chloro groups in a tetragonal arrangement around Mn^{II} [10].

References:

[1] Ohyoshi, E. (Polyhedron **5** [1986] 1165/70).
[2] Basak, A. K.; Banerjea, D. (J. Indian Chem. Soc. **55** [1978] 853/6).
[3] Erlenmeyer, H.; Griesser, G.; Prijs, B.; Sigel, H. (Helv. Chim. Acta **51** [1968] 339/48).
[4] Kowala, C.; Murray, K. S.; Swan, J. M.; West, B. O. (Australian J. Chem. **24** [1971] 1369/75).
[5] Van Landschoot, R. C.; van Hest, J. A. M.; Reedijk, J. (J. Inorg. Nucl. Chem. **38** [1976] 185/90).
[6] Srivastava, S. K.; Gupta, A. (Acta Chim. [Budapest] **118** [1985] 249/54).
[7] Srivastava, S. K.; Gupta, A. (Acta Chim. [Budapest] **118** [1985] 255/9).
[8] Srivastava, S. K.; Gupta, A. (Chim. Acta Turc. **12** [1984] 369/75).
[9] Srivastava, S. K.; Kudesia, V. P.; Kaur, K. H. (Acta Ciencia Indica Ser. Chem. **6** [1980] 85/8).
[10] Zaitsev, B. E.; Davydov, V. V.; Palishkin, M. V.; Kazanskii, L. P.; Sheban, G. V.; Kukalenko, S. S.; Novikova, G. A.; Shestakova, S. I. (Zh. Neorgan. Khim. **31** [1986] 947/54; Russ. J. Inorg. Chem. **31** [1986] 539/43).

35.19.6 With Derivatives of 4,5-Dihydro-thiazole or Thiazolidine

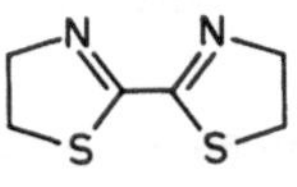

ligand 1 (= $C_6H_8N_2S_2$)

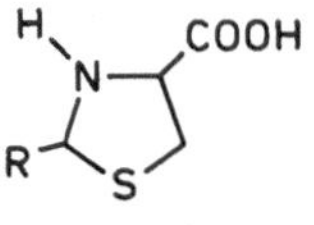

ligands 2 to 6

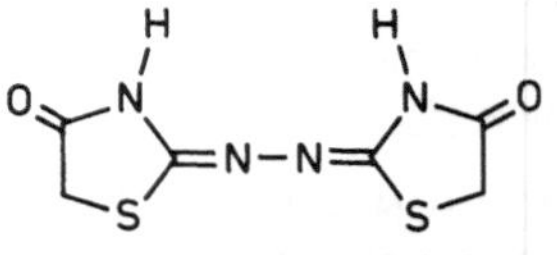

ligand 7 (= $C_6H_6N_4O_2S_2$)

ligand	2	3	4	5	6
R	H	CH_3	$(CH_3)_2CH$	C_6H_5	
formula	$C_4H_7NO_2S$	$C_5H_9NO_2S$	$C_7H_{13}NO_2S$	$C_9H_{11}NO_2S$	$C_8H_9NO_3S$

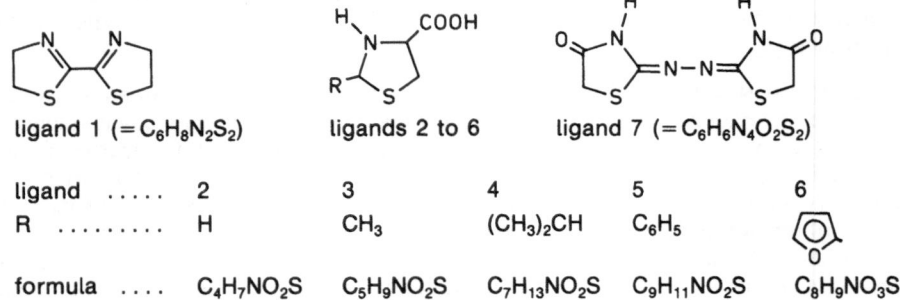

Remark. Manganese(II) complexes with 2-thiazolidinethione are described on p. 61.

Complexes in Solution. Stability constants of $Mn^{II}L^+$ complexes with HL = ligands 2 and 4 to 6 were determined potentiometrically under nitrogen in aqueous solution at ionic strength I = 0.1 M and various temperatures:

complex	$Mn(C_4H_6NO_2S)^+$				
t in °C	37	25	30	40	50
log K_1*)	1.904	3.10	3.03(3.16)	2.94	2.84
Ref.	[1]	[2]	[2]	[2]	[2]

complex	$Mn(C_7H_{12}NO_2S)^+$	$Mn(C_9H_{10}NO_2S)^+$	$Mn(C_8H_8NO_3S)^+$
t in °C	30	30	30
log K_1*)	2.67 (2.79)	3.25 (3.42)	3.32 (3.43)
Ref.	[3]	[3]	[3]

*) In parentheses: log K_1 at zero ionic strength.

The log K_1 values 2.79, 2.93, 3.04, and 3.25 for $Mn(C_7H_{12}NO_2S)^+$, 3.38, 3.54, 3.70, and 3.96 for $Mn(C_9H_{10}NO_2S)^+$, 3.40, 3.51, 3.64, and 3.88 for $Mn(C_8H_8NO_3S)^+$ were determined at 30°C and I = 0.1 in aqueous solution containing 10, 20, 30, or 50% ethanol, respectively. The stability constants are directly dependent on the mole fraction of ethanol and not on the dielectric constant of the solvent mixtures. Since water solvates Mn^{II} to a greater extent than ethanol does, the less solvated Mn^{II} attracts the ligand more than a heavily solvated Mn^{II} would [3]. Thermodynamic parameters for the formation of complexes with ligands 2 [2] and 4 to 6 [3] at 30°C and I = 0.1 M were calculated as follows:

complex	$Mn(C_4H_6NO_2S)^+$	$Mn(C_7H_{12}NO_2S)^+$	$Mn(C_9H_{10}NO_2S)^+$	$Mn(C_8H_8NO_3S)^+$
ΔH in kcal/mol	−4.20	−3.66	−4.53	−4.63
ΔG in kcal/mol	−4.22	−3.72	−4.34	−4.34
ΔS in $cal \cdot mol^{-1} \cdot K^{-1}$	0.07	0.21	0.63	0.95

The reactions are exothermic and spontaneous in nature as evidenced by the negative values of ΔH and ΔG. Both parameters have been separated into their electrostatic and nonelectrostatic components, which indicates that the nonelectrostatic forces are stronger, corresponding to a more covalent Mn^{II}–ligand bond [2, 3]. It is suggested that the ligand is attached through the ring nitrogen atom and the monodentate carboxylato group [1 to 3].

^1H NMR studies confirm that ligand 2 is bonded to Mn^{II} via the carboxylato group and the ring nitrogen atom. The transverse relaxation time for protons in the presence of Mn^{II} has been determined from NMR line broadening measurements on D_2O solutions at 25°C. Analysis of the data yields an effective radius of 4 Å for the complex. Parameters for the ligand exchange in the region of fast exchange are the activation energy, $E_a = 4.6$ kcal/mol, and the relaxation time, $\tau = 6.5 \times 10^{-11}$ s at 298 K. Comparison of the measured and experimentally determined proton–Mn^{II} distances for the two possible conformations of the heterocyclic ring show that the ring has the A conformation [4].

$Mn^{II}(C_4H_6NO_2S)_2$ was prepared by reaction of manganese(II) acetate (5 mmol) dissolved in 200 mL ethanol and ligand 2 (10 mmol) dissolved in 20 mL water at 80°C. The mixture was stirred at room temperature for 2 h. The pale pink precipitate was washed with water, then methanol, and dried in vacuum. The magnetic moment of the complex is $\mu_{eff} = 6.01\ \mu_B$. The main bands in the IR spectrum of the complex, compared with band positions of the free ligand (in parentheses), are assigned as follows: ν(NH) at 3310, 3275 (3050), δ(NH) at 1630 sh, 1605 (1625), ν_{as}(OCO) at 1590 (1555), ν_s(OCO) at 1410 (1435) cm^{-1}. The ligand is tridentate, bonded to Mn^{II} by the ring nitrogen and one carboxylic oxygen atom in the plane and by the second oxygen atom at the apex of an octahedron. The second oxygen atom is proposed to be bridging in a polymeric arrangement. The complex is insoluble in water and common organic solvents [5].

$Mn^{II}(C_5H_8NO_2S)_2 \cdot 2H_2O$ was obtained by the reaction of freshly prepared manganese(II) hydroxide and ligand 3 (mole ratio 1:2) in aqueous solution at 70 to 80°C. The beige precipitate was washed with ethanol, recrystallized from water, and dried over P_4O_{10}. The IR spectrum of the complex shows prominent bands in the 3460 to 3340 and 1640 to 1560 cm^{-1} regions and at 1400, 1360, and 675 cm^{-1}. The spectral results support the coordination of the ligand to Mn^{II} through the oxygen atom of the carboxylato group and the ring nitrogen atom [6].

$Mn^{II}(C_6H_4N_4O_2S_2)$ was prepared by digesting an aqueous reaction mixture of manganese(II) salt and the disodium salt of ligand 7 (mole ratio 1:1) on a water bath for 1 h. The precipitate was washed several times with water, then with ethanol, and dried in vacuum [7, 8]. The magnetic moment of the light green complex at room temperature is $\mu_{eff} = 5.50\ \mu_B$. The IR spectrum of the compound in KBr, compared with that of the free ligand, shows the ν(C=O) vibration at 1720 cm^{-1} and the ν(CS) mode unaffected by complexation, where as the acyclic ν(C=N) vibration at ~1600 cm^{-1} has been shifted to lower frequencies. The complex is probably a polymer in which the ligand acts bis-bidentately through both acyclic nitrogens and both ring nitrogens. In the complex the thiazolidine nitrogen atoms of the ligand are *trans* to each other [8].

$Mn^{II}(C_6H_8N_2S_2)_2(NCS)_2$ obtained by dropwise addition of a hot ethanolic solution of $Mn(NCS)_2$ to a hot solution of ligand 1 in the same solvent. The precipitate was recrystallized from methanol. The crystals are suitable for co-crystallization with the two crystalline forms of $Fe(C_6H_8N_2S_2)_2(NCS)_2$. For ESR studies of the Mn^{II} doped crystals, see the paper [9].

References:

[1] Huang, Z.; May, P. M.; Williams, D. R. (Inorg. Chim. Acta **56** [1981] 41/4).
[2] Ramesh, P.; Ram Reddy, M. G. (Acta Ciencia Indica Chem. **9** [1983] 138/42; C.A. **100** [1984] No. 216615).
[3] Ramesh, P.; Vinod Kumar, P.; Ram Reddy, M. G. (J. Indian Chem. Soc. **60** [1983] 231/3).
[4] Fazakerley, G. V.; Jackson, G. E. (J. Chem. Soc. Perkin Trans. II **1975** 567/71).
[5] Yang, W.; Chen, B. (Wuji Huaxue **2** [1986] 85/94; C.A. **107** [1987] No. 59419).
[6] Catrina, E. (Rev. Roumaine Chim. **21** [1976] 81/7).
[7] Satpathy, K. C.; Mishra, H. P.; Mahana, T. D. (J. Indian Chem. Soc. **56** [1979] 248/50).
[8] Satpathy, K. C.; Mahana, T. D.; Mishra, K. C. (J. Indian Chem. Soc. **57** [1980] 1232/3).
[9] Ozarowski, A.; McGarvey, B. R.; Sakar, A. B.; Drake, J. E. (Inorg. Chem. **27** [1988] 628/35).

35.19.7 With Penicillins

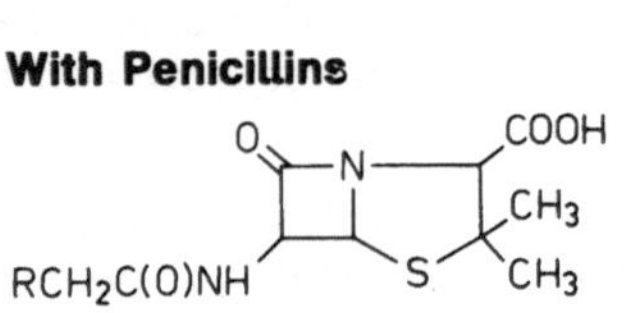

ligand 1 with $R = C_6H_5$ $(= C_{16}H_{18}N_2O_4S)$
ligand 2 with $R = C_6H_5O$ $(= C_{16}H_{18}N_2O_5S)$

ligand 3 with R =

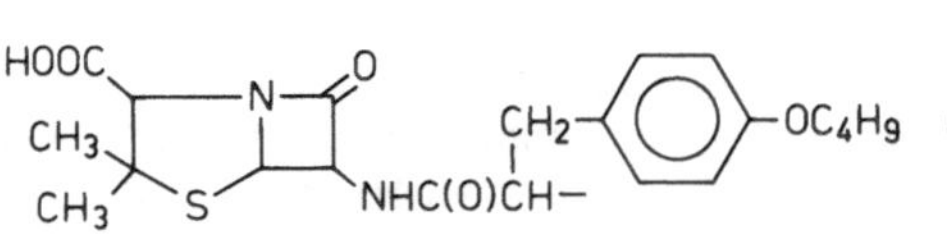

$(= C_{31}H_{40}N_4O_9S_2)$

Complexes in Solution. Stability constants of **$Mn^{II}L^+$** and **$Mn^{II}L_2$** with HL = ligand 1 (penicillin G) and ligand 2 (penicillin V) were determined in 50% aqueous acetone solutions at I = 0.1 M ($NaClO_4$) and various temperatures. The table shows average values of log K_1 and log K_2, obtained by the half $\bar{n}$ and least squares methods:

ligand	1	1	1	2	2	2
t in °C	20	30	40	20	30	40
log K_1	4.96	4.75	4.51	4.50	4.27	3.85
log K_2	4.73	4.20	3.49	4.05	3.78	3.50

The thermodynamic parameters (ΔH and ΔG in kcal/mol, ΔS in $cal \cdot mol^{-1} \cdot K^{-1}$) were determined at 30°C: $\Delta H_1 = -11.40$, $\Delta H_2 = -11.20$, $\Delta G_1 = -6.63$, $\Delta G_2 = -5.78$, $\Delta S_1 = 15.83$, $\Delta S_2 = -17.80$ for the stepwise formation of the complexes with ligand 1; $\Delta H_1 = -15.44$, $\Delta H_2 = -15.10$, $\Delta G_1 = -5.96$, $\Delta G_2 = -5.28$, $\Delta S_1 = -31.30$, $\Delta S_2 = -32.40$ for the complexes with ligand 2. The high negative values of ΔH indicate a considerable degree of covalency in the Mn^{II} ligand bond, which presumably involves coordination of the ring nitrogen atom [1]. Coordination of ligand 1 to Mn^{II} through the oxygen atom of the carboxylato group and the ring nitrogen atom was confirmed by ^1H NMR measurements in D_2O solutions at 25°C and pH 5.5. The transverse relaxation time for protons in the presence of Mn^{II} was determined from line-broadening measurements. The B conformation of the thiazolidine ring in the coordinated penicillin G molecule is slightly favored over conformation A. In the B type the sulfur atom is located nearer to Mn^{II} than in A type [2, 3].

$Mn^{II}L_2 \cdot 2H_2O$ complexes with HL = ligand 1 or 2 were precipitated upon raising the pH of the 50% aqueous acetone solution of an Mn^{II} salt [1, 4]. The IR spectra of the solid compounds in KBr show a shift of the vibration assigned to the tertiary nitrogen atom and the thiazolidine nucleus and a second shift ($\Delta\nu \approx 100\ cm^{-1}$) of the ν(CO) vibration. This indicates bonding of the ligand to Mn through the carboxylate oxygen and the ring nitrogen atoms. Coordination of the water molecules is indicated by bands occurring at ~3200, ~1600, and ~800 cm^{-1}. The octahedral coordination at Mn^{II} is supported by the thermal behavior, both complexes lose water on heating at 110°C [1].

$Mn^{II}(C_{31}H_{38}N_4Na_2O_9S_2)Cl_2$ was obtained by reaction of $MnCl_2$ and the disodium salt of ligand 3, a semisynthetic penicillin [5] in ethanol solution. The stable compound is slightly soluble in heptane, benzene, or carbon tetrachloride. It is proposed that the ligand coordinates to Mn^{II} via the carboxylate oxygen and the thiazolidine nitrogen atoms [6].

References:

[1] Chakrawarti, P. B.; Tiwari, A.; Sharma, H. N. (Indian J. Chem. A **21** [1982] 200/1).
[2] Fazakerley, G. V.; Jackson, G. E. (J. Chem. Soc. Perkin Trans. II **1975** 567/71).
[3] Fazakerley, G. V.; Jackson, G. E. (J. Inorg. Nucl. Chem. **37** [1975] 2371/5).
[4] Tiwari, A.; Sharma, H. N.; Chakrawarti, P. B. (Indian J. Chem. A **19** [1980] 83/5).
[5] Mudzhoyan, A. L.; Grigoryan, M. T.; Vardanyan, S. O.; Nazaryan, V. O.; Ter-Zakharyan, Yu. Z.; Oganyan, Sh. G.; Zhuravleva, L. P.; Mudzhoyan, Sh. L. (Khim. Farm. Zh. **10** [1976] 40/4; C.A. **85** [1976] No. 46542).
[6] Avakyan, S. N.; Pogosyan, L. E.; Avakyan, A. S.; Minasyants, M. Kh. (Mater. 4th Resp. Soveshch. Neorgan. Khim., Yerevan 1975 [1976], pp. 64/5; C.A. **91** [1979] No. 133333).

35.19.8 With Benzothiazole or Its Derivatives

Remark. Complexes with benzothiazolylazo compounds are described in "Manganese" D 5, 1987, pp. 316/7. For additional data, see [1]. Complexes with Schiff bases, hydrazones, and bisthiosemicarbazones containing the benzothiazolyl group are described in "Manganese" D 6, 1988, pp. 246/8, 251/4, 285/7, and 351. A benzothiazolesulfonamide is described on p. 125, a benzothiazole derivative of thiourea on p. 197, and a 2(3H)-benzothiazolethione on p. 61 of the present volume.

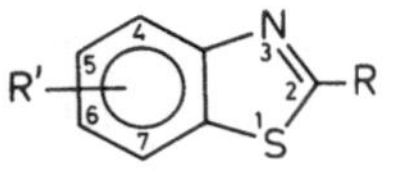

ligands 1 to 7

ligand	R	R′	formula	
1	H	H	C_7H_5NS	(= L)
2	H	4-OH	C_7H_5NOS	(= HL)
3	H	4-OH, 7-SO_3H	$C_7H_5NO_4S_2$	(= H_2L)
4	NH_2	4-OH	$C_7H_6N_2OS$	(= HL)
5	NH_2	6-CH_3	$C_8H_8N_2S$	(= L)
6	NHC(O)CH_3	6-CH_3	$C_{10}H_{10}N_2OS$	(= L)
7	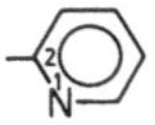	H	$C_{12}H_8N_2S$	(= L)

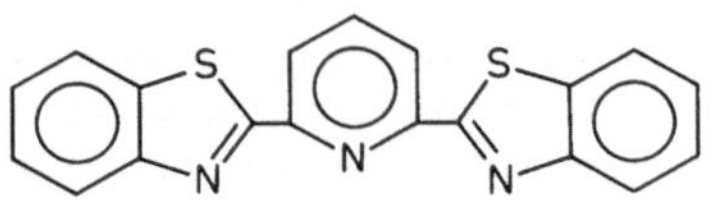

ligand 8 (= $C_{19}H_{11}N_3S_2$ = L)

Complexes in Solution. Potentiometric studies of solutions containing Mn^{2+} and ligand 2 or 4 reveal the formation of MnL^+ and MnL_2 complexes, while ligand 3 forms MnL and MnL_2^{2-} species. Stability constants of the complexes in 50 vol% aqueous dioxane at 25°C are summarized below:

ligand	ionic strength	log K_1	log K_2	Ref.
2	—	5.4	4.9	[2]
3	0.1 (NaCl)	5.1	3.9	[3]
4	0.1 (NaCl)	6.2	5.2	[3]

The stabilities are compared to those of analogous complexes of various metals such as Zn^{II}, Cd^{II}, Pb^{II}, Co^{II}, and Cu^{II} [2, 3].

[$Mn^{II}(C_7H_4NOS)_2$] was prepared by the reaction of ligand 2 in slight excess and manganese(II) perchlorate in an aqueous ethanol solution. The pH of the solution was raised slowly with continuous stirring until precipitation occurred. The reaction mixture was warmed on a water bath until the precipitate coagulated. The mixture was cooled, and the isolated compound washed thoroughly with water, and dried in a vacuum desiccator. The precipitate contained a large amount of water that was difficult to remove. The IR spectrum of the complex in Nujol shows characteristic bands at 1562, 1314, 1274, 1183, 1143, 945, 767, and 730 cm^{-1}. The bands at 1562 (1570) and 1314 (1308) cm^{-1} were assigned to the ν(C–O) vibration mode (free ligand bands in parentheses). A considerable shift in this vibration, as compared to the ligand, indicates binding to the metal by the oxygen atom of the uninegative ligand.

The complex is decomposed by heating at 105°C. It is insoluble in all common solvents [2].

[$Mn^{II}(C_8H_8N_2S)_4I_2$] was prepared by the reaction of hot ethanolic solutions of MnI_2 and ligand 5 in a 1:4 mole ratio. The reaction mixture was refluxed for 3 to 4 h and then evaporated to dryness. The residue was washed with water and 10% aqueous ethanol and dried in vacuum over anhydrous $CaCl_2$. The complex melts at 85°C. The magnetic moment, $\mu_{eff} = 6.10\ \mu_B$, resulting from susceptibility measurements at room temperature, indicates a high-spin octahedral $Mn^{II}(d^5)$ complex. IR spectral results for several different Mn^{II} complexes with ligand 5 (see p. 235) suggest that the neutral ligand is bonded to the Mn^{II} via the NH_2 group. The molar electrical conductivity of a 10^{-3} M acetate solution, $\Lambda = 0.04\ cm^2 \cdot \Omega^{-1} \cdot mol^{-1}$, demonstrates that the complex is a nonelectrolyte. It is soluble in methanol, acetone, and dioxane [4].

[$Mn^{II}L_2X_2$]. The complexes with ligand 1 or 5, [$Mn(C_7H_5NS)_2Cl_2$] and [$Mn(C_8H_8N_2S)_2I_2$], the chloro complex with ligand 6, [$Mn(C_{10}H_{10}N_2OS)_2Cl_2$], and [$Mn(C_{12}H_8N_2S)_2X_2$] complexes with ligand 7 (X = Cl, Br, I, NCS) were prepared by reaction of the MnX_2 salts with stoichiometric amounts of the corresponding benzothiazole. The [$Mn(C_8H_8N_2S)_2(NCS)_2$] complex was prepared from hot ethanolic solutions of $MnCl_2$, NH_4NCS, and ligand 5 in a 1:2:2 mole ratio [5]. The reaction mixture was refluxed for 2 to 4 h and then concentrated and cooled [6], or it was evaporated to dryness [4, 5, 7]. The product was washed with water and 10% aqueous ethanol and dried in vacuum over anhydrous $CaCl_2$ [4 to 7]. The [$Mn(C_7H_5NS)_2Cl_2$] complex was finally washed with light petroleum [8]. The complex with ligand 5, [$Mn(C_8H_8N_2S)_2I_2$] melts at 108°C [4], the complex with ligand 6, [$Mn(C_{10}H_{10}N_2OS)_2Cl_2$] at 59°C [7].

Magnetic moments of the [MnL_2X_2] complexes from susceptibility measurements at room temperature are listed below:

ligand L	5	6	7	7	7	7
X	I	Cl	Cl	Br	I	NCS
μ_{eff} in μ_B	5.90	6.02	5.86	5.87	5.93	5.88
Ref.	[4]	[7]	[6]	[6]	[6]	[6]

The IR spectra (Nujol mulls) of the complexes with ligand 5 show a considerable lowering in the frequencies due to the vibration of the exocyclic NH_2 group, whereas the cyclic ν(C–S–C) and ν(C=N) bands remain unchanged compared to the spectrum of the free ligand. These data suggest coordination of the neutral ligand to the metal through the NH_2 group [4, 5]. Bands at 2080 and around 2075 cm^{-1} in the spectra of the isothiocyanato complexes with ligand 5 or 7 were attributed to the ν(Mn–NCS) vibration mode [5, 6]. Important bands (in cm^{-1}) in the IR spectrum of [$Mn(C_{10}H_{10}N_2OS)_2Cl_2$] were assigned as follows (free ligand in parentheses): ν(NH) at 3120 (3200), ν(CO) at 1680 (1705), ν_{ring}(C=N) at 1525 (1545), ν(Mn–N) at 550, ν(Mn–O) at 240, ν(Mn–Cl) at 210, and amide I to V at 1680 (1705), 1495 (1525), 1300 (1295), 715 (710), 580 (575), respectively. A negative shift in the amide I and amide II bands and a positive shift in the amide III band indicate bonding through oxygen of the amide group. The nonligand bands at 550, 240, and 210 cm^{-1} suggest that the neutral bidentate ligand 6 is six-coordinated to manganese(II) via the ring nitrogen, the carbonyl oxygen, and the chlorine atoms [7]. As with the complexes of ligand 7, containing only one ligand molecule (see p. 236), the IR spectra of the [$Mn(C_{12}H_8N_2S)_2X_2$] complexes show coordination through pyridine and benzothiazole nitrogens [6]. The electronic absorption spectrum of [$Mn(C_{10}H_{10}N_2OS)_2Cl_2$] exhibits maxima at 10416, 14206, 18518, 23809, 34482, and 36364 cm^{-1} [7].

Thermal decomposition of [$Mn(C_7H_5NS)_2Cl_2$] was studied by differential scanning calorimetry in a nitrogen atmosphere with a heating rate of 16 K/min. The complex decomposes in two steps: losing 1.67 mol of ligand 1 at 455 to 510 K ($\Delta H = 23.1 \pm 0.1$ kcal/mol) and then 0.33 mol at 540 to 595 K ($\Delta H = 12.4 \pm 0.4$ kcal/mol). The specific heat of the complex can be calculated in the temperature range 320 to 400 K using $C_p = a + bT$, where a = 16.0 cal/mol and $b = 0.23\ cal \cdot mol^{-1} \cdot K^{-1}$ [8]. In boiling water the complexes with ligand 7 dissociate into Mn^{II} salts and free ligand [6].

The complexes are nonelectrolytes. [$Mn(C_8H_8N_2S)_2I_2$] is soluble in methanol, dioxane, and acetone [4]. The complexes with ligand 7 are fairly soluble in ethanol, methanol, and DMF, but are almost insoluble in water [6].

[$Mn^{II}(C_{12}H_8N_2S)_2(H_2O)_2](NO_3)_2$ was prepared by the method outlined for [$Mn^{II}L_2X_2$] complexes with ligand 7, described above. It was separated from the reaction mixture by adding an equal volume of ether. The magnetic moment is $\mu_{eff} = 5.90\ \mu_B$ at room temperature. The ν(OH) band at 3450 and the δ(OH) band at 840 cm^{-1} in the IR spectrum indicate coordination of the water molecules. The complex shows nitrate bands at 1381, 1270, and 812 cm^{-1} [6].

[$Mn^{II}LX_2$] complexes with ligand 5, [$Mn(C_8H_8N_2S)Br_2$], and ligand 7, [$Mn(C_{12}H_8N_2S)X_2$] (X = Cl, Br) were prepared by the reaction of hot ethanolic solutions of MnX_2 and the ligand in a 1:1 mole ratio [6, 9]. The reaction mixture was refluxed for a few hours and the [$Mn(C_{12}H_8N_2S)X_2$] complexes that separated were filtered off, washed with ethanol, and dried over $CaCl_2$ [6]. In the case of the [$Mn(C_8H_8N_2S)Br_2$] complex, the reaction mixture was evaporated to dryness, the residue washed with water, then 10% ethanol, and dried in vacuum over anhydrous $CaCl_2$ [9]. Yellow crystals of **[$Mn(C_{19}H_{11}N_3S_2)Cl_2$]·0.5$H_2O$** were obtained by the reaction of ligand 8 in hot $CHCl_3$ and a slight excess of ethanolic $MnCl_2$. The complex crystallized almost immediately, was isolated, washed with hot $CHCl_3$, then ethanol followed by diethyl ether, and dried in vacuum over P_4O_{10}. The magnetic moment of [$Mn(C_{19}H_{11}N_3S_2)Cl_2$]·0.5$H_2O$ is $\mu_{eff} = 5.92\ \mu_B$ at 298 K [10]. Magnetic moments, $\mu_{eff} = 5.88$ and 5.92 μ_B at 309 K, determined for [$Mn(C_{12}H_8N_2S)Cl_2$] and [$Mn(C_{12}H_8N_2S)Br_2$], indicate an octahedral geometry. The IR spectrum (KBr disks) of free ligand 7 exhibits bands at 1612 and 992 cm^{-1}, assigned to ($\nu_{C=N} + \nu_{C=C}$) and the pyridine breathing mode, respectively. A negative shift of −10 to −12 cm^{-1} in the first band and a positive shift of 8 to 10 cm^{-1} in the second band indicate the complexes coordinated through the pyridine and benzothiazole nitrogens. A six-coordinated polymeric structure with bridging halogen atoms was therefore proposed [6]. From magnetic moment measurements and IR spectral results (bands not given in the original), it was assumed that the neutral ligand 5 is bonded to the manganese(II) through the NH_2 group in a tetrahedral geometry [9]. The IR spectrum of [$Mn(C_{19}H_{11}N_3S_2)Cl_2$]·0.5$H_2O$ in Nujol mulls shows the ν(Mn–N) band at 365 cm^{-1}, which is characteristic for the tris-ligand complexes, $[M(N\text{–}N)_3]^{2+}$ (M = Ni^{II}, Co^{II}, Fe^{II}). The bands at 313, 295, and 273 cm^{-1} were assigned to the ν(Mn–Cl) vibration mode. A five-coordinate high-spin Mn^{II} complex with a neutral tridentate ligand chelated via a pyridine and two benzothiazole nitrogens was proposed. The X-ray powder diffraction patterns of the complex (not given in the original) are consistent either with a polymeric six-coordinate or five-coordinate structure [10].

The complex with ligand 5 is soluble in methanol, acetone, and chloroform, and it is a nonelectrolyte in each [9]. The low solubility of the [$Mn(C_{12}H_8N_2S)X_2$] complexes in ethanol, methanol, and acetone indicates their polymeric nature [6]. The complex with ligand 8 is virtually insoluble in water and common organic solvents [10].

References:

[1] Shimidzu, N.; Uno, T. (Chem. Pharm. Bull. [Tokyo] **25** [1977] 2942/7).
[2] Feng, Pao Kuo; Fernando, Q. (J. Am. Chem. Soc. **82** [1960] 2115/8).
[3] Feng, Pao Kuo; Fernando, Q. (Inorg. Chem. **1** [1962] 426/7).
[4] Chaurasia, M. R.; Shukla, Prema; Singh, N. K. (Def. Sci. J. **32** [1982] 75/9).
[5] Chaurasia, M. R.; Shukla, Prema (J. Indian Chem. Soc. **60** [1983] 1011/3).
[6] Chaturvedi, A. P.; Tiwari, H. N.; Mishra, L. K. (J. Indian Chem. Soc. **64** [1987] 306/7).
[7] Chaurasia, M. R.; Shukla, Prema; Singh, N. K. (Chem. Petro-Chem. J. **12** [1981] 3/8).
[8] Mortimer, C. T.; McNaughton, J. L. (Thermochim. Acta **6** [1973] 269/74).
[9] Chaurasia, M. R.; Shukla, Prema (Indian J. Phys. Nat. Sci. **2** [1982] 1/5).
[10] Livingstone, S. E.; Nolan, J. D. (J. Chem. Soc. Dalton Trans. **1972** 218/23).

35.19.9 With Derivatives of 2,3-Dihydro-benzothiazole

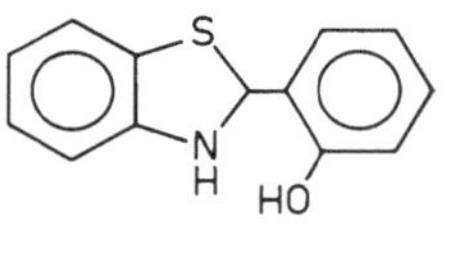

ligand 1 (= $C_{13}H_{11}NOS$ = HL)

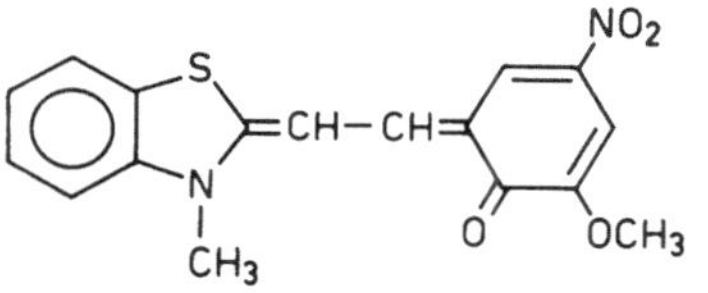

ligand 2 (= photomerocyanine = $C_{17}H_{14}N_2O_4S$ = L)

Complexes in Solution. Formation constants of the complexes with ligand 1, $Mn^{II}(C_{13}H_{10}NOS)^+$ and $Mn^{II}(C_{13}H_{10}NOS)_2$ in aqueous dioxane (50 vol%) were determined potentiometrically (glass electrode) at 25°C: log K_1 = 5.88, log K_2 = 4.56(?). The stabilities are compared to analogous complexes of other metals ($Cu^{II} > Co^{II} > Fe^{II} > Zn^{II} > Pb^{II} > Ni^{II} > Mn^{II}$), and to those with 8-quinolinol, 2-(2-benzoxazolyl)phenol, and 2-(2-benzothiazolyl)phenol [1, 2].

$[Mn^{II}(C_{13}H_{10}NOS)_2]$ was prepared by the reaction of an aqueous ethanol solution of $Mn(CH_3COO)_2$ with ligand 1 (mole ratio 1:2). The reaction mixture was stirred to yield a granular precipitate. The orange-yellow product was isolated, washed with aqueous ethanol, and dried in vacuum. The magnetic moment of the solid is μ_{eff} = 5.84 μ_B at 304 K. The IR spectral results suggest that the uninegative ligand is coordinated to manganese(II) through the phenolate oxygen and the benzothiazole nitrogen. The complex is a nonelectrolyte in DMF [3] and decomposes at 130°C [2]. It is only slightly soluble in ethanol and methanol, but is appreciably soluble in DMF [3].

$[Mn^{II}(C_{17}H_{14}N_2O_4S)Br_2]$ was obtained on stirring a suspension of appropriate amounts of $MnBr_2$ with an open spiropyrane form (photomerocyanine) in anhydrous acetone for 10 to 15 h. The red solid, which crystallized poorly, is isomorphous with the analogous complexes of Zn^{II} and Co^{II}. The IR spectrum shows one band at 428 cm^{-1}, assigned to ν(Mn–O), and two bands at 239 and ~200 cm^{-1}, due to ν(Mn–Br). There is no indication of a bond between the metal and the heterocyclic nitrogen atom; the heterocyclic merocyanine is dipolar, carrying a positive charge on the heterocyclic N atom and a negative charge on the carbonyl oxygen atom. It was proposed that the metal ion is surrounded by the oxygen atoms of the CO and OCH_3 groups in addition to the two Br^- ions in a pseudo-tetrahedral arrangement. The complex has limited solubility in all organic solvents [4].

References:

[1] Freiser, H. (Analyst **77** [1952] 830/45).
[2] Charles, R. G.; Freiser, H. (Anal. Chim. Acta **11** [1954] 1/11).
[3] Das, K.; Gupta, S. K.; Prasad, U. S.; Mishra, L. K. (J. Indian Chem. Soc. **59** [1982] 334/5).
[4] Le Baccon, M.; Guglielmetti, R. (J. Chem. Res. S **1979** 154; J. Chem. Res. M **1979** 1801/9).

35.19.10 With Derivatives of Thiadiazole

Remark. A thiadiazolylazo complex was described in "Manganese" D 5, 1987, p. 317, complexes with Schiff bases containing thiadiazolyl groups in "Manganese" D 6, 1988, pp. 41/2, 72, and 88. For complexes with sulfathiazole or a thiadiazolylsulfanylacetic acid, see pp. 117/21 and 84, respectively, in the present volume.

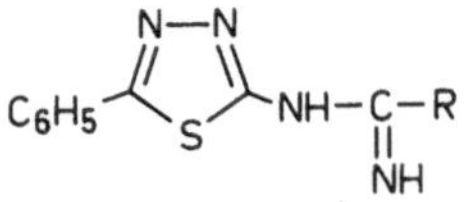

ligand 1 with R = CH_3 (= $C_{10}H_{10}N_4S$ = HL)
ligand 2 with R = C_6H_5 (= $C_{15}H_{12}N_4S$ = HL)

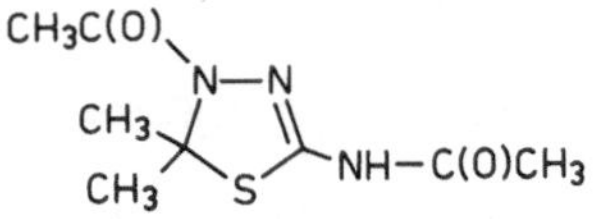

ligand 3
(= $C_8H_{13}N_3O_2S$ = L)

$[Mn^{II}(C_{10}H_9N_4S)_2(H_2O)_2]$ and **$[Mn^{II}(C_{15}H_{11}N_4S)_2(H_2O_2]$** complexes with ligand 1 or 2 were prepared by the reaction of an Mn^{II} salt and the sodium salt of the appropriate ligand in ethanol. The reaction mixture was stirred and refluxed for 3 to 4 h, then cooled. The solid was washed with water, followed by ethanol, and dried in a vacuum desiccator. The complexes do not melt up to 250°C. The magnetic moments, μ_{eff} = 5.94 μ_B for $[Mn(C_{10}H_9N_4S)_2(H_2O)_2]$ and 5.92 μ_B for $[Mn(C_{15}H_{11}N_4S)_2(H_2O)_2]$, result from susceptibility data at room temperature. The electronic spectra of the complexes in Nujol mull exhibit a charge transfer band at 35000 cm^{-1} and two d-d bands at 25100 and 16000 cm^{-1}, which were assigned to the transitions $^6A_{1g} \rightarrow {}^4A_{1g}(G)$, $^4E_g(G)$, $\rightarrow {}^4T_{2g}(G)$, $^4T_{1g}(G)$, respectively, in the distorted octahedral ligand field. The IR spectra of the complexes in KBr were discussed together with the spectra of the analogous complexes of Cd^{II}, Ni^{II}, Fe^{II}, and Cu^{II}. The disappearance of the free ligand ν(NH) band at 3400 cm^{-1} and a negative shift of 25 to 30 cm^{-1} in the ν(NN) band of the thiadiazole ring suggest that the uninegative bidentate ligand is coordinated to the octahedral Mn^{II} via the ring and exocyclic nitrogen atoms. The presence of coordinated water is confirmed by the bands in the 650 to 880 cm^{-1} range. The ν(CS) band at 718 or 730 cm^{-1} in the spectrum of free ligand 1 or 2, respectively, shows a positive shift of ca. 20 cm^{-1} on complexation. Coordination through the amide nitrogen is also confirmed by the ^1H NMR spectra of the complexes, which show a downfield shift of the NH resonance relative to the free ligand. The complexes are fungitoxic [1].

$[Mn^{II}(C_8H_{13}N_3O_2S)_2(H_2O)_2]Cl_2$ was prepared by the reaction of equal volumes of ethanol solutions of $MnCl_2$ and ligand 3 in a 1:2 mole ratio. The pH of the reaction mixture was increased to 7.0 by adding dilute ammonia. The reaction mixture was digested on a water bath for about 3 h and then cooled. The precipitate that forms was isolated, washed with ethanol, and dried in vacuum. The solid does not melt up to 350°C. A magnetic moment of 5.93 μ_B results from measurements of the magnetic susceptibility (Gouy's method) at room temperature. The electronic spectrum of the complex exhibits the same d-d and charge transfer bands as the complex above. The IR spectrum (CsI pellets) was discussed together with those of the analogous complexes (M^{II} = Ni, Co, and Fe). The free ligand bands of ν(NH), ν(C=N), and ν(NN) at 3410, 1630, and 915 cm^{-1}, respectively, remain at the same positions in the spectra of the complexes, suggesting that none of the nitrogen atoms take part in bonding with the metal. A downward shift of the ν(CO) vibration from 1680 to ~1650 cm^{-1} indicates coordination through the carbonyl oxygen atom. The ν(CS) vibration is shifted from 655 to 620 cm^{-1} on complex formation, showing coordination through the ring sulfur atom. The new bands appearing at 550 to 520 cm^{-1} and 380 to 335 cm^{-1} were assigned to ν(M–O) and ν(M–S), respectively. Bands of the coordinated water molecules are present in the 1000 to 800 cm^{-1} region. The molar electrical conductivity of a 10^{-3} M solution in nitromethane, Λ = 4.0 $cm^2 \cdot \Omega^{-1} \cdot mol^{-1}$, indicates a nonelectrolytic nature for the complex. The complex is soluble in acetone, methanol, nitromethane, DMF, DMSO, acetonitrile, and chloroform. It shows fungitoxic activity [2].

References:

[1] Srivastava, R. S. (Inorg. Chim. Acta **55** [1981] L71/L74).
[2] Thimmaiah, K. N.; Chandrappa, G. T.; Lloyd, W. D. (Inorg. Chim. Acta **107** [1985] 281/4).

35.19.11 With 1,4-Oxathiane

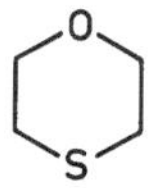

(= C_4H_8OS)

$[Mn^{II}(C_4H_8OS)_2(H_2O)_4](ClO_4)_2$ was prepared by the reaction of an excess of the ligand (mole ratio 6:1) with a solution of dehydrated manganese(II) perchlorate, made by warming (at 50°C) the hydrated salt in a 1:1 mixture of triethyl orthoformate and ethanol and stirring for 1 h. The reaction mixture was evaporated and the residue treated with a little ligand to yield an off-white crystalline product. The complex was isolated under a dry nitrogen atmosphere, washed with anhydrous diethyl ether, and dried in an evacuated desiccator over P_4O_{10} or $CaCl_2$. The magnetic moment of the solid is $\mu_{eff}=6.13\ \mu_B$ at 300 K. The IR spectrum of the complex (in Nujol) was investigated, but no satisfactory resolution could be obtained. Bonding of the monodentate ligand only through the sulfur atom was assumed. Bands of the ionic perchlorate groups were observed at 1100 to 1080 cm^{-1}. The presence of coordinated water is confirmed by the ν(OH) band at 3500 to 3400 cm^{-1} and the δ(HOH) band at 1640 to 1620 cm^{-1}. The water molecules apparently form strong covalent bonds with the metal ion, since they cannot be removed by prolonged desiccation. A distorted octahedral structure is assumed for the complex cation. The molar electrical conductivity of a 0.001 M solution of the compound in nitromethane, $\Lambda=154\ cm^2\cdot\Omega^{-1}\cdot mol^{-1}$, indicates a 1:2 electrolyte. The atmosphere-stable complex is soluble in the parent ethers and many organic solvents [1].

$Mn^{II}(C_4H_8OS)X_2\cdot CH_3CN$ (X = Cl, Br) were prepared as follows: The manganese halide was treated with methyl cyanide, the excess CH_3CN was removed, and an excess of ligand in dichloromethane was added. After standing for one day at room temperature, the solvent was removed to give a pink residue. Magnetic moments (μ_{eff} in μ_B) and important IR bands (in cm^{-1}) of the complexes and the ligand in Nujol mulls are listed below:

compound	μ_{eff}	ν_{as}(COC)	ν_s(COC)	ν_{as}(CSC)	ν_s(CSC)	ν(CN)*)	Ref.
X = Cl	5.65	1104, 1075	821, 815	693	666	2310, 2287	[2]
X = Br	5.97	1078	823	—	665	2305, 2275	[2]
C_4H_8OS	—	1102, 1050	830	693	664	—	[1, 2]

*) Uncoordinated CH_3CN reveals bands at 2253 and 2288 cm^{-1}.

The IR spectrum suggests coordination of the monodentate oxathiane to manganese only through the oxygen atom, and of methyl cyanide through the nitrogen atom. The diffuse reflectance electronic absorption spectra show maxima at ~7000, 15600, 20400, 30400, and 44000 cm^{-1} (for X = Cl), at ~8400, 26600, 30200, 34700, 34800, and 44400 cm^{-1} (for X = Br). The insolubility of the complexes implies a polymeric structure with the manganese atom possessing an octahedral configuration [2].

References:

[1] Karayannis, N. M.; Mikulski, C. M.; Speca, A. N.; Cronin, J. T.; Pytlewski, L. L. (Inorg. Chem. **11** [1972] 2330/5).
[2] Anagnostopoulos, A. (J. Inorg. Nucl. Chem. **37** [1975] 268/9).

35.19.12 With Thiomorpholin-3-one

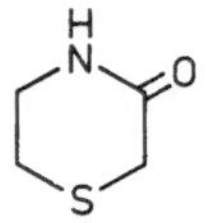

(= C_4H_7NOS = L)

The Mn^{II} complexes listed in the table below were prepared by reaction of the appropriate manganese(II) salt with a small excess of the molten ligand. The compounds were purified by repeated washing with diethyl ether. The table lists color, decomposition range, magnetic moment (μ_{eff} in μ_B) at room temperature, and main IR bands (ν in cm^{-1}) in KBr or Nujol:

compound	color	t_{dec} in °C	μ_{eff}	ν(NH)	ν(CO)	ν_{as}(CS)	ν_s(CS)	ν(Mn–O)
$[MnL_6](ClO_4)_2 \cdot 2H_2O$	white	83 to 85	—	3340	1595	700	650	364 355
$[MnL_6](BF_4)_2$	white	148 to 150	6.2	3320	1625	700	650	365 355
$[MnL_4Cl_2] \cdot 2H_2O$	pink-white	84 to 86	5.9	3290	1625	698	650	359
$[MnL_4Br_2]$	pink-white	78 to 80	5.9	3300	1610	695	648	358
$[MnL_4I_2]$	white-yellow	132 to 134	5.9	3280	1620	690	650	359
$[MnL_4SO_4] \cdot H_2O$	white	85 to 87	6.0	3280	1625	700	650	361 354
$[MnL_2(CH_3COO)_2] \cdot 2H_2O$	white	82 to 84	6.4	3400	1600	695	650	353
C_4H_7NOS = L*)	—	—	—	3400	1660	690	650	—

*) IR data from chloroform solution.

The IR spectra indicate that the ligand is bonded to Mn^{II} through the oxygen atom. Coordination of the chloride, bromide, or iodide ions is indicated by ν(Mn–X) vibration modes observed at 228, 208, and 154 cm^{-1}, respectively. The sulfato complex shows bands typical of bidentately coordinated sulfato groups in a C_{2v} symmetry. Bands of ν_{as}(COO) at 1510 and ν_s(COO) at 1410 cm^{-1} indicate monodentate coordination of the carboxylato group. The perchlorate and fluoroborate groups are not coordinated to Mn. Solid state electronic band maxima assigned to transitions from the $^6A_1$ ground state, and the calculated ligand field parameters (in cm^{-1}) are:

compound	$^6A_1 \rightarrow {}^4T_{1g}(G)$	$\rightarrow {}^4A_{1g}, {}^4E_g(G)$	$\rightarrow {}^4E_g(D)$	$\rightarrow {}^4T_{1g}(P)$	$\rightarrow {}^4T_{1g}(F)$	Dq	B	C
$[MnL_6](ClO_4)_2 \cdot 2H_2O$	20320	24880	28570	34245	39800	740	493	3990
$[MnL_6](BF_4)_2$	20490	24810	28330	34850	39840	723	537	3880
$[MnL_4Cl_2] \cdot 2H_2O$	20410	24690	28490	33670	39700	656	543	3852
$[MnL_4Br_2]$	20920	24270	28010	32360	39215	594	534	3786
$[MnL_4I_2]$	21100	23920	27626	33110	37880	508	529	3726
$[MnL_4SO_4] \cdot H_2O$	19920	24700	28410	35080	39680	770	530	3880
$[MnL_2(CH_3COO)_2] \cdot 2H_2O$	20410	24510	29070	33100	39850	589	651	3600

The data agree with an octahedral environment at Mn^{II}. In 10^{-3} M nitromethane solution the perchlorate and fluoroborate complexes are 1:2 electrolytes, consistent with electrical conductivities of Λ=167.8 and 177.6 $cm^2 \cdot \Omega^{-1} \cdot mol^{-1}$. All the other compounds behave as non-electrolytes. They are soluble in nitromethane and dimethylformamide. In many other (coordinating) solvents in which the solubility is relatively high, dissociation is accompanied by extensive solvolysis. The perchlorate complex is hygroscopic.

Reference:

Preti, C.; Tosi, G. (Australian J. Chem. **29** [1976] 543/9).

35.19.13 With 10H-Phenothiazine

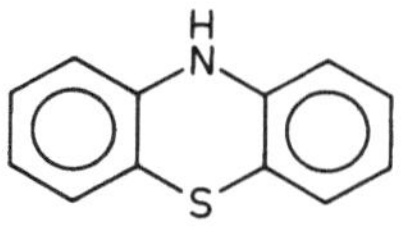

(= $C_{12}H_9NS$)

Remark. Mixed ligand complexes of Mn containing the 3,7-diaminophenothiazin-5-ium cation are described, together with complexes of dithiol compounds, see p. 49.

The complex formation of phenothiazine was examined spectroscopically in the surfactant micellar solutions of bivalent metal dodecyl sulfates, $M(C_{12}H_{25}OSO_3)_2$, with M^{II} = Mg, Mn, Ni, Co, and Cu, where the bivalent metals constitute the counter ions of micelles. Formation of a manganese complex was indicated by absorption maxima at 660 nm ($\varepsilon = 6.15 \times 10^2$ $L \cdot mol^{-1} \cdot cm^{-1}$) and 457 nm.

Reference:

Moroi, Y.; Saito, M.; Matuura, R. (Nippon Kagaku Kaishi **1980** 482/5; C. A. **92** [1980] No. 186531).

35.19.14 With a Macrocyclic Ligand

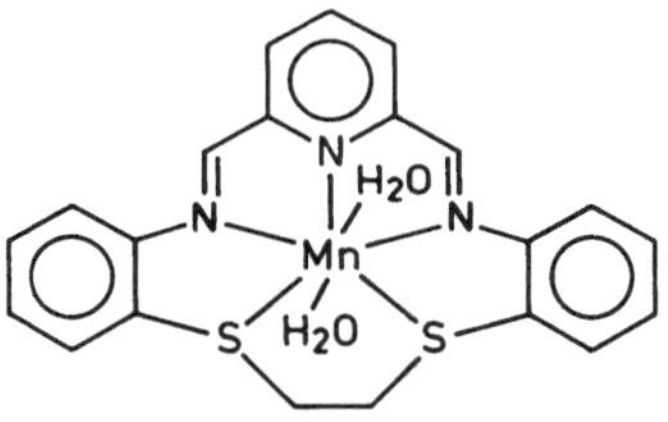

(= $C_{21}H_{17}N_3S_2$)

$[Mn(C_{21}H_{17}N_3S_2)(H_2O)_2](ClO_4)_2$ was prepared by a template reaction of 2,6-pyridinedicarbaldehyde, 2,2′-(1,2-ethanediyl)bis(sulfanyl)bis(benzenamine), and a slight excess of manganese(II) perchlorate in methanol or acetonitrile. The reaction mixture was kept at room temperature for 24 h. The reaction in methanol yielded an amorphous orange precipitate, which was washed with a little cold methanol and dried under reduced pressure. The reaction in acetonitrile resulted in an orange solution from which crystals were obtained by slow evaporation under nitrogen. It is assumed that the three nitrogen and two sulfur atoms of the 15-membered ligand ring are coordinated to Mn in an approximately planar equatorial arrangement, while the water molecules are axially bonded [1]; preliminary report [2].

References:

[1] Liles, D. C.; McPartlin, M.; Tasker, P. A. (J. Chem. Soc. Dalton Trans. **1987** 1631/6).
[2] Alcock, N. W.; Liles, D. C.; McPartlin, M.; Tasker, P. A. (J. Chem. Soc. Chem. Commun. **1974** 727/8).

36 Complexes with Ligands Containing Selenium or Tellurium

36.1 With Diphenylselane or Diphenyltellane

$(C_6H_5)_2Se$ ($=C_{12}H_{10}Se$)

$(C_6H_5)_2Te$ ($=C_{12}H_{10}Te$)

[Mn(NO)$_3$(C$_{12}$H$_{10}$Se)] and [Mn(NO)$_3$(C$_{12}$H$_{10}$Te)]. To an n-pentane solution of [Mn(NO)$_3$THF], obtained by UV irradiation of equimolar amounts of $Mn_2(CO)_{10}$ and $[Co(NO)_2Cl]_2$ in THF, a threefold molar quantity of $(C_6H_5)_2Se$ or $(C_6H_5)_2Te$ was added to replace the THF. The reaction with diphenyl selenide was not quantitative. The green to brown solutions are stable at room temperature. The NMR spectra investigated at 16.0 MHz in CH_2Cl_2 show chemical shifts, relative to saturated aqueous $KMnO_4$, of $\delta(^{55}Mn)=-1250$ ppm for [Mn(NO)$_3$(C$_{12}$H$_{10}$Se)] and -1130 ppm for [Mn(NO)$_3$(C$_{12}$H$_{10}$Te)]. The shielding of the Mn nucleus in the complex, quantified by the δ parameter, does not increase, as expected, on going down the chalcogen series. (For the corresponding complex with methylphenylsulfane, [Mn(NO)$_3$(C$_7$H$_8$S)], a chemical shift of $\delta(^{55}Mn)=-1130$ ppm was observed.)

Reference:

Rehder, D.; Ihmels, K.; Wenke, D.; Oltmanns, P. (Inorg. Chim. Acta **100** [1985] L11/L12).

36.2 With 8-Quinolineselenol

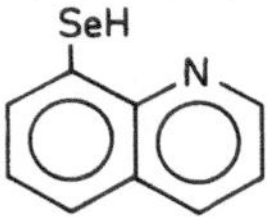

($=C_9H_7NSe=HL$)

Mn^{II}(C$_9$H$_6$NSe)$_2$ was prepared by addition of the sodium salt of the ligand (in 50% excess) to an aqueous buffered manganese(II) salt solution at 65 to 75°C and pH 5.9 [1] or 5.2 [2]. After 10 min the gray-brown product was collected, washed with hot (80°C) filtrate adjusted to the desired pH, then with hot water, and dried over silica gel. All operations were carried out under nitrogen [1]. Precipitation of a brown complex at pH 3 to 6 was reported in [3]. The effect of acidity on the precipitation of various bivalent metal complexes has been studied. The minimum pH value required for favorable precipitation of the 1:2 complexes increases in the order $Zn^{II}<Cd^{II}<Ni^{II}<Co^{II}<Pb^{II}<Cu^{II}<Mn^{II}$ [1].

The IR spectra recorded in KBr disks or Nujol mulls in the 4000 to 400 cm^{-1} region exhibit strong bands at 1487, 975, 776, 746, and 660 cm^{-1}, which may originate from the metal chelate. Bands at 962, 788, 761, 757, and 642 cm^{-1} may arise from the diselane, $(C_9H_6NSe)_2$, since the ligand is easily oxidized to the diselane by air, especially in alkaline solution. The percentage of the regular metal chelate is estimated to be about 40% for the compound obtained at pH 5.2, and about 10% for the sample obtained at pH 10. Thus, even at the pH value most favorable for complete precipitation, the percentage of regular Mn^{II} chelate is the lowest among the examined metal complexes of 8-quinolineselenol. This indicates a rather low stability for $Mn(C_9H_6NSe)_2$ [4]. The complex is easily converted into a white-brown solid even on gentle heating, and is decomposed by heating with a mixture of HNO_3 and $HClO_4$ [1].

References:

[1] Sekido, E.; Fujiwara, I.; Masuda, Y. (Talanta **19** [1972] 479/87).
[2] Mido, Y.; Fujiwara, I.; Sekido, E. (J. Inorg. Nucl. Chem. **36** [1974] 537/41).

[3] Schneeweis, G.; König, K.-H. (Z. Anal. Chem. **316** [1983] 16/22).
[4] Mido, Y.; Fujiwara, I.; Sekido, E. (J. Inorg. Nucl. Chem. **36** [1974] 1003/10).

36.3 With Dimethyl Selenoxide

$(CH_3)_2SeO$ ($=C_2H_6OSe$)

$[Mn^{II}(C_2H_6OSe)_6](ClO_4)_2$ was prepared by adding dropwise an acetone solution of hydrated $Mn(ClO_4)_2$ containing triethyl orthoformate to an acetone solution of dimethyl selenoxide (mole ratio 1:8) under dry nitrogen. The slightly pink crystalline compound was dried under reduced pressure at room temperature. In contrast to its components the complex is not hygroscopic. The IR spectrum (Nujol mull) shows bands of ν(SeO) + ν(SeC) at 1430 and 1420 cm^{-1}, ν(SeO) at 787 cm^{-1}, ν_{as}(SeC) at 593 cm^{-1}, ν_s(SeC) at 580 cm^{-1}, and ν(Mn–O) at ~400 cm^{-1}. A considerable negative shift of the ν(SeO) vibration mode, as compared to the free ligand ($\Delta\nu = -38$ cm^{-1}), is typical of coordination through the oxygen atom of the ligand. The magnetic moment, $\mu_{eff} = 5.99$ μ_B, as well as the electronic spectrum (not completely evaluated), support a strong octahedral MnO_6 structure. The ligand field parameters derived from the spectral data are: B = 800 cm^{-1} and $\beta = 0.90$.

Reference:

Paetzold, R.; Bochmann, G. (Z. Anorg. Allgem. Chem. **368** [1969] 202/10).

36.4 With Diphenyl Selenoxide or Diphenyl Telluroxide

$(C_6H_5)_2SeO$ ($=C_{12}H_{10}OSe$) $(C_6H_5)_2TeO$ ($=C_{12}H_{10}OTe$)

$Mn^{II}Cl_2 \cdot nC_{12}H_{10}OSe$ (n = 2,3) and **$Mn^{II}Cl_2 \cdot 2C_{12}H_{10}OTe$.** The complex $MnCl_2 \cdot 2C_{12}H_{10}OSe$ was prepared by adding dropwise a warm solution of 3.17 mmol diphenyl selenoxide in 25 mL CCl_4 to one of anhydrous $MnCl_2$ (1.47 mmol) in a mixture of ethanol (5 mL) and CCl_4 (10 mL). The almost colorless crystals which precipitated after short refluxing were collected, washed with benzene, and dried under reduced pressure. To prepare $MnCl_2 \cdot 3C_{12}H_{10}OSe$, a solution of anhydrous $MnCl_2$ (2.16 mmol) in 15 mL of ethanol was added dropwise with stirring to the benzene solution of the ligand (13 mmol in 20 mL). The initially colorless mixture gradually became red, and the slightly green crystals, which deposited after ca. 12 h, were washed with benzene and dried [1]. $MnCl_2 \cdot 2C_{12}H_{10}OTe$ is reported to form within 24 h from $MnCl_2$ and diphenyl telluroxide in ethanol solution at 5°C. The complex, recrystallized from ethanol, melts at 246 to 247°C [2]. The IR spectra of the diphenyl selenoxide complexes recorded in Nujol mulls show absorption bands assignable to ν(SeO) at 798 and 809 cm^{-1} for $MnCl_2 \cdot 2C_{12}H_{10}OSe$ and at 800 cm^{-1} for $MnCl_2 \cdot 3C_{12}H_{10}OSe$. The considerable negative shift of ν(SeO), compared to the ligand with ν(SeO) = 831 cm^{-1}, indicates coordination of the ligand molecules through the selenoxide oxygen atoms [1].

References:

[1] Paetzold, R.; Vordank, P. (Z. Anorg. Allgem. Chem. **347** [1966] 294/303).
[2] Khandelwal, B. L.; Jain, S. K. (Inorg. Chim. Acta **59** [1982] 193/6).

36.5 With Benzeneseleninic Acid or Its Derivatives

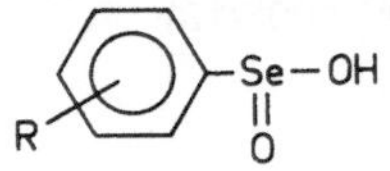

(=HL)

ligand	R	formula
1	H	$C_6H_6O_2Se$
2	4-Cl	$C_6H_5ClO_2Se$
3	3-Cl	$C_6H_5ClO_2Se$
4	4-Br	$C_6H_5BrO_2Se$
5	3-Br	$C_6H_5BrO_2Se$
6	$4-CH_3$	$C_7H_8O_2Se$

$Mn^{II}L_2$ complexes with ligands 1 to 6 were obtained by heating the corresponding aqua complexes (see p. 245) at 170 to 180°C for 4 h. Color, magnetic moments, characteristic IR bands of the complexes (Nujol mulls or KBr disks, 4000 to 50 cm^{-1} range), and bands observed in the solid state electronic spectra from 45000 to 4000 cm^{-1} are tabulated below [1]:

No.	complex	color	μ_{eff} in μ_B	IR data in cm^{-1} ν_{as}(SeO)	ν_s(SeO)	ν(SeC)	δ(OSeC)	ν(Mn–O)
1	$Mn(C_6H_5O_2Se)_2$	ivory	5.92	748	727	660	400, 350	421
2	$Mn(C_6H_4ClO_2Se)_2$	brown	5.94	808	730	695(sh)	395, 358	431
3	$Mn(C_6H_4ClO_2Se)_2$	light brown	5.89	780	730	648	385, 348	440
4	$Mn(C_6H_4BrO_2Se)_2$	light brown	5.91	796	725	708	390, 345	425
5	$Mn(C_6H_4BrO_2Se)_2$	pink-white	6.12	775	720	641	380, 335	433
6	$Mn(C_7H_7O_2Se)_2$	ivory	6.14	800	745	698	385, 380	432

No.	electronic spectra, transitions from $^6A_{1g}$ and ligand field parameters in cm^{-1} $\rightarrow{}^4T_1(^4G)$	$\rightarrow{}^4E, {}^4A_1(^4G)$	$\rightarrow{}^4E(^4D)$	$\rightarrow{}^4T_1(^4P)$	$\rightarrow{}^4T_1(^4F)$	10Dq	B	C	Z*
1	19800	23950	27600	32570	—	6270	521	3748	0.56
2	20500	25000	28985	34250	39700	5680	569	3862	0.70
3	20200	24570	28410	33670	39000	5740	549	3816	0.64
4	20400	25840	29670	33600	—	6000	547	4074	0.64
5	20000	24750	28570	33600	39370	6040	546	3858	0.63
6	20000	24700	28570	34010	39700	6310	553	3834	0.66

The magnetic moments in the range 5.89 to 6.14 μ_B are consistent with high-spin $Mn^{II}(d^5)$ complexes. The 10Dq values decrease in the order $R = 4-CH_3 > H > 3-Br > 4-Br > 3-Cl > 4-Cl$ and are lower than those of the corresponding aqua complexes (see p. 245). The IR bands due to the SeO_2 group are shifted toward higher wavenumbers in comparison with those of the free ligand. Coordination of the ligand anion to manganese through both oxygen atoms is suggested. The observed differences between ν_{as}(SeO) and ν_s(SeO) vibrations (more pronounced than in the tetragonally-distorted octahedral dihydrates) are indicative of a pseudotetrahedral structure. The ν(Mn–O) vibrational modes were observed at higher wavenumbers than in the aqua complexes, in accordance with the fact that the change in the stereochemistry accompanied by a decrease in coordination number causes an increase in the metal-ligand stretching vibration. The Z* values in the 0.56 to 0.70 range, the nephelauxetic parameters (β) in the 0.59 to 0.64 range, and the nonconductivity in DMF indicate the covalent nature of the metal-ligand bond. The compounds are microcrystalline or powder-like, stable in the atmosphere, and soluble in most common organic solvents [1, 2].

[$Mn^{II}L_2(H_2O)_2$]. An aqueous solution of a manganese(II) salt was added dropwise to an aqueous solution of the sodium salt of the ligand (mole ratio 1:2). The mixture was stirred for 5 to 10 h at room temperature, the crystalline precipitate then collected, washed with water, acetone, and diethyl ether, and dried in vacuum over P_4O_{10}. Color, magnetic moments, μ_{eff}, characteristic IR bands, and bands observed in the solid state electronic spectra are tabulated below [3]:

No.	complex	color	μ_{eff} in μ_B	IR data in cm^{-1} ν_{as}(SeO)	ν_s(SeO)	ν(SeC)	δ(OSeC)	ν(Mn–O)	ν(Mn–OH_2)
1	[$Mn(C_6H_5O_2Se)_2(H_2O)_2$]	pink-white	5.85	751	735	667	390sh, 350	420	328
2	[$Mn(C_6H_4ClO_2Se)_2(H_2O)_2$]·$H_2O$	pink-white	6.13	798	765	—	400sh, 367	430	333
3	[$Mn(C_6H_4ClO_2Se)_2(H_2O)_2$]	white	5.90	777	749	647	—, 342	435	318
4	[$Mn(C_6H_4BrO_2Se)_2(H_2O)_2$]	white	5.96	795	748	705	390sh, 352	423	320
5	[$Mn(C_6H_4BrO_2Se)_2(H_2O)_2$]	white	6.38	780	750	641	384, 342	430	330
6	[$Mn(C_7H_7O_2Se)_2(H_2O)_2$]·2$H_2O$	pink-white	6.40	807	789	702	378, 342	430	320

No.	electronic spectra, transitions from $^6A_{1g}$ and ligand field parameters in cm^{-1} →$^4T_{1g}(^4G)$	→^4E_g,$^4A_{1g}(^4G)$	→$^4E_g(^4D)$	→$^4T_{1g}(^4P)$	→$^4T_{1g}(^4F)$	10Dq	B	C	Z*
1	20830	25125	29585	34000	41320	7050	637	3751	0.94
2	20490	25050	29670	34485	40980	7010	660	3690	1.03
3	20240	24875	29325	34250	40650	7110	636	3703	0.94
4	20530	25250	29600	35200	40800	7030	621	3808	0.88
5	20000	24630	29240	33780	40500	7080	659	3608	1.03
6	20410	24940	29410	34780	40600	7110	639	3710	0.95

Shifts of ν_s(SeO) and ν_{as}(SeO) toward higher wavenumbers in comparison to those of the free ligand suggest coordination through both oxygens of the ligand ions. The IR spectra also show bands characteristic of coordinated water with well-resolved ν_{as}(OH) bands in the 3400 to 3360 cm^{-1} range, ν_s(OH) bands in the 3300 to 3230 cm^{-1} region, and δ(HOH) around 1655 cm^{-1}. New bands appearing in the 970 to 935 and 665 to 600 cm^{-1} regions can be assigned to rocking, twisting, and wagging modes of coordinated water. These observations are also confirmed by a far-IR band in the 333 to 318 cm^{-1} range, assignable to ν(Mn–OH_2). A distorted octahedral geometry of the aqua complexes was proposed, which is supported by the results of the electronic spectra and by the high-spin magnetic moments. The considerably low Z* values in the 0.88 to 1.03 range, as well as the nonconductivity in methanol and ethanol, indicate the covalent nature of these compounds. The complexes with ligands 2 and 6 possess, respectively, one and two additional water molecules of crystallization.

The complexes are very stable and are soluble in methanol and ethanol [3]; see also [2, 4].

References:

[1] Candrini, G.; Malavasi, W.; Preti, C.; Tosi, G.; Zannini, P. (Spectrochim. Acta A **39** [1983] 635/9).

[2] Candrini, G.; Preti, C.; Tosi, G. (16th Congr. Nazl. Chim. Inorg. Atti, Ferrara, Italy, 1983, pp. 154/6; C.A. **100** [1984] No. 28858).

[3] Preti, C.; Tosi, G. (Australian J. Chem. **33** [1980] 1203/11).

[4] Preti, C.; Tosi, G.; Zannini, P. (13th Congr. Nazl. Chim. Inorg. Atti, Camerino, Italy, 1980, pp. 171/3; C.A. **94** [1981] No. 218838).

36.6   With Selanyl or Tellanyl Carboxylic Acids and Their Derivatives

36.6.1   With Arylselanylacetic Acids

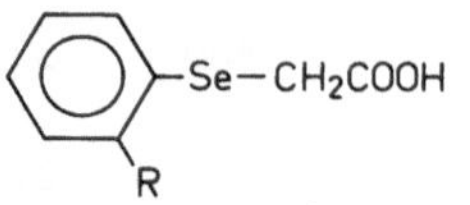

$\langle\bigcirc\rangle$—Se—CH$_2$COOH   ligand 1 with R = H   (= $C_8H_8O_2Se$)

ligand 2 with R = 2-CH$_3$O (= $C_9H_{10}O_3Se$)

Complexes in Solution. Formation constants of MnL$^+$ complexes in aqueous solution were determined using a method based on equilibria in which Ag$^+$ and Mn^{2+} ions compete for coordination to the ligands. Potential measurements were made by silver-silver chloride and glass electrodes versus a saturated mercury-mercurous sulfate electrode. At 25°C and ionic strength I = 0.1M (KNO$_3$) the formation constants are K_1 = 2.1(2) and 3.1(2) L/mol for the complexes Mn($C_8H_7O_2Se$)$^+$ and Mn($C_9H_9O_3Se$)$^+$, respectively. The complexes are of low stability and are not in line with the Irving-Williams order.

Reference:

Ford, G. J.; Gans, P.; Pettit, L. D.; Sherrington, C. (J. Chem. Soc. Dalton Trans. **1972** 1763/5).

36.6.2   With Selanediyl or Tellanediyl Dicarboxylic Acids

Se $\begin{array}{c} \diagup R-COOH \\ \diagdown R-COOH \end{array}$ Te $\begin{array}{c} \diagup CH_2CH_2COOH \\ \diagdown CH_2CH_2COOH \end{array}$

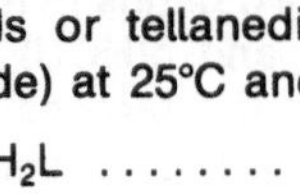

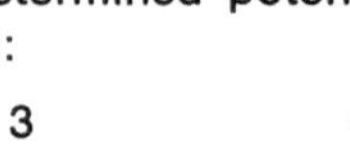

ligand 1 with R = CH$_2$ (= $C_4H_6O_4Se$)   ligand 4 (= $C_6H_{10}O_4Te$)

ligand 2 with R = CH$_2$CH$_2$ (= $C_6H_{10}O_4Se$)

ligand 3 with R = CH(CH$_3$) (= $C_6H_{10}O_4Se$)

Complexes in Solution. Formation constants of MnL complexes with selanediyl dicarboxylic acids or tellanediyl dicarboxylic acid have been determined potentiometrically (glass electrode) at 25°C and ionic strength I = 0.1M (KNO$_3$) [1]:

ligand H$_2$L	1	2	3	4
log K$_1$	2.02(1)	1.50(4)	2.02(1)	1.2(1)

A stability constant, log K_1 =1.6, was determined at 25°C and I = 0.1M (NaClO$_4$) for the complex with ligand 1, (Mn($C_4H_4O_4Se$) [2]. For the formation of the species Mn($C_4H_5O_4Se$)$^+$ with the monodeprotonated ligand 1 a stability constant, log K_{MnHL}^{Mn} = 0.88(8), was obtained [1]. On comparison with other dicarboxylic acids of the X(RCOOH)$_2$ type the stability of the manganese complexes was found to decrease in the order X = O > S > Se > CH$_2$ > Te [1, 2]. Coordination of selenium or tellurium to manganese was assumed by [2].

References:

[1] Laing, D. K.; Pettit, L. D. (J. Chem. Soc. Dalton Trans. **1975** 2297/301).
[2] Suzuki, K.; Yamasaki, K. (J. Inorg. Nucl. Chem. **28** [1966] 473/80).

36.7 With O-Pentyl Carbonodiselenoate

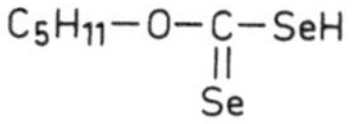

$$C_5H_{11}-O-\underset{\underset{Se}{\|}}{C}-SeH \quad (=C_6H_{12}OSe_2=HL)$$

$Mn^{III}(C_6H_{11}OSe_2)_3$ is reported to form as a red-brown oil on reaction of potassium carbono-diselenoate and a manganese(III) salt in a mixture of water and benzene at ca. 0°C. The complex is soluble in gasoline and acts as an effective antiknock agent in nonleaded gasoline.

Reference:

Loder, W. R., Jr.; Fay, P. L.; Veatch, F. (U.S. 3976440 [1976]; C.A. **86** [1977] No. 158134).

36.8 With N, N-Dialkylselenothiocarbamic Acids or -diselenocarbamic Acids

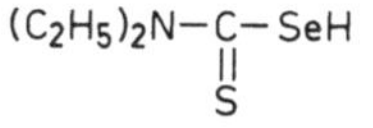

$$(C_2H_5)_2N-\underset{\underset{S}{\|}}{C}-SeH \qquad (C_4H_9)_2N-\underset{\underset{Se}{\|}}{C}-SeH$$

$$(=C_5H_{11}NSSe=HL) \qquad (=C_9H_{19}NSe_2=HL)$$

$Mn^{III}(C_5H_{10}NSSe)_3$. A methanolic or aqueous solution of manganese(II) chloride or nitrate (0.005 mol in 25 mL) was added with stirring to a methanolic solution of ligand 1 (0.015 mol in 50 mL). The black manganese(III) complex, precipitating due to air oxidation of the solution, was sucked off, and was recrystallized from a mixture of CH_2Cl_2 and petroleum ether. It melts at 80°C with decomposition. The IR spectrum (KBr disks) exhibits an absorption band of $\nu(C-N)$ at 1500 cm^{-1}. The magnetic moment, $\mu_{eff}=5.05$ μ_B at 23°C, is indicative of four unpaired d electrons. The electronic spectrum of the complex in $CHCl_3$ solution shows a d-d band at 18700 cm^{-1} (log $\varepsilon=3.41$). The complex is less stable to air oxygen than the corresponding compounds of most other metals [1].

$Mn^{III}(C_9H_{18}NSe_2)_3$ was prepared, similarly to $Mn^{III}(C_5H_{10}NSSe)_3$, by air oxidation of a metha-nolic solution containing $MnCl_2$ (0.005 mol) and ligand 2 (0.01 mol). The precipitate was redis-solved in 30 mL CH_2Cl_2. After addition of methanol, black leaflets were obtained which melt at 59°C. A magnetic moment of $\mu_{eff}=5.03$ μ_B was found at 23°C. The tris-chelate is unstable in $CHCl_3$ and less stable to air oxygen than the corresponding compounds of other metal ions [2].

References:

[1] Heber, R.; Kirmse, R.; Hoyer, E. (Z. Anorg. Allgem. Chem. **393** [1972] 159/67).
[2] Lorenz, B.; Kirmse, R.; Hoyer, E. (Z. Anorg. Allgem. Chem. **378** [1970] 144/51).

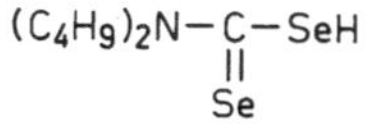

36.9 With Selenourea

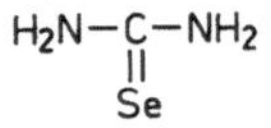

($=CH_4N_2Se$)

$Mn^{II}Cl_2 \cdot nCH_4N_2Se$ (n = 1, 4). Equilibrium studies of the ternary systems MnX_2–CH_4N_2Se–H_2O by the solubility method revealed that temperature has a significant effect on the composition of the complexes formed. In the systems $MnCl_2$–CH_4N_2Se–H_2O [1] and $MnSO_4$–CH_4N_2Se–H_2O [2] at 30°C only the solid phases selenourea and $MnCl_2 \cdot 4H_2O$ or $MnSO_4 \cdot 4H_2O$ are present. At 50°C the $MnCl_2 \cdot 4CH_4N_2Se$ complex appeared as the third binary solid phase [1], which could be crystallized by isothermal evaporation [3]. At 70°C $MnCl_2 \cdot CH_4N_2Se$ was formed in addition to the solid phases $MnCl_2 \cdot 4CH_4N_2Se$, $MnCl_2 \cdot 2H_2O$, and CH_4N_2Se [4].

References:

[1] Grekova, N. D.; Sulaimankulov, K.; Nogoev, K. (Zh. Neorgan. Khim. **18** [1973] 1385/91; Russ. J. Inorg. Chem. **18** [1973] 734/7).

[2] Grekova, N. D.; Nogoev, K. (Geterogen. Ravnovesiya Sist. Neorgan. Org. Soedin. **1974** 30/3; Ref. Zh. Khim. **1975** No. 12 B 912; C.A. **83** [1975] No. 153393).

[3] Grekova, N. D.; Nogoev, K.; Sulaimankulov, K. (V Sb. XI Mendeleevsk. S'ezd. po Obshch. i Prikl. Khim. Ref. Dokl. i Soobshch. **1975** No. 1, p. 61; Ref. Zh. Khim. **1976** No. 2 V 189; C.A. **84** [1976] No. 157615).

[4] Grekova, N. D.; Nogoev, K.; Sulaimankulov, K. (Zh. Neorgan. Khim. **18** [1973] 3107/9; Russ. J. Inorg. Chem. **18** [1973] 1653/4).

36.10 With the Selenosemicarbazone of 2-Acetylpyridine

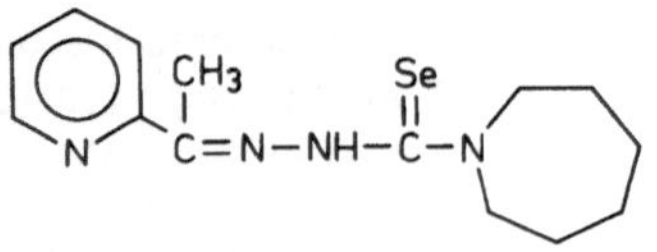

($=C_{14}H_{20}N_4Se = HL$)

$Mn^{II}(C_{14}H_{19}N_4Se)_2$ was prepared by the reaction of hot ethanol solutions of $MnCl_2 \cdot 4H_2O$ and the ligand in a 1:2 mole ratio. The reaction mixture was gently warmed for 1 h with stirring. The orange precipitate was collected, washed with ethanol and diethyl ether, and finally dried over P_4O_{10} in vacuum. The complex reveals a magnetic moment of 5.9 μ_B at 298 K. The main bands (in cm^{-1}) in the IR spectrum taken from KBr or CsI disks in the 4000 to 400 cm^{-1} range were assigned as follows (free ligand bands in parentheses): 1590(1576) to ν(C=N) + ν(C=C), 901(963) to ν(CSe), 668(716) to ν(NN), and 476 to ν(Mn–N). An octahedral structure is assumed for the complex; manganese is bonded to the azomethine nitrogen, the pyridine nitrogen, and the selenium of the deprotonated ligand. It was concluded that each set of two identical donor atoms has a *trans* configuration. The electronic spectrum of the complex in DMF shows band maxima at 16600, 26300, and 30300 cm^{-1}. They were assigned to the electronic transitions from $^6A_{1g}$ ground state to the $^4T_{1g}(^4G)$, $^4E_2$, $^4A_{1g}(^4G)$, and $^4E_g(^4D)$ excited states, respectively, in agreement with an octahedral geometry at the manganese atom. The complex also exhibits charge-transfer bands of high intensity. The ESR X-band spectrum of the solid compound shows a very sharp signal with g = 1.993 and 2.005 at 298 and 77 K, respectively, and a broad signal corresponding to g ≈ 6. The DMF solution exhibits a six-line spectrum with a broad signal corresponding to g = 2.209 at room temperature. At 77 K the values of g = 2.018 and $A_0 = 94 \times 10^{-4}$ cm^{-1} were obtained, consistent with an octahedral coordination of the Mn atom. The molar electrical conductivity of a 10^{-3} DMF solution, $\Lambda = 9.7$ $cm^2 \cdot \Omega^{-1} \cdot mol^{-1}$, indicates a non-electrolytic nature for the complex.

Reference:

Garg, Bhagwan S.; Kurup, M. R. Prathapachandra; Jain, Satendra K.; Bhoon, Yudhvir K. (Transition Metal Chem. [Weinheim] **13** [1988] 92/5).

36.11 With 1,5-Diphenylselenocarbazone

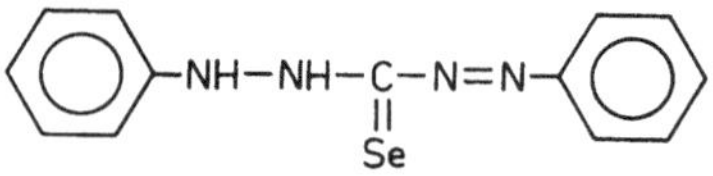

(= $C_{13}H_{12}N_4Se$ = selenazone)

Formation of a strongly colored brown-red complex was observed on reaction of Mn^{2+} with the ligand. It can be extracted into $CHCl_3$ from aqueous solution at pH 6.8.

Reference:

Ramakrishna, R. S.; Irving, H. M. N. H. (Anal. Chim. Acta **48** [1969] 251/69).

36.12 With 1-(Selenophene-2-yl)-1,3-alkanediones

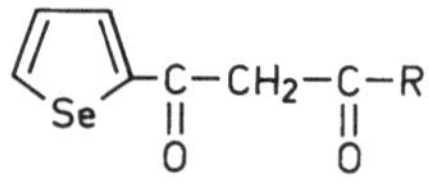

ligand 1 with R = CH_3 (= $C_8H_8O_2Se$ = HL)

ligand 2 with R = 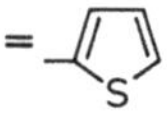 (= $C_{11}H_8O_2SSe$ = HL)

$Mn^{II}(C_8H_7O_2Se)_2 \cdot 2L'$ (L′ = H_2O, py) and **$Mn^{II}(C_{11}H_7O_2SSe)_2 \cdot 2L'$** (L′ = H_2O, py, 3-, or 4-methylpyridine). The dihydrates were prepared by mixing saturated ethanolic solutions of manganese(II) chloride and the ligand (mole ratio 1:2) and adding dilute ammonia up to pH 8 with constant stirring. The adducts with L′ = pyridine, 3-, and 4-methylpyridine were obtained by reaction of $MnCl_2$ with the ligands HL and the amine L′ (mole ratio 1:2:5) in ethanol, after addition of dilute ammonia up to pH 7 to 8. The yellow to orange complexes, precipitating in the form of prisms, were washed with ethanol and ether, and dried in air [1 to 3]. The IR spectra of $Mn(C_{11}H_7O_2SSe)_2 \cdot 2L'$ complexes recorded in Nujol mulls in the 4000 to 400 cm^{-1} region indicate coordination of the β-diketone to manganese through both oxygen atoms, forming a chelate ring. The L′ ligands are probably *trans* to each other [2]. As shown by thermal analyses (DTA, TG, DTG), $Mn(C_8H_7O_2Se)_2 \cdot 2H_2O$ loses two water molecules at 65 to 90°C in an endothermic reaction. The second (exothermic) step between 215 and 300°C indicates the loss of one ligand molecule [1]. Thermal analyses (DTA, TG, DTG) in air at a heating rate of 5°C/min up to 400°C reveal two DTG peaks for each $Mn(C_{11}H_7O_2SSe)_2 \cdot 2L'$ adduct. The first peak of the dihydrate and dipyridine adducts was observed at 100°C, the second one at 240 and 230°C, respectively. It was assumed that $Mn(C_{11}H_7O_2SSe)_2 \cdot 2py$ loses the pyridine in two stages with formation of $Mn(C_{11}H_7O_2SSe)^+$ in the second stage. 3-Methylpyridine was lost at 240°C and 4-methylpyridine at 210°C in the first stage; in the second stage (at 280°C for both complexes) $Mn(C_{11}H_7O_2SSe)^+$ was formed. The adducts are less stable than the corresponding complexes of Co^{II} and Ni^{II} [1 to 3]. The air-stable $Mn(C_8H_7O_2Se)_2 \cdot 2H_2O$ compound is readily soluble in methanol, ethanol, acetone, and ether [1]. The $Mn(C_{11}H_7O_2SSe)_2 \cdot 2L'$ adducts are stable in air, insoluble in water, and highly soluble in methanol and ethanol. The solutions are comparatively stable, but nicotinamide and 2-methylpyridine readily displace the L′ molecules [2].

The complexes are stabilizers for polyurethane adhesives [2, 4] and effective catalysts for the formation of polyurethanes [4]. Polymerizations in the presence of the complexes were studied kinetically. The catalytic activity is higher for the complexes with ligand 1 and decreased along the series $Mn^{II} > Cu^{II} > Co^{II} > Ni^{II}$. The presence of *cis-trans* isomerization in the structure of the complexes has a significant effect on their catalytic activity and reaction mechanism, which, probably occurs through the formation of an intermediate four-centered bimolecular complex [4].

References:

[1] Balan, V. T.; Byrke, A. I.; Fedoseev, M. S.; Kaptar, K. G.; Fedoseeva, A. M. (Koord. Soedin. Perekhodnykh Elem. Vopr. Khim. Khim. Tekhnol. **1983** 16/21; C.A. **99** [1983] No. 168425; Ref. Zh. Khim. **1983** No. 16 V 179).

[2] Byrke, A. I.; Kharitonov, Yu. Ya.; Shafranskii, V. N.; Balan, V. T.; Fedoseev, M. S.; Kaptar, K. G. (Koord. Khim. **9** [1983] 51/8; Soviet J. Coord. Chem. **9** [1983] 44/51).

[3] Balan, V. T.; Byrke, A. J.; Kaptar, K. G. (Koord. Soedin. Perekhodnykh Elem. Vopr. Khim. Khim. Tekhnol. **1983** 11/6; C.A. **99** [1983] No. 186334; Ref. Zh. Khim. **1983** No. 16 V 180).

[4] Fedoseev, M. S.; Kharitonov, Yu. Ya.; Balan, V. T.; Surkov, V. D.; Byrke, A. I.; Shafranskii, V. N. (Koord. Khim. **13** [1987] 1299/304; Soviet J. Coord. Chem. **13** [1987] 733/8).

Ligand Formula Index

The ligands treated in this volume are arranged in the index according to the system of Hill, A. (J. Am. Chem. Soc. **22** [1900] 478/94). In this system, the first criterion for the location of a ligand is the number of carbon atoms. The second is the number of hydrogen atoms, and finally the number of atoms of the other elements, in alphabetical order.

The first column contains the empirical formulas of the ligands. The second column shows their linearized structural formulas. Additional ligands of mixed ligand complexes and of adducts are placed in subheadings. The last column lists the pertinent pages.

The formulas of ligands coordinated as deprotonated acids are given in their nondeprotonated form. Ligands occurring in tautomeric equilibria are presented only in the form commonly known. The numbering of locants for substituents corresponds to IUPAC rules.

List of abbreviations for ligands or solvents used in the volume:

bpy	2,2′-bipyridine	α-pic	2-methylpyridine
dmf = DMF	dimethylformamide	β-pic	3-methylpyridine
DMSO	dimethyl sulfoxide	γ-pic	4-methylpyridine
en	ethylenediamine	py	pyridine
Hacac	acetylacetone	terpy	2,2′:6′,2″-terpyridine
phen	1,10-phenanthroline	THF	tetrahydrofuran

	and $C_6H_2Cl_4S_2$	44
	$C_6H_2Cl_4S_2$ = 1,2-$(HS)_2C_6Cl_4$	
	and $C_{12}H_{10}N_3S^+$	49
	$C_{12}H_{10}N_3S^+$ = 3,7-$(H_2N)_2C_{12}H_6NS^+$ = 3,7-Diamino-phenothiazin-5-ium	
$C_4H_2O_2S_2$	$(HS)_2C_4(=O)_2$	
	3,4-Dimercapto-3-cyclobutene-1,2-dione	33, 40
	and C_6H_6S	41
	C_6H_6S = HSC_6H_5	
$C_4H_2S_4$	$(HS)_2C_4(=S)_2$	
	3,4-Dimercapto-3-cyclobutene-1,2-dithione	55/7
$C_4H_3N_3O_3S$	$(HON=)C_4H_2N_2(=O)_2(=S)$	
	or Tautomeric compounds; 5-Hydroxyimino-2-thioxo-dihydro-4,6(1H,3H)-pyrimidine dione = Thiovioluric acid	76/8
$C_4H_4N_2OS$	$C_4H_4N_2(=O)(=S)$	
	2-Thioxo-2,3-dihydro-4(1H)-pyrimidinone = Thiouracil	75, 78/9
$C_4H_4N_2O_2S$	$C_4H_4N_2(=O)_2(=S)$	
	or Tautomeric compounds; 2-Thioxo-dihydro-4,6(1H,3H)-pyrimidinedione = Thio-barbituric acid	75/8
$C_4H_4N_2S$	$C_4H_4N_2(=S)$	
	2(1H)-Pyrimidinethione	75, 78/9
$C_4H_6N_2S$	$(CH_3)C_3H_3N_2(=S)$	
	1-Methyl-1,3-dihydro-2H-imidazole-2-thione	58
	$(H_2N)C_3HNS(CH_3)$	
	4-Methyl-2-thiazolamine	228/9
$C_4H_6O_2S_2$	$HOOC-C_3H_5S_2$	
	1,2-Dithiolane-3-carboxylic acid = Tetranorlipoic acid	226
$C_4H_6O_2S_4$	$HSC(=S)OCH_2CH_2OC(=S)SH$	133/4
$C_4H_6O_4S$	$HOOCCH(SH)CH_2COOH$	50/3
	$HOOCCH_2SCH_2COOH$	84/6
$C_4H_6O_4S_2$	$HOOCCH_2SSCH_2COOH$	90/1
$C_4H_6O_4Se$	$HOOCCH_2SeCH_2COOH$	246/7

$C_4H_{10}O_3S$	$HOCH_2CH_2S(=O)CH_2CH_2OH$	107
	$HO_3SC_4H_9$	114/5
$C_4H_{10}O_6S_2$	$HO_3S(CH_2)_4SO_3H$	114/5
$C_4H_{12}N_2S$	$HSCH_2CH_2NHCH_2CH_2NH_2$	28
C_5		
C_5H_4BrNOS	$(HO)(Br)C_5H_3N(=S)$	
	5-Bromo-1-hydroxy-3-methyl-2(1H)-pyridinethione	64/5
$C_5H_4N_2O$	$CH_3C(=O)CH(CN)_2$	14
	and C_5H_5N	15
	C_5H_5N = Pyridine = py	
	and C_6H_7N	151
	$C_6H_7N = (CH_3)C_5H_4N$ = 2-Methyl-pyridine = α-pic	
	and C_6H_7N	15
	$C_6H_7N = (CH_3)C_5H_4N$ = 4-Methyl-pyridine = γ-pic	
$C_5H_4O_2S$	$HOOC\text{-}C_4H_3S$	
	2-Thiophenecarboxylic acid	222
C_5H_5NOS	$(HO)C_5H_4N(=S)$	
	1-Hydroxy-2(1H)-pyridinethione	64/5
	3-Hydroxy-2(1H)-pyridinethione	64
	$HON{=}CH\text{-}C_4H_3S$	
	2-Thiophenecarbaldehyde oxime	222/4
$C_5H_5N_5S$	$(H_2N)C_5H_3N_4(=S)$	
	or Tautomeric compounds; 2-Amino-1,7(or 1,9)-dihydro-6H-purine-6-thione	76/8
$C_5H_6N_2OS$	$C_4H_3O\text{-}C(=S)NHNH_2$	
	or Enethiol form; C_4H_3O = 2-Furanyl	218/9
	$C_4H_3S\text{-}C(=O)NHNH_2$	
	2-Thiophenecarbohydrazide	222
$C_5H_6N_2O_2S$	$(CH_3)C_4H_3N_2(=O)_2(=S)$	
	or Tautomeric compounds; 1-Methyl-2-thioxo-dihydro-4,6(1H,3H)-pyrimidinedione = 1-Methyl-thiobarbituric acid	75/8

$C_5H_{11}NS_2$	$HSSCN(C_2H_5)_2$	
	and C_9H_7NO	143/4
	$C_9H_7NO = (HO)C_9H_6N$ = 8-Quinolinol	
	and $C_{10}H_8N_2$	142/3
	$C_{10}H_8N_2 = C_5H_4N\text{-}C_5H_4N$ = 2,2'-Bipyridine = bpy	
	and $C_{12}H_8N_2$	142/3
	$C_{12}H_8N_2$ = 1,10-Phenanthroline = phen	
	$HSSCNHC_4H_9$	136/7
$C_5H_{12}O_2S_2$	$CH_3S(=O)CH_2CH_2CH_2S(=O)CH_3$	108
$C_5H_{12}O_3S$	$HO_3SC_5H_{11}$	114/5
$C_5H_{12}O_3S_4$	$HSCH_2CH(SH)CH_2SCH_2CH_2SO_3H$	49/50
$C_5H_{12}O_4S_3$	$HSCH_2CH(SH)CH_2OCH_2CH_2SO_3H$	49/50
$C_5H_{12}O_5S_4$	$HSCH_2CH(SH)CH_2S(=O)_2CH_2CH_2SO_3H$	49/50
$C_5H_{12}O_6S_2$	$HO_3S(CH_2)_5SO_3H$	114/5
$C_5H_{14}N_2S$	$HSCH_2CH_2NHCH_2CH_2CH_2NH_2$	28
C_6		
$C_6H_2Cl_4S_2$	1,2-$(HS)_2C_6Cl_4$	33/7, 44/5, 48/9
	and $C_4H_2N_2S_2$	44
	$C_4H_2N_2S_2 = HSC(CN){=}C(CN)SH$	
$C_6H_5BrO_2Se$	1-$HO_2SeC_6H_4Br$-3	244/6
	1-$HO_2SeC_6H_4Br$-4	244/6
$C_6H_5ClO_2Se$	1-$HO_2SeC_6H_4Cl$-3	244/6
	1-$HO_2SeC_6H_4Cl$-4	244/6
$C_6H_5ClO_3S$	1-$HO_3SC_6H_4Cl$-4	114/5
	and C_5H_5N	114
	C_5H_5N = Pyridine = py	
C_6H_5NOS	$C_6H_5N{=}S{=}O$	107
$C_6H_5NO_2$	$C_6H_5NO_2$	22

C_8

$C_8H_8O_3S$	$CH_3OOCC(SH){=}CH\text{-}C_4H_3O$	
	C_4H_3O = 2-Furanyl	50/2
	$HOOCC(SH){=}C(CH_3)\text{-}C_4H_3O$	
	C_4H_3O = 2-Furanyl	50/2
	$HOOCCH_2S({=}O)C_6H_5$	107
C_8H_9NOS	$HSCH_2C({=}O)NHC_6H_5$	54
	$HON(CH_3)C({=}S)C_6H_5$	215/7
$C_8H_9NO_2S$	$HOOCCH_2S(1\text{-}C_6H_4NH_2\text{-}2)$	82/3
	$HOOCCH_2S(1\text{-}C_6H_4NH_2\text{-}4)$	82/3
	$HONHC({=}S)(1\text{-}C_6H_4OCH_3\text{-}4)$	215/6
$C_8H_9NO_3S$	$HOOC\text{-}C_3H_5NS\text{-}C_4H_3O$	
	2-(2-Furanyl)-4-thiazolidinecarboxylic acid	230/1
$C_8H_9NS_2$	$HSSCN(CH_3)C_6H_5$	139, 141, 146/7, 149/50, 153/6
$C_8H_{10}N_2O_3S$	$(4\text{-}H_2NC_6H_4\text{-}1)S({=}O)_2NHC({=}O)CH_3$	
	or Tautomeric compounds; Albucid	116/8, 121
$C_8H_{10}N_2S_2$	$C_6H_5CH_2SC({=}S)NHNH_2$	185
$C_8H_{10}N_4O_2S$	$(2\text{-}HOC_6H_4\text{-}1)C({=}O)NHNHC({=}S)NHNH_2$	
	or Tautomeric compounds	207/8
$C_8H_{10}O_8S$	$HOOCCH_2CH(COOH)SCH(COOH)CH_2COOH$	88/9
$C_8H_{10}S_2$	$1,2\text{-}(HS)_2C_6H_2(CH_3)_2\text{-}4,5$	33
	and C_5H_5N	33/8, 45/8
	C_5H_5N = Pyridine = py	
$C_8H_{12}N_2O_3S$	$HOOCCH_2CH_2\text{-}C_5H_7N_2S({=}O)$	
	Bisnorbiotin = Hexahydro-2-oxo-[3aS-(3aα,4β,6aα)]-1H-thieno[3,4-d]-imidazole-4-propanoic acid	227
$C_8H_{12}N_2S$	$HSCH_2CH_2NHCH_2\text{-}C_5H_4N$	
	C_5H_4N = 2-Pyridinyl	28
$C_8H_{12}O_4S_2$	$(HS)_2C{=}C(COOC_2H_5)_2$	31/2
	and $C_{10}H_8N_2$	32
	$C_{10}H_8N_2 = C_5H_4N\text{-}C_5H_4N$ = 2,2′-Bipyridine = bpy	

$C_9H_6Cl_6OS_2$	$HSC(=S)OCH_2$-$C_7H_3Cl_6$	
	$C_7H_3Cl_6$ = 1,4,5,6,7,7-Hexachloro-bicyclo[2.2.1]hept-5-en-2-yl	130/1
C_9H_6FNS	$(HS)(F)C_9H_5N$	
	5-Fluoro-8-quinolinethiol	66, 71/2
	6-Fluoro-8-quinolinethiol	66, 71/2
	7-Fluoro-8-quinolinethiol	66, 71/2
C_9H_6INS	$(HS)(I)C_9H_5N$	
	5-Iodo-8-quinolinethiol	66, 71/2
	7-Iodo-8-quinolinethiol	66, 71/2
$C_9H_6NNaO_3S_2$	$(HS)C_9H_5N(SO_3Na)$	
	8-Mercapto-5-quinolinesulfonic acid, Na-salt	66/8, 71
$C_9H_7BrN_2S$	$(H_2N)C_3HNS(1$-C_6H_4Br-4)	
	4-(4-Bromophenyl)-2-thiazolamine	228/9
$C_9H_7ClN_2S$	$(H_2N)C_3HNS(1$-C_6H_4Cl-4)	
	4-(4-Chlorophenyl)-2-thiazolamine	228/9
$C_9H_7FN_2S$	$(H_2N)C_3HNS(1$-C_6H_4F-4)	
	4-(4-Fluorophenyl)-2-thiazolamine	228/9
C_9H_7NS	$(HS)C_9H_6N$	
	8-Quinolinethiol = Thioxine	66/9, 71/2
$C_9H_7NS_2$	HSSC-C_8H_6N	
	1H-Indole-1-carbodithioic acid	166, 175
C_9H_7NSe	$(HSe)C_9H_6N$	
	8-Quinolineselenol	242/3
$C_9H_7N_3S_3$	HSSCNH-$C_2N_2S(C_6H_5)$	
	(5-Phenyl-1,3,4-thiadiazol-2-yl)carbamodithioic acid	136, 138
$C_9H_8N_2OS$	$(CH_3)C_8H_5N_2(=O)(=S)$	
	or Tautomeric compounds; 3-Methyl-2-thioxo-2,3-dihydro-4(1H)-quinoxalinone	76/8
$C_9H_8N_2S$	$(CH_2{=}CH)C_7H_5N_2(=S)$	
	1-Vinyl-1,3-dihydro-2H-benzimidazole-2-thione	58/9
	$C_7H_5N_2$-$SCH{=}CH_2$	
	2-Vinylsulfanyl-1H-benzimidazole	80/1

$C_{10}H_8Cl_6OS_2$	$HSC(=S)OCH_2$-$C_7H_2Cl_6(CH_3)$ $C_7H_2Cl_6(CH_3)$ = 1,4,5,6,7,7-Hexachloro-3-methyl-bicyclo[2.2.1]hept-5-en-2-yl	130/1
$C_{10}H_8N_2O_2S$	$(C_6H_5)C_4H_3N_2(=O)_2(=S)$ or Tautomeric compounds; 1-Phenyl-2-thioxo-dihydro-4,6(1H,3H)-pyrimidinedione = 1-Phenyl-thiobarbituric acid	75/8
$C_{10}H_8N_2O_2S_2$	C_4H_3O-C(=S)NHNHC(=S)-C_4H_3O or Enethiol form; C_4H_3O = 2-Furanyl	218/9
$C_{10}H_8N_2O_3S$	C_4H_3O-C(=O)NHNHC(=S)-C_4H_3O or Enethiol form; C_4H_3O = 2-Furanyl	218/9
$C_{10}H_8O_2S$	$HO_2S(1$-$C_{10}H_7)$	109/10
$C_{10}H_9NOS$	$(HS)(CH_3O)C_9H_5N$	
	4-Methoxy-8-quinolinethiol	66/7, 71/2
	5-Methoxy-8-quinolinethiol	66, 71/2
	6-Methoxy-8-quinolinethiol	66/7, 71
$C_{10}H_9NO_3S_2$	$(HS)(CH_3)C_9H_4N(SO_3H)$ 8-Mercapto-2-methyl-5-quinolinesulfonic acid	66, 68, 71
$C_{10}H_9NS$	$(HS)(CH_3)C_9H_5N$	
	2-Methyl-8-quinolinethiol	66, 71/2
	4-Methyl-8-quinolinethiol	66/7, 71/2
	6-Methyl-8-quinolinethiol	66/7, 71/2
	7-Methyl-8-quinolinethiol	66/7, 71/2
$C_{10}H_9NS_2$	$(HS)(CH_3S)C_9H_5N$	
	4-Methylsulfanyl-8-quinolinethiol	66/7, 71/2
	5-Methylsulfanyl-8-quinolinethiol	66/7, 71/2
	7-Methylsulfanyl-8-quinolinethiol	66, 71/2
$C_{10}H_9N_3O_2S_2$	$HOOCCH_2S$-$C_2N_2S(NHC_6H_5)$ $C_2N_2S(NHC_6H_5)$ = 5-Phenylamino-1,3,4-thiadiazol-2-yl	82/4
$C_{10}H_9N_5O_2S$	$(HS)(HO)(H_2N)C_4N_2$-N=N(1-C_6H_4OH-2) or Tautomeric compounds; 6-Amino-5-(2-hydroxyphenylazo)-2-mercapto-4-pyrimidinol	75, 77
$C_{10}H_{10}N_2OS$	$CH_3C(=O)NH$-$C_7H_3NS(CH_3)$ N-(6-Methyl-2-benzothiazolyl)acetamide	234/5

$C_{11}H_{12}N_4O_3S$ — $(4\text{-}H_2NC_6H_4\text{-}1)S(=O)_2NH\text{-}C_4H_2N_2(OCH_3)$
or Tautomeric compounds; $C_4H_2N_2(OCH_3)$ = 2-Methoxy-4-pyrimidinyl 116/7, 119/20
or Tautomeric compounds; $C_4H_2N_2(OCH_3)$ = 6-Methoxy-4-pyrimidinyl; Sulfamonomethoxine 116/8, 121
or Tautomeric compounds; $C_4H_2N_2(OCH_3)$ = 6-Methoxy-3-pyridazinyl; Sulfapyridazine = Sulfamethoxypyridazine 116/22

$C_{11}H_{17}N_3O_3S$ — $(4\text{-}H_2NC_6H_4\text{-}1)S(=O)_2NHC(=O)NHC_4H_9$
or Tautomeric compounds; Bucarban 116/8, 121

$C_{11}H_{18}N_2O_3S$ — $HOOC(CH_2)_5\text{-}C_5H_7N_2S(=O)$
Homobiotin = Hexahydro-2-oxo-[3aS-(3aα,4β,6aα)]-1H-thieno[3,4-d]-imidazole-4-hexanoic acid 227

$C_{11}H_{23}NS_2$ — $HSSCN(C_5H_{11})_2$ 139, 146/7, 149/50
$HSSCN(C_5H_{11}\text{-}i)_2$ 139, 146/7, 149/50

C_{12}

$C_{12}H_4N_4$ — $(NC)_2C{=}(1\text{-}C_6H_4\text{-}4){=}C(CN)_2$ 14, 16/8
and $C_{10}H_{24}N_4$ 18
$C_{10}H_{24}N_4$ = 1,4,8,11-Tetraazacyclotetradecane
and $C_{12}H_8N_2$ 18
$C_{12}H_8N_2$ = 1,10-Phenanthroline = phen

$C_{12}H_6Cl_4O_2S$ — $(1\text{-}HO(3,4\text{-}Cl_2)C_6H_2\text{-}2)S(2\text{-}C_6H_2(Cl_2\text{-}3,4)OH\text{-}1)$ 81

$C_{12}H_8Cl_2O_2S$ — $(1\text{-}HO(4\text{-}Cl)C_6H_3\text{-}2)S(2\text{-}C_6H_3(Cl\text{-}4)OH\text{-}1)$ 81

$C_{12}H_8N_2S$ — $C_7H_4NS\text{-}C_5H_4N$
2-(2-Pyridinyl)-benzothiazole 234/6

$C_{12}H_9NS$ — $C_{12}H_9NS$
10H-Phenothiazine 241

$C_{12}H_{10}N_2O_2S$ — $(2\text{-}HOC_6H_4\text{-}1)CH{=}NNHC(=S)\text{-}C_4H_3O$
or Enethiol form; C_4H_3O = 2-Furanyl 218/9

$C_{12}H_{10}N_2O_3S$ — $(2\text{-}HOC_6H_4\text{-}1)C(=O)NHNHC(=S)\text{-}C_4H_3O$
C_4H_3O = 2-Furanyl 218

C_{15}

$C_{15}H_{12}OS$ — $C_6H_5C(=O)CH_2C(=S)C_6H_5$ 55/7
and C_5H_5N 57
C_5H_5N = Pyridine = py

$C_{15}H_{14}N_2O_2S$ — $C_6H_5C(=O)NHC(=S)NH(1\text{-}C_6H_4OCH_3\text{-}2)$
or Tautomeric compounds 198/9

$C_{15}H_{14}N_4O_2S$ — $(4\text{-}H_2NC_6H_4\text{-}1)S(=O)_2NH\text{-}C_3H_2N_2(C_6H_5)$
or Tautomeric compounds; $C_3H_2N_2(C_6H_5)$ = 1-Phenyl-1H-pyrazol-5-yl; Sulfaphenazole ... 116/8

$C_{15}H_{15}NS_2$ — $HSSCN(CH_2C_6H_5)_2$ 139, 142, 146/7, 149/50, 153/6, 159

$C_{15}H_{17}NO_2S$ — $(CH_3)_2C_6H_5(=O)_2\text{-}C(=S)NHC_6H_5$
or Enol form; 4,4-Dimethyl-2,6-dioxo-N-phenyl-1-cyclohexanecarbothioamide 213/4

$C_{15}H_{24}S$ — $1\text{-}HSC_6H_2(C_3H_7\text{-}i)_3\text{-}2,4,6$ 24, 27

C_{16}

$C_{16}H_{11}N_3O_3S$ — $(HON=)(C_6H_5)_2C_4N_2(=O)_2(=S)$
or Tautomeric compounds; 5-Hydroxyimino-1,3-diphenyl-2-thioxo-dihydro-4,6(1H,3H)-pyrimidinedione = 1,3-Diphenyl-thiovioluric acid 76/8

$C_{16}H_{14}Cl_2N_6O_2S_2$ — $(4\text{-}ClC_6H_4\text{-}1)NHC(=S)NHNHC(=O)C(=O)NHNHC(=S)NH(1\text{-}C_6H_4Cl\text{-}4)$
or Tautomeric compounds 202/3

$C_{16}H_{14}N_2O_2S$ — $(4\text{-}CH_3C_6H_4\text{-}1)S(=O)_2NH\text{-}C_9H_6N$
or Tautomeric compounds; C_9H_6N = 8-Quinolinyl 124

$C_{16}H_{16}N_4O_5S$ — $HOOCCH{=}CHC(=O)NH(1\text{-}C_6H_4\text{-}4)S(=O)_2NH\text{-}C_4HN_2(CH_3)_2$
or Tautomeric compounds; $C_4HN_2(CH_3)_2$ = 4,6-Dimethyl-2-pyrimidinyl 123

$C_{16}H_{18}N_2O_4S$ — $HOOC\text{-}C_5H_3NS(CH_3)_2(=O)\text{-}NHC(=O)CH_2C_6H_5$
Penicillin G = 3,3-Dimethyl-7-oxo-6-[(phenylacetyl)amino]-[2S-(2α,5α,6β)]-4-thia-1-azabicyclo[3.2.0]heptane-2-carboxylic acid 232/3

$C_{16}H_{18}N_2O_5S$ — $HOOC\text{-}C_5H_3NS(CH_3)_2(=O)\text{-}NHC(=O)CH_2OC_6H_5$
Penicillin V = 3,3-Dimethyl-7-oxo-6-[(phenoxyacetyl)amino]-[2S-(2α,5α,6β)]-4-thia-1-azabicyclo[3.2.0]heptane-2-carboxylic acid 232/3

$C_{16}H_{34}OS$ — $CH_3S(=O)C_{15}H_{31}$ 101

C_{17}

$C_{17}H_{14}N_2OS$ — $(CH_3)(C_6H_5)C_3HN_2(=O)\text{-}C(=S)C_6H_5$
5-Methyl-2-phenyl-4-(phenylthioxomethyl)-2,4-dihydro-3H-pyrazol-3-one 55/6
$(CH_3)(C_6H_5)C_3HN_2(=S)\text{-}C(=O)C_6H_5$
(3-Methyl-1-phenyl-5-thioxo-4,5-dihydro-1H-pyrazol-4-yl)phenylmethanone 59

$C_{17}H_{14}N_2O_4S$ — $(O=)(CH_3O)(O_2N)C_6H_2=CHCH=C_7H_4NS(CH_3)$
2-Methoxy-6-[(3-methyl-2,3-dihydro-2-benzothiazolylidene)ethylidene]-4-nitro-2,4-cyclohexadien-1-one 237

$C_{17}H_{28}O_8S_4$ — $(HOOCCH(CH_3)S)_2CH(CH_2)_3CH(SCH(CH_3)COOH)_2$ 89/90

C_{18}

$C_{18}H_{15}N_3O_3S$ — $(HON=)(2\text{-}CH_3C_6H_4\text{-}1)_2C_4N_2(=O)_2(=S)$
or Tautomeric compounds; 5-Hydroxyimino-2-thioxo-di-o-tolyl-dihydro-4,6(1H,3H)-pyrimidinedione = 1,3-Di-o-tolyl-thiovioluric acid 76/8
$(HON=)(3\text{-}CH_3C_6H_4\text{-}1)_2C_4N_2(=O)_2(=S)$
or Tautomeric compounds; 5-Hydroxyimino-2-thioxo-di-m-tolyl-dihydro-4,6(1H,3H)-pyrimidinedione = 1,3-Di-m-tolyl-thiovioluric acid 76/8
$(HON=)(4\text{-}CH_3C_6H_4\text{-}1)_2C_4N_2(=O)_2(=S)$
or Tautomeric compounds; 5-Hydroxyimino-2-thioxo-di-p-tolyl-dihydro-4,6(1H,3H)-pyrimidinedione = 1,3-Di-p-tolyl-thiovioluric acid 76/8

C_{19}

$C_{19}H_{11}N_3S_2$ — $C_7H_4NS\text{-}C_5H_3N\text{-}C_7H_4NS$
2,2′-(2,6-Pyridinediyl)bis(benzothiazole) 234, 236

$C_{19}H_{18}N_4O_2S$ — $C_6H_5C(=O)NHC(=S)NH\text{-}C_3N_2(CH_3)_2(C_6H_5)(=O)$
or Tautomeric compounds; $C_3N_2(CH_3)_2(C_6H_5)(=O)$ = 1,5-Dimethyl-3-oxo-2-phenyl-2,3-dihydro-1H-pyrazol-4-yl 198

$C_{19}H_{19}Cl_2N_3S_2$ — $HSSCN(CH_2CH_2N=CH(1\text{-}C_6H_4Cl\text{-}2))_2$ 164/5
$HSSCN(CH_2CH_2N=CH(1\text{-}C_6H_4Cl\text{-}3))_2$ 164/5
$HSSCN(CH_2CH_2N=CH(1\text{-}C_6H_4Cl\text{-}4))_2$ 164/5

$C_{19}H_{21}N_3S_2$ — $HSSCN(CH_2CH_2N=CHC_6H_5)_2$ 164/5

$C_{19}H_{27}NS_2$ — $(HS)(C_{10}H_{21}S)C_9H_5N$
5-Decylsulfanyl-8-quinolinethiol 66, 71

C_{20}–C_{31}

Physical Constants and Conversion Factors

Avogadro constant N_A (or L) = 6.02214×10^{23} mol^{-1}
Faraday constant F = 9.64853×10^{4} C/mol
molar gas constant R = 8.31451 $J \cdot mol^{-1} \cdot K^{-1}$
molar volume (ideal gas) $V_m = 2.24141 \times 10^{1}$ L/mol
(273.15 K, 101325 Pa)

Planck constant $h = 6.62608 \times 10^{-34}$ $J \cdot s$
elementary charge $e = 1.60218 \times 10^{-19}$ C
electron mass $m_e = 9.10939 \times 10^{-31}$ kg
proton mass $m_p = 1.67262 \times 10^{-27}$ kg

1 kg = 2.205 pounds
1 m = 3.937×10^{1} inches = 3.281 feet
1 m^3 = 2.642×10^{2} gallons (U.S.)
1 m^3 = 2.200×10^{2} gallons (Imperial)

Force	N	dyn	kp
1 N	1	10^{5}	1.019716×10^{-1}
1 dyn	10^{-5}	1	1.019716×10^{-6}
1 kp	9.80665	9.80665×10^{5}	1

Pressure	Pa	bar	kp/m^2	at	atm	Torr	lb/in^2
1 Pa = 1 N/m^2	1	10^{-5}	1.019716×10^{-1}	1.019716×10^{-5}	9.86923×10^{-6}	7.50062×10^{-3}	1.450378×10^{-4}
1 bar = 10^6 dyn/cm^2	10^{5}	1	1.019716×10^{4}	1.019716	9.86923×10^{-1}	7.50062×10^{2}	1.450378×10^{1}
1 kp/m^2 = 1 mm H_2O	9.80665	9.80665×10^{-5}	1	10^{-4}	9.67841×10^{-5}	7.35559×10^{-2}	1.422335×10^{-3}
1 at (technical)	9.80665×10^{4}	9.80665×10^{-1}	10^{4}	1	9.67841×10^{-1}	7.35559×10^{2}	1.422335×10^{1}
1 atm = 760 Torr	1.01325×10^{5}	1.01325	1.033227×10^{4}	1.033227	1	7.60×10^{2}	1.469595×10^{1}
1 Torr = 1 mm Hg	1.333224×10^{2}	1.333224×10^{-3}	1.359510×10^{1}	1.359510×10^{-3}	1.315789×10^{-3}	1	1.933678×10^{-2}
1 lb/in^2 = 1 psi	6.89476×10^{3}	6.89476×10^{-2}	7.03069×10^{2}	7.03069×10^{-2}	6.80460×10^{-2}	5.17149×10^{1}	1

Work, Energy, Heat	J	kW·h	kcal	Btu	eV
1 J = 1 W·s = 1 N·m = 10^7 erg	1	2.778×10^{-7}	2.39006×10^{-4}	9.4781×10^{-4}	6.242×10^{18}
1 kW·h	3.6×10^6	1	8.604×10^2	3.41214×10^3	2.247×10^{25}
1 kcal	4.1840×10^3	1.1622×10^{-3}	1	3.96566	2.6117×10^{22}
1 Btu (British thermal unit)	1.05506×10^3	2.93071×10^{-4}	2.5164×10^{-1}	1	6.5858×10^{21}
1 eV	1.602×10^{-7}	4.450×10^{-14}	3.8289×10^{-11}	1.51840×10^{-10}	1

1 cm^{-1} = 1.239842×10^{-4} eV   1 Hz = 4.135669×10^{-15} eV

1 hartree = 27.2114 eV   1 eV ≙ 23.0578 kcal/mol

Power	kW	hp	$kp \cdot m \cdot s^{-1}$	kcal/s
1 kW = 10^3 J/s	1	1.35962	1.01972×10^2	2.39006×10^{-1}
1 hp (horsepower, metric)	7.3550×10^{-1}	1	7.5×10^1	1.7579×10^{-1}
1 $kp \cdot m \cdot s^{-1}$	9.80665×10^{-3}	1.333×10^{-2}	1	2.34384×10^{-3}
1 kcal/s	4.1840	5.6886	4.26650×10^2	1

References:

International Union of Pure and Applied Chemistry, Manual of Symbols and Terminology for Physicochemical Quantities and Units, Pergamon, London 1979; Pure Appl. Chem. **51** [1979] 1/41.

The International System of Units (SI), National Bureau of Standards Spec. Publ. 330 [1972].

Landolt-Börnstein, 6th Ed., Vol. II, Pt. 1, 1971, pp. 1/14.

ISO Standards Handbook 2, Units of Measurement, 2nd Ed., Geneva 1982.

Cohen, E. R., Taylor, B. N., Codata Bulletin No. 63, Pergamon, Oxford 1986.

Key to the Gmelin System of Elements and Compounds

HCl ZnCl$_2$

System Number	Symbol	Element
1		Noble Gases
2	H	Hydrogen
3	O	Oxygen
4	N	Nitrogen
5	F	Fluorine
6	**Cl**	**Chlorine**
7	Br	Bromine
8	I	Iodine
8a	At	Astatine
9	S	Sulfur
10	Se	Selenium
11	Te	Tellurium
12	Po	Polonium
13	B	Boron
14	C	Carbon
15	Si	Silicon
16	P	Phosphorus
17	As	Arsenic
18	Sb	Antimony
19	Bi	Bismuth
20	Li	Lithium
21	Na	Sodium
22	K	Potassium
23	NH_4	Ammonium
24	Rb	Rubidium
25	Cs	Caesium
25a	Fr	Francium
26	Be	Beryllium
27	Mg	Magnesium
28	Ca	Calcium
29	Sr	Strontium
30	Ba	Barium
31	Ra	Radium
32	**Zn**	**Zinc**
33	Cd	Cadmium
34	Hg	Mercury
35	Al	Aluminium
36	Ga	Gallium

CrCl$_2$ ZnCrO$_4$

System Number	Symbol	Element
37	In	Indium
38	Tl	Thallium
39	Sc, Y La–Lu	Rare Earth Elements
40	Ac	Actinium
41	Ti	Titanium
42	Zr	Zirconium
43	Hf	Hafnium
44	Th	Thorium
45	Ge	Germanium
46	Sn	Tin
47	Pb	Lead
48	V	Vanadium
49	Nb	Niobium
50	Ta	Tantalum
51	Pa	Protactinium
52	**Cr**	**Chromium**
53	Mo	Molybdenum
54	W	Tungsten
55	U	Uranium
56	Mn	Manganese
57	Ni	Nickel
58	Co	Cobalt
59	Fe	Iron
60	Cu	Copper
61	Ag	Silver
62	Au	Gold
63	Ru	Ruthenium
64	Rh	Rhodium
65	Pd	Palladium
66	Os	Osmium
67	Ir	Iridium
68	Pt	Platinum
69	Tc	Technetium[1]
70	Re	Rhenium
71	Np, Pu . . .	Transuranium Elements

Material presented under each Gmelin System Number includes all information concerning the element(s) listed for that number plus the compounds with elements of lower System Number.

For example, zinc (System Number 32) as well as all zinc compounds with elements numbered from 1 to 31 are classified under number 32.

[1] A Gmelin volume titled "Masurium" was published with this System Number in 1941.

A Periodic Table of the Elements with the Gmelin System Numbers is given on the Inside Front Cover